The Systematics Association Special Volume Series 72

# Reconstructing the Tree of Life

## Taxonomy and Systematics of Species Rich Taxa

T0204017

# The Systematics Association Special Volume Series

Series Editor

Alan Warren

*Department of Zoology, The Natural History Museum,*
*Cromwell Road, London SW7 5BD, UK.*

The Systematics Association promotes all aspects of systematic biology by organizing conferences and workshops on key themes in systematics, publishing books and awarding modest grants in support of systematics research. Membership of the Association is open to internationally based professionals and amateurs with an interest in any branch of biology including palaeobiology. Members are entitled to attend conferences at discounted rates, to apply for grants and to receive the newsletters and mailed information; they also receive a generous discount on the purchase of all volumes produced by the Association.

The first of the Systematics Association's publications *The New Systematics* (1940) was a classic work edited by its then-president Sir Julian Huxley, that set out the problems facing general biologists in deciding which kinds of data would most effectively progress systematics. Since then, more than 70 volumes have been published, often in rapidly expanding areas of science where a modern synthesis is required.

The *modus operandi* of the Association is to encourage leading researchers to organize symposia that result in a multi-authored volume. In 1997 the Association organized the first of its international Biennial Conferences. This and subsequent Biennial Conferences, which are designed to provide for systematists of all kinds, included themed symposia that resulted in further publications. The Association also publishes volumes that are not specifically linked to meetings and encourages new publications in a broad range of systematics topics.

Anyone wishing to learn more about the Systematics Association and its publications should refer to our website at www.systass.org.

Other Systematics Association publications are listed after the index for this volume.

**The Systematics Association Special Volume Series 72**

# Reconstructing the Tree of Life

## Taxonomy and Systematics of Species Rich Taxa

Edited by

## Trevor R. Hodkinson

## John A. N. Parnell

Department of Botany
School of Natural Sciences
Trinity College Dublin
Dublin, Ireland

**CRC Press**
Taylor & Francis Group
Boca Raton London New York

CRC Press is an imprint of the
Taylor & Francis Group, an **informa** business

Outside to inside of image: water ermine moth, UK (*Spilosoma urticae*); barley, UK (*Hordeum distichon*); fossilised sea urchins, Tunisia (*Mecaster* spp.); seeds, unknown origin (Bignoniaceae); purple sea snails, worldwide (*Janthina janthina*); fossilised shark teeth, USA (*Isurus* sp.); and sea urchin, Greece (*Arbacia lixula*)

artwork by: Diccon Alexander (diccona@hotmail.com)

CRC Press
Taylor & Francis Group
6000 Broken Sound Parkway NW, Suite 300
Boca Raton, FL 33487-2742

First issued in paperback 2019

© 2007 by Taylor & Francis Group, LLC
CRC Press is an imprint of Taylor & Francis Group, an Informa business

No claim to original U.S. Government works

ISBN-13: 978-0-8493-9579-6 (hbk)
ISBN-13: 978-0-367-38958-1 (pbk)

---

**Library of Congress Cataloging-in-Publication Data**

---

Reconstructing the tree of life : taxonomy and systematics of species rich taxa / editors, Trevor R.
    Hodkinson and John A.N. Parnell.
        p. cm. -- (The Systematics Association special volume series)
    Includes bibliographical references and index.
    ISBN 0-8493-9579-8 (alk. paper)
    1. Biology--Classification. I. Hodkinson, Trevor R. II. Parnell, John A. N.

    QH83.R43 2006
    578.01'2--dc22                                                                2006048341

---

Visit the Taylor & Francis Web site at
http://www.taylorandfrancis.com

and the CRC Press Web site at
http://www.crcpress.com

# Preface

The twenty chapters of this book are based on the theme of the plenary session of the Fourth Biennial Conference of the Systematics Association, held at Trinity College Dublin (TCD), Ireland, in August 2003, namely the systematics of species rich taxa. During the five-day conference, there were stimulating presentations, posters and discussions, covering a broad sample of the 'tree of life'; these also influenced the shape and content of this volume. Papers were contributed by a number of conference delegates and by others subsequently invited to broaden the book's scope or address particular theoretical issues.

Consideration of the book's theme and content began at a conference planning meeting at TCD in early 2003 with the local conference organiser, Steve Waldren of TCD, and Gordon Curry, the honorary treasurer of the Systematics Association. These were refined further in discussions with Alan Warren, the Systematics Association special volumes series editor, and Chris Humphries, the president of the Systematics Association. We are grateful to all of them for their input and encouragement, particularly our colleague, Steve. Two anonymous book proposal reviewers also provided valuable content guidance. We are particularly grateful for the manuscript preparation input of Sandra Velthuis of Whitebarn Consulting, who has worked long and hard to proofread chapters and standardise their format, and to the production team, especially Gail Renard, Pat Roberson and John Sulzycki, at CRC Press, who have been highly supportive and professional. We also thank Diccon Alexander for the superb cover artwork. Finally we thank all 51 contributing authors to the book, many of whom also peer reviewed other chapters. We encourage all readers to support the activities of the Systematics Association (www.systass.org).

**Trevor R. Hodkinson**
**John A.N. Parnell**
*Department of Botany*
*School of Natural Sciences*
*Trinity College Dublin*
*Ireland*

# The Editors

**Dr Trevor Hodkinson** is Senior Lecturer in the Department of Botany, School of Natural Sciences, Trinity College Dublin (TCD), Ireland. He is head of the Molecular Laboratory and specialises in the research fields of molecular systematics, genetic resources and taxonomy (http://www.tcd.ie/Botany/Staff/THodkinson.html).

**Professor John Parnell** is also from the Department of Botany at TCD. He is curator of the herbarium and his research interests are mainly in the fields of taxonomy and systematics (http://www.tcd.ie/Botany/Staff/JParnell.html).

# Contributors

**T.G. Barraclough**
Division of Biology and NERC Centre
  for Population Biology
Imperial College London, UK

**E. Biffin**
Division of Botany and Zoology
Australian National University
Canberra, Australia

**O.R.P. Bininda-Emonds**
Institut für Spezielle Zoologie und
  Evolutionsbiologie mit Phyletischem
  Museum
Friedrich-Schiller-Universität Jena
Jena, Germany

**K.E. Black**
School of Forest Resources
Penn State University
University Park, Pennsylvania, USA

**Y. Bouchenak-Khelladi**
Department of Botany
School of Natural Sciences
Trinity College Dublin
Dublin, Ireland

**J. Brodie**
Botany Department
The Natural History Museum
London, UK

**G. Cassis**
Research and Collections Branch
Australian Museum
Sydney, Australia

**M.W. Chase**
Jodrell Laboratory
Royal Botanic Gardens, Kew
Richmond, UK

**J.J. Clarkson**
Jodrell Laboratory
Royal Botanic Gardens, Kew
Richmond, UK

**J.A. Cotton**
Zoology Department
The Natural History Museum
London, UK

**L.A. Craven**
Australian National Herbarium
Centre for Plant Biodiversity Research
Canberra, Australia

**C.J. Creevey**
European Molecular Biology Laboratory
EMBL Heidelberg
Heidelberg, Germany

**T.J. Davies**
Department of Biology
University of Virginia
Charlottesville, Virginia, USA

**R.P.J. de Kok**
Herbarium
Royal Botanic Gardens, Kew
Richmond, UK

**D.A. Fitzpatrick**
Conway Institute
University College Dublin
Dublin, Ireland

**G. Fusco**
Department of Biology
University of Padova
Padova, Italy

**M. Geerts**
Burg. Heynenstraat 11
Swalmen, The Netherlands

**K.W. Hilu**
Department of Biological Sciences
Virginia Polytechnic Institute and State
  University
Blacksburg, Virginia, USA

**T. R. Hodkinson**
Department of Botany
School of Natural Sciences
Trinity College Dublin
Dublin, Ireland

**K.D. Hyde**
Centre for Research in Fungal Diversity
Department of Ecology and Biodiversity
The University of Hong Kong
Hong Kong, China

**S.W.L. Jacobs**
National Herbarium
Royal Botanic Gardens
Sydney, Australia

**M.S. Kinney**
Department of Botany
School of Natural Sciences
Trinity College Dublin
Dublin, Ireland

**A.F. Konings**
Cichlid Press
El Paso, Texas, USA

**J.O. McInerney**
Department of Biology
National University of Ireland Maynooth
Maynooth, Ireland

**K.R. McKaye**
Appalachian Laboratory
University of Maryland System
Frostburg, Maryland, USA

**A. Minelli**
Department of Biology
University of Padova
Padova, Italy

**E. Negrisolo**
Department of Public Health, Comparative
  Pathology and Veterinary Hygiene
University of Padova
Legnaro, Italy

**M.J. O'Connell**
Department of Biochemistry
University College Cork
Cork, Ireland

**J.A.N. Parnell**
Department of Botany
School of Natural Sciences
Trinity College Dublin
Dublin, Ireland

**G. Petersen**
Botanical Garden and Museum
The Natural History Museum of Denmark
Copenhagen, Denmark

**D.E. Pisani**
Department of Biology
National University of Ireland Maynooth
Maynooth, Ireland

**G. Reid**
Botany Department
The Natural History Museum
London, UK

**N. Rønsted**
Jodrell Laboratory
Royal Botanic Gardens, Kew
Richmond, UK

**N. Salamin**
Department of Ecology and Evolution
University of Lausanne
Lausanne, Switzerland

**V. Savolainen**
Jodrell Laboratory
Royal Botanic Gardens, Kew
Richmond, UK

**F.R. Schram**
Department of Biology
University of Washington
Seattle, Washington, USA

**R.T. Schuh**
American Museum of Natural History
Division of Invertebrate Zoology
New York, New York, USA

**O. Seberg**
Botanical Garden and Museum
The Natural History Museum of Denmark
Copenhagen, Denmark

**B.D. Shenoy**
Centre for Research in Fungal Diversity
Department of Ecology and Biodiversity
The University of Hong Kong
Hong Kong, China

**A. Stamatakis**
Swiss Federal Institute of Technology
School of Computer and Communication
  Sciences
Lausanne, Switzerland

**J.R. Stauffer, Jr.**
School of Forest Resources
Penn State University
University Park, Pennsylvania, USA

**M. Steel**
Biomathematics Research Centre
University of Canterbury
Christchurch, New Zealand

**A.M.C. Tang**
Centre for Research in Fungal Diversity
Department of Ecology and Biodiversity
The University of Hong Kong
Hong Kong, China

**K. Turk**
Jodrell Laboratory
Royal Botanic Gardens, Kew
Richmond, UK

**T.M.A. Utteridge**
Herbarium
Royal Botanic Gardens, Kew
Richmond, UK

**M.A. Wall**
Department of Entomology
San Diego Natural History Museum
San Diego, California, USA

**W.C. Wheeler**
Division of Invertebrate Zoology
American Museum of Natural History
New York, New York, USA

**M. Wilkinson**
Zoology Department
The Natural History Museum
London, UK

**D.M. Williams**
Botany Department
The Natural History Museum
London, UK

**E. Yektaei-Karin**
Jodrell Laboratory
Royal Botanic Gardens, Kew
Richmond, UK

**G.C. Zuccarello**
School of Biological Sciences
Victoria University of Wellington
Wellington, New Zealand

# Contents

## SECTION C  Taxonomy and Systematics of Species Rich Groups (Case Studies)

# Section A

Introduction and General Context

# 1 Introduction to the Systematics of Species Rich Groups

*T. R. Hodkinson and J. A. N. Parnell*
Department of Botany, School of Natural Sciences, Trinity College Dublin

## CONTENTS

## ABSTRACT

To completely document the world's diversity of species we need to undertake some simple but mountainous tasks; above all we need to tackle its species rich groups. We need to collect them, name and classify them, and position them on the tree of life. We need to do this systematically across all groups of organisms, and because of the biodiversity crisis we need to do it quickly. A qualitative approach to defining a species rich taxon — such as a species rich genus, family, order, class or phylum — appears more broadly applicable than a quantitative definition, but combining such categories of definition also appears useful. We define a species rich group as: 'a group with a relatively high number of species in comparison to other groups of the same, and comparable, taxonomic rank'. This chapter introduces, with examples, the concept of species rich groups and discusses how these groups are central to efforts to document the world's diversity of species and to help address the biodiversity crisis. Naming and describing species rich groups is the first step in placing them on the phylogenetic tree of life. Phylogenetic trees are becoming bigger (supersized) and methods are being developed to deal with the computational complexity of such trees. This paper also outlines the wider context of the book and papers presented herein. With species rich taxa, evolution has set taxonomists and systematists a difficult, but not unattainable, challenge that must be addressed as a matter of urgency.

## 1.1  INTRODUCTION

It may be a surprise to many readers that biologists cannot answer two seemingly simple yet fundamental questions: 'how many species are there in the world?' and 'how do the world's species relate to one another in an evolutionary context?'. The first question is a basic challenge for taxonomists who list, describe and classify the world's organisms. The second is a challenge for systematists/phylogeneticists who try to place organisms in an evolutionary framework by inferring a tree of life such as that shown in Figure 1.1. Activities of both groups of workers are critically impeded by species rich taxa, as they are often poorly sampled and described, yet make up a high proportion of total global species richness.

There is a huge variance in the published estimates of the total number of species on Earth. It could lie anywhere in the region of 4 million to 100 million[1-3]. We cannot even accurately count the number of species that have so far been described because of synonymy (the same species unwittingly recorded under different names by different researchers, that is, duplication). For example, 1.7 million species have been described but levels of synonymy could be in the range of 20–50%[4-6] (but see Cassis et al., *Chapter 13*, for a higher value). Even for a particular species rich group, estimates can vary enormously. For example, in the insects with approximately 1 million described species, estimates of the total number of species have varied from 1.8 million by Hodkinson and Casson[7] to 80 million by Stork[8]. An intermediate 10 million, proposed by Ødegaard et al.[9], may well be more appropriate, but such estimates are often based on crude methods (Cassis et al., *Chapter 13*). Furthermore, for many species rich groups, only a low proportion of the total estimated number of species has been described. For example, approximately 100,000 fungi have been described but 1.5 million species may exist (Tang et al., *Chapter 15*), only 15,000–20,000 diatom species (heterokont algae) have been described but up to 200,000 may exist (Williams and Reid, *Chapter 19*) and approximately 5,800 red algae (Rhodophyta) have been described but 20,000 may exist (Brodie and Zuccarello, *Chapter 20*).

Why do estimates of the number of species in the world vary by an order of magnitude or more, and why is there such uncertainty? Some of the reasons are covered in the chapters of this book, particularly *Chapter 2* (Schram) and *Chapter 3* (Seberg and Petersen), but one problem stands out above all others, namely that of the species rich groups. It is probably fair to say that taxonomists have collected representatives of most of the major lineages (groups) of life and that the discovery of new major branches is a rare event meriting high publicity; for example that surrounding a new species, *Symbion pandora,* discovered feeding on the mouth of the Norway lobster and assigned to a new phylum, Cycliophora[10]. However, there is now a need to fill in the gaps to find and characterise, in an evolutionary framework, all the other representatives belonging to those groups and particularly, in the context of this book, its species rich taxa.

Species diversity is not evenly distributed across the range of life forms that have existed on Earth. If species were distributed evenly between and within major groups of organisms, and if the taxonomic units were strictly comparable, we could simply and accurately count the number of species in one section of the tree (Figure 1.2a) and multiply up by the number of comparable sections so that the whole tree is represented. However, this pattern is not seen in nature, and we find striking examples of imbalance. Some evolutionary lineages have succeeded while others have perished. For example the hexapods, a group including the insects, are a species rich group compared to their closest relatives the myriapods, crustaceans, cheliceriformes and tardigrades (Figure 1.2b) and all other eukaryotic life (Cassis et al., *Chapter 13*). Furthermore, there may be as many as 200,000 diatoms (heterokont algae), but their sister group has recently been recognised as a group of tiny flagellates, Bolidophyceae, which has no more than three to five currently recognised species[11,12] (see also Williams and Reid, *Chapter 19*). Therefore, speciation and extinction are not random processes; some groups of organism have speciated to a staggering degree, while others have not. The factors leading to such imbalance are discussed throughout this book but especially in *Chapter 10* (Davies and Barraclough), *Chapter 11* (Hilu) and *Chapter 17* (Hodkinson et al.).

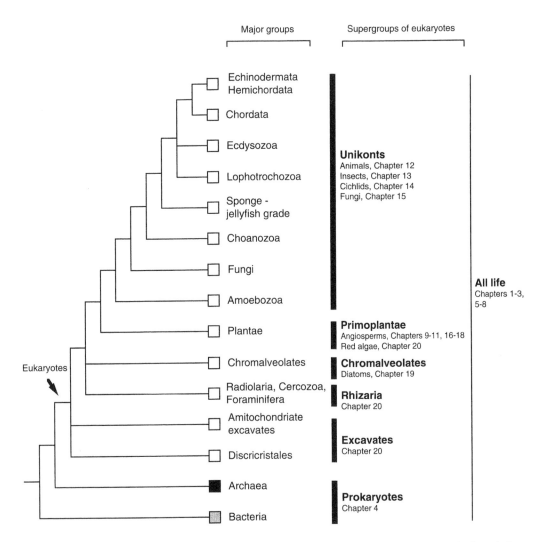

**FIGURE 1.1** Tree of life. Chapters within the book that relate to specific species rich taxa are indicated. Open squares represent eukaryotes, the black square represents archaea and the hatched square represents bacteria. Representatives of the major groups include (1) *Bacteria:* hydrogenobacteria, blue-green bacteria, green-sulphur bacteria, spirochaetes; (2) *Archaea:* korarchaeotes, crenarchaeotes, euryarchaeotes; (3) *Discricristales:* euglenids, trypanosomes, acrasid slime moulds; (4) *Amitochondriate excavates:* parabasalids, diplomonads; (5) *Radiolaria:* radiolarians; (6) *Cercozoa:* cercomonads; (7) *Foraminifera:* foraminiferans; (8) *Chromalveolates:* diatoms, brown algae, oomycetes (water moulds), ciliates, dinoflagellates; (9) *Plantae:* angiosperms (flowering plants), gymnosperms, ferns, liverworts, mosses, green algae; (10) *Amoebozoa:* slime moulds, lobose amoebae (mycetozoans); (11) *Fungi:* microsporidians, zygomycetes, basidiomycetes, ascomycetes; (12) *Choanozoa:* choanoflagellates, ichthyosporeans; (13) *Sponge — jellyfish grade:* siliceous 'sponges', calcareous 'sponges', corals, jellyfish, aceolomorphs; (14) *Lophotrochozoa:* gastropods (snails), bivalves (clams), platyhelminths, rotifers, brachiopods; (15) *Ecdysozoa:* nematodes, insects, centipedes, crabs, barnacles, spiders, velvet worms; (16) *Chordata:* humans, birds, lizards, fish, lancelets, tunicates; (17) *Echinodermata:* sea urchins, sea cucumbers; and (18) *Hemichordata:* acorn worms. (Major groups and representatives adapted from Pennisi[2] and super-groups of eukaryotes from Baldauf[27].)

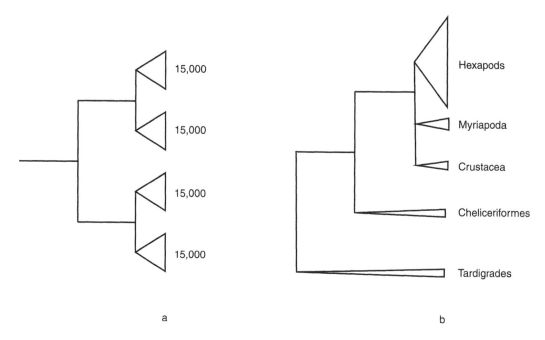

a                                                        b

**FIGURE 1.2** Species richness of phylogenetic groups is not evenly distributed. (a) If speciation and extinction had proceeded in a stochastic manner we would not expect to see significant levels of variation from the model shown (in a fully resolved and bifurcating tree). Triangles are drawn in proportion to species richness in that clade (15,000 species in all clades of Figure 1.2a). (b) An example of imbalance in species diversification within the animal group comprising the insects. Insects belong to the hexapods and account for three quarters of all described animal diversity. The hexapod clade is much larger in terms of species number than any of its sister groups of same taxonomic rank (Mriapoda, Crustacea and Cheliceriformes). (Figure 1.2b adapted from Cassis et al., *Chapter 13*.)

This book concerns the taxonomy and systematics of species rich groups; it is about how to collect, document, describe and classify them. It is also about the inextricably linked phylogenetic studies that try to position species rich taxa on the tree of life and represent their diversity. This introduction defines species rich groups, highlights examples of major species rich groups, introduces the concept of the tree of life and discusses the problems and prospects of dealing with species rich groups. It unashamedly focuses on species rich groups. Species poor groups are obviously important components of world species diversity, but they lie outside the aims and scope of this volume.

## 1.2  WHAT IS A SPECIES RICH GROUP?

### 1.2.1  QUANTITATIVE AND OBJECTIVE DEFINITIONS

Surprisingly, there is little literature on what the essential properties of a species rich group are, nor much discussion of how such groups might be defined. Rather it seems assumed that a species rich group will always be easily and universally recognisable as such and therefore needs no formal definition. We disagree and believe that it is important to attempt to define what constitutes a species rich group. Such a definition could be quantitative or qualitative, or both. In a quantitative or objective approach we might try to define a species rich group in everyday numerical terms. We could, for example, simply give a numerical threshold, which the size of a group must exceed, before it can be classified as species rich or 'big'. Frodin[13] takes this approach for plant genera and defines a 'big genus' as one containing 500 or more species. The same argument could be applied at different taxonomic ranks; that is, we could identify suitable thresholds that could be considered as big. For example, a large family could be defined as containing at least 5,000 species or a large order as containing at least 20,000 species. This approach may work within some groups such as the angiosperms or insects and for comparisons between them. For example, the grass family (Hilu, *Chapter 11* and Hodkinson et al., *Chapter 17*) and the insect bug family Miridae (Cassis et al., *Chapter 13*) both contain approximately 10,000 species and both can be considered, under this definition, to be species rich families. These families can also be considered big in that they usually present a mountainous challenge to systematists specialising in the group.

Therefore, a quantitative approach can sometimes work, but it soon runs into difficulties if used in a wider context. For example, the threshold value given above could not be used to sensibly describe the largest families of mammal because no mammal families or genera would be considered big under such a definition; there are only an estimated 5,500 mammal species in approximately 1,000 genera which themselves tend to be small. The largest mammal order, Rodentia, contains 2,000–3,000 species, but the largest mammal family, Muridae (including mice, rats and gerbils), has approximately 600 species[14–17]. Likewise this threshold figure could not be used for the fish suborder, Labroidei, containing the cichlids (Stauffer et al., *Chapter 14*), a group with approximately 1,800 species. Clearly this is unsatisfactory, as the cichlids, in most biologists' minds, are species rich (850 species of cichlid have been found in the African Great Lake Malawi alone).

A further complication in trying to numerically define a species rich group is that there are no quantitative ways of defining a particular taxonomic rank. Taxonomic ranks are clearly defined in a relative hierarchical sense (a genus is a collection of species; a family a collection of genera, and so on) but not in any absolute numerical sense. Without such common yardsticks, taxonomists can recognise species and classify them in different ways, and because of this, a taxonomic group in one rank does not necessarily represent the same degree of distinction (evolutionary divergence) as that in another taxonomic group of the same rank. For this reason it is often not possible to make meaningful comparisons from one taxonomic group to another even if they are from the same rank.

The size of taxonomic groups can also be quantified using a phylogenetic approach and sister clade comparisons. A clade may be large in comparison to its sister clade(s). For example, Hexapoda in Figure 1.2b are much more species rich than Myriapoda and Crustacea. This approach allows us to get a relative measure for comparative purposes but is not widely applicable beyond the sister clades in question. For example, both Myriapoda and Crustacea can be considered large in comparison to many other animal groups of the same taxonomic rank. This quantitative method is also open to the same problems of transferability between taxonomic groups as is the basic quantitative definition of a species rich group discussed above. Thresholds must be chosen in order to say how big a group has to be to be regarded as species rich in comparison to its sister groups.

### 1.2.2 QUALITATIVE AND SUBJECTIVE DEFINITIONS

If we recognise that 'big' for one group is 'small' for another, then we may prefer a qualitative (that is, relative or subjective) definition of a species rich group. A species rich group could therefore be defined as: 'a group with a relatively high number of species in comparison to other groups of the same, and comparable, taxonomic rank'. The caveat 'comparable' has been added to the definition to avoid the problem introduced by the wide taxonomic comparisons discussed above and the lack of common yardsticks in taxonomy.

Clearly using a qualitative approach such as that suggested above immediately leads to the well known 'hollow curve' of Willis[18,19] discussed by many other authors (including Hilu, *Chapter 11,* and to a lesser degree Parnell et al., *Chapter 16*). Therefore, to some extent we are here entering the realm of Dial and Marzluff[20] who argued that an index of dominance (the ratio of $N_{Max}/N_{Tot}$, where $N_{Max}$ is the number of subtaxa in the largest taxon and $N_{Tot}$ is the total number of subtaxa) could be used to characterise the size distributions of taxa. Clearly, as this index is not dependent on the absolute value of $N_{Tot}$, high values of the index are comparable across different taxonomic groups and so could be used to define a taxon rich group. But what value of the index should be chosen? We further discuss

**TABLE 1.1**
**Top Five Species Rich Orders of Insects**

| Orders | Species | % of All Insect Species |
| --- | --- | --- |
| Coleoptera (beetles) | 350,000 | 35.0 |
| Lepidoptera (butterflies and moths) | 150,000 | 15.0 |
| Hymenoptera (bees and wasps) | 125,000 | 12.5 |
| Diptera (flies) | 120,000 | 12.0 |
| Hemiptera (true bugs, cicadas, leafhoppers, aphids) | 90,000 | 9.0 |
| Total | 835,000 | 83.5 |

*Note:* The five largest orders, representing 6.4% percent of all insect orders, contain approximately 83.5% of all insect species.

*Source:* Cassis et al., *Chapter 13*, and references therein.

the hollow curve in Section 1.2.3 below, where we attempt to combine quantitative and qualitative definitions, and in Section 1.3.4, where patterns and processes are tackled. The following examples serve to illustrate the qualitative definition of species rich groups, namely the species rich insects and the species rich angiosperms and various subgroups within each.

Within the species rich hexapods (Cassis et al., *Chapter 13;* Figure 1.2b), the insects dominate, and so far approximately 1 million species have been described and divided into 31 orders. Insects also make up approximately three quarters of all animal species that have been described. The insects are, therefore, clearly species rich hexapods and species rich animals. Within the insects, the vast majority of species are found in one of five orders (Coleoptera, Diptera, Hymenoptera, Lepidoptera and Hemiptera). These represent 835,000 of the species and over 80% of all insect species diversity (Table 1.1). They can without difficulty be called species rich orders. The top five families account for 21% of the species (Table 1.2) despite representing less than 1% of all insect families, and 20 insect families account for almost 45% of the insects; these can all legitimately be termed species rich families. A number of species rich genera can also be identified, such as *Agrilus* (Coleoptera) with over 8,000 species, *Camponotus* (Hymenoptera) with over 1,500 species, and *Megaselia* (Diptera) also with over 1,500 species (Wall, personal communication).

Such a pattern of uneven species distribution holds true across all major groups of life. For example, within the angiosperms (more than 250,000 species in 13,185 genera[21]), five families

**TABLE 1.2**
**Top Five Species Rich Families of Insects**

| Families | Species | % of All Insect Species |
| --- | --- | --- |
| Curculionidae (weevils and snout beetles) | 50,000 | 5.4 |
| Staphylinidae (rove beetles) | 47,000 | 5.1 |
| Cerambycidae (long horned beetles) | 35,000 | 3.8 |
| Chrysomelidae (leaf beetles, flea beetles, root worms) | 35,000 | 3.8 |
| Carabidae (ground beetles) | 30,000 | 3.2 |
| Total | 197,000 | 21.3 |

*Note:* The five largest families (all beetles), representing less than 1% of all insect families, contain approximately 21% of all insect species.

*Source:* Cassis et al., *Chapter 13*, and references therein.

**TABLE 1.3**
**Top Five Species Rich Families of Angiosperms**

| Families | Species | % of All Species | Genera | % of All Genera |
|---|---|---|---|---|
| Asteraceae (daisies) | 22,750 | 9.1 | 1,528 | 11.6 |
| Orchidaceae (orchids) | 18,500 | 7.4 | 788 | 6.0 |
| Fabaceae (beans) | 18,000 | 7.2 | 624 | 4.7 |
| Rubiaceae (coffees) | 10,200 | 4.1 | 630 | 4.8 |
| Poaceae (grasses) | 9,500 | 3.8 | 668 | 5.1 |
| Total | 78,950 | 31.6 | 4,238 | 32.1 |

*Note:* The largest five families, representing just 1% of all angiosperm families, contain 31.6% of all angiosperm species.

*Source:* Data from Mabberley[21].

(beans, coffees, daisies, grasses, orchids) account for 31.6% of the species and 32.1% of the genera (Table 1.3), so these can be legitimately defined as species rich families. The top 10 angiosperm genera all have more than 1,000 species and account for 7% of all angiosperm species, the largest 15 for 9.3% (Table 1.4) and the largest 50 for 19.8% (data not shown) despite representing only

**TABLE 1.4**
**Top 15 Species Rich Angiosperm Genera**

| Rank | Genus (Family) | Number of Species |
|---|---|---|
| 1 | *Astragalus* (Fabaceae) | 3,270 |
| 2 | *Bulbophyllum* (Orchidaceae) | 2,032 |
| 3 | *Psychotria* (Rubiaceae) | 1,951 |
| 4 | *Euphorbia* (Euphorbiaceae) | 1,836 |
| 5 | *Carex* (Cyperaceae) | 1,795 |
| 6 | *Begonia* (Begoniaceae) | 1,484 |
| 7 | *Dendrobium* (Orchidaceae) | 1,371 |
| 8 | *Acacia* (Fabaceae) | 1,353 |
| 9 | *Solanum* (Solanaceae) | 1,250 |
| 10 | *Senecio* (Asteraceae) | 1,250 |
| 11 | *Croton* (Euphorbiaceae) | 1,223 |
| 12 | *Pleurothallis* (Orchidaceae) | 1,120 |
| 13 | *Eugenia* (Myrtaceae) | 1,113 |
| 14 | *Piper* (Piperaceae) | 1,055 |
| 15 | *Ardisia* (Myrsinaceae) | 1,046 |
| | Total | 23,149 |

*Note:* All top 10 angiosperm genera have at least 1,000 species, and together they contain 7% of the angiosperm species despite only representing 0.075% of the genera. The top 15 largest genera contain 23,149 species (9.3% of all angiosperms) despite representing 0.1% of all angiosperm genera. *Syzygium* (Mrytaceae) ranks 16th with 1,041 species (but see Parnell et al., *Chapter 16*), and *Ficus* ranks 31st with 750 species and is the topic of *Chapter 9* (Rønsted et al.).

*Source:* Figures from Frodin[13] and percentages calculated from total angiosperm species and genus numbers in Mabberley[21].

0.4% of all angiosperm genera (calculated from values given in Frodin[13] and Mabberley[21]). These can all be defined as species rich genera. The taxonomy and systematics of two species rich angiosperm genera are explored in more detail within this book; *Syzygium* with between 1,000 and 1,500 species (Parnell et al., *Chapter 16*) and *Ficus* with 750 species (Rønsted et al., *Chapter 9*). Whilst we believe that the pattern of uneven distribution does allow for the construction of a qualitative definition of a species rich group it is somewhat unsatisfactory in that it is largely subjective. How are the defining percentages to be set?

### 1.2.3 COMBINING OBJECTIVE AND SUBJECTIVE DEFINITIONS

It is clear that both quantitative and qualitative definitions are, to some extent, problematic. However, it is possible to combine these categories and to define a species rich group using a combination of qualitative and quantitative criteria.

As shown in *Chapter 11* (Hilu) and *Chapter 16* (Parnell et al.), the distribution of subtaxa within taxa follows a hollow curve distribution. Of the taxa so distributed, this volume is concerned with those that are large relative to the rest. Cronk[22] pointed out the asymmetry of the size distributions of taxa and indicated that the variance of the lognormal distribution is the best general descriptor of the hollow curve. Scotland and Sanderson[23] compared a number of hypothetical distributions with curves generated from real data. They concluded that none of their tested hypothetical distributions matched their real hollow curves very well. However, as only four real hollow curves were tested, the transferability of their conclusions remains unclear. In this book, we seek to define species rich groups and differentiate them from groups that are not species rich. In other words, we need to define where to stop as we slide down the hollow curve towards its tail; we require a stopping rule (or some other way of deciding how to partition the curve). Jackson's discussion of stopping rules applicable for ecological ordination[24] is, we believe, of relevance, although his favoured solutions cannot be applied, partly because of the findings of Scotland and Sanderson[23]. However, it appears to us that an extension of the concept of the scree plot, despite its disadvantages[24], does offer an opening first approximation to a definition satisfactorily combining subjective and objective criteria.

The idea underlying use of the scree plot in the context of Jackson[24] is that there is a break point in a curve, where its slope flattens out, and that values in the flat part of the curve may be disregarded. In our case, we are interested not in this particular point on the curve but in the concept of a break point. In particular, is there a break point in the tail of the curve; that is, is the tail continuous or fragmented towards its tip? Examination of the tails of a number of published hollow curves or of the data used to construct them, generally does show a break towards the end of the tail. For example, there are break points visible in tails of all the curves published by Scotland and Sanderson[23]. We understand that there may well be cases where the break point is not obvious nor possibly even singular. However, in general for the curves and data we have seen, there does appear to be an obvious gap. For example, in the case of Myrtaceae (Parnell et al., *Chapter 16*) our method yields three species rich genera in the family — *Syzygium*, *Eucalyptus* and *Eugenia*. Such a procedure seems to yield a relatively small number of truly exceptionally taxa that are exceptionally subtaxa rich; other workers may wish to extend the concept of species rich groups further into the flat part of the curve, perhaps defining species rich groups in terms of the uppermost quartile or some other non-arbitrary concept.

### 1.2.4 LARGE TAXONOMIC GROUPS

The basis of the concept for a qualitative (or combined qualitative and quantitative definition) of a species rich group can be extended to other large taxonomic assemblages. For this reason we can distinguish between 'large taxa' and 'subtaxa rich taxa'. Not all large groups are subtaxa rich, but most will be. By recognising other large taxonomic groups we accommodate the taxonomic hierarchy. For example, a family can be considered a large group because it contains a large number of genera, with no reference to the number of species. So we can use terms such as 'genus rich family', 'family rich order' or 'order rich class'.

## 1.3 RECONSTRUCTING AND USING THE TREE OF LIFE

### 1.3.1 THE TREE OF LIFE

Naming all the world's species is just a first step in understanding them. We need to know how each of these organisms relates to one another and how they are positioned on the tree of life. The tree of life model is one of the most enduring and powerful tools at the disposal of an evolutionary biologist. It is a hypothesis or statement of inferred relationships between organisms displayed in a graphical form approaching a branching, tree-like structure. Such trees, also known as phylogenetic trees, have evolved from the early attempts of Charles Darwin[25] and Ernst Haeckel[26] to more sophisticated and presumably accurate trees such as Figure 1.1. This tree has been divided up into seven major subgroups (Unikonts, Primoplantae, Chromalveolates, Rhizaria, Excavates, Archaea and Bacteria, following Baldauf[27]) and 15 minor subgroups (following Pennisi[2]). For more detailed trees, with more subgroups, see Pennisi[2], Baldauf[27], Cracraft and Donoghue[28], and Palmer et al.[29].

Despite the power of phylogenetic trees, there has been much debate about whether a tree of life exists and whether life can be accurately represented using a phylogenetic tree model or a combination of trees[30–34] (see also McInerney et al., *Chapter 4*). The answer is yes and no. Life is unlikely to be fully represented by an all-encompassing and unambiguous tree of life model. Endosymbiosis, genome fusion, horizontal gene transfer, hybridisation, polyploidy and reticulation are all substantive issues that sometimes make simple trees unrealistic approximations of phylogeny[33,35] (see also Rønsted et al., *Chapter 9*). However, the evolution of life is likely to have taken, at least within eukaryotes, a tree-like pattern (Figure 1.1 and Figure 1.3a). Within most eukaryotes there is little reason to suggest that such processes occur commonly enough to prevent the recovery of a tree of life except possibly near some reticulating tips[29]. Steel (*Chapter 7*) adds weight to this argument by showing, with a simple mathematical result, that an underlying tree of life can always be defined (and exists) even in the presence of complications such as reticulation. He shows how the notion of a tree of life can be rigorously defended but recognises that such a tree defined in the presence of complications such as reticulation will miss much of the detail and richness of evolutionary history and will be largely unresolved in places. He goes on to discuss methods for better representing and studying reticulate evolution. However, within the prokaryotes horizontal gene transfer and genome fusions have been more common, and this begs the question of whether an underlying tree structure exists and is recoverable[29,32] (see also McInerney et al., *Chapter 4*).

There are numerous other models that can be used to describe the evolutionary history of organisms, and some of these are particularly apt for prokaryotes where horizontal gene transfer and endosymbiosis have played a more significant role in evolution. They may also be more appropriate for closely related species where frequent hybridisation is known to occur. Therefore, it is clear that a basic three-domain tree of all life is an oversimplification[30], a network of some sort with a 'universal ancestor', or network of gene trees, may better explain the pattern[31,32] (see also Steel, *Chapter 5*). Zimmer[36] has coined this concept the 'mangrove of life' (Figure 1.3b). A more recent concept to emerge is the 'ring of life' (Rivera and Lake[34]). This concept has the potential to represent prokaryotic evolution and the origin of eukaryotes. Rivera and Lake's ring of life (Figure 1.3c) is based on the analysis of hundreds of genes and a method called 'conditioned reconstruction' that uses shared genes as a measure of genome similarity and allows horizontal gene transfer to be used in assessing genome based phylogeny. It resolves the dual nature of eukaryotic genomes that sit simultaneously on an eubacterial lineage (bacteria) and an archaebacterial lineage (archaea). This is what seals the ring (Martin and Embley[33]). Their model supports

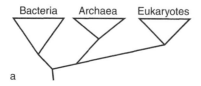

a

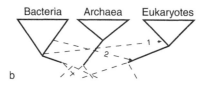

b

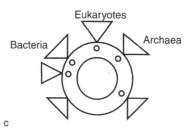

c

**FIGURE 1.3** Tree of life models. (a) A standard phylogenetic tree showing the three domains of life; within the triangles a standard tree like branching pattern is seen. (b) A network tree incorporating reticulation; reticulations are seen by endosymbiotic events (fusion of genomes) and by exchange of genes in gene trees. For example one event involved bacteria giving rise to chloroplasts (1) and another event involved bacteria giving rise to mitochondria (2). (c) A ring of life, a model used to depict evolutionary pattern especially useful for the prokaryotes and origin of the bigenomic eukaryotes. Small circles within the ring represent defining ancestors of the major groups. (Figure 1.3a adapted from Woese[30], 1.3b from Zimmer[36]; 1.3c from Rivera and Lake[34].)

the idea that a union has occurred between achaebacterial and eubacterial genomes, likely to be an endosymbiotic association between two prokaryotes. The evolution of prokaryotes and the notion of a prokaryotic tree are discussed further in McInerney et al. (*Chapter 4*).

The debate about the shape of the tree, network or ring of life is essential and stimulating. However, all models are by definition imperfect, but many are good enough to work from, and a simple tree is as good a place to start as anywhere else (as it is the simplest of the models). Even though a network or other model may better explain these patterns, they may not have the same analytical power or simplicity of a tree (or combination of trees).

### 1.3.2 BIG TREE RECONSTRUCTION FOR SPECIES RICH GROUPS: ARE LARGE PHYLOGENETIC TREES ACCURATE?

Most phylogenetic studies have included relatively few species, and only a few studies have included the large numbers of taxa required for detailed understanding of species rich groups or other large tree of life problems[37,38]. The next decade will see the rise of supersized phylogenetic trees (Hodkinson et al., *Chapter 17*) because DNA sequencing has become a standard laboratory technique and costs have dropped. Advances in DNA sequencing techniques are also envisaged. Phylogenetic analyses will therefore include more characters and more species.

One major concern is whether methods of phylogenetic reconstruction can accommodate large datasets. The first step in the production of phylogenetic trees often involves applying a method of phylogenetic reconstruction such as maximum likelihood, parsimony analysis or Bayesian inference. The second step in maximum likelihood and parsimony (but not Bayesian inference) involves the assessment of internal support, via resampling methods such as bootstrapping and the jacknife[39,40] so that the investigator can discriminate between groups with clear phylogenetic signals and those needing more investigation or more data to resolve[44]. The production of large phylogenetic trees and assessing internal support via resampling methods are mathematical and computational challenges because they involve searches of tree space (the total set of possible trees for the relevant set of taxa), and the number of possible trees grows more than exponentially with the number of taxa on the tree. This means that, as the number of taxa increases, the job of accurately finding the optimal trees under some objective function becomes relatively much more difficult due to the increase in tree space. We must therefore ask whether existing methods can, or will ever be able to, accurately reconstruct the phylogeny of species rich groups with several thousands and possibly hundreds of thousands of taxa. These are the topics explored by Wilkinson and Cotton (*Chapter 5*), Bininda-Emonds and Stamatakis (*Chapter 6*) and Steel (*Chapter 7*) and to a lesser extent by Wheeler (*Chapter 8*) and Hodkinson et al. (*Chapter 17*). Despite the scale of the problem there is cause for optimism. Increasing the number of characters in a dataset[42–44] and the number of taxa sampled[45–47] (see also Hodkinson et al., *Chapter 17*) generally results in more reliable phylogenetic inferences, if not limited significantly by computational issues. At some point the computational complexity of the problem must, however, outweigh the benefits of adding taxa (Bininda-Emonds and Stamatakis, *Chapter 6*).

Empirical and theoretical studies show that existing methods perform relatively well with large datasets[47–49]. For example, Salamin et al.[44] have shown, using Monte Carlo simulations, good accuracy of parsimony and neighbour joining methods to retrieve model trees with taxon numbers up to 13,000 (the number of angiosperm genera and close to the number of species in a large angiosperm family such as the grasses) if sequences of sufficient length (number of nucleotides) were used (see Hodkinson et al., *Chapter 17*). Testing the reliability of phylogenetic inference using, for example, resampling methods is also a major challenge with large DNA matrices[41]. However, existing methods and shortcuts perform relatively well[41], and we expect that advances in tree search methods will facilitate this process.

Better and more powerful phylogenetic methods are being developed and tested for analysing large computationally demanding phylogenetic datasets. These methods can be categorised into supermatrix and supertree methods[50–52] (see also Bininda-Emonds and Stamatakis, *Chapter 6;* Steel, *Chapter 7*). Supermatrix and supertree approaches are not mutually exclusive, as supertrees are essential in many formal divide-and-conquer analysis methods of single datasets (supermatrices). These divide-and-conquer strategies seek to break down the problem into smaller subproblems (a process known as decomposition) that are computationally easier to solve (Wilkinson and Cotton, *Chapter 5*). The results from these subproblems are then combined to provide an answer for the initial global problem. Large analyses may incorporate divide-and-conquer search strategies such as quartet puzzling and disk covering. These methods are likely to become increasingly important for analyses of large data sets as well as for searches of smaller data sets using more complex and computationally demanding optimality criteria.

Wilkinson and Cotton (*Chapter 5*) discuss advances in supertree methodology as part of a divide-and-conquer strategy. They explore the issue of effective taxon overlap and how it may be achieved via suitable decomposition, and they present a new fast supertree method. Bininda-Emonds and Stamatakis (*Chapter 6*) further discuss theoretical issues surrounding the reconstruction of large phylogenetic trees. They investigate the potential to reconstruct phylogenies for species rich groups and ever-larger portions of the tree of life using a range of methods; they explore the scalability of phylogenetic accuracy with respect to species number. Their results show that taxon number itself, especially with the implementation of disk covering methods, may not be the constraining factor in these analyses but that the strategy used to sample taxa may have a larger impact on both accuracy and analysis time.

### 1.3.3 Characters and Homology

Accurate phylogenetic analysis is critically based on the input of high quality phylogenetically informative characters (that is, 'good' characters), and these can be of many types but are predominantly molecular, morphological or anatomical. Obviously, different types of data are useful for the study of different evolutionary processes and at different levels of evolutionary divergence/taxonomic rank. For example, nucleotide substitutions within single-locus nuclear genes are proving highly valuable for studies of closely related species. Likewise, combinations of different genes including nuclear, plastid and mitochondrial genomes are utilised for studies of hybridisation, introgression and polyploidy in such closely related species (discussed in depth by Rønsted et al., *Chapter 9*). Morphological characters are essential for many analyses including those of extinct fossil species[29] and are vital in investigations of evolution and development (evo-devo). Minelli et al. (*Chapter 12*) outline the importance of morphology in evo-devo studies and show how it can help with phylogenetic reconstruction in general.

The use of DNA sequence data in phylogenetic analysis requires assumptions to be made about the homology of characters (positional homology of nucleotides within aligned sequences). This is often an overlooked problem and is particularly important for analyses of large datasets. Wheeler (*Chapter 8*) explores this critical issue of homology assessment and describes the various solutions to the problem. Sequence availability is also an issue and is discussed in Hodkinson et al. (*Chapter 17*).

### 1.3.4 Patterns and Processes of Diversity and Understanding the Hollow Curve

Large phylogenetic trees can be used for the study of pattern and processes in evolution but also a whole list of other biological questions. Dobzhansky's statement that 'nothing in biology makes sense except in the light of evolution'[52] has almost become a cliché but remains highly relevant and pertinent.

One of the most commonly used applications of large trees is for classification and taxonomy. However, they also have wider application to a host of biological and evolutionary questions[53,54]. Large trees have convenience from a statistical perspective (Steel, *Chapter 7*) and there are many theoretical reasons for using large trees[46,55–57]. For example, they are required for accurate inferences of macro-evolutionary processes because in such studies it is desirable to sample most of the diversity within a study group to reduce the risk of incorrect phylogenetic tree reconstruction and to allow meaningful comparisons to be made or hypotheses to be tested[40,53].

Large phylogenetic trees of species rich taxa are useful tools for detecting diversification rate variation, extinction and exploring the processes that may have led to the diversity of the group. We may, for example, wish to know why some groups have become species rich and others have either failed to diversify or have perished. The distribution of species richness within a phylogenetic tree, even between closely related groups of organisms, can vary enormously.

As discussed above, the hollow curve[18,19] has been used to describe patterns of diversification where few taxonomic groups are species rich while the majority are species poor. There may be, for example, an inverse relationship of large to small genera (that is, lots of small genera and few large ones). Within the angiosperms the frequency distribution of genera containing increasing numbers of species (number of species in a genus plotted against the number of genera) approximates to the logarithmic hollow curve, although the first term is always larger than expected. Because of this, classifications are generally strongly polarised, having some 80% of the genera smaller than average but some 80% of the species concentrated in genera larger than average[58,59]. Age of a genus, species richness of genera and geographical area that the genus occupies tend to be correlated, although there are opposing views as to how that correlation maps out. Cronk[22] considers large genera to be recent blooms of evolution, whereas Willis interpreted big genera as being old (for further discussion see Hilu, *Chapter 11*). Modern phylogenetic reconstruction allows these alternative hypotheses to be tested. Widespread genera are often larger than continental genera[60,61]. Clayton and Renvoize[61] suggest that there may be a dichotomy in evolutionary strategies between large genera speciating in a wide variety of niches and small genera in labile environments subject to continuing processes of disruption and replacement. These are hypotheses that require detailed analysis and testing. The properties of the hollow curve and processes leading to it are discussed in detail by Hilu (*Chapter 11*) and Parnell et al. (*Chapter 16*).

A number of tests using the temporal and/or topological properties of phylogenetic trees exist to determine if diversification variation is statistically significant[62–65]. In the species rich angiosperms, for example, diversification can vary by over several orders of magnitude between clades (Davies and Barraclough, *Chapter 10*). Furthermore, within any particular angiosperm family, such as the grasses, diversification rates have also been shown to vary (Hodkinson et al., *Chapter 17*). Factors including key biological traits, coevolution, geography and environmental variables may have contributed to the variation that exists in net diversification between clades[62,65]. Davies and Barraclough (*Chapter 10*) review studies to explore diversification in flowering plants using large scale phylogenetic trees. They also discuss further statistical tests to explore these processes. Rønsted et al. (*Chapter 9*) also discuss the coevolution and cospeciation of *Ficus* with hymenopteran wasps belonging to the species rich insect family Agonidae.

## 1.4 TAXONOMY OF SPECIES RICH GROUPS

If we are going to document and understand the diversity of species in the world, and that of species rich groups in particular, we need to make sure that some basic tasks, including the collecting, naming, describing and classifying of those organisms, are undertaken. We need to complete these tasks systematically across all groups of organisms, and because of the currently high rate of ablation of biodiversity (the biodiversity crisis), we need to complete these tasks soon. We are facing a potentially massive episode of extinction, so it is essential that such studies are carried out as quickly as possible so that conservation policies and strategies are based on the best possible information.

### 1.4.1 COLLECTING

Collecting trips need to avoid unnecessary duplication and ensure that the maximum species diversity is sampled. They also need to be shown to be good value for money. Collecting is one of the main rate determining steps in documenting the world's species and further characterising them. The topics of how we should focus and prioritise our collecting efforts to maximise new species discovery are covered in *Chapter 18* (Utteridge and de Kok) and to a lesser degree in *Chapter 2* (Schram) and *Chapter 3* (Seberg and Petersen). Collecting is a slow and expensive process. For example, over 100 grasses were collected in a recent two-week period in New South Wales and Queensland, Australia, by the first author of this chapter and Surrey Jacobs, a highly cooperative and experienced grass taxonomist, from the Royal Botanic Gardens, Sydney. One of the grass species, *Alexfloydia repens,* is only known from one location in the world. The second, *Homopholis belsonii,* is very rare and endangered (Jacobs, personal communication). Both species took close to a day to track down and collect, entailing considerable financial expense, not to mention leech attacks, tick infestations and mosquito bites (bloody biodiversity!). Beyond such anecdotal statements, others have tried to quantify the pace of collecting in an attempt to estimate the scale of the task. Parnell's quantification of the costs of collecting[66] showed that about 85% of the costs of collecting a specimen for a number of expeditions were salary associated, with 63% being direct salary costs. Surprisingly, he showed that expenses such as travel, local living and postage for a collecting expedition, which is the part external agencies are most likely to be asked to fund (and without which the expedition simply cannot occur), constituted only about 12–17% of the total costs. Seberg and Petersen (*Chapter 3*) and Cassis et al. (*Chapter 13*) have tried to quantify the effort required to sample species rich groups by doing some simple calculations based on the number of people days it will take to collect all remaining species of a species rich group.

Such estimates allow us to see the scale and potential cost of the problem, but we should also remember this is only part of the process. It covers the resources required for collection, but not the additional resources needed for describing and classifying the organisms (that could amount to the same or more again). In reality these figures are also likely to be underestimates because geographical areas will need to be resampled many times, at different times of the year, with different methods (with specialist and generalist collectors; see Utteridge and de Kok, *Chapter 18*) before we can be sure that we are close to collecting all species in an area.

### 1.4.2 Naming, Describing and Classifying

The process of naming, describing and classifying organisms is sometimes known as alpha taxonomy (Williams and Reid, *Chapter 19*); it is time consuming and requires highly qualified staff. For some taxa, the shortage of specialists is an issue, leading to huge delays in identification. Therefore, ensuring some degree of evenness of taxonomic coverage is an important issue. Taxonomy needs to be done across the board, not just for well known organisms (we have provided examples in the latter section of this book for a range of taxonomic groups). The focus tends to be on well known groups and may be excessive. Working on the relatively small genus *Cyclamen* (c. 30 species) Compton et al.[67] indicate that the 'differing infrageneric classifications produced in *Cyclamen* result from varying taxon sampling, differing interpretation of morphological data, changes in the sources and analysis of data, and inconsistent application of names'. They conclude that 'extensive subdivision of small genera in the absence of adequate data that could provide evidence for consistent patterns of relationship is premature and leads to a proliferation of names'. Clearly, large or species rich genera offer far more potential for inappropriate subdivision, a topic briefly discussed in Parnell et al. (*Chapter 16*). Concentration on relatively well known groups may occur at the expense of the less well known ones, a real problem if those less well known groups are also big, a topic discussed in Schram (*Chapter 2*).

The taxonomic coverage of papers in this book spans the tree of life and can be seen in Figure 1.1. *Chapter 2* (Schram) and *Chapter 3* (Seberg and Petersen) introduce general issues, and more specific discussions are given for insects (Cassis et al., *Chapter 13*), fish (Stauffer et al., *Chapter 14*), fungi (Tang et al., *Chapter 15*), angiosperms (Rønsted et al., *Chapter 9*; Parnell et al., *Chapter 16*; Hodkinson et al., *Chapter 17*), diatoms (Williams and Reid, *Chapter 19*) and algae (Brodie and Zuccarello, *Chapter 20*). Many of the chapters discuss the advances made in electronic resources that make 'taxonomic information readily available at the click of a mouse' (Bisby et al.[68]). Such systems will involve 'terascale taxonomy', having to handle enormous volumes of information including data, literature and images, on behaviour, classification, ecology, genome, geography, morphology, nomenclature, ontogeny, phylogeny and physiology (Wheeler et al.[69]). Considering an estimated world species number of 10 or more million, this will ultimately result in trillions of observations associated with specimens in natural history collections[69]. Digital databasing has started[68] and is making good progress. It will certainly facilitate taxonomic work and make information globally available by linking institutions such as museums, herbaria, universities and their taxon specialists. For specialist species rich groups there are several existing high quality database systems that can be used as models (Schram, *Chapter 2*). There is therefore, as Schram explains, no need to reinvent the wheel, although a review of such systems could be useful. Experiences with some model groups such as the plant bugs (Cassis et al., *Chapter 13*) and grasses (Hodkinson et al., *Chapter 17*) should be evaluated and recommendations made on how best develop other systems. We must also remember that the digital interface is only a tool and cannot replace well trained taxonomists or physical resources such as herbaria and museums. These resources will only work with international cooperation. Such coordinated action at an international level is also needed to reach consensus over taxonomic nomenclature and accepted names.

The DNA revolution has offered huge potential to taxonomy and systematics, but as with the digital revolution, we should take care. Obviously we should be prepared to embrace the methods where they can offer real help. For example, a recent development that may help with the taxonomy and systematics of species rich groups is DNA barcoding and DNA taxonomy. The slow pace of species description and taxonomy has led some to call for a modern DNA based taxonomy[70-72]. In this method, DNA sequences are used to identify the organism. Sequences are generated and compared to sequences found in a database that have known identity and are linked to real, accurately identified specimens in institutions such as herbaria and museums. The appeal of this fully automated approach is that anybody should be able to identify an organism without specialist knowledge of the group. It also offers the potential to develop futuristic tools that can instantaneously identify an organism by sampling its DNA and making a comparison to a database of sequences. This would have particular advantages in species rich groups where taxon identification is often a problem and synonymy a big issue.

However, there are a number of issues with this technology, especially if interpreted in the strict sense, including concerns about sequence quality, insufficient sampling within and amongst species, pseudogenes, herbarium specimen quality and availability, type specimen use and common occurrence of hybridisation and introgression and associated DNA exchange (capture) between closely related species. Seberg and Petersen (*Chapter 3*) discuss the pitfalls of DNA technology and highlight the danger of using it inappropriately as a shortcut in taxonomy. DNA barcoding is seen by many as a better alternative in that it uses DNA sequences to aid identification but is not all prevailing when it comes to identification. DNA can certainly facilitate and improve taxonomy. DNA sequences have the added bonus that they have high potential for phylogenetics, classification and for providing a phylogenetic framework for developing a meaningful monographic study (Hodkinson et al., *Chapter 17*), although caveats may apply[73]. Phylogenetics, molecular systematics and taxonomy are therefore inextricably linked.

## 1.5   CONCLUSIONS: BLAME EVOLUTION AND POLITICIANS

This book is concerned with species rich groups. By concentrating on such groups we do not mean to suggest that species poor groups should be ignored. Far from it, but they are outside the scope of this book. We divide this book into three sections:

- Introduction and general context
- Reconstructing and using the tree of life
- Taxonomy and systematics of species rich groups (case studies)

To document and characterise the world's species rich groups is one of the largest challenges of biology and needs financial and political support. The reason this challenge has not been adequately addressed is partly because evolution has set us an enormous task and partly because politicians have not prioritised the problem sufficiently highly; we should therefore blame both evolution and politicians. However, the task is achievable. Schram, in the next chapter, outlines his vision of how this could be achieved. Readers may not agree with all his points but will hopefully find some common ground on most of them. It will require the meshing together of phylogenetics and taxonomy, considerable advances in informatics, improved and increased collecting, training of taxonomists and significant financial support. We hope that this book goes some way to help achieve that aim.

## ACKNOWLEDGEMENTS

We thank Gerry Cassis, Nicolas Salamin and Michael Wall for comments on this manuscript.

## REFERENCES

1. Blackmore, S., Biodiversity update: progress in taxonomy, *Science,* 298, 365, 2002.
2. Pennisi, E., Modernizing the tree of life, *Science,* 300, 1692, 2003.
3. Wheeler, Q.D., Taxonomic triage and the poverty of phylogeny, *Phil. Trans. R. Soc. Lond. B,* 359, 571, 2004.
4. Gaston, K.J. and May, R.M., The taxonomy of taxonomists, *Nature,* 356, 281, 1992.
5. Solow, A.R., Mound, L.A., and Gaston, K.J., Estimating the rate of synonymy, *Syst. Biol.,* 44, 93, 1995.
6. May, R.M., The dimensions of life on earth, in *Nature and Human Society: The Quest for a Sustainable World,* National Academy of Sciences Press, Washington DC, 2000.
7. Hodkinson, I.D. and Casson, D., A lesser predilection for bugs — Hemiptera (Insecta) diversity in tropical rain-forests, *Biol. J. Linn. Soc.,* 43, 101, 1991.
8. Stork, N.E., Insect diversity: facts, fiction and speculation, *Biol. J. Linn. Soc.,* 35, 321, 1988.
9. Ødegaard, F., Diserud, O.H., and Ostbye, K., The importance of plant relatedness for host utilization among phytophagous insects, *Ecol. Lett.,* 8, 612, 2005.
10. Funch, P. and Kristensen, R.M., Cycliophora is a new phylum with affinities to Entoprocta and Ectoprocta, *Nature,* 378, 711, 1995.
11. Guillou, L. et al., *Bolidomonas:* a new genus with two species belonging to a new algal class, the Bolidophyceae (Heterokonta), *J. Phycol.,* 35, 368, 1999.
12. Kühn, S., Medin, M., and Eller, G., Phylogenetic position of the parasitoid nanoflagellate *Pirsonia* inferred from nuclear-encoded small subunit ribosomal DNA and a description of *Pseudopirsonia* n. gen. and *Pseudopirsonia mucosa* (Drebes) comb. nov., *Protist,* 155, 143, 2004.
13. Frodin, D.G., History and concepts of big plant genera, *Taxon,* 53, 753, 2004.
14. Vaughan, T.A., Ryan, J.M., and Capzaplewski, N.J., *Mammalogy,* 4th ed., Saunders College Publishing, 2000.
15. Michaux, J., Reyes, A., and Catzeflis, F., Evolutionary history of the most speciose mammals: molecular phylogeny of muroid rodents, *Molec. Biol. Evol.,* 17, 280, 2001.
16. O'Leary, M.A. et al., Building the mammalian sector of the tree of life: combining different data and a discussion of divergence times for placental mammals, in *Assembling the Tree of Life,* Cracraft, J. and Donoghue, M.J., Eds., Oxford University Press, Oxford, 2004, 490.
17. Wilson, D.E. and Reeder, D.M., Eds., *Mammal Species of the World,* 3rd ed., Johns Hopkins University Press, 2005.
18. Willis, J.C., *Age and Area,* Cambridge University Press, Cambridge, 1922.
19. Willis, J.C., The birth and spread of plants, *Boissera,* 8, 1949.
20. Dial, K.P. and Marzluff, J.M., Nonrandom diversification within taxonomic assemblages, *Syst. Zool.,* 38, 26, 1989.
21. Mabberley, D.J., *The Plant Book,* 2nd ed., Cambridge University Press, Cambridge, 1997.
22. Cronk, Q., Measurement of biological and historical influences on plant classifications, *Taxon,* 38, 357, 1989.
23. Scotland, R.W. and Sanderson, M.J., The significance of few versus many in the tree of life, *Science,* 303, 643, 2004.
24. Jackson, D.A., Stopping rules in principal components analysis: a comparison of heuristical and statistical approaches, *Ecology,* 74, 2204, 1993.
25. Darwin, C., *On the Origin of Species by Means of Natural Selection, or the Preservation of Favoured Races in the Struggle for Life,* John Murray, London, 1859.
26. Haeckel, E., *Generale Morphologie der Organismen,* Verlag von Georg Reimer, Berlin, 1866.
27. Baldauf, S.L., The deep roots of eukaryotes, *Science,* 300, 1703, 2003.
28. Cracraft, J. and Donoghue, M.J., *Assembling the Tree of Life,* Oxford University Press, Oxford, 2004.
29. Palmer, J.D., Soltis, D.E., and Chase, M.W., The plant tree of life: an overview and some points of view, *Amer. J. Bot.,* 91, 1437, 2004.

30. Woese, C.R., Kandler, O., and Wheelis, M.C., Towards a natural system of organisms: proposal for the domains Archaea, Bacteria and Eucarya, *Proc. Natl. Acad. Sci. USA,* 87, 4576, 1990.
31. Woese, C.R., The universal ancestor, *Proc. Natl. Acad. Sci. USA,* 95, 6854, 1998.
32. Doolittle, W.F., Phylogenetic classification and the universal tree, *Science,* 284, 2124, 1999.
33. Martin, W. and Embley, T.M., Early evolution comes full circle, *Nature,* 431, 134, 2004.
34. Rivera, M.C. and Lake, J.A., The ring of life provides evidence for a genome fusion origin of eukaryotes, *Nature,* 431, 152, 2004.
35. Linder, C.R. and Rieseberg, L.H., Reconstructing patterns of reticulate evolution in plants, *Amer. J. Bot.,* 91, 1700, 2004.
36. Zimmer, C., *Evolution: The Triumph of an Idea,* William Heinemann, London, 2002, 101.
37. Savolainen, V. and Chase M.W., A decade of progress in plant molecular phylogenetics, *Trends Genet.,* 19, 717, 2003.
38. Sanderson, M.J. and Driskell, A.C., The challenge of constructing large phylogenetic trees, *Trends Plant Sci.,* 8, 374, 2003.
39. Efron, B., Bootstrap methods: another look at the jackknife. *Ann., Stat.,* 7, 1, 1979.
40. Felsenstein, J., Phylogenies and the comparative method, *Am. Nat.,* 125, 1, 1985.
41. Salamin N., et al., Assessing internal support with large phylogenetic DNA matrices, *Molec. Phylogenet. Evol.,* 27, 528, 2003.
42. Erdos, P.L. et al., A few logs suffice to build (almost) all trees: part II, *Theor. Comp. Sci.,* 221, 77, 1999.
43. Bininda-Emonds, O.R.P., et al., Scaling of accuracy in extremely large phylogenetic trees, in *Pacific Symposium on Biocomputing 6,* Altman, R.B., et al., Eds., World Scientific Publishing Company, River Edge, New Jersey, 2001, 547.
44. Salamin, N., Hodkinson T.R., and Savolainen, V., Towards building the tree of life: a simulation study for all angiosperm genera, *Syst. Biol.,* 54, 183, 2005.
45. Hillis, D.M., Inferring complex phylogenies, *Nature,* 383, 130, 1996.
46. Hillis, D.M., Taxonomic sampling, phylogenetic accuracy, and investigator bias, *Syst. Biol.,* 47, 3, 1998.
47. Källersjö, M. et al., Simultaneous parsimony jackknife analysis of 2538 *rbcL* DNA sequences reveals support for major clades of green plants, land plants, seed plants and flowering plants, *Pl. Syst. Evol.,* 213, 259, 1998.
48. Soltis, P.S., Soltis, D.E., and Chase, M.W., Angiosperm phylogeny inferred from multiple genes as a tool for comparative biology, *Nature,* 402, 402, 1999.
49. Savolainen, V. et al., Phylogeny reconstruction and functional constraints in organellar genomes: plastid versus animal mitochondrion, *Syst., Biol.,* 51, 638, 2002.
50. Salamin, N., Hodkinson T.R., and Savolainen, V., Building supertrees: an empirical assessment using the grass family (Poaceae), *Syst. Biol.,* 51, 136, 2002.
51. Wilkinson, M. et al., The shape of supertrees to come: tree shape related properties of fourteen supertree methods, *Syst. Biol.,* 54, 419, 2005.
52. Dobzhansky, T., Nothing in biology makes sense except in the light of evolution, *Am. Biol. Teach.,* 35, 125, 1973.
53. Purvis, A., Using interspecies phylogenies to test macroevolutionary hypotheses, in *New Uses for New Phylogenies,* Harvey, P.H. et al., Eds., Oxford University Press, Oxford, 1996, 153.
54. Harvey, P.H. et al., Eds., *New Uses for New Phylogenies,* Oxford University Press, Oxford, 1996.
55. Rannala, B. et al., Taxon sampling and the accuracy of large phylogenies, *Syst. Biol.,* 47, 702, 1998.
56. Källersjö, M., Albert, V.A., and Farris, J.S., Homoplasy increases phylogenetic structure, *Cladistics,* 15, 91, 1999.
57. Hillis, D.M. et al., Is sparse taxon sampling a problem for phylogenetic inference? *Syst. Biol.,* 52, 124, 2003.
58. Clayton, W.D., Some aspects of the genus concept, *Kew Bull.,* 27, 281, 1972.
59. Clayton, W.D., The logarithmic distribution of angiosperm families, *Kew Bull.,* 29, 271, 1974.
60. Clayton, W.D., Chorology of the genera of Gramineae, *Kew Bull.,* 30, 111, 1975.
61. Clayton, W.D. and Renvoize, S.A., *Genera Graminum: Grass Genera of the World,* Her Majesty's Stationery Office, London, 1986.
62. Barraclough, T.G. and Nee, S., Phylogenetics and speciation, *Trends Ecol. Evol.,* 16, 391, 2001.
63. Chan, K.M.A. and Moore B.R., Whole-tree methods for detecting differential diversification rates, *Syst. Biol.,* 51, 855, 2002.

64. Chan, K.M.A. and Moore B.R., SYMMETREE: whole-tree analysis of differential diversification rates, *Bioinformatics*, 21, 1709, 2004.

65. Moore, B.R., Chan, K.M.A., and Donoghue, M.J., Detecting diversification rate variation in supertrees, in *Phylogenetic Supertrees: Combining Information to Reveal the Tree of Life*, Bininda-Emonds, O.R.P., Ed., Kluwer Academic Publishers, Dordrecht, 2004, 487.

66. Parnell, J.A.N., The monetary value of herbarium collections, in *Biological Collections and Biodiversity*, Rushton, B.S., Hackney, P., and Tyrie, C.R., Eds., Linnean Society of London Special Publication 3, England, 2001, 271.

67. Compton, J.A., Clennett, J.C.B., and Culham, A., Nomenclature in the dock. Overclassification leads to instability: a case study in the horticulturally important genus *Cyclamen, Bot. J. Linn. Soc.*, 146, 339, 2004.

68. Bisby, F.A. et al., Taxonomy, at the click of a mouse, *Nature*, 418, 367, 2002.

69. Wheeler, Q.D., Lipscomb, D., and Platnick, N., Terascale taxonomy: cyber-infrastructure and the Linnaean legacy, in *Proc. of the Fourth Biennial Conference of the Systematics Association*, Trinity College Dublin, Ireland, 2003.

70. Tautz, D. et al., DNA points the way ahead in taxonomy, *Nature*, 418, 479, 2002.

71. Tautz D. et al., A plea for DNA taxonomy, *Trends Ecol. Syst.*, 18, 70, 2003.

72. Lipscomb, D., Platnick N., and Wheeler, Q., The intellectual content of taxonomy: a comment on DNA taxonomy, *Trends Ecol. Syst.*, 18, 65, 2003.

73. Stace, C.A., Plant taxonomy and biosystematics: does DNA provide all the answers? *Taxon*, 54, 999, 2005.

# 2 Taxonomy/Systematics in the Twenty-First Century

*F. R. Schram*
Department of Biology, University of Washington, Seattle, USA
Formerly of Zoological Museum, University of Amsterdam, The Netherlands

## CONTENTS

*And out of the ground the Lord God formed every beast of the field, and every fowl of the air; and brought them unto Adam to see what he would call them: and whatsoever Adam called every living creature, that was the name thereof. And Adam gave names to all cattle, and to the fowl of the air, and to every beast of the field ... (Genesis 2:19–20)*

## ABSTRACT

Taxonomy/systematics has had a history extending back to the 1880s, with Cassandras issuing dire warnings about the future of the science, but little hard data exist to document these warnings. Some institutions have done well, while others have endured severe cutbacks or even disappeared. Meanwhile, the need for effective biodiversity knowledge is increasing exponentially. The numbers of species in many groups is truly staggering, and the use of information technology to manage terascale volumes of data in the science of taxonomy is inarguably essential. The tools to effectively move on this need to be developed, and online models for specific groups of organisms including species rich groups need to be made available. Some unfortunate decisions and trends in the management of natural history museums and universities have occurred in the recent past. Human capital and mobility need to be enhanced. The biodiversity crisis is real. Rivalries must be put aside, and true cooperation must occur if the crisis is to be addressed. An action plan is needed to: (1) establish an international structure to deal with issues vital to furthering a healthy taxonomy/ systematics community, including a czar to spearhead the plan; (2) increase spending with funding levels targeted on per capita population; (3) approach staffing needs in universities with proactive arguments for replacing retiring staff with taxonomists; (4) channel people and training into the study of understudied groups of organisms; (5) direct training and education at enhancing human capital in systematics in developing countries; (6) require and facilitate international cooperation of networks and institutions; and (7) apply information technology on a large scale with the establishment of super computing centres.

## 2.1 HISTORICAL WAILINGS

Taxonomy/systematics is the world's oldest profession. Every culture, no matter how 'primitive', has developed systems of nomenclature to catalogue the plants and animals in its environment. However, modern systematists have been somewhat lonely professionals until recently. Systematists are often perceived as sitting quietly in front of the microscope, preparing specimens and describing new species. This kind of science has historically been low-cost research that could be done just about anywhere with some basic equipment, laboratory supplies, and a carefully assembled col- lection of monographs and reprints of earlier work. I used to tell my students that it was the kind of work with which you could still carry on an active research and intellectual life even if you ended up at a mythical remote place like North West North Dakota State Teachers College. This productivity would not be possible if you specialised in a field such as immunochemistry or some other science that would require high technology, high cost hardware. Furthermore, systematics is the kind of science that yields a lot of 'bang for the buck'; for example, research grant proposals can, for modest sums, promise yields thick in revisionary monographs and many individual species descriptions. However, times have changed. This kind of low-intensity effort is no longer adequate. Systematics as a science has metamorphosed in a generation from having 'just sufficient' funding sources, to having 'not nearly enough'.

We have read recently about the dire straits in which the science of taxonomy/systematics currently finds itself. Wheeler[1] paints a bleak picture of staffing and funding of taxonomy even while he rallies us to reshape our science and take our own fate into our hands. Wheeler et al.[2] lay a lot of the blame for the state of taxonomy at the feet of taxonomists themselves, pointing out that, for all but a few taxa, the data in museums attached to specimens is outdated, incomplete, or otherwise unreliable; and many specimens are misidentified, while others are undescribed. Carvalho et al.[3] took issue with some of the points of Wheeler et al.[2], especially with their Amero-Eurocentric viewpoint. Nevertheless, they argue that the problem with systematics is not within the science, but within the greater community of scientists and policy makers at large, and that all might be well (or at least better) if we had more jobs, more money, and more of a commitment to sustain museums and systematic collections. Whilst all this would be helpful, I believe that their outlook is too narrow and a little naïve.

I have sympathy with all these positions but have read this all before; their comments merely echo the observations of, as an example, Manning[4] and Feldmann and Manning[5], who also pleaded for more money, for more jobs and for better support for museum collections. Moreover, they in turn admitted that their comments merely reflected the same observations of K. P. Schmidt 40 years earlier[6], and those remarks were no different than the wailings of Waldo Schmitt[7], who in turn merely echoed the same issues expressed by Sir E. Ray Lankester during the 1880s. Indeed, one might question that after more than 125 years of this kind of purported attrition in the field, why are there any systematists left at all, let alone institutions in which they work? Are modern-day taxonomists the equivalent of Cassandras to whom we are not listening? Or are they merely crying wolf?

Hard data is often difficult to come by, as Parnell[8] pointed out in connection with his observations concerning the state of tropical systematic botany. Many of the negative observations are anecdotal. We can point to institutions such as the Field Museum in Chicago, USA, or the Zoological Museum in Copenhagen, Denmark, which have successfully maintained curatorial staff numbers for decades. Nevertheless, there are also institutions such as the Zoological Museum in Amsterdam, the Netherlands, that have been decimated, where there were 12 curators on the staff in the late 1980s–early 1990s, but where at the time of writing there are only three full-time curators and one half-time curator. Another seriously affected institution is the Philadelphia Academy of Science, USA, which recently sustained some significant cuts in curatorial staff[9].

At the very least, the apparent general reduction in overall staff numbers in natural history institutions will make it difficult to meet the target of the Convention on Biological Diversity. Are we now reaping the result of over a century of de-emphasis of organismic biology in favour of the more stylish and 'modern' molecular studies? I would hope not, since the integration of molecular studies holds promise for a great stimulus for progress in taxonomy/systematics. We stand on the threshold of a real renaissance in biodiversity studies. However, we cannot refuse to recognise past problems. If a living dynamic science is to grow, then steps have to be taken to correct these conditions.

## 2.2   USING TECHNOLOGY

One can make a case with some justification as to whether the science of taxonomy is evolving fast enough in the face of the demands being placed upon it by the public and by government policy makers in connection with the so-called biodiversity crisis. Quentin Wheeler and his colleagues[1,2] have argued repeatedly in recent years for a terascaling of taxonomy as a science. Potentially vast amounts of data concerning species must be digitised and manipulated in order to expedite research in taxonomy/systematics. In other words, systematic data can no longer be sampled in discrete titbits, but rather must be swallowed in huge chunks. Feldmann and Manning[5] furthermore pointed out that there was little excuse to delay (even back in 1992). Many grocery store chains and megamarketers such as Wal-Mart can keep track of millions of pieces of merchandise in thousands of stores around the world, and link these to databases of 'just in time' suppliers and backup suppliers to keep their shelves full, check out items being purchased, while producing bills for payment automatically. Yet, instead of using this technology, many systematists continue to organise meetings and workshops to plan for the day when we might barcode, or when we might link collection records, or when we might digitise specimen images. Maybe instead of meetings, we should take that money and hire some receiving and stock managers from the local Wal-Mart outlets to show museum people how to do it.

There is no denying that the task is daunting. At present, amongst the crustaceans, a species rich group with which I am most familiar, there are some 10,184 species of isopods described, in excess of 7,910 species of amphipods, between 15,000 and 20,000 species of decapods (10,500 of those species are of crabs alone), 13,000 living ostracodes with another 65,000 fossil species and at least 12,000 species of copepods. These are staggering numbers, though they do not rank amongst the largest for all groups of organisms. For example, David Williams and Geraldine Reid (*Chapter 19*) from the Natural History Museum in London, England, announce that there are only an estimated 10,000 to 12,000 species of diatoms described in the literature out of a total diversity that may possibly approach 100,000 to 200,000 species! My mind can barely grasp this claim, yet I believe them. Frank Almeda of the California Academy of Sciences, San Francisco, USA, and his colleagues claim that there are over 1,500 species in a single plant genus, *Miconia*[10]. Khidir Hilu (*Chapter 11*) and Trevor Hodkinson et al. (*Chapter 17*) have displayed trees that estimate the higher level phylogenetic relationships of 10,000 species of grasses. In addition, often there are curious disjunctions in our knowledge. Nina Rønsted et al. (*Chapter 9*) admit that, while entomologists have a good understanding of the phylogenetic relationships of fig wasps, the 750 species of figs upon which the wasps are symbionts are only now being sorted out.

These numbers are startling enough. However, one example can serve to illustrate what is involved with terascaling. I have published an inventory of the species of mantis shrimp, or Stomatopoda[11]. This is a relatively small group of crustaceans, comprising only about 482 fossil and living species. Even with the advantage of beginning with an established database it, nonetheless, took me almost two person years to track down all the valid species and available names. I also assembled information on species distributions; most of it not yet in geographic coordinates let alone in any Geographical Information System (GIS) format. That is, much of it was only general designations of approximate localities, depth ranges, general habitat (not always available I discovered), colour (an important element in the biology for mantis shrimp) and size ranges for both sexes. I had only mixed success in this effort despite the fact that this is a group that has had intensive scholarly attention over the last 40 years. Some 25% of the species of mantis shrimp are known from little more than their original morphological description based on a single specimen of only one of the sexes in museum collections; in other words, with little or no data yet available on depth, details of locality and habitat preferences. Whole genera of the mantis shrimps are incompletely understood due to these parameters.

Now, multiply the work load involved for the assembling of base data for the 482 mantis shrimp by whatever factor you need to scale up to any of the figures given above, for example, the 10,000 species of grasses. One can clearly see that to arrive at just a basic species list for some groups will entail intense effort. Why is this so? I never cease to be amazed at how much alpha taxonomic science has been done sloppily, an earmark perhaps of its cottage industry tradition. For example, a major effort in assembling the above mentioned catalogue of stomatopods entailed tracing down the location of, and catalogue numbers for, the type specimens. A significant amount of such material was never clearly designated in the original taxonomic descriptions (and many of these are late twentieth-century papers), and a significant number of the total array of mantis shrimp type specimens remain 'lost'. This little group of crustaceans is not unique. Moreover, assembling the basic species list of life is only the beginning. Wheeler has pointed out that to terascale our knowledge of the diversity of life, that is, to be able to link the basic species list to information about nomenclatural history, ecology, behaviour, development and geographic distribution (with proper GIS coordinates) will be a daunting task. This is not to say we should not undertake this effort; we must! However, it will not be easy. It will take time and money, as well as patience and persistence. This effort will also distract taxonomists and slow down their efforts to describe new species.

Modern information technology can achieve some amazing things DNA barcoding is one approach that derives directly from inventory control in consumer marketing. Recent discussions at conferences have even proposed the possibility that one day inventory barcoding might be combined with science fiction so that StarTrek-like 'tricorders' aimed at specimens in the field are linked by satellite to central databases of previously identified DNA sequences. This would certainly be an exciting new tool if it could be perfected, but Wheeler[1] warns that, unless it is handled in a rational way, barcoding could do more harm than good (see also Seberg and Petersen, *Chapter 3*). This does not mean we should not undertake barcoding, we just need to be careful that we make clear to people or agencies demanding this kind of service what the limitations of the tool are. In this case, the technology might serve to belittle real species as discovered in nature. Barcoded DNA sequences are rather dry data, but majestic 100-metre redwood trees or dazzling male peacocks in full colour are real. As Wheeler[1] asks, "Why forego all that is intellectually engaging and aesthetically beautiful to settle for what is clinically efficient?". There is wisdom in these words. There is a value in standing in front of actual species in nature; it is the principle upon which pilgrimage, and indeed mass tourism, is based.

Information technology tools must be practically focused. Just because we can do something does not mean we have to do something. Too many recently funded projects and programmes involving information technology have been done because they could be done, that is, they were undertaken because money was available, and someone from the 'top' enticed the scientists 'below' to undertake it. As an example, a European Union funded programme was recently completed to produce a species list of the terrestrial and freshwater species in Fauna Europaea (FE) (http://www.faunaeur.org). Some  4,000,000 over four years was expended. The stated target audience included the public agencies that might have call to have this data, such as customs officers who might want to know if someone is trying to import an endangered species. Much intense argument occurred amongst cooperating taxonomists within the programme concerning what sorts of data were to be included in the checklist. Furthermore, an unexpected amount of time was needed to develop the software to enter the data and to provide online access. However, what was achieved in the end works better for specialists rather than the non-scientist public servants on whom the development of the database was originally justified.

This programming problem is a critical point. A similar earlier project for the European Register of Marine Species (ERMS) ran into the same difficulties, which in the end were never resolved. ERMS settled on merely putting the faunal lists online (http://www.marbef.org/data/erms.php). This was not necessarily all bad. ERMS has the advantage of ease of use, allowing users to see whole lists of species without having to know specific Linnaean binomens before the website is further engaged. On the other hand, FE has a rather sophisticated website, but it assumes that any user will know a name to begin with before any data can be accessed, and of course the slightest misspelling (not uncommon with Latin binomens) will yield nothing.

I do not find these types of top down databases as useful as the bottom up sites that have been assembled by scientists 'in the trenches' on an as needed basis. As an example, the Crayfish Home Page is amongst the nicest and most useful (http://crayfish.byu.edu). Here one can find all sorts of practical and useful data including a complete taxonomy of the group, regional faunal lists, biogeographic distributions (with maps), photographs of species, and links to other sites. It is a well designed and easy to use site. Its utility is a reflection of its bottom up genesis. The Crayfish Home Page shows every sign of evolving into the kind of online monograph for which Godfray[12] called. Slightly more complex is Antbase (http://www.antbase.org), which focuses on the 11,000 species of ants. Antbase is a bit more difficult to navigate than the Crayfish Home Page until one becomes accustomed to the layout, but it does provide very up-to-date information and other material, including scanned literature for old species descriptions. At the other end of the spectrum, the Turbellarian Taxonomic Database (http://devbio.umesci.maine.edu/styler/turbellaria) is simplicity itself to use.

Naturally, these kinds of websites are not easy to develop, but fortunately there are people who specialise in doing such technical things. It merely requires that the people with the taxonomic expertise pair with a person with the necessary information technology expertise, and if done correctly the team can emerge with a very useful tool that not only advances and facilitates research in that group, but is also attractive, easy to navigate and informative to the general public. We should look forward to the day that there is a website for every taxonomist, or better yet for every cooperating network of taxonomists.

## 2.3  INSTITUTIONAL ISSUES

Museum systematic collections are, and should remain, at the heart of taxonomic/systematic research. Yet, natural history museums in my opinion are probably the most misunderstood institutions in the current cultural milieu. They grew out of the cabinets of Enlightenment savants, who assembled noteworthy collections on their own, such as the Teylers Museum in Haarlem, the Netherlands, or grew out of official patronage of major benefactors or government largesse, for example, the Field Museum in Chicago and any of the national museums of natural history in places like London, Paris and Washington. These collections and institutions were scholarly in nature and intellectual in their focus. They were responsible for the first great blooming of science and natural history in the 1700s. They persisted in this vein for almost 200 years with the addition of exhibits in glass cases of excess materials for the edification and education of the public.

All of this began to quickly fall by the wayside on the occasion of a particularly famous royal visit in 1977, the King Tut Exhibit. This blockbuster travelled the world drawing huge crowds to institutions in a few weeks that were only accustomed to similar numbers of entrants spread out over several years. Suddenly, there was not a museum director or board of trustees in the world that was not asking 'Why not us? Why not here?'. The blockbuster exhibit suddenly became a permanent fixture in the calendars of museum events. This happened without any serious debate on what museums used to be and the 'infotainment' venues they became. There was a practical downside as well. Carpeting in exhibition halls that was to last years, toilets that were to survive decades, had to be replaced after King Tut moved on to his next venue, immediately consuming some of the profits generated from the blockbuster. For natural history museums this was further exacerbated by the appearance a few years after Tut's show by the marvelously clever dinorobots that in many places eventually became parts of permanent exhibits.

I enjoy blockbuster exhibitions like everyone else. What I bewail is the impact they have had on institutions that had formed an important part of fundamental science and real education of the public. Institutions have handled the impact in different ways. In the mid-1960s, when I first went to work as a research assistant in the Field Museum, my boss took me down to the personnel and finance department to be signed on to the payroll. It was located in a single large room on the ground floor and about six people, including the department head, handled the entire operation. Today, the departments of personnel, financial operations and development occupy significant portions of that museum. Conversely, what was a rather modest exhibits department in the mid 1960s, is now staffed by a considerable number of people well housed in quarters quite expanded from what they were. As a scientist, I cannot complain about what happened at that particular museum. Significant improvements were made to the collections storage facilities in the ensuing decades, and the number of active scientist/curators has been maintained and perhaps somewhat expanded from what they were. I do regret that the amount of exhibition space for the general public has decreased while needs for more office and marketing space were assuaged. However, not all the museums of natural history in the world have fared so well. Many smaller collections have been orphaned in the last 40 years, and whilst their demise was one thing, their impact on the larger institutions that have taken on the curation of the orphaned collections has been significant[13].

There have been some notable good and useful things happening in the world of natural history museums. The creation of virtual museums online and their increasing linkage through initiatives such as the Global Biodiversity Information Facility (GBIF) have been very helpful (http://www.gbif.org). In connection with my stomatopod catalogue mentioned above, I was able to make good use of online data for the verification of location of type specimens, saving me weeks of time relying on regular surface post or actual physical visits to collections. There is a slowly growing development of online collection records. Some of these are marvellously constructed for ease of use. The American Museum of Natural History has an excellent site that is relatively easy to use (http://research.amnh.org/informatics) with an hierarchical layout that allows one to navigate the collections without having to necessarily know Linnaean binomens. The Smithsonian National Museum of Natural History is another example (http://www.mnh.si.edu/rc/db/colldb.html), although some sections seem to have organised their data in a more user friendly format than others. The Royal Botanic Gardens Kew's epic site is another good example (http://www. rbgkew.org.uk/ epic/index.htm) that will soon allow linked online access to images created for the African Plants Initiative, one of the most exciting digitisation projects in herbaria worldwide (John Parnell, personal communication). In addition, the Chicago area institutions that cooperated in setting up Plant Base have done a good job in creating relatively easy access to their herbarium and botanical garden holdings. If all museums, herbaria, live culture collections and botanic gardens did similar jobs, the ease of taxonomic research would be increased considerably.

## 2.4  HUMAN CAPITAL

There are institutions, nevertheless, where serious staff reductions have occurred; for example, in the Netherlands, biodiversity institutions such as the Zoological Museum in Amsterdam mentioned previously, Naturalis in Leiden and the various centres of the Dutch National Herbarium. Institutions in Eastern Europe are in dire straits.

The situation in universities in Europe and North America also merits some discussion. On the one hand, I have noted in the last few years the appearance of more and more job adverts for invertebrate and vertebrate zoologists. This is due to the ongoing need to replace the retiring generation of professors who began their careers in the late 1950s and 1960s under the impetus to expand and support science education in Europe and America as a response to the Russian sputnik. On the other hand, there has been both a disappearance of taxonomy/systematics courses in the curricula, and as noted above, a reduction in staff numbers in many institutions. Our Cassandras for the field of taxonomy/systematics have repeatedly bewailed through the decades staff number reductions in the universities, yet I have seen very little real data to support or refute this. This is not to deny that reductions have occurred, but it does indicate that a lot of the evidence is anecdotal.

Nevertheless, there have been several special programmes developed in the last decade to introduce and teach advanced systematics in many institutions. For example, Imperial College and the Natural History Museum in London have one such programme, and the University of Amsterdam created a M.Sc. programme in Systematics and Evolution. What is noteworthy about these efforts is that they have been aimed not at just a national population, but have been deliberately directed at an international clientele. Again, hard data are lacking, but I do know that several graduates of the Amsterdam programme returned to their home countries, such as Brazil, China and Indonesia, and took up positions in museums and other agencies there, and some of these people were sufficiently stimulated by their experience to start investigating Ph.D. programmes. On a pan-European level, the Erasmus Network in Systematic Biology was founded in the early 1990s under the aegis of the European Union Fourth Framework with the express purpose of introducing and spreading practical and theoretical knowledge about systematic biology within the participating European countries. Originally small with only about 10 institutions, the network grew in the ensuing decade to some 25 institutions directly and indirectly involved. Eventually the network faded away not only because funding priorities in Brussels changed, but also because it had achieved its original goals as thriving new communities of systematists took hold across Europe.

What seems to be lacking in all of this is concrete evidence of any real increase in the numbers of permanent full-time positions in Europe and North America. All too often, many institutions have remained satisfied with promises of flat funding, rather than seeking real increases in new funds not only to keep up with inflation, but also to provide real career paths for the newly trained scientists in the field. Oddly, it appears that the developing world is doing better than the Western world in employing people involved in taxonomy/systematics. Brazil must remain a role model in this regard[3], but recently other countries such as Indonesia have shown signs of expanding opportunities in our field. The developed countries can help by increasing their efforts to train young people from abroad. However, recent attempts by many countries to control the flow of people and information across borders seem to be acting against the free exchange of human capital and technical knowledge that is so desperately needed in the face of the biodiversity crisis, and this has to be strongly resisted.

## 2.5  THE BIODIVERSITY CRISIS

The word 'biodiversity', when it first appeared in the media, was castigated by scientists with comments like 'Ugh! Buzzword! It means nothing!'. This happened even though Professor E.O. Wilson[14], one of the most respected figures in the field, coined the term. However, it did mean something because it is a comfortable word that anyone can, and does, understand (politicians as

well as the general public). Eventually the scientists surrendered. Now even the hardened professionals use the word biodiversity without a second thought, but it took years of persistence to get
this acceptance. The biodiversity crisis arises because world events threaten to destroy much of the
evidence of evolution of biological diversity before it can be discovered and described. The
biodiversity crisis is real. Nevertheless, whilst the current crisis is largely attributed to man-made
changes, biodiversity crises have occurred repeatedly in the history of the planet. This is not offered
as an excuse to do nothing, but it does offer an opportunity to examine why the time imperative
is so critical now.

How systematists treat each other is another great problem. All too often, these interactions
entail rancour. Good, old fashioned argument and debate in science is healthy; it makes for progress.
Some sciences are known for it. Particle physics is a case in point, where my physicist friends tell
me that the arguments at meetings and conferences can be intense, but when everyone retires and
drinks some beer, all is harmony. Systematists do this too. The difference is that when particle
physicists return home and receive colleagues' grant proposals to review, they generally see the
benefit for the field as a whole and provide a good evaluation. All too often when systematists get
proposals to review, they remember the arguments at the meetings rather than the beer that was
drunk afterwards, see not the good of the field but rather take the opportunity to even scores, and
provide bad or cool reviews to colleagues' proposals. Admittedly, not everyone does this, but enough
do, and I am reliably informed that it remains a problem for maintaining a viable level of funding
for the field of systematics as a whole around the world. Meanwhile, important aspects of the
biodiversity crisis go unaddressed.

Aside from reacting more kindly to each other, there are many things that taxonomist/system-
atists can do in their own research that would relieve the workload. Mark Wilkinson and James
Cotton (*Chapter 5*) demonstrate that there are tricks we can employ to deal with terascale data
while the software and hardware available to systematics are being upgraded. This field, however,
has sustained much argument of late, both in journals and in scientific meetings, as to whether it
is best to use either supertrees or supermatrices (total evidence). This is again an example of one
of those pointless and suicidal tendencies in systematics. Neither method is perfect; each has both
strong and weak points; each is a way to organise and process very large chunks of information.
So, why argue and why not do both?

Systematists have to learn to talk with each other, and listen. Systematists are sometimes like
people on islands, isolated by deep and dangerous waters from the other islands (other systematists).
What taxonomy/systematics needs to develop is a 'polder model' for our science. The Dutch replaced
many of their islands with polders. Early in their history, the Dutch discovered they had to sit together
and come to a consensus as to how, and where, to build dams and dikes to sequester areas of the
landscape out of which they could pump the water and unite the high terps into good, productive
polder land. Cooperation in this effort was seen as a benefit to all; so too it should be with systematics.
The purpose of our work is not to pointlessly argue about 'methodologies', a term I have come to
loathe, but to discover biodiversity, describe it, and try to understand it! Why is it a do or die proposition
if one uses method A versus method B in one's work? Use both and get on with it! All too often
there is little sense of balance, or even common sense, in the debates of taxonomists.

## 2.6 WHAT TO DO?

The question is, will this chapter become just another in that long series of articles extending back
to E. Ray Lankester? I hope not, and to this end I would offer the following as a battle plan.

### 2.6.1 ORGANISATIONAL STRUCTURE

First, we need something to get these efforts underway and keep them going. All the ideas discussed
above, and the points I am going to put forth below, have been proposed in one way or another

many times in the past. Yet, we still read articles every few years bewailing the state of taxonomy/ systematics. We can keep doing this, or we can get organised. What I think we need is an international coordinating council to push these issues and keep pushing. It should be formed out of the taxonomically oriented societies and institutions. Maybe this body should be attached to GBIF, or maybe it should come out of the international commissions; whatever, it needs to be formed urgently. What we need in effect is a taxonomic czar with dedicated cadres who know where the buttons are to push on the political and institutional scene and when to push them. The objective of this commission would be to advance the points to follow.

### 2.6.2 INCREASED SPENDING

Second, we need more money. I do not mean 'some' money, I mean more, real cash. I am going to put forth here something that many may believe is decidedly controversial. The world spends hundreds of billions of dollars on biomedical research. Although this work is noble in its aim, by extending human life spans and reducing infant mortality, it merely exacerbates the biodiversity crisis whose central cause is too many people. If, however, we could get just 10% of the money spent on biomedical research allocated for biodiversity studies, we could have some hope of documenting diversity in the biosphere and perhaps come to grips with the issue of sustainability.

Where is this money to come from? Quentin Wheeler, formerly seconded to the National Science Foundation in Washington, USA, relates that the Foundation has responded to this need in recent years, and currently has about US$50 million a year going into various programmes that support systematic research, such as Partnerships for Enhancing Expertise in Taxonomy (PEET), Assembling the Tree of Life (ATOL), Biological Surveys and Inventories (BSI) and other initiatives. Although the government of the United States has been justly criticised the world over for its policies concerning the Rio Accords and the Kyoto Protocols, it nevertheless remains true that few other countries support the science of systematic biology in such a substantial and sustained way. As a point of comparison, the Sixth Framework European Union funding cycle is now underway, and while initial results seemed encouraging, in the end systematics will be lucky to achieve programme funding of 50 million. This is a tidy sum but one which will have to stretch over a five- to six-year period, that is, one fifth to one sixth the level of current funding in the USA.

However, numbers can be misleading. The government of Sweden, largely under the influence of the Green Party, has recently committed the equivalent of some 15 million a year to various biodiversity initiatives. This seems like a small sum compared to the ones just mentioned above. The figure is not so paltry when one considers that Sweden has only about 5 million people; that is, Sweden will spend about 3 per person in this effort. The USA, with some 285 million people will only be spending about US$1.75 per person, and the European Union with over 400 million citizens will only be allocating about 0.15 per person.

Somehow, all first-world countries will have to find ways to pick up the slack concerning funding taxonomy/systematics programmes. Ideally, this will be done in ways that will constructively involve our developing world colleagues. There are nascent communities of good scientists in places like Africa, Asia and Latin America who study biodiversity and who must be brought into the global mainstream.

### 2.6.3 JOBS

Third, we do need more jobs; permanent positions for taxonomists and systematists the world over. This has been a repeated theme through the decades. Jobs mean money to hire people. How this may be achieved is problematic. One avenue meriting further consideration concerns recruitment in universities in North America and Europe. If job adverts are any indication, many of these positions are calling for floristic and faunistic specialists. While subspeciality is often open, many are intended for ecologists and behaviourists. It would seem to me, however, that many of these positions should go to morphologists and taxonomists with some proactive persuasion being exerted from colleagues and scientific societies when the positions become available.

### 2.6.4 CHANNEL STAFF

Fourth, it is not enough to add positions to staff lists. Channelling people and resources is required. As positions in museums and universities become available, there should be some assessment of what taxonomic specialities are needed in a world context. I do not want to denigrate the study of beetles and butterflies, but it would seem to me that the last thing we need is more coleopterists or lepidopterists, when there could be a real need to see some young people undertake careers in the study of earwigs, or some other little studied group of organisms. Why do we need so many ornithologists when whole phyla of organisms have no available expertise? Likewise, we do not suffer from a lack of botanists who study angiosperms, but there is a real need to train people conversant in mosses. Even so, I am told by my botany colleagues that there are whole families of flowering plants for which there are no experts available at all in the world. Somehow, we have to slow down the production of taxonomists specialising in well studied groups and encourage the training of workers in groups that have been neglected. Professional societies should conduct detailed inventories and then become proactive in this channelling. They might start with a continuing census of groups and the numbers of specialists who work on them. The overlooked or understudied groups, many of which are species rich taxa, could then be targeted, and when positions become available, strong cases should be made for filling them with someone working on one of the neglected groups.

### 2.6.5 TRAINING AND EDUCATION

Fifth, education has to become a priority, and linked with this, jobs for students once they are trained. However, the training and job placement of people has to be globalised and addressed on a world scale. We need to train more taxonomists certainly, but these people need not necessarily obtain positions in the developed countries. There is a desperate need for taxonomists in the developing world in all groups of organisms. Granting agencies can help here immensely by making available targeted sums to specifically train people from developing countries at the Masters, Ph.D., and postdoctoral levels in Europe and North America with the idea that these people would go back to positions in museums, herbaria, botanic gardens and universities in their home countries. Thus, the agencies with funds, such as the National Science Foundation and the European Union, would coordinate their funding not only with the host institutions, but also the sending countries. The idea is that advanced training will be supported financially in Europe and North America for people from developing countries, if the mother country will insure a position after the degree is obtained. Hence, the need for trained taxonomists is addressed and we facilitate technology transfer on a global scale.

### 2.6.6 INSTITUTIONAL COOPERATION

Sixth, natural history museums and collections should be helped and encouraged to undertake networking. Several efforts are now underway, as noted above, but the faster this takes place the better it will be for the field as a whole. In addition, some model systems already being implemented are marvels of ease of use, such as ITIS (http://www.itis.usda.gov). Furthermore, museums should be encouraged to adopt pre-existing information technology systems rather than waste time and money reinventing the wheel and developing their own. Encouraging signs of facilitating cooperative programmes are emerging in the ATOL program in pan-European funding initiatives.

### 2.6.7 INFORMATICS

Seventh, the adoption of technology needs to be encouraged and supported even more than it is now. We need to move quickly from the model project stage to the actuated system stage. This could be abetted by the International Commission of Zoological Nomenclature and International Commission of Botanical Nomenclature proactively pushing for web-based revisions, bottom up generated web pages on all groups of organisms, and technologically friendly modes of publication. We do not and should not wait until the perfect worldwide system is available. We need to get moving on this now and work the 'bugs' out later. This should be coordinated with GBIF in Copenhagen, which is working well, but the national and regional nodes need to do more — stabilise and make permanent the staff needed to achieve well coordinated goals.

As I have suggested before[15], serious discussion and planning should take place within the taxonomic community, funding agencies and private foundations to set up across the world regional, or continental, supercomputer facilities. If we are serious about terascaling biodiversity data, then we have to have the hardware, that is, supercomputers dedicated to biodiversity work, available for the exclusive use of the biodiversity research and databasing community.

## 2.7  CONCLUDING REMARKS

Our czar will not have an easy job of it. The science of taxonomy/systematics is encountering hard times, but if past publications are any indication, times have always been hard. And, even if this were the case, this does not mean that the supposed last taxonomists have to turn out the lights. We as a scientific community need to assume a proactive, planned and coordinated stance to push the issues that are critical to us, that is, push these issues not just to taxonomists, but to biology and the world as a whole. As Wheeler[1] pointed out, we have the technology, and we have the absolute need to use it. Because of the biodiversity crisis, it is now or never.

## ACKNOWLEDGEMENTS

Dr. Ronald Sluys, University of Amsterdam, offered some useful comments on the manuscript. He and Dr. Cees Hof, University of Amsterdam, and Prof. Koen Martens, now of the Royal Belgian Institute, Brussels, have offered debate and discussion about many of these issues in the past, from which I have drawn ideas and inspiration.

## REFERENCES

1. Wheeler, Q.D., Taxonomic triage and the poverty of phylogeny, *Phil. Trans. R. Soc. Lond. B,* 359, 571, 2004.
2. Wheeler, Q.D., Raven, P.H., and Wilson, E.O., Taxonomy: impediment or expedient? *Science,* 303, 285, 2004.
3. Carvalho, M.R. et al., Revisiting the taxonomic imperative, *Science,* 307, 353, 2005.
4. Manning, R.B., The importance of taxonomy and museums in the 1990s, *Memoirs of the Queensland Museum,* 31, 205, 1991.
5. Feldmann, R.M. and Manning, R.B., Crisis in systematic biology in the 'age of biodiversity', *J. Paleontol.,* 66, 157, 1992.
6. Anonymous, *Conference on the Importance of Systematics in Biology, April 22 1953,* National Academy of Science, National Research Council, Washington, DC, 1953.
7. Schmitt, W.L., The study of scientific material in the museum, *The Museum News,* 8, 8, 1930.
8. Parnell, J., Plant taxonomic research, with special reference to the tropics: problems and potential solutions, *Conserv. Biol.,* 7, 809, 1993.
9. Kaiser, J., Philadelphia institution forced to cut curators, *Science,* 307, 28, 2005.
10. Almeda, F. et al., *Miconia,* 1531 species names, 1061 readily distinguishable entities, in *Systematics, Fourth Biennial Conference of the Systematics Association, Program and Abstracts,* Trinity College, Dublin, 2003, 15.

11. Schram, F.R. and Müller, H.G., *Catalog and Bibliography of the Fossil and Recent Stomatopoda,* Backhuys Publ., Leiden, 2004.
12. Godfray, H.C.J., Challenges for taxonomy, *Nature,* 417, 17, 2002.
13. West, R.M., Endangered and orphaned natural history and anthropology collections in the United States and Canada, *Collection Forum,* 4, 65, 1988.
14. Wilson, E.O., The biological diversity crisis, *Bioscience,* 35, 700, 1985.
15. Schram, F.R., The truly new systematics: megascience in the information age, *Hydrobiologia,* 519, 1, 2004.

# 3 Assembling the Tree of Life: Magnitude, Shortcuts and Pitfalls

*O. Seberg and G. Petersen*
Botanical Garden and Museum, The Natural History Museum of Denmark, Copenhagen

## CONTENTS

## ABSTRACT

Assembling the tree of life is 'big science', and this chapter discusses the magnitude of the task. It also discusses in greater depth DNA taxonomy and DNA barcoding, two recent shortcuts, that have been proposed to achieve the goal of assembling the tree of life. Whilst DNA taxonomy is largely a futile exercise, DNA barcodes, short standardised portions of the genome, may become helpful tools in species identification, especially in species rich taxa. However, it is less likely that barcodes will significantly speed up the discovery of new species.

## 3.1 INTRODUCTION

Phylogenetic information is central to biology[1] and has proven useful in many fields, such as choosing experimental systems for biological research, tracking the origin and spread of emerging diseases and their vectors, bioprospecting for pharmaceutical and agrochemical products, preserving germplasm, targeting biological control of invasive species and evaluating risk factors for species conservation and ecosystem restoration[2].

Acknowledging that many branches in the tree of life remain unanalysed and unresolved, particularly the species rich groups, and accepting that we have only limited information about most species on Earth, have been significant factors behind the USA National Science Foundation's recent initiative to launch a major programme aimed at assembling the tree of life (http://www.nsf.gov/bio/progdes/ bioatol.htm). Assembling the tree of life is 'big science', and its planetary scope makes it mandatory that all countries realise their responsibility for adding to this endeavour.

## 3.2  THE SCALE OF THE PROBLEM

The estimated number of extant species varies enormously, ranging from three million to 50 million or even more[3]. According to Groombridge and Jenkins[4] the most likely estimate is in the vicinity of 14 million. Another recent estimate[5] places the number closer to seven million. Of these between 1.5[5] and 1.75 million[4] have already been described. The uncertainty of the number of described species is aggravated by synonymy. Hammond[3] has proposed a generally applicable average figure of 20% synonymy, but recently, Scotland and Wortley[6] have predicted that almost 80% of the already published names in the angiosperms are synonyms. However, in the present context Groombridge and Jenkins' numbers are accepted as realistic estimates.

Consequently, the systematic community is faced with a tremendous task even to obtain the basic material for assembling the tree of life. It has been suggested that new species are described at a rate of approximately 10,000 per year by May[5] and 7,000 per year by Wheeler[11]. Given the numbers for several major taxa in Table 3.1 it seems more probable to estimate that the figure is in the vicinity of 13,000 per year. This is close to the estimate by Stork[12] and in agreement with the estimate of Hammond[3]. Within these annual rates of new taxonomic descriptions a 1,000 fold rate increase would be necessary for the scientific community to describe all undescribed taxa in one year. Put in a different way, if there are still 12,250,000 undescribed species out there, and if we describe them at a rate of 13,000 each year, it will take us slightly more than 940 years to describe them all.

**TABLE 3.1**
**The Number of New Species Described per Year, the Total Number of Accepted/Described Species, and the Estimated Total Number of Species in Selected Groups**

|             | New         | Accepted/Described[4,6] | Estimated Total |
|-------------|-------------|-------------------------|-----------------|
| Bacteria    | 120[9]      | 10,000                  | 1,000,000       |
| Fungi       | 1,700[8]    | 72,000                  | 1,500,000       |
| Algae       | Unknown[10] | 40,000                  | 400,000         |
| Plants      | 1,700[9]    | 270,000                 | 320,000         |
| Nematodes   | 365[8]      | 25,000                  | 400,000         |
| Mandibulata | 7,200[8]    | 963,000                 | 8,000,000       |
| Birds       | 5[8]        | 9,750                   | —               |
| Mammals     | 26[8]       | 4,630                   | —               |
| Total       | c.13,000    | 1,750,000               | 14,000,000      |

*Note:* In some sources, for example Groombridge and Jenkins[4] and Scotland and Wortley[6], the terms 'accepted species' and 'described species' are used interchangeably. The total number of new species of Mandibulata described covers insects only. Estimates for diatom species numbers range from 6,000 to 1,000,000 according to Guiry in Parnell[10].

*Source:* Data from Groombridge and Jenkins[4]; Hall and Hawksworth[7]; Hammond[8]; Hawksworth[9].

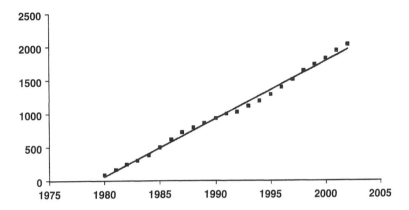

**FIGURE 3.1** Cumulative number of publications in WebSpirs (version 5.02) since 1980 that cite the term taxonomic revision in the title or abstract ($r^2 = 0.99$, $P = 1.0$)

All these estimates ignore the fact that one has to find the new species first[13,14] and that there is a paucity of field naturalists. If the accumulation of type specimens at the Royal Botanic Gardens, Kew, UK, and the US National Herbarium is an indication of taxonomic activities, the description of new plant species has decreased dramatically since 1909[15]. Even revolutions in taxonomic methods[5] will not solve this problem[13]. For this reason alone, it seems overly optimistic to assume that the task of describing all species will be done in only 25 years as the All Species Foundation (http://www.all-species.org) hopes to do[16]. It is pertinent to reiterate Raven's recent statement: "Finally, nothing will substitute for the activities of the field naturalist. No matter how much we speak about instant identification through DNA analysis, hand-held keys or other modern approaches, unless there are very many people who can recognise organisms, find them, go into the field and find them again, whether they be in the tropical moist forests of the Congo or the chalk grasslands on the South Downs of England, nothing will work"[17].

In marked contrast to the nearly exponential increase in number of papers that deal with molecular phylogenies[1,18], the number of published taxonomic revisions has been remarkably constant over the last 20 years and has steadily been in the region of 90 per year (Figure 3.1). Acknowledging the bias of WebSpirs, this figure is verging on the insignificant and is unlikely to increase in the near to middle future.

It has been suggested that somewhere between 50,000 and 80,000 species[19,20] are currently placed in a phylogenetic analysis. Accordingly, the task of gathering the overwhelming number of basic building blocks to assemble the tree of life will require an effort that will be orders of magnitude larger than the work that went into sequencing the human genome. To speed up the process of describing unknown biodiversity and to make it possible to assemble the tree of life within reasonable time, it is tempting to develop shortcuts to reach the goals. Two such recent proposals are DNA taxonomy[21,22] and supertrees[23–25]. Supertrees construct new trees based on the topologies of trees obtained from existing phylogenetic trees. In this way they substitute the collection of new data with analyses of trees derived from different data sources (occasionally analysed with incompatible methods) and at least partially different taxon sampling. Supertrees are not dealt with further here, as their advantages and disadvantages are discussed elsewhere (Hodkinson and Parnell, *Chapter 1*; Wilkinson and Cotton, *Chapter 5*; Bininda-Emonds and Stamatakis, *Chapter 6*).

## 3.3 SHORTCUTS IN SYSTEMATICS: DNA TAXONOMY

The basic procedure of DNA taxonomy is simple[22]; a tissue sample is taken from an individual collection and DNA is extracted. One or more gene regions are sequenced and an unambiguous link between the collected individual and the sequence (or sequences) is made. The sequence serves as a first approximation, an identification tag, for the species from which the DNA sample was extracted. The sequence is compared against existing sequences and made available to the scientific community through appropriate databases together with other types of information, ideally including its taxonomic status. The sequence is a standard for future reference and should ideally be linked with the type specimen and the DNA preparation. As a prerequisite, existing Linnaean names should be matched with appropriate DNA sequences. However, many or most existing type specimens are not useful for this purpose. In such instances, DNA preparations should be based mainly on sequences from newly collected individuals, preferably from the type locality, and identified by experienced taxonomists; it is even suggested that these should have the nomenclatural status of neotypes and hence should replace the existing types[22]!

Even if theoretically possible, DNA taxonomy seems to pay no attention to the fact that many species are known only from the type collection. Although the 'neotypifications' proposed by Tautz et al.[22] seem straightforward, it is not necessarily a simple matter given, for example, that 50–70% of all arthropod species only turn up as one or two specimens in most surveys, approximately 40% of all beetles are only known from one locality[5], and roughly half of the 38,000 known spiders were originally described on the basis of a single specimen[26]. Generally, it will be extremely difficult and labour intensive to find neotypes in all the megadiverse groups. Perhaps even worse, given the postulated lack of taxonomic experience (and the complete lack of expertise in many fields) one may wonder who the experienced taxonomists are that should undertake the task of verifying existing species identifications and identify the new samples in the endless number of instances of 'neotypification'? Although it is unlikely to happen, consider the havoc that would ensue when mixing the types of *Homo sapiens* and *Pan troglodytes*. There are innumerable examples far less likely to be spotted).

Whilst in principle acknowledging the importance of morphological information, it emerges that in DNA taxonomy sequences will have preference over all other types of data, even if it involves total destruction of the specimen (or type) and replacement by a photograph. Tautz et al.[22] suggest that the routine identification of specimens collected during ecological studies should be done by high through-put DNA sequencing facilities. Such facilities "could routinely handle c. 1,000 samples per day" at a cost of 5 per sample (a calculation that of course disregards all other expenses, such as equipment, technical and scientific staff).

According to its proponents, DNA taxonomy will solve a number of pertinent problems, each of which is also relevant to DNA barcoding, and these are discussed in the following sections:

- The identification problem
- Instability of Linnaean names
- Taxonomic bias
- The 'taxonomic impediment'
- The inadequacy of taxonomic data and standards in existing databases

## 3.4 THE IDENTIFICATION PROBLEM

### 3.4.1 DNA BARCODING

The notion of DNA taxonomy has largely been rejected by the taxonomic community[27–29] as an ill-founded attempt to monopolise taxonomy, whereas a growing sympathy for the reality of the identification problem is emerging (but see [30–34]), as epitomised in the concept of DNA barcoding. In contrast to DNA taxonomy, DNA barcoding is to be viewed as a new and exciting addition to the taxonomists' toolbox and can go some way to filling the knowledge void left when experienced taxonomists retire. Although there are obvious similarities between DNA taxonomy and DNA barcoding and some confusion still prevails[14,35–37], the fundamental difference between DNA taxonomy and DNA barcoding resides in the very different use and consequences of the collected data.

**DNA barcoding** is a technique for characterizing species of organisms using a short DNA sequence from a standard and agreed-upon position in the genome. DNA barcodes are therefore useful to taxonomists who are trying to discover, distinguish and describe new species, and to anyone who is trying to assign an unidentified specimen to a known species. DNA barcodes can be a powerful addition to the traditional methods we use to discover new species and identify specimens. They can be used by people who are not experts on a particular group of organisms, and can be obtained from specimens that are hard or impossible to identify with traditional methods (like damaged, incomplete, or immature specimens). (Consortium for the Barcode of Life, CBOL, http://barcoding.si.edu/index_detail.htm).

In striking contrast, DNA taxonomy clearly goes beyond species identification and puts prime importance on the sequence[22]. The widespread acceptance of DNA barcoding, even within the usually very conservative taxonomic community, and the willingness of larger institutions to use their resources for barcoding is evident from the success of the newly founded CBOL, which was created as an international initiative in 2004, and in December 2005 the consortium counted 93 member organisations including many botanic gardens, herbaria, natural history museums and zoos.

The major advantages and applications of barcoding may well lie outside the taxonomic community, in applied taxonomy[34] where it will enable nonspecialists in governmental and intergovernmental agencies, NGOs and other users of taxonomic information, to produce fast and reliable species identifications. Within the taxonomic realm its strength lies in its ability to provide researchers with a new way of identifying potentially new species[38–41]. Even for the experienced taxonomist, species identification from material at all life history stages (or from fragments of specimens) may be difficult if not impossible. In such cases barcoding offers tremendous help[40].

Obviously, finding a new barcode sequence does not mean a new species has been discovered[30,31,42]; the new sequence may just add to the variation of an already described species, but it directs attention to the potentially undescribed. It is not the intention, and certainly not necessary or desirable, to use mtDNA divergence (or divergence of whatever sequence is being used) as "a primary criterion for recognizing species boundaries"[35] (see also Sites and Marshall[43]). Similarly, different species may have identical barcode sequences, depending on the choice of sequences used.

For barcoding to work, a number of problems must be solved or minimised to the largest possible extent. Ideally, a simple, short DNA sequence should be able to identify all known and unknown species. Thus, a sequence of only 15 base pairs (bps) is in theory able to distinguish four to the power of 15 (more than a billion) species; a figure that by far exceeds even the most unrealistic estimates of the Earth's biodiversity. However, it is evident that no such single, universally applicable sequence exists. In the animal kingdom, there has been a strong focus and increased consensus on using a small portion (c. 650 bps) of the mitochondrial COI (cytochrome *c* oxidase subunit I) gene[38,39], although small subunit ribosomal sequences (SSU) are also in use[44,45].

Among higher plants, mitochondrial sequences are unlikely to be useful as barcodes simply because they are far too invariable. Several attempts to find other suitable regions including nonprotein coding plastid regions are under way[36,46]. Finding suitable DNA regions depends on finding short sequences with sufficient variability to discern even closely related species, but at the same time intraspecific variation should be minimal (if possible nonexistent)[35]. In some animal groups COI meets these criteria, but in others it does not, and it may be necessary to look for different sequences in different taxonomic groups. In the higher plant community an approach that rests either on simultaneous use of different barcode sequences, or on a hierarchical system of identification, seems far more acceptable than among zoologists and is perhaps the only way forward. It appears likely that the widely used plastid gene *rbcL* (the large subunit of ribulose-1,5-biphosphate carboxylase/oxygenase) will in the majority of cases be able to identify plant species to genus or family level, but other sequences will be required for identifications at the species level, most likely even different sequences in different taxa. Obviously, in species identification there will be instances where barcode sequences behave just as poorly or worse than morphology. Recently evolved species may lack clear sequence divergence, as has occurred in many island environments[47]. Equally, organelle sequences (which are in most instances inherited from single parents) are, despite their practical advantages, unable to allow the identification of hybrids, instances of introgression and recent polyploidisation, all of which may blur species boundaries.

For most taxonomists DNA barcodes may never become anything more than a new gizmo in the toolbox. For researchers in other fields of biology they may become another identification tool akin to a good illustrated field guide. Obviously, in instances when organisms have no or limited morphological variation or when they are unable to be cultured, such as most prokaryotes and nematodes, the only option may be to collect DNA sequence data and invent a classification[48,49], which reflects sequence similarity only. This is of course a caricature of a classification and has resulted from a simple need for recognition/identification. Needless to say, microbiologists and other scientists working with such groups would prefer to know considerably more about their organisms than this, and they know that they can do far better if the organisms can be cultured[28]. To use this parody of a classification as an argument for revolutionising taxonomy seems bizarre[50].

However, once made, DNA barcodes may obviously be used for purposes other than identification. Even though the length and variation level of most barcode sequences make them largely inappropriate for phylogenetic analyses, they do include information that makes them amenable to rough and dirty estimates of phylogeny[22], and they are a resource that may be combined with other data.

It is obvious that DNA barcodes may be viewed as a first step on the slippery slope to reach the goals of DNA taxonomy[34], but this is a route that we wholeheartedly advise against[28,29], and CBOL has wisely avoided it. Implementation of DNA taxonomy and the widespread, naïve attitudes towards classification as articulated by, for example, Felsenstein in his recent textbook[51] in which he takes a stance on 'the irrelevance of classification', will only serve to take us further away from assembling the tree of life, and represent an unwarranted arrogance and ignorance towards one of the central issues in biology.

### 3.4.2 PRACTICAL PROBLEMS OF DNA-BASED METHODS

If one accepts the rationale behind DNA taxonomy and DNA barcoding, the sheer task of assembling data points is formidable; in April 2005 there were sequences from approximately 160,000 classified organisms in GenBank, and of these approximately 107,000 are classified to species level and a further 16,000 or so also to infraspecific categories. These sequences are from widely different parts of the genome(s), but let us for the sake of argument consider them all relevant to DNA taxonomy or barcoding. Hence, we need a further 13,840,000 sequences to have just one sequence from all species (undescribed as well as described) and just 1,590,000 sequences to have one sequence for all described species.

With respect to DNA barcoding, there were in April 2005 close to 12,000 COI sequences from eukaryotes in GenBank that were not classified as barcodes and nearly 1,000 sequences annotated as such, but in the independently created Barcode of Life Database (BOLD; http://www.barcodinglife.com), more than 32,000 barcode sequences are deposited representing over 12,000 species. Thus a theoretical maximum of 24,000 species have so far been barcoded. BOLD is a special-purpose application for barcodes, and it is the explicit intention of BOLD that GenBank will act as the primary sequence repository for barcodes. Hence, it is to be expected that the latter sequences will eventually be deposited in GenBank.

Ideally, it would be advantageous if all species could be identified by a single, unique, short DNA sequence, but this is far from possible. The task of producing sequences for all species, even only for those described thus far, is as difficult for barcoding as it is for DNA taxonomy. Initially most barcode projects have therefore been restricted to geographically and/or taxonomically restricted groups of organisms, for instance the flora of Costa Rica and the birds of North America[40]. However, the latter project is now being extended to cover the c. 10,000 bird species of the world; the All Birds Barcoding Initiative (http://barcoding.si.edu/AllBirds.htm). A similar network project, Fish-BOL, seeks to barcode the 23,000 known species of fish and is estimated to involve the barcoding of 500,000 specimens (http://www.fishbol.org).

If we imagine that all the described species are valid (that is, no synonymy) and are immediately available for sequencing, it would require approximately seven years (given 225 work days, of eight hours per day, per year) to produce a single sequence for each, given the availability of a single high-throughput DNA sequencing facility, that could routinely handle c. 1,000 samples per day, as estimated by Tautz et al.[22]. However, just producing the sequences is insufficient, as they also have to be checked and read. Even if we assume absolutely perfect sequences of c.1,000 bps and a minimal handling time of five minutes per sequence, it would take approximately 10.5 years to handle just one year's worth of the generated sequences. This amounts to more than 80 years for all known species and around 640 years for the estimated total of species, at a cost of approximately €8 and €70 million, respectively (neglecting all other expenses such as equipment and salaries for technical and scientific staff).

Evidently, a much higher level of automation is possible at all stages in the process than is standard, but sequencing such an enormous diversity of species is not a straightforward process, and DNA extraction (which has not been included in the calculations) is far from trivial. In GenBank, the overall acquisition rate of sequence(s) from new species has been constant in the period 1995–2003 at 2,088 ($r^2 = 0.99$, $P = 0.95$) new species per year. This also applies to green plants (Viridiplantae = Chlorobiota) where there has been a constant acquisition rate of 764 ($r^2 = 0.95$, $P = 0.77$) new species per year. In comparison, the accumulation of sequence from the very widely used *rbc*L gene has also been constant but a quarter of the total species acquisition rate. However, given the widespread commitment to barcoding, there is every reason to believe that barcodes will be obtained at a significantly increased rate.

Existing collections are of course a potentially valuable source of material both for DNA taxonomy and barcoding, and as stated by Tautz et al.[22] and Blaxter[44] it may be possible to nondestructively sample large animals, insects, most plants and fungi in existing collections. However, the present, often well justified, reluctance of many curators to accept destructive sampling of many collections (for example, in the case of small insects or when only the type or very few collections are known) makes it difficult to believe that such practice will ensue more widely in groups that are less suitable for DNA extraction. It is an additional complication that a very large fraction of the existing types are in excess of 100 years old, which would significantly increase the error rates during PCR and complicate matters when assembling the small fractions of sequence that may be amplified from low-quality DNA.

Additionally, a large number of specimens are preserved under conditions that makes successful large-scale recovery of DNA unrealistic, such as animals pickled in formalin and plants dried with alcohol. All specimens have interesting features that go beyond their contents of nucleic acids, and it would be ill advised to replace them, if successfully extracted, with a DNA sample or a photograph. Knowing that, if the quality of the extraction and the sequence(s) are far from perfect, all we might be left with is one or more lousy sequences and a photograph. Every practicing taxonomist is aware of the limited value of types that exist only as descriptions or drawings. For example, although the actual taxonomic status of the new shrike species, *Laniarius liberatus* Smith et al.[52] was based on extensive studies of morphology and behaviour, only a minimalist holotype (photographs, moulted feathers and minute samples of blood) were used to designate the type. This created considerable furore among ornithologists[53–55], and the decision not to preserve a complete specimen seems ill founded. However, it is important to stress that neither the zoological, nor the botanical, code preclude the inclusion of DNA data in the diagnosis or description of species, or in the designation of tissue samples as types.

One of the alleged advantages of DNA taxonomy and DNA barcoding is that sequence information is digital and not influenced by subjective assessment. This is unquestionably true from the very moment the sequence is entered into the computer, but a series of decisions leading up to this particular moment are certainly not. Errors occur due to misincorporation of bases, misreading of chromatograms, mistyping of results and miscommunication of the sequences to the database. Some of these errors can be quantified, like the misincorporation of bases by the polymerase, but others are difficult to estimate[56]. However, the problems are all largely technical, and one way around them is to make it possible to quantify error rates by including trace files with the sequences (as implemented in BOLD). These trace files will in turn make it possible to calculate a probability score for each base call, and archive them with each trace file for every barcode sequence.

Additional problems may be caused by the analysis of the sequences themselves such as alignment and distinguishing orthologous from paralogous sequences. Quantifying similarity between a new barcode sequence and existing sequences from other specimens may not be a trivial matter, and the standard practice involves a series of difficult steps, namely sequence alignment, calculation of pairwise similarities or dissimilarities, and clustering. These problems are theoretical and not easily solved. A wide range of different alignment algorithms, (dis)similarity measures and clustering methods are in existence, and phenetics is surely not the only available or potential methodology. Although the actual species limits are open to interpretation, the 10 groupings, provisionally recognised as cryptic species recovered in the Skipper Butterfly, *Astraptes fulgerator*, complex using Neighbour Joining by Herbert et al.[40] based on COI sequences, are robust when subjected to a parsimony approach (personal observation). However, the interspecific sequence variation in some of these putative cryptic species is comparable to the intraspecific variation, and some of the groupings gain their credibility from other types of data, such as correlation with the colour patterns of the caterpillars and their food plants.

Our ability to identify species using barcodes may also be hampered by paralogous sequences, which may often have a higher level of sequence divergence than the orthologous sequences; contrary to widely held beliefs, orthology is not an emergent property of a sequence, but a testable hypothesis. Putative paralogous sequences, nuclear mitochondrial pseudogenes (NUMTS), are of widespread occurrence in eukaryotes[57] and are also known from COI[58].

## 3.5  INSTABILITY OF LINNAEAN NAMES

The assumed instability of the Linnaean naming system seems to be a major concern not only of Tautz et al.[22], but also to adherents of the Phylocode (however, see [59–63]). Admittedly, it is a nuisance for everyone that names, which have been in use for a long time, are suddenly no longer used. However, it is difficult to see how linking a name with a sequence will solve this problem. The instability of the current system is caused by changes in knowledge; for example, previously recognised species are split into two species, or subsumed into others. Hypotheses about relationships are always subject to revision as new information becomes available, or existing data are reinterpreted. In fact the goal of phylogenetic systematics is hypothesis testing. The taxonomy of species is not fixed[4]. In taxonomy 'stability is ignorance'[64], and the mere idea behind creating a unitary taxonomy[50] runs counter to scientific practice[65]. If *Pan paniscus* and *P. troglodytes* are subsumed into *Homo*, which in light of the low sequence divergence between *Homo* and *Pan* is an option, two species names disappear no matter whether they were linked to physical types or electronically stored sequences.

## 3.6  TAXONOMIC BIAS

The undue emphasis on charismatic taxonomic groups, such as vertebrates, insects and flowering plants, is beyond doubt a problem for our scientific understanding of life on Earth. To express it differently, it is a general problem that the typical bird or mammal species on the average will be mentioned in one scientific paper per year, whereas the average invertebrate species will only be mentioned in 0.1 to 0.01 scientific papers per year[5]. It is difficult to disagree with Knapp et al.[66] that the taxonomic inflation and instability in species names postulated by Isaac et al.[67] largely pertains to these charismatic groups and, in general, remains a negligible contribution to our estimate of biodiversity. It would certainly be advantageous if more scientists turned their research interests towards lesser known groups, but we must not neglect how important groups like birds, butterflies and whales are to ecology and the general public. To a large extent these are the kind of organisms that shape public opinion on threats to biodiversity. However, it is difficult to see how DNA technology can divert more scientific attention to orphan groups[68]. Would it be any more interesting to study mite sequences than mite morphology? It is difficult to persuade anyone that we will know more about mites because we have a sequence and a photograph of one, than if we know something about its morphology and life history.

## 3.7  THE TAXONOMIC IMPEDIMENT

It seems very difficult to imagine how the introduction of DNA-based technologies will contribute significantly towards alleviating the 'taxonomic impediment' (for example, reducing the shortage of trained taxonomists and curators, especially in developing countries[14]). There are better ways of doing this than to become ignorant of the organisms we work with. One is the training of a new generation of taxonomists[69] and the other is funding. For further details, see the Darwin Declaration of the Global Taxonomy Initiative (http://www.biodiv.org/programmes/cross-cutting/taxonomy/dar-win-declaration.asp). It is perhaps no coincidence that a recent attempt to centralise registration of plant names was rejected by the International Botanical Congress in 1999. The move to reject this notion was lead by taxonomists from developing nations fearing that wealthier countries would monopolise taxonomic information.

## 3.8  INADEQUACY OF TAXONOMIC DATA AND STANDARDS IN EXISTING DATABASES

Evidently there is only limited control on the taxonomic standards of the submissions to GenBank, or most other molecular databases. The biggest efforts are concentrated on keeping nomenclature up to date, which is a noble effort in its own right. Apart from the large genome projects, which work with well known model organisms, most information on the origin of individuals is deposited in GenBank by people (mostly taxonomists) who hopefully know the kinds (and indeed the type) of organisms upon which they work. It may be an exaggeration[70–72] and not a generally applicable estimate "that up to 20% of publicly available, taxonomically important DNA sequences for three randomly chosen groups of fungi may be incorrectly named, chimeric, of poor quality or too incomplete for reliable comparison"[73] (see also Bridge et al.[74]), but the identity of specimens is certainly a serious problem that many researchers have confronted, and which may seriously confound the conclusions drawn.

It would be naïve and against scientific practice to think that one can create an authoritative body of taxonomists to supervise taxonomic aspects of sequence submission, and it is evidently beyond the responsibility of GenBank to act as such. In fact NCBI has a disclaimer to the contrary (http://www.ncbi.nlm.nih.gov/Taxonomy/Browser/wwwtax.cgi): "The NCBI taxonomy database is not an authoritative source for nomenclature or classification — please consult the relevant scientific literature for the most reliable information".

It may be that there are no solid taxonomic standards in current repositories of sequence data, but neither is there (as indicated above) solid control of the sequences themselves. As a natural consequence of this, Harris[75] has recently appealed for a much greater authentication of the sequences deposited in GenBank. There is every reason to be cautious when three out of 16 sequences from the mitochondrial cytochrome *b* gene from different reptile genera contain either 'stop codons' or indels that disrupt the reading frame[76]. Indeed, Noor and Larkin[77] also failed to confirm, by resequencing, any of 22 previously published polymorphisms in mitochondrial 12S rRNA genes from *Drosophila pseudoobscura*. By resequencing, checking vouchers and adding four new sequences from the same genus, Kristiansen et al.[78] have recently unequivocally shown that both *rbcL* sequences in GenBank from *Oxychloë andina* (Juncaceae), which have caused considerable confusion in the phylogeny of Cyperaceae and Juncaceae, are erroneous, at least one of them being chimeric. The quality of taxonomy and sequences in GenBank relies solely on the quality and thoroughness of the researchers. Neither problem will be solved by DNA taxonomy or barcoding.

To what extent initiatives like GenBank should enforce standards that require the linking of taxa with sequences is a contentious issue. In the majority of cases it suffices to require that traditional voucher information is available, either directly or indirectly through the journal in which the studies are published. This is of course the responsibility of the editors of refereed journals, who should never allow publications of sequence-based studies without simultaneous submission of the sequences to a public database and should never allow publication of such data unless the necessary voucher information is available. The cornerstone of scientific inquiry is repeatability[79]. It can hardly be stressed enough that specimens used in scientific investigations should be catalogued and vouchered in publicly accessible sites such as culture collections, herbaria and museums, ensuring that species identification can be checked[79–82]. To exclude voucher information from the printed issues of journals and publish it on web pages or other ephemeral media is a highly detrimental and unfortunate practice that some journals like the *American Journal of Botany* have started to implement[82].

In connection with DNA barcoding, very strict rules for voucher information have to be followed, requiring that the barcode sequences must be linked to voucher specimens. At this stage, the NCBI has accepted that barcode sequences submitted by members of CBOL should carry the keyword 'barcode', and discussions of further improvements are ongoing.

## 3.9  CONCLUSIONS

DNA data certainly have a major role to play in taxonomy and for documenting and understanding species rich groups. However, if we know nothing else about the organisms than a tiny part of their DNA, there are few interesting observations to make, apart from sequence similarities. For millennia humans have been fascinated and puzzled by morphological diversity; this is a large part of what needs to be explained, but of course not the only part. Hence, the potential of DNA taxonomy to relegate taxonomy to a high tech service industry centred around a few DNA sequences will turn taxonomy away from being an intellectually stimulating hypothesis-driven science into a purely technical, metaphysical discipline[2,28], or at best, into a cataloguing device for other biologists.

It is often erroneously supposed that taxonomy is a descriptive science, but as emphasised by Wheeler[15], taxonomy is hypothesis driven. "The conclusion that the distribution of a homologous attribute qualifies it as a character of a species or a synapomorphy of a higher taxon is a hypothesis. A species is a hypothesis. Every clade at every Linnaean rank is a hypothesis … [a] species name is an effective shorthand notation for an explicit hypothesis about the distribution of attributes among populations of organisms".

In contrast, it is the goal of DNA taxonomy to reinvent taxonomy and give the sequence prime importance over all other types of data. Wisely, BOLD has strongly emphasised that barcodes are not substitutes for, or an attempt to supplant, existing taxonomic practice (http://phe.rockefeller.edu/BarcodeConference). However, these identifications will never be better than the available taxonomy. Hence, DNA barcodes may be used to create or test hypotheses about species, as in the recent investigation on the neotropical skipper butterfly (*Astraptes fulgerator*)[40], but barcodes are not the arbiter of species status[30,31,35,42]. Although the studies of the *Astraptes fulgerator* complex is viewed as a scholarly implementation of the barcoding technique, it is based on more than 25 years of experiments and has involved the rearing of more than 2,500 caterpillars caught in the wild, and is supplemented by meticulous studies of the morphology of both caterpillars and imagos.

It is difficult to disagree with Wheeler et al.[83] that molecular data, abundant and inexpensive as they are, have revolutionised phylogenetics but not diminished the importance of traditional work. The need for this research has largely been masked because molecular researchers have been able to draw on centuries of banked morphological knowledge.

Likewise the views of May[5] are pertinent. "The task of inventorying is sometimes mistaken for 'stamp collecting' by thoughtless colleagues in the physical sciences [and, sadly, one might add, amongst ecologists and microbiologists]. But such information is a prerequisite to the proper formulation of evolutionary and ecological questions, and essential for rational assignment of priorities in conservation biology. Lacking basic knowledge about the underlying taxonomic facts, we are impeded in our efforts to understand the structure and dynamics of food webs, patterns in the relative abundance of species, or, ultimately, the causes and consequences of biological diversity".

Although we fundamentally agree with the dire need for trained taxonomists and with the controversial but simple fact that a barcode sequence in itself is of limited value[14,84,85], we do not envisage DNA barcoding as a replacement of classical taxonomy but recognise it as a means to revitalise it. DNA barcodes are data, and "the future for systematics and biodiversity research is integrative taxonomy, which uses a large number of characters including DNA and many other types of data, to delimit, discover, and identify meaningful, natural species and taxa at all levels"[34]. However, DNA barcoding at least has the potential to raise public awareness of an increased need for taxonomy expertise, and hence become a major benefit to the taxonomic community[86,87].

## ACKNOWLEDGEMENTS

This manuscript has benefited from comments from Chris Humphries (Natural History Museum, London), Nikolaj Scharff (Natural History Museum of Denmark), Dennis W. Stevenson (New York Botanical Garden) and Dave Williams (Natural History Museum, London).

## REFERENCES

1. Pagel, M., Inferring the historical patterns of biological evolution, *Nature,* 401, 877, 1999.
2. Cracraft, J.M. et al., Eds., *Assembling the Tree of Life,* American Museum of Natural History, New York.
3. Hammond, P.M., The current magnitude of biodiversity, in *Global Biodiversity Assessment,* Heywood, V.H., Ed., Cambridge University Press, Cambridge, UK, 1995, 113.
4. Groombridge, B. and Jenkins, M.D., *World Atlas of Biodiversity: Earth's Living Resources in the 21st Century,* University of California Press, Berkeley, Los Angeles, 2002.
5. May, R.M., The dimensions of life on Earth, in *Nature and Human Society: The Quest for a Sustainable World,* Raven, P.H. and Williams, T., Eds., The National Academy of Sciences, Washington, DC, 1999, 30.
6. Scotland, R.W. and Wortley, A.H., How many species of seed plants are there? *Taxon,* 52, 101, 2003.
7. Hall, G.S. and Hawksworth, D.L., Resources for microbial biosystematics in Europe, in *Systematics Agenda 2000: The Challenge for Europe,* Blackmore, S. and Cutler, D., Eds., Samara Press for the Linnean Society of London, London, 1996, 5.
8. Hammond, P.M., Species inventory, in *Global Biodiversity: Status of the Earth's Living Resources,* Groombridge, B., Ed., Chapman and Hall, London, 1992, 17.
9. Hawksworth, D.L., Orphans in 'botanical' diversity, *Muelleria,* 10, 111, 1991.
10. Parnell, J., European systematics and the European flora, in *Systematics Agenda 2000: The Challenge for Europe,* Blackmore, S. and Cutler, D., Eds., Samara Press for the Linnean Society of London, London, 1996, 31.
11. Wheeler, Q.D., Systematics, the scientific basis for inventories of biodiversity, *Biodivers. Conserv.,* 4, 476, 1995.
12. Stork, N.E., Measuring global biodiversity and its decline, in *Biodiversity II: Understanding and Protecting Our Biological Resources,* Reaka-Kudla, M.L., Wilson, D.E. and Wilson, E.O., Eds., Joseph Henry Press, Washington, DC, 1997, 41.
13. May, R.M., Tomorrow's taxonomy: collecting new species in the field will remain the rate-limiting step, *Phil. Trans. R. Soc. Lond. B,* 359, 733, 2004.
14. Scotland, R. et al., The Big Machine and the much-maligned taxonomist. *Sys. Biodiv.,* 1, 139, 2003.
15. Wheeler, Q.D., Taxonomic triage and the poverty of phylogeny, *Phil. Trans. R. Soc. Lond. B.,* 359, 571, 2004.

16. Wilson, E.O., The encyclopedia of life, *Trends Ecol. Evol.* 18, 77, 2003.
17. Raven, P.H., Taxonomy: where are we now? *Phil. Trans. R. Soc. Lond. B,* 359, 720, 2004.
18. Hillis, D.M., The tree of life and the grand synthesis of biology, in *Assembling the Tree of Life,* Cracraft, J. and Donoghue, M.J., Eds., Oxford University Press, Oxford, 2004, 545.
19. Cracraft, J., The seven great questions of systematic biology: an essential foundation for conservation and sustainable use of biodiversity, *Ann. Missouri Bot. Gard.,* 89, 127, 2002.
20. Pennisi, E., Modernizing the tree of life, *Science,* 300, 1692, 2003.
21. Tautz, D. et al., DNA points the way ahead in taxonomy, *Nature,* 418, 479, 2002.
22. Tautz, D. et al., A plea for DNA taxonomy, *Trends Ecol. Evol.,* 18, 70, 2003.
23. Bininda-Emonds, O.R.P., Ed., *Phylogenetic Supertrees: Combining Information to Reveal the Tree of Life,* Kluwer Academic Publishers, Dordrecht, 2004.
24. Bininda-Emonds, O.R.P., Gittleman, J.L. and Steel, M.A., The (super)tree of life: procedures, problems, and prospects, *Annu. Rev. Ecol. Syst.,* 33, 265, 2002.
25. Sanderson, M.J., Purvis, A. and Henze, C., Phylogenetic supertrees: assembling the tree of life, *Trends Ecol. Evol.,* 13, 105, 1998.
26. Coddington, J.A. and Levi, H.W., Systematics and evolution of spiders (Araneae), *Annu. Rev. Ecol. Syst.,* 22, 565, 1991.
27. Andersen, N.M., Publishing in systematic entomology: present and future, *Insects Syst. Evol.,* 34, 1, 2003.
28. Lipscomb, D., Platnick, N. and Wheeler, Q.D., The intellectual content of taxonomy: a comment on DNA taxonomy, *Trends Ecol. Evol.,* 18, 65, 2003.
29. Seberg, O. et al., Shortcuts in systematics? A commentary on DNA-based taxonomy, *Trends Ecol. Evol.,* 18, 63, 2003.
30. Funk, D.J. and Olmland, K.E., Species-level paraphyly and polyphyly: frequency, causes, and consequences, with insights from animal mitochondrial DNA, *Annu. Rev. Ecol. Syst.,* 34, 397, 2003.
31. Sperling, F., DNA barcoding: deus ex machina, *Newsl. Biol. Surv. Canada (Terrestrial Arthropods),* 22, 50, 2003.
32. Will, K.W. and Rubinoff, D., Myth of the molecule: DNA barcodes for species cannot replace morphology for identification and classification, *Cladistics,* 20, 47, 2004.
33. Wheeler, Q.D., Losing the plot: DNA 'barcode' and taxonomy, *Cladistics,* 21, 405, 2005.
34. Will, K.W., Mishler, B.D. and Wheeler, Q.D., The perils of DNA barcoding and the need for integrative taxonomy, *Syst. Biol.,* 54, 844, 2005.
35. Moritz, C. and Cicero, C., DNA barcoding: promise and pitfalls, *PLoS Biology,* 2, 1529, 2004.
36. Chase, M.W. et al., Land plants and DNA barcodes: short-term and long-term goals, *Phil. Trans. R. Soc. B.,* 360, 1889, 2005.
37. Savolainen, V. et al., Towards writing the encyclopaedia of life: an introduction to DNA barcoding, *Phil. Trans. R. Soc. B.,* 360, 1805, 2005.
38. Hebert, P.D.N. et al., Biological identification through DNA barcodes, *Proc. R. Soc. Lond. B.,* 270, 313, 2003.
39. Herbert, P.D.N., Ratnasingham, S. and de Waard, J.R., Barcoding animal life, cytochrome *c* oxidase subunit 1 divergences among closely related species, *Proc. R. Soc. London. B. (Suppl.),* 270, 1, 2003.
40. Herbert, P.D.N. et al., Ten species in one: DNA barcoding reveals cryptic species in the neotropical skipper butterfly *Astraptes fulgerator, Proc. Nat. Acad. Sci. USA,* 101, 14812, 2004.
41. Stoeckle, M., Taxonomy, DNA, and the barcode of life, *BioScience,* 23, 2, 2003.
42. Sperling, F., Butterfly molecular systematics: from species definition to higher-level phylogenies, in *Butterflies: Ecology and Evolution Taking Flight,* Boggs, C.L., Watt, W.B. and Ehrlich, P.R., Eds., The University of Chicago Press, Chicago, 2003, 431.
43. Sites Jr., J.W. and Marshall, J.C. Delimiting species: a renaissance issue in systematic biology, *Trends Ecol. Evol.,* 18, 462, 2003.
44. Blaxter, M.L., The promise of DNA taxonomy, *Phil. Trans. R. Soc. Lond. B.,* 359, 669, 2004.
45. Blaxter, M.L., Elsworth, B. and Daub, J., DNA taxonomy of a neglected animal phylum: an unexpected diversity of tardigrades, *Proc. R. Soc. Lond. B. (Suppl.),* 271, S189, 2003.
46. Kress W.J. et al, Use of DNA barcodes to identify flowering plants, *Proc. Nat. Acad. Sci. USA,* 102, 8369, 2005
47. Givnish, T.J., Adaptive plant evolution on islands: classical patterns, molecular data, new insights, in *Evolution on islands,* Grant, P.R., Ed., Oxford University Press, Oxford, 1998, 281.
48. Blaxter, M. et al., Defining operational taxonomic units using DNA barcode data, *Phil. Trans. R. Soc. Lond. B.,* 360, 1889, 2005.

49. Markmann, M. and Tautz, D., Reverse taxonomy: an approach towards determining the diversity of meiobenthic organisms based on ribosomal RNA signature sequences, *Phil. Trans. R. Soc. Lond. B.,* 360, 1917, 2005.

50. Godfray, H.C.J., Challenges for taxonomy, *Nature,* 417, 17, 2002.

51. Felsenstein, J., *Inferring Phylogenies,* Sinauer Associates, Sunderland, Massachusetts, 2004.

52. Smith, E.F.G. et al., A new species of shrike (Laniidae, Laniarius) from Somalia, verified by DNA-sequence data from the only known individual, *Ibis,* 133, 227, 1991.

53. Hughes, A.L., Avian species described on the basis of DNA only, *Trends Ecol. Evol.,* 7, 2, 1992.

54. Hughes, A.L., Reply from Austin Hughes (to Peterson, A.T. and Layon, S.M., 1992), *Trends Ecol. Evol.,* 7, 168, 1992.

55. Peterson, A.T. and Layon, S.M., New bird species: DNA studies and type specimens, *Trends Ecol. Evol.,* 7, 167, 1992.

56. Clark, A.G. and Whittam, T.S., Sequencing errors and molecular evolutionary analysis, *Mol. Biol. Evol.,* 9, 744, 1992.

57. Bensasson, D. et al., Mitchondrial pseudogenes: evolution's misplaced witnesses, *Trends Ecol. Evol.,* 16, 314, 2001.

58. Bucklin, A. et al., Taxonomic and systematic assessment of planktonic copepods using mitochondrial COI sequence variation and competitive, species-specific PCR, *Hydrobiologia,* 401, 230, 1999.

59. Carpenter, J.M., Critique of pure folly, *Bot. Rev.* 69, 79, 2003.

60. Keller, R.A., Boyd, R.N. and Wheeler, Q.D., The illogical basis of phylogenetic nomenclature, *Bot. Rev.,* 69, 93, 2003.

61. Nixon, K.C., Carpenter, J.M. and Stevenson, D.W., The Phylocode is fatally flawed, and the 'Linnaean' system can easily be fixed, *Bot. Rev.,* 69, 111, 2003.

62. Flann, C., Phylocode — may the force be with us: an attempt to understand, *The Systematist,* 24, 9, 2005.

63. Pickett, K.M., The new and improved Phylocode, now with types, ranks, and even polyphyly: a conference report from the First International Phylogenetic Nomenclature Meeting, *Cladistics,* 21, 79, 2005.

64. Gaffney, E.S., An introduction to the logic of phylogenetic reconstruction, in *Phylogenetic Analysis and Paleontology,* Cracraft, J. and Eldredge, N., Eds., Columbia University Press, New York, 1979, 79.

65. Vane-Wright, R.I., Indifferent philosophy versus almighty authority: on consistency, consensus and unitary taxonomy, *Syst. Biodiv.,* 1, 3, 2003.

66. Knapp, S., Lughadha, E.N. and Paton, A., Taxonomic inflation, species concepts and global species lists, *Trends Ecol. Evol.,* 20, 7, 2005.

67. Isaac, N.J.B., Mallet, J. and Mace, G.M., Taxonomic inflation: its influence on macroecology and conservation, *Trends Ecol. Evol.,* 19, 464, 2005.

68. Hyde, K.D., Who will look after the orphans? *Muelleria,* 10, 139, 1997.

69. Schram, F.R. and Los, W., Training systematists for the 21st century, in *Systematics Agenda 2000: The Challenge for Europe,* Blackmore, S. and Cutler, D., Eds., Samara Press for the Linnean Society of London, London, 1996, 89.

70. Holst-Jensen, A., Vrålstad, T. and Schumacher, T., On reliability, *New Phytol.,* 161, 11, 2004.

71. Hawksworth, D.L., 'Misidentifications' in fungal DNA sequence databanks, *New Phytol.,* 161, 13, 2004.

72. Vilgalys, R., Taxonomic misidentifications in public DNA bases, *New Phytol.,* 160, 4, 2003.

73. Bridge. P.D. et al., On the unreliability of published DNA sequences, *New Phytol.,* 160, 43, 2003.

74. Bridge, P.D., Spooner, B.M. and Roberts, P.J., Reliability and use of published sequence data, *New Phytol.,* 161, 15, 2004.

75. Harris, D.J., Can you bank on GenBank? *Trends Ecol. Syst.,* 18, 317, 2003.

76. Harris, D.J., Reassessment of comparative genetic distance in reptiles from mitochondrial cytochrome b genes., *Herp. J.,* 12, 85, 2002.

77. Noor, M.A.F. and Larkin, J.C., A re-evaluation of 12S ribosomal RNA variability in *Drosophila pseudoobscura, Mol. Biol. Evol.,* 17, 938, 2000.

78. Kristiansen, K.A. et al., DNA taxonomy: the riddle of *Oxychloë* (Juncaceae), *Syst. Bot.,* 30, 284, 2005.

79. Ruedas, L.A. et al., The importance of being earnest: what, if anything, constitutes a 'specimen examined', *Mol. Phyl. Evol.,* 17, 129, 2000.

80. Agerer, R. et al., Always deposit vouchers, *Mycol. Res.,* 104, 642, 2000.
81. Barkworth, M.E. and Jacobs, S.W.L., Valuable research or short stories: what makes the difference? *Hereditas,* 135, 263, 2001.
82. Funk, V.A. et al., The importance of vouchers, *Taxon,* 54, 127, 2005.
83. Wheeler, Q.D., Raven, P.H. and Wilson, E.O., Taxonomy: impediment or expedient? *Science,* 303, 285, 2004.
84. Dunn, C.P., Keeping taxonomy based in morphology, *Trends Ecol. Evol.,* 18, 270, 2003.
85. Ebach, M.C. and Holdrege, C., DNA barcoding is no substitute for taxonomy, *Nature,* 434, 697, 2005.
86. Gregory, T.R., DNA barcoding does not compete with taxonomy, *Nature,* 434, 1067, 2004.
87. Schindel, D.E. and Miller, S.E., DNA barcoding a useful tool for taxonomists, *Nature,* 435, 17, 2005.

# Section B

Reconstructing and Using
the Tree of Life

# 4 Evolutionary History of Prokaryotes: Tree or No Tree?

*J. O. McInerney and D. E. Pisani*
Department of Biology, National University of Ireland Maynooth, Ireland

*M. J. O'Connell*
Department of Biochemistry, University College Cork, Ireland

*D. A. Fitzpatrick*
Conway Institute, University College Dublin, Ireland

*C. J. Creevey*
European Molecular Biology Laboratory, EMBL Heidelberg, Germany

## CONTENTS

## ABSTRACT

Prokaryotes are likely to be the most numerous and species rich organisms on the planet[1], occupying a more diverse set of ecological niches than eukaryotes. Knowledge of prokaryote diversity is severely limited by our inability to recreate the conditions in the laboratory that are needed to cultivate the majority. Discrepancies between direct microscopical counts and the numbers of colony-forming units can be as much as 100-fold, leading to speculation concerning how much we really know about prokaryotes. In contrast, genomic studies of prokaryotes are advanced. So, while on one hand we know that we have a poor overview of prokaryotic life on the planet, we have, paradoxically, succeeded in obtaining more completed genomic sequences of prokaryotes than of eukaryotes. Therefore, even though taxon sampling has been restricted, we have now reached the stage where we can evaluate whether there is a meaningful prokaryotic phylogenetic tree or taxonomy. Questions remain as to whether the history of prokaryotic life has been overwritten by continuous and random interspecies gene transfer and occasional genome fusions, or whether these events have only been minor contributors, thereby enabling prokaryotic evolutionary history to be adequately described by a tree.

## 4.1 A BRIEF HISTORY OF PROKARYOTIC SYSTEMATICS

Haeckel formalised the concept of using a phylogenetic tree in order to depict the relationships between all the life forms on the planet (see below). The metaphor of the tree seemed to work quite well, and indeed, in Charles Darwin's magnum opus of 1859[2], the only diagram that was used was one depicting a phylogenetic tree. For botanists and zoologists the concept of a phylogenetic tree with large trunks giving rise to smaller branches and then to leaves had so many attractive properties that its position as a central metaphor is almost unshakeable. For microbiologists, however, phylogenetic trees of the prokaryotes have always been problematic; even the definition of prokaryotic and eukaryotic taxa was not satisfactorily resolved until the 1960s[3].

The ranges of morphological characters that have been the subject of analysis in animal and plant groups simply do not exist in the prokaryotes. Cell morphology in many prokaryotes can be described using adjectives as simple as 'rod shaped' or 'round'; nothing approaching the rich lexicon that can be used by botanical or zoological systematists to describe their study taxa. The description of the prokaryotes (called Monera, at the time) given by Haeckel is perhaps the most colourful. He described them as: "... not composed of any organs at all, but consist entirely of shapeless, simple homogeneous matter ... nothing more than a shapeless, mobile, little lump of mucus or slime, consisting of albuminous combination of carbon"[4].

Stanier and van Niel finally settled on a definition of the prokaryotes that included three traits that they lacked: absence of true nuclei, absence of sexual reproduction and absence of plastids[5]. This lack of morphological diversity resulted in a situation where microbiologists settled for classification systems that were taxonomically based, rather than phylogenetically based. Naturally, this led to the downgrading of microbial phylogenetics, and with students of microbiology being presented with nothing more than lists of species names, prokaryotic systematics proceeded at a very much slower pace than was seen in plants and animals. The definitive authority on prokaryotic species, *Bergey's Manual of Determinative Bacteriology,* was published first in 1923 and made no attempt at presenting the prokaryotes in a hierarchical manner based on common ancestry, and indeed, the most recent version still does not[6]. While this approach is changing, with *Bergey's Manual of Systematic Bacteriology*[7] presenting the prokaryotes in a phylogenetic context, the absence of a phylogenetic paradigm in earlier editions of Bergey's manual was reflective of the prevailing attitude that the natural history of the prokaryotes was not knowable at that time and perhaps even that it was not important.

While the phylogenetic relationships between the prokaryotes did not receive much attention in the early part of the last century, it was becoming increasingly clear that metabolic diversity in the prokaryotes was extensive[8]. Prokaryotes could live at a wider range of temperatures than eukaryotes, could live on a very diverse range of diets and produced an almost endless range of secondary metabolites. The source of this metabolic diversity was obviously the result of differences in the genetic composition of the organisms. However, there was no reason at that stage to suggest that microorganisms varied enormously in their genomic composition; after all, they all needed to replicate, carry out transcription and translation and other housekeeping functions. Perhaps small numbers of genes were responsible for this huge amount of metabolic variation?

## 4.2  THE RIBOSOMAL RNA REVOLUTION

In a seminal paper in 1965, Zuckerkandl and Pauling compared the degree of divergence between α-globin proteins of various animals and the separation times of these animals as judged by the fossil record[9]. The result was a generally linear increase in protein divergence with time. The implications were that cellular macromolecules could be used to make inferences concerning historical events, and if this was so, then these molecules could potentially be used to infer phylogenetic relationships, and ultimately, the tree of life might be inferred using these data.

By the early 1970s, manipulation of the macromolecules of the cell became more tractable, and this led Woese and coworkers to the development of classification systems based on ribosomal RNA oligonucleotide cataloguing[10]. Within a few years, enough information was available for the first really big change in our views concerning prokaryotic evolutionary relationships. This change in perception centred on the discovery that prokaryotes could be divided into two groups, with neither group being particularly closely related to each other and certainly no more closely related to each other than either was to eukaryotes[11]. Suddenly, a complete revolution took place. The pace of change in molecular biology facilitated some of this revolution. Rapid DNA sequencing technologies would develop over the following fifteen years[12,13], resulting in the sequencing of tens of thousands of ribosomal RNA molecules from prokaryotes and eukaryotes. In part, the renewed interest was driven by one of the most comprehensive and incisive manuscripts to have ever been written on the subject of bacterial evolution[14], and in part, it seemed that microbiologists were making up for lost time. By the late 1980s, ribosomal RNA phylogenetic trees became the gold standard for inferring evolutionary relationships across all levels.

Quite ironically, Darwin had cautioned that "The importance, for classification, of trifling characters, mainly depends on their being correlated with several other characters of more or less importance. The value indeed of an aggregate of characters is very evident in natural history"[2]. Even though Darwin knew nothing of genes and the cellular macromolecules, he was clear that classification systems should be based on a broad spectrum of traits whose functions were also diverse. Woese[14], to his credit, accepted that it was possible to see differences between the phylogenies that were inferred using small subunit ribosomal RNA molecules and the phylogenies that were being inferred using Cytochrome C genes. Quite likely, these differences were due to interspecies gene transfer, a form of prokaryotic sex, first described by Lederberg and Tatum in 1946[15]. Woese also noted that the phylogenetic trees were otherwise almost identical and concluded that it was "safe to assume" that there was a unique prokaryotic evolutionary history and that some of the cellular macromolecules would have recorded this history[14].

Molecular biology continued to advance, and with the arrival of automated sequencing methods[16] the first genome sequence of a prokaryote, that of *Haemophilus influenzae,* became available[17]. The genome was followed soon afterwards by the genome sequence of an archaeon, *Methanosarcina janaschii*[18], and the genome sequence of the smallest known autonomously replicating organism, *Mycoplasma genitalium*[19]. The genome sequence of *Escherichia coli* K12 was a relatively late arrival[20], given that it was the first organism for which a genome sequencing effort had started. However, when three 'strains' of this species were sequenced completely, the full extent of the nature of gene transfer in prokaryotes was seen[20–22]. These three genomes have no more than 39% of their genes in common and vary in sequence length by almost one million base pairs. Clearly, if any pair of plants or animals differed in genome content by more than 20%, they would not be considered to be the same species; however, in prokaryotes, the standard taxonomic tools had grouped these organisms together as a species. The underlying cause of this genome content difference appears to be the independent acquisition of large numbers of genes in the process known by varying terms including lateral gene transfer (LGT), horizontal gene transfer, or simply gene transfer. This presents us with a problem for inferring phylogenetic relationships. If this pattern is replicated throughout the prokaryotic world, then perhaps the inference of phylogenies based on genomic data may not be possible.

## 4.3  CONFLICTING TREES

An examination of many phylogenetic trees derived from single-copy genes reveals that there is a considerable degree of similarity across these trees. Consider the situation in Figure 4.1. These two trees are constructed from a ribonuclease gene (left) and a DNA polymerase gene family (right). There is a large degree of similarity between these two trees. The differences are to be seen in the absence of an ortholog for the ribonuclease gene in *Xylella fastidiosa* and *Vibrio cholera*, an alternative resolution of the branching order within the *Escherichia coli* strains and a resolution of the branching order within the *Neisseria meningitidis* strains. Overall, these two trees generally suggest highly congruent, but not identical, evolutionary histories. These minor differences are probably attributable to errors in phylogeny reconstruction and to lineage specific gene loss.

By 1998, the first large-scale comparative genome analyses were being carried out, and one of the first findings was that 755 of the identified 4,288 open reading frames in the *E. coli* genome (547.8 Kb) were introduced by LGT in at least 234 lateral transfer events since its divergence from *Salmonella* approximately 100 million years before present[23]. If this was true and these LGT events were stable, then it was relatively easy to conclude that LGT was indeed a major feature, perhaps the most important feature, of prokaryotic evolution. The implication also was that ribosomal RNA phylogenetic trees were no more than gene trees and did not reflect organismal phylogeny.

In 1999, Doolittle, writing in *Science,* made the statement that "If 'chimerism' or 'lateral gene transfer' cannot be dismissed as trivial in extent or limited to special categories of genes, then no hierarchical universal classification can be taken as natural"[24]. The reason for making a statement such as this, which was a radical departure from the questions that were being asked at the time (such as "What is the shape of the universal tree and how should we try to infer this shape?"), had to do with what genomic data was beginning to tell us. Increasingly, ortholog-derived trees were being produced that were not in agreement with the ribosomal RNA tree and were not in agreement with one another. In another paper at the time, Doolittle suggested that it was more appropriate to visualise the evolutionary history of life on the planet as a web[25]. This would reflect the central role of LGT in life's evolution and would be more accurate. This caused controversy and was seen in some quarters as an effort to hark back to the dark days when it was accepted that a prokaryotic phylogeny was unknowable.

Defending the phylogenetic tree concept (tree thinking), Kurland and coworkers[26] refuted the suggestion that LGT was the "essence of phylogeny". They pointed out the difficulties of incorporating a new gene into a genome, particularly when there may be an incumbent gene that is performing a similar or identical function. Their conclusion was that stable integration of a new gene into a genome is at such a low rate that it has little or no influence on the idea of a core phylogenetic tree uniting all organisms.

Woese had already put forward the 'genetic annealing model' of organismal evolution[27]. In this model, Woese suggested that, prior to organismal diversification, the planet was populated by 'progenotes', and gene transfer between these progenotes was high. Subsequently, gene transfer became more difficult, and currently there are high barriers to LGT. Interestingly, Woese stated that "By now, it is obvious that what we have come to call the universal phylogenetic tree is no conventional organismal tree. Its primary branchings reflect the common history of central components of the ribosome, components of the translation apparatus, and a few other genes. But that is all. In its deep branches, the tree is merely a gene tree"[27]. Subsequently, Woese extended his hypothesis, stating explicitly that the very importance of LGT is in part evidenced by the universality of the genetic code; if it was not universal, then LGT would not be possible[28]. However, Woese, in sticking with his doctrine of espousing the view that vertical inheritance is the most important mode of organismal evolution, defined the 'Darwinian Threshold' as the critical point that is reached when vertical inheritance becomes more important than horizontal transfer. According to Ge and coworkers, the evolutionary history of life is somewhat like a great tree with occasional cobwebs joining branches[29]. They estimate the extent of LGT to be 2% per genome.

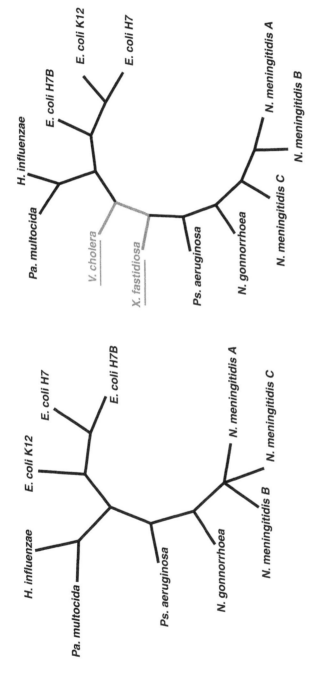

**FIGURE 4.1** Phylogenetic trees of *Escherichia, Haemophilus, Neisseria, Pseudomonas* (Ps), *Pasteurella* (Pa), *Vibrio* and *Xylella*. On the left is a phylogenetic tree derived using orthologs from the ribonuclease family. On the right is a phylogenetic tree derived using orthologs of DNA polymerase III. The completed genomes of these species were searched for orthologs, and all available orthologs were used. For the DNA polymerase III family, there was no ortholog present in the genomes of *V. cholera* and *X. fastidiosa*.

Therefore, the early part of this century has resulted in the formation of two camps, one that emphasises evolution by vertical inheritance and focuses on the identification of 'core' genomic components that tend to be inherited together, using this information to define prokaryotic relationships (tree thinkers), and the other that emphasises LGT and attempts to accommodate it (net thinkers or web thinkers).

## 4.4 METHODOLOGICAL DEVELOPMENTS

Initially, the post genome era debate was being fought on philosophical grounds with ad hoc invocation of analyses of small amounts of data; the finding of the extraordinary level of plasticity in the *E. coli* genome being the most highlighted case to argue in favour of the pan transfer advocates[20–23].

At the moment, the ground on which the debate is being fought is becoming technical, and arguments are being made on the basis of sophisticated methods of analysis and large amounts of data. In general, the sequences of large numbers of presumed orthologous genes are being collected, trees constructed, and phylogenetic hypotheses based on these trees postulated. The LGT debate centres on the analysis of these trees, and usually, though not always, the analysis method of choice involves some kind of consensus or supertree approach. We will briefly review supertree methods, before describing the kinds of findings that these methods are producing.

The use of supertree methods in phylogenetics can trace its origin back to Gordon's classic paper[30], although the earliest supertree algorithm actually predates Gordon's work[31]. Supertree approaches seek to amalgamate the information contained in a set of phylogenetic trees (that is, dendrograms or cladograms), the only requirement being that they overlap in a specific way. Whilst there is no requirement for any given tree to contain the entire set of leaves, there is a requirement that the combined trees can be linked to one another through common subsets of their leaves. Supertrees cannot be constructed from sets of input trees with disjoint leaf sets. The output from a supertree analysis is a supertree, that is, a tree summarising, according to a defined set of rules, the information contained in the input trees. Different supertree methods are based on different sets of rules. From this point of view, supertrees are no different from standard consensus methods, and supertrees can be considered generalisations of consensus tree methods[32]. However, in contrast to consensus methods, supertrees combine partially overlapping trees. This can provide an inference based on the information contained in the input trees and may result in clades being present in the supertree that do not appear in the input set. In any case, the relationships that are present in the supertree but not in the input trees must be implied by some of the trees in the input set and should never be contradicted by all input trees[33,34].

Many supertree methods exist, and a classification of these methods is now difficult, in part because there are such a diverse range of methods and in part because, in some cases, a method could be said to belong to more than one kind of approach. Broadly speaking, two categories can be distinguished, that of the strict/semistrict supertree methods, and that of the liberal supertree methods. Strict and semistrict supertree methods do not allow conflict among the input trees to be resolved, while liberal supertree methods allow for conflict resolution[35]. Strict and semistrict methods are generally not used in practical studies because they tend to return artificially highly unresolved supertrees. The most frequently used supertree methods are the liberal ones, and amongst these the most common involve the generation of matrices that are representations of the input treesy[36]. Alternative matrix representation-based methods are characterised by the way the trees are recoded (for example, as sets of splits or quartets) and by the optimality criterion used to analyse these matrices, such as parsimony[36] compatibility[37], or the minimum number of flips (state changes) necessary to eliminate all the incompatibilities from the matrix representation of a set of trees (Min Flip supertrees)[38]. In any case, it is important to note that all supertree methods can be defined in terms of the tree-to-tree distance they use as an optimality criterion; for example tree length in the case of Matrix Representation using Parsimony (MRP), the Robinson-Foulds distance[39] in the case

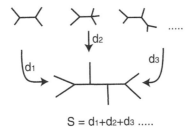

$$S = d_1 + d_2 + d_3 \dots$$

**FIGURE 4.2** Outline of the procedure for evaluating a supertree using **DFIT** or **SFIT** measures as implemented in CLANN. For each input tree, its similarity to an appropriately pruned supertree is measured. The overall score for the supertree is either the sum or average distance computed for all input trees. The difference between the **DFIT** and the **SFIT** measures is to be found in the way in which the distance is computed. S = supertree score, d = distance between the input tree and the appropriately pruned supertree.

of Split Fit, or a flip distance in the case of Min Flip[40]. An alternative is to use a path length distance-based approach to infer the optimal supertree[41]. Approaches using path length distances include the Distance Fit (DFIT) method[42], and the Average Consensus[43], the latter having the potential advantage that it can use branch length information if available.

All of these methods (with the exclusion of Min Flip) are implemented in the program CLANN[44], which also implements a fast Neighbour Joining Average Consensus (NJAC) procedure, and Quartet Fit (QFIT). For the DFIT approach, a supertree can be proposed for the dataset; this supertree can be randomly generated, or an initial rapid supertree construction method such as NJAC can be used to provide a starting tree. The proposed supertree is compared with any input tree, even when the input tree only contains a subset of the total complement of leaves. This can be achieved by pruning the supertree appropriately. Once the pruned supertree and the input tree have the same leaf set, a simple comparison can be made to evaluate their similarity (see Figure 4.2). The DFIT approach involves the calculation of a path length distance from every taxon to the others. The distance is simply the number of nodes that separates the taxa on the tree. If the pruned supertree and the input tree are identical, then the distance matrix that is derived from the pruned supertree and the distance matrix derived from the input tree will also be identical. If the two are different, then the distance matrices will be different, and with increasing dissimilarity in tree shape, there will be increasing dissimilarity in the distances derived from the trees. The supertree that is chosen is therefore the one that is most similar to the input trees.

Other methods like QFIT and Split Fit (SFIT), although originally thought as matrix representation based methods[34,41], can be similarly derived. QFIT involves breaking up the pruned supertree and the input trees into the quartets they entail. Naturally, the two collections of quartets will be identical in terms of leaf content. Again, if both the pruned supertree and the input tree have identical topologies, their quartets will be identical. However, increasing dissimilarity in tree shape will result in fewer quartets with identical topologies. Therefore for QFIT, the score of any given supertree will be proportional to the number of quartets that it contains that have identical topologies to those found in the input trees. SFIT involves breaking up the pruned supertree and the input trees into the splits they entail. SFIT can then be seen as comparing an appropriately pruned supertree with each input tree. The measure of similarity in this case will be the Robinson-Foulds distance[39], and the best supertree will be the one minimising the distance between it and the input trees.

The first large-scale supertree that was constructed for prokaryotes was constructed by Daubin and coworkers[45]. The dataset included a total of 33 prokaryotes and four eukaryotes. They indicated that they could produce a robust supertree when they used ortholog trees with a broader taxon sampling, that is when they avoided using gene trees with small numbers of leaves, and they also indicated that this genome phylogenetic tree was very much in agreement with the ribosomal RNA trees. Subsequently, this work was followed up with an analysis of differences between these ortholog trees, using a multivariate analysis method to identify a core of gene trees with similar topologies and then using these gene trees in order to construct a MRP supertree. For many of the groups on this supertree there is strong support (support being assessed using the bootstrap method); however, the spine of the tree appeared to only have low to medium levels of support.

## 4.5   AN EMERGING CONSENSUS?

Recently Creevey et al.[42] carried out an analysis of completed genome information in a supertree context. The question that was being addressed was whether or not it was possible to identify a robust phylogenetic tree among the deepest branches of the prokaryotic domains. This was to be compared and contrasted with an analysis of a similarly sized dataset that spanned a relatively well-characterised but less ancient group of prokaryotes, the γ-proteobacteria. The analysis involved using the single-gene families from completed genomes, inferring the phylogenetic relationships between these gene sequences and only retaining the inferences that, according to the most widely used methods of analysis, were the most robust. These phylogenetic hypotheses were combined using the MSSA (DFIT) supertree approach implemented in CLANN[44], and the input trees were compared with the supertree to evaluate the goodness of fit of the data to the tree. The results were interesting. While the gene trees derived from the γ-proteobacteria were in good agreement with each other and were in good agreement with the supertree, the trees that spanned the deepest branches of the prokaryotes strongly conflicted with each other. This was not a simple case of lacking signal; it was a case that the gene trees were strongly supported but conflicting. In fact, a comparison of the congruence across these trees and congruence across randomised sets of trees, using the YAPTP test[42], showed that the trees derived from the data were no more in agreement with one another than trees that were completely random. The conclusion from this work was that the prokaryotic phylogeny inferred is strongly supported in parts and not so in other parts. The γ-proteobacteria were also examined using an entirely different approach[46], but the conclusions were the same: congruence across different gene trees is excellent. Lerat et al.[46] recorded that for 205 gene trees examined, there was concordance across 203. In another study using the same methods, we examined the relationships within the α-proteobacteria and found, once again, that there was good agreement between the individual gene trees[47]. However, these analyses have only concentrated on gene families where there are no apparent paralogs. Obviously, if duplicated genes were also taken into consideration, there would be much more data to examine. In addition, taxonomic sampling is sparse and it remains to be seen if the conclusions still hold when sampling is improved.

In a recent report by Beiko and coworkers[48], 'highways' of gene sharing between prokaryotic groups were identified. Their analysis centered on using edit distances to transform ortholog derived trees into a topology that is consistent with a supertree. The finding was that vertical inheritance of genes was in the majority, but the patterns of LGT could not be ignored, and that LGT mostly took place between closely related organisms (presumably using homologous recombination as a means of integrating new genetic material) or between distantly related organisms that occupied the same environment (presumably using illegitimate recombination as the means of integration). The frequency with which each category of genes was transferred was not uniform, with genes involved in 'metabolism', and 'cellular processes' being significantly more frequently involved in a LGT event.

This leads to the question of whether there is or there is not a meaningful prokaryotic phylogenetic tree. If there is, then the paradigm of a tree still stands; if there is not, then the paradigm falls and we need to revert to the descriptive taxonomy of yesteryear, and any evolutionary indications would refer to some subset of the organism's genes, but not the organism. There is clearly an emerging lack of consensus. We can easily find instances where congruence is excellent, and we can find instances where congruence is impossibly poor.

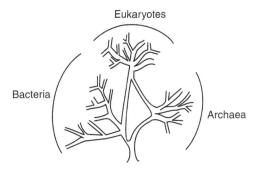

**FIGURE 4.3** A stylised outline of how the evolutionary history of cellular life could be represented using the ring of life theory.

## 4.6  THE PROKARYOTIC INFLUENCE ON THE EUKARYOTE

The ribosomal RNA tree of all cellular life is a metaphor that is very widely recognised. The three main divisions of life, Bacteria, Archaea and Eucarya, are widely recognised, and even though there are disagreements about the importance of this classification system, there is general agreement that these three life forms are very different from one another. The ribosomal RNA tree also has an important inference, that the first microorganisms on the planet were prokaryotes, but eukaryotes evolved from this prokaryote world. The ribosomal RNA tree suggests that there was some kind of discrete event that led to the development of the nucleus early in eukaryote evolution. The ribosomal RNA tree also led to the conclusion that mitochondria were α-proteobacteria-like and had evolved via some kind of symbiosis[14].

Challenges to this dogma have been in circulation for some time, but recently, the first evidence has been produced for a discrete event that suddenly resulted in the development of the eukaryotic cell. Using a new method of genome analysis, conditioned reconstruction, Lake and coworkers have suggested that the eukaryotic cell was created as a result of a fusion of the genome of a bacterium and an archaeum[49–51]. They then suggest that there is no tree of life; if anything, there is a ring of life (see Figure 4.3 for an illustration of what is inferred). If this analysis proves correct, then the consequences for prokaryotic systematics are profound. This would mean that neither the bacteria nor the archaea are monophyletic and both would have the eukaryotic lineage as one of their descendents. This could also mean that the development of our ideas concerning prokaryotic evolution may be incorrect. If true, it also begs the question concerning whether or not there are other 'rings' of life.

## 4.7  CONCLUSIONS, FUTURE DIRECTIONS AND OPEN QUESTIONS

The consensus at the moment is that the prokaryote phylogeny is more tree like than random. There are clear instances of groups of prokaryotes where the agreement across their ortholog phylogenies is high[42,47]. Speciation in prokaryotes is not well understood; however, it is likely that inheritance patterns are generally divergent, and in that respect, the evolutionary history of the prokaryote cells are tree like. What is at question is whether there are groups of genes that make the inference of this history deviate from a tree like pattern. Various metaphors have been used to describe the evolutionary history of prokaryotes such as tree, web, ring or cobweb. However, it is clear that one single description is insufficient to describe the entire history of the group. Future work will centre on more precise descriptions of prokaryote genes, genome and cellular evolution.

## ACKNOWLEDGEMENTS

This work was supported by a Science Foundation Ireland Research Frontiers Programme grant to James McInerney, and a Marie Curie Intra European Fellowship to Davide Pisani (contract number MEIF-CT-2005-01002). The authors would like to thank the two referees for their helpful advice.

## REFERENCES

1. Whitman, W.B., Coleman, D.C., and Wiebe, W.J., Prokaryotes: the unseen majority, *Proc. Natl. Acad. Sci. USA*, 95, 6578, 1998.
2. Darwin, C., *On the Origin of Species by Means of Natural Selection*, John Murray, London, 1859.
3. Sapp, J., The prokaryote-eukaryote dichotomy: meanings and mythology, *Microbiol. Mol. Biol. Rev.*, 69, 292, 2005.
4. Haeckel, E., *The History of Creation*, Trench and Co., London, 1883.
5. Stanier, R.Y. and van Niel, C.B., The concept of a bacterium, *Arch. Mikrobiol.*, 42, 17, 1961.
6. Holt, J.G., *Bergey's Manual of Determinative Bacteriology*, Williams and Wilkins, Baltimore, 1994.
7. Garrit, G.M., Ed., *Bergey's Manual of Systematic Bacteriology*, Springer, New York, 2001.
8. Breed, R.S., Murray, E.G.D., and Hitchens, A.P., *Bergey's Manual of Determinative Bacteriology.* The Williams and Wilkins Company, 1948.
9. Zuckerkandl, E. and Pauling, L., Molecules as documents of evolutionary history, *J. Theor. Biol.*, 8, 357, 1965.
10. Sogin, S.J., Sogin, M.L., and Woese, C.R., Phylogenetic measurement in procaryotes by primary structural characterization, *J. Mol. Evol.*, 1, 173, 1971.
11. Woese, C.R. and Fox, G.E., Phylogenetic structure of the prokaryotic domain: the primary kingdoms, *Proc. Natl. Acad. Sci. USA*, 74, 5088, 1977.
12. Sanger, F., Nicklen, S., and Coulson, A.R., DNA sequencing with chain-terminating inhibitors, *Proc. Natl. Acad. Sci. USA*, 74, 5463, 1977.
13. Smith, L.M. et al., Fluorescence detection in automated DNA sequence analysis, *Nature*, 321, 674, 1986.
14. Woese, C.R., Bacterial evolution, *Microbiol. Rev.*, 51, 221, 1987.
15. Lederberg, J. and Tatum, E., Gene recombination in *Escherichia coli*, *Nature*, 158, 558, 1946.
16. Wilson, R.K. et al., Development of an automated procedure for fluorescent DNA sequencing, *Genomics*, 6, 626, 1990.
17. Fleischmann, R.D. et al., Whole-genome random sequencing and assembly of *Haemophilus influenzae* Rd, *Science*, 269, 496, 1995.
18. Bult, C.J. et al., Complete genome sequence of the methanogenic archaeon, *Methanococcus jannaschii*, *Science*, 273, 1058, 1996.
19. Fraser, C.M. et al, The minimal gene complement of *Mycoplasma genitalium*, *Science*, 270, 397, 1995.
20. Blattner, F.R. et al., The complete genome sequence of *Escherichia coli* K-12, *Science*, 277, 1453, 1997.
21. Hayashi, T. et al., Complete genome sequence of enterohemorrhagic *Escherichia coli* O157:H7 and genomic comparison with a laboratory strain K-12, *DNA Res.*, 8, 11, 2001.
22. Welch, R.A. et al., Extensive mosaic structure revealed by the complete genome sequence of uropathogenic *Escherichia coli*, *Proc. Natl. Acad. Sci. USA*, 99, 17020, 2002.
23. Lawrence, J.G. and Ochman, H., Molecular archaeology of the Escherichia coli genome, *Proc. Natl. Acad. Sci. USA*, 95, 9413, 1998.
24. Doolittle, W.F., Phylogenetic classification and the universal tree, *Science*, 284, 2124, 1999.
25. Doolittle, W.F., Lateral genomics, *Trends Cell. Biol.*, 9, M5, 1999.
26. Kurland, C.G., Canback, B., and Berg, O.G., Horizontal gene transfer: a critical view, *Proc. Natl. Acad. Sci. USA*, 100, 9658, 2003.
27. Woese, C.R, The universal ancestor, *Proc. Natl. Acad. Sci. USA*, 95, 6854, 1998.
28. Woese, C.R., On the evolution of cells, *Proc. Natl. Acad. Sci. USA*, 99, 8742, 2002.
29. Ge, F., Wang, L.S., and Kim, J., The cobweb of life revealed by genome-scale estimates of horizontal gene transfer, *PLoS Biol.*, 3, e316, 2005.
30. Gordon, A.D., Consensus supertrees: the synthesis of rooted trees containing overlapping sets of labeled leaves, *J. Classif.*, 3, 31, 1986.

31. Aho, A.V. et al., Inferring a tree from lowest common ancestors with an application to the optimisation of relational expressions, *SIAM J. Comput.*, 10, 405, 1981.
32. Semple, C. and Steel, M., A supertree method for rooted trees., *Discrete Appl. Math.*, 105, 2000.
33. Pisani, D.E. and Wilkinson, M., Matrix representation with parsimony, taxonomic congruence and total evidence, *Syst. Biol.*, 51, 151, 2002.
34. Wilkinson, M. et al., Measuring support and finding unsupported relationships in supertrees, *Syst. Biol.*, 54, 823, 2005.
35. Wilkinson, M. et al., Some desiderata for meta-analytical supertrees, in *Phylogenetic Supertrees: Combining Information to Reveal the Tree of Life.*, Bininda-Emonda, O.R.P., Ed., Kluwer Academic, Dordrecht, 2004, 227.
36. Ragan, M.A., Phylogenetic inference based on matrix representation of trees, *Mol. Phylogenet. Evol.*, 1, 53, 1992.
37. Ross, H.A. and Rodrigo, A.G., An assessment of matrix representation with compatibility in supertree construction, in *Phylogenetic Supertrees: Combining Information to Reveal the Tree of Life,* Bininda-Emonda, O.R.P., Ed., Kluwer Academic, Dordrecht, 2004, 35.
38. Burleigh, J.G et al., MRF supertrees, in *Phylogenetic Supertrees: Combining Information to Reveal the Tree of Life,* Bininda-Emonds, O.R.P., Ed., Kluwer Academic, 2004, 65.
39. Robinson, D. and Foulds, L., Comparison of phylogenetic trees, *Math. Biosci.*, 53, 131, 1981.
40. Chen, D. et al., *Flipping: A Supertree Construction Method,* American Mathematical Society, Providence, Rhode Island, 2003, 135.
41. Steel, M. and Penny, D., Distributions of tree comparison metrics—some new results, *Syst. Biol.*, 42, 126, 1993.
42. Creevey, C.J. et al., Does a tree-like phylogeny only exist at the tips in the prokaryotes? *Proc. R. Soc. Lond. B. Biol. Sci.*, 271, 2551, 2004.
43. Lapointe, F.-J. and Cucumel, G., The average consensus procedure: combination of weighted trees containing identical or overlapping sets of taxa, *Syst. Biol.*, 46, 306, 1997.
44. Creevey, C.J. and McInerney, J.O., Clann: investigating phylogenetic information through supertree analyses, *Bioinformatics,* 21, 390, 2005.
45. Daubin, V., Gouy, M., and Perriere, G., Bacterial molecular phylogeny using supertree approach, *Genome Inform. Ser. Workshop Genome Inform.*, 12, 155, 2001.
46. Lerat, E., Daubin, V., and Moran, N.A., From gene trees to organismal phylogeny in prokaryotes: the case of the gamma-Proteobacteria, *PLoS Biol.*, 1, e19, 2003.
47. Fitzpatrick, D.A., Creevey, C.J., and McInerney, J.O., Genome phylogenies indicate a meaningful {alpha}-proteobacterial phylogeny and support a grouping of the mitochondria with the Rickettsiales, *Mol. Biol. Evol.*, 23, 74, 2006.
48. Beiko, R.G., Harlow, T.J., and Ragan, M.A., Highways of gene sharing in prokaryotes, *Proc. Natl. Acad. Sci. USA,* 102, 14332, 2005.
49. Rivera, M.C. and Lake, J.A., The ring of life provides evidence for a genome fusion origin of eukaryotes, *Nature,* 431, 152, 2004.
50. Lake, J.A. and Rivera, M.C., Deriving the genomic tree of life in the presence of horizontal gene transfer: conditioned reconstruction, *Mol. Biol. Evol.*, 21, 681, 2004.
51. McInerney, J.O. and Wilkinson, M., New methods ring changes for the tree of life, *Trends Ecol. Evol.*, 20, 105, 2005.

# 5 Supertree Methods for Building the Tree of Life: Divide-and-Conquer Approaches to Large Phylogenetic Problems

*M. Wilkinson and J. A. Cotton*
Zoology Department, The Natural History Museum, London, UK

## CONTENTS

## ABSTRACT

Reconstructing the tree of life will require fast methods for building very large phylogenetic trees from patchy data. The leading candidates for such an approach employ supertree methods as part of a divide-and-conquer strategy. Here, we discuss two aspects of a phylogenetic divide-and-conquer method: the decomposition of the tree into subproblems and the recombining of these into an overall solution. In particular, we highlight and explore the issue of effective taxon overlap, how it might be achieved via suitable decomposition and how it might be used to guide the setting of priorities for additional data acquisition, and we show how some knowledge of phylogeny is vital in both contexts. Last, we show that quartet puzzling, the best known phylogenetic divide-and-conquer method, can perform poorly when not all quartets are available, and we present a new fast supertree method designed to perform better in this context. Whilst a great deal of work remains, such an approach has great potential as part of a divide-and-conquer method for reconstructing large phylogenies on the scale of the tree of life or for large subsets of species rich taxa.

## 5.1  INTRODUCTION

Reconstructing the complete history of life is an ultimate goal of biology[1], and recent interest in constructing the phylogenetic tree of life reflects the central role of phylogenetic trees in understanding evolutionary history[2,3]. Notwithstanding that most living species may be as yet undescribed, there are major methodological challenges in realising the tree of life. Whilst there has been some debate[4-8], we generally expect accurate reconstruction of phylogenetic trees to become more difficult as trees become larger. The main reason for this is computational complexity. As is well known, the number of possible trees grows more than exponentially with the number of taxa on the tree, so as we seek to identify optimal trees under some objective function the size of tree space (the set of trees for the relevant set of taxa) in which we hope to locate them becomes impossibly large. Exact methods such as exhaustive searches or branch-and-bound algorithms are prohibitively time consuming for all but the smallest phylogenetic problems, and for any substantial problem we are forced to rely on heuristics to guide a limited search of tree space in the hope of finding good (optimal or near optimal) trees. As the size of tree space grows, such searches will become more and more difficult, as they search for one or more needles in an ever expanding haystack. Building trees as large as the tree of life (of the order of millions of taxa) using any known heuristic will be unfeasible, and new approaches will be needed.

A second difficulty in reconstructing the tree of life is the patchy availability of data for different leaves. DNA sequences have become the principal source of data for phylogenetic reconstruction and are accumulating at a rapid rate. As the number of leaves increases, however, it becomes increasingly unlikely that a single gene or single source of data is available for all the taxa, or that a single gene will be effective in reconstructing their relationships. Thus we can expect some information to be unavailable for some taxa ('missing data'). Extensive nonrandom missing data may complicate or compromise analyses[9]. Furthermore, if different markers are needed for different taxa, then accurate analysis will probably need to account for heterogeneity between these markers[10,11]. Modelling this heterogeneity can be complex and will tend to make methods for analysing such data slow.

A solution to both of these problems may be to use divide-and-conquer approaches in which large phylogenetic problems are decomposed into subproblems and the solutions of these subproblems combined to give a global solution. Such approaches reflect the expectation that subproblems can be more easily analysed separately because they are smaller in size and because they can include just those taxa for which a particular type of data is available, reducing the problem of missing data and allowing the process of evolution for particular data to be more accurately modelled. A decomposition might be a natural one, such as dividing a large molecular dataset into data from individual orthologous sequences, or could be designed to yield subproblems that should be more easy to solve accurately and that are readily recombined. The problem of combining a set of phylogenetic trees into a single estimate of phylogeny is addressed by supertree methods, which are therefore integral to any divide-and-conquer approach to building large phylogenetic trees. For example, quartet puzzling (QP)[12] is perhaps the best known divide-and-conquer approach to phylogenetic inference, and the puzzling step is a heuristic supertree method.

In this chapter our main aims are to draw attention to the problem of achieving effective overlap between subproblems and to outline a new fast supertree method. Speed is an obvious important consideration in building large phylogenetic trees, but the importance of effective overlap and how it might most efficiently be achieved has been less widely appreciated. We begin with an overview of divide-and-conquer methods.

## 5.2  DIVIDE-AND-CONQUER METHODS

Divide-and-conquer is a standard approach to solving difficult computational problems by splitting them into smaller, easier (often trivial) subproblems which can be independently solved and then combined to give a global solution[13]. A number of well known algorithms use a divide-and-conquer approach, such as merge sort and quick sort and the fast Fourier transform. The efficiency of solving the subproblems and the efficiency of this merging process will determine how effective a divide-and-conquer approach can be. Whilst classical divide-and-conquer algorithms provide globally optimal solutions to problems, this is probably an unrealistic aim for phylogenetic methods, given that most optimisation-based phylogenetic problems are known to be, or likely to be, NP-complete[14] (NP = nondeterministic polynomial time) so there is very unlikely to be a polynomial time algorithm to solve them (NP-completeness has been shown for parsimony[15], compatibility[16], distance metrics[17] and at least one likelihood problem[18]). Thus we should expect phylogenetic divide-and-conquer strategies to be heuristic rather than exact algorithms. There is no guarantee that solutions to subproblems will be accurate and thus no guarantee that they will all be compatible or readily combinable. Even apparently disjoint subproblems may be incompatible (while quartets with less than three leaves in common must be pairwise compatible, three such quartets can be incompatible). In the phylogenetic context, there may be choices to be made about the order in which subproblems are combined, and even about which sets of subproblems to consider at all.

Most uses of supertree methods have been to build larger phylogenetic trees from sets of previously published trees. Whilst this is divide-and-conquer analysis of a sort (the published trees can be thought of as the results of a given decomposition of the overall problem), we prefer to view divide-and-conquer approaches more narrowly as those in which a designed decomposition of the problem is integral to the analysis. Here, divide-and-conquer analyses offer methods for inferring trees from large datasets rather than from sets of previously inferred trees, and the criterion for choosing among alternative inferences, be it parsimony, likelihood or something else, need be no different from that used by other methods of analysing large datasets.

An important advantage of divide-and-conquer approaches is that they may be relatively computationally efficient[19]. Whilst only two supertree methods have been studied in the divide-and-conquer setting[20], the structure of divide-and-conquer algorithms suggest that many existing supertree methods are perhaps unsuitable for such use. In particular, optimisation supertree methods such as matrix representation with parsimony (MRP) require time-consuming heuristic searches of trees, and combining subproblems for large sets of taxa using these methods will take just as long as solving the problem in a single step (it may or may not be more accurate). These optimisation methods will not be suitable for the amalgamation step in a divide-and-conquer strategy that hopes to be quicker than a conventional analysis: we need faster supertree methods that take a length of time proportional to some polynomial in the number of input taxa. MinCut[21] and modified MinCut[22] supertree methods are both polynomial time approaches, as is the strict consensus merger (SCM[23]). However, whilst saving time is important, accuracy is paramount. MinCut supertrees have been shown to be less accurate than trees constructed using other methods in simulation studies[24] and show a significant bias with respect to shape[25] that might be correlated with poor accuracy. There is clearly scope for new, fast supertree methods in the context of divide-and-conquer approaches (see below).

Exact divide-and-conquer algorithms generally break a problem down into the smallest, trivial subproblems. Similarly, QP breaks a phylogeny problem into the smallest meaningful (unrooted) problem of quartets of taxa. Solving quartets is possible very quickly, as only three different quartets need to be compared for a four-taxon subproblem. However, some quartets may be difficult to accurately infer, and a heuristic analysis of larger subproblems might be quicker or more accurate, leading to an optimal 'granularity' of the decomposition for particular problems.

## 5.3  EFFECTIVE OVERLAP

It is widely recognised that the efficacy of supertree construction is contingent upon which taxa are shared between different input trees (their overlap), but what distinguishes effective and ineffective overlap? To simplify matters we consider the special case of compatible input trees. If two (or more) trees are compatible, then there exists at least one supertree that displays both (or all)

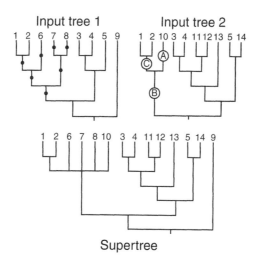

**FIGURE 5.1** Two compatible input trees and their strict consensus supertree. Polytomies on the supertree show where there is no effective overlap between the input trees. Dots indicate the seven different positions in which leaf 10 could occur on input tree 1 while the two input trees remain compatible. Letters indicate the same positions for taxa 6, 7 and 8 on tree 2. Sampling leaf 10 for the tree 1 gene would produce a fully resolved supertree if it was placed in any of these positions. The improvement in overlap with sequencing leaves 6, 7 or 8 for tree 2 depends on where the taxa appear on this tree. (Adapted from Gordon[26].)

the input trees. The set of supertrees that displays all the input trees is termed the span of the input trees and denoted <S>. The strict component consensus of <S> is referred to here as the consensus supertree. Figure 5.1 gives an example of two compatible input trees and their consensus supertree in which there is a mixture of effective and ineffective overlap[26]. Their span <S> includes seven fully resolved supertrees that differ only in the placement of leaf 10 with respect to leaves 1, 2, 6, 7 and 8. Although it mostly does a good job of combining the information in the two input trees, the consensus supertree conveys much less information about the relationships of these leaves than does tree 1. Dealing with compatible trees allows a natural definition of effective and ineffective overlap: effective overlap occurs when the consensus supertree displays all of the input trees, whilst overlap is ineffective to the extent that information present in the input trees is not present in their consensus supertree. In this example, it is easy to see that there is mostly good overlap, but that there is not sufficient information in the input trees to determine the relationships of leaves 6–8 from tree 1 to leaf 10 from tree 2.

### 5.3.1 The Importance of Phylogeny

Overlap has mostly been considered only in terms of the number of leaves in common (for example[27]). Existing approaches to designing meta-analyses of sequence data have focused on maximising this overlap, either by identifying datasets for which all genes are available for all taxa[28] so that there are no missing data, or by minimising the number of gene taxon pairs that are missing from a dataset, with no consideration of the phylogenetic relationship between taxa[29]. The first approach leads to very conservative datasets in which most of the available data need to be discarded, while neither approach directly addresses the effectiveness of the overlap. Given a pair of unrooted trees, the minimum requirement for effective overlap is three common leaves (two leaves plus the root in rooted trees). However, as our example shows, effective overlap depends upon both the number of common leaves and their relationships to each other (see also Wilkinson et al.[30]).

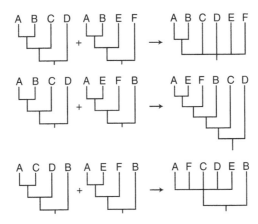

**FIGURE 5.2** Three pairs of four taxon input trees together with their strict consensus supertrees. The phylogenetic position of the two shared leaves has a profound effect on the effectiveness of the overlap between the two trees.

We can show the importance of the phylogeny of the input leaves with some simple examples. The three simple cases shown in Figure 5.2 each consist of a pair of trees with two leaves in common, but differ greatly in the relative positions of the common leaves. These examples show that the common taxa occurring as sister taxa in both trees give no effective overlap between the trees, and neither do the two taxa occurring widely separated on both trees. The optimal situation appears to be when a small clade in one tree spans a deep split in the second tree, as in the second example. Figure 5.3 shows this result more generally. We can define the separation of any pair of taxa (and mean separations of any sets of taxa) on a given tree by counting the number of internal edges separating them. This quantity might be helpful in weighting simple co-occurrence metrics used to assess overlap; if the above result holds in more complicated cases, then it should be optimal to have taxa with highly different mean separations across different input trees. Measures developed in different contexts might also prove useful here, such as the proportion of phylogenetic history sampled by a set of taxa[31].

### 5.3.2 The Importance for Experimental Design and Future Sampling

Achieving effective overlap has been an important practical issue in supertree construction. With the exception of QP trees, most published supertrees have been constructed from input trees harvested from the literature. Although such use of existing phylogenetic inferences has allowed the assembly of several large-scale phylogenetic trees without the need to reanalyse primary data[32–34], achieving effective overlap has sometimes required the use of comprehensive taxonomic hierarchies, interpreted as phylogenies[34]. Faced with needing to rely upon taxonomies or confronted with ineffective overlap, the obvious question is how best to improve overlap. Answering this question has obvious importance for guiding the prioritisation and targeting of additional data acquisition so as to most efficiently bridge the gaps. Our aim here is more to highlight the importance of this question than to address it, but a few preliminary comments may be worthwhile.

If we have non-overlapping data and trees for genes X and Y, with some ineffective overlap identified or suggested by a poorly resolved supertree, then we can ask how the tree for X impinges upon the choice of which leaves in Y should be sequenced for gene X. Consider again the example from Gordon[26] shown in Figure 5.1, which we can think of as showing two trees constructed from two different genes. There are a number of possible choices of additional sampling of genes for

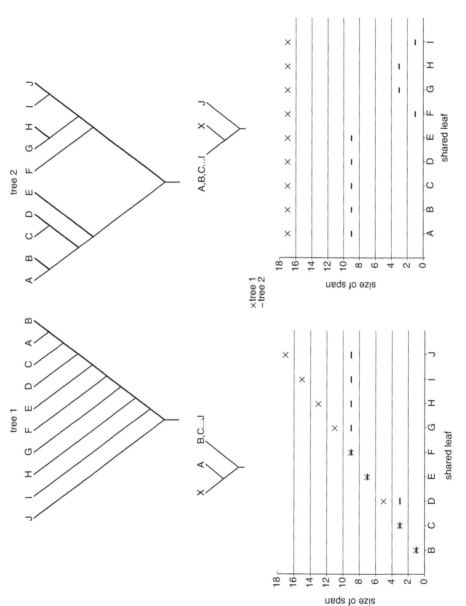

**FIGURE 5.3** The size of the span of supertrees inferred from two different input trees combining with two larger trees. The two small input trees overlap by two taxa (one of which is varied) with the larger trees and have one unique leaf (X). The effectiveness of overlap between the two trees (measured by the size of the span) varies greatly with which leaves are shared and with the topology of the two trees.

particular taxa that might improve overlap sufficiently to allow the consensus supertree to display both input trees, and some selections that would be less likely to help. For example, assuming that the new sequences introduce no conflict in the input trees, then obtaining additional data for leaves 11–14 so as to include them in tree 1 would obviously produce no practical improvement in overlap. An obvious choice that would provide completely effective overlap would be to sample 10 to include it in tree 1, but we could alternatively sample from 6, 7 and 8 for the tree 2 gene. Other things being equal we would target whichever (10 or 6, 7, 8) were most convenient or least expensive to obtain. If we are constrained by the availability of samples or other resources (for example, if we had sufficient funding to sample a single gene for a single taxon and gene 1 cannot be sampled for taxon 10), we can ask which of the remaining targets should be our priority. Thus we would ask whether the available phylogenetic information suggests that additional data for one of 6, 7 and 8 would be most likely to provide effective overlap. Sampling taxon 6 and adding it to tree 2 on branches A, B or C gives a fully resolved supertree if on branch A or B but not on C, while sampling 7 (or 8) on this tree gives a fully resolved supertree on branch C, but not on either A or B. Sampling taxon 6 is more likely to resolve the problem and so might be an optimal choice for this gene.

Similar issues need to be addressed when designing a divide-and-conquer algorithm. Different subproblems need to overlap to some extent if they are to be combined in a global solution, but too much overlap will lead to the same relationships being inferred many times, making the algorithm inefficient. The problem of designing a decomposition of a particular tree so as to provide effective overlap is trivial. Simultaneously providing effective overlap and easily solvable subproblems is more challenging, particularly without knowledge of the tree. Previous workers have designed decompositions that give subproblems that are easily solved, and are even provably easy to solve. Huson et al. created the original disk-covering method (DCM) to produce subproblems of minimal 'evolutionary diameter' in that taxa in a particular subproblem have small pairwise sequence divergence[23]. Distance-based methods such as neighbour joining are known to be accurate for such data, and this allowed Huson et al. to prove a number of theorems about the accuracy of analysis using their decomposition together with these methods. This DCM decomposition did not attempt to control the degree of overlap, however, and so performs poorly in the more general supertree context[19,20]. A second method, DCM2, identifies a single set of taxa which have bounded diameter and which produce subproblems of bounded diameters such that the largest subproblem is as small as possible[35]. Both DCM2 and a related recursive alternative (Rec-I-DCM3) have been shown to be remarkably effective divide-and-conquer algorithms[20,36] (see also Bininda-Emonds and Stamatakis, *Chapter 6*), but it remains unclear whether a strategy in which all subproblems share a set of common taxa is preferable to one in which pairs of subproblems have different shared taxa. We note in passing that consensus efficiency[37], which is the ratio of the cladistic information content[38] of a consensus to that of a set of trees (such as the span), provides a potential measure of the efficacy of overlap which could be used to compare different decompositions of the same dataset.

If different taxonomic groups are studied using different molecular markers, as they inevitably will be to some extent, then the tree of life can only be inferred by combining individual studies. It would be helpful to be able to give some guidance to molecular systematists as to how to design such studies (focusing on their particular taxonomic group of interest) to be easily combinable in this context. For example, it might be best to sequence a marker for a few closely related taxa (as is currently done for outgroup rooting for instance), or it might be better for every study to include a few of a selected set of systematic 'model organisms' which might, but need not, coincide with the model organisms of molecular biologists, for many of which complete sequence data are already available. Tentatively, it seems that the first solution is likely to be better, and we might encourage systematists to sample closely related sequential sister groups to their clade of interest in molecular studies (Figure 5.4). Further work is needed, and our limited discussion and exploration of the simplest examples is intended simply to highlight this need.

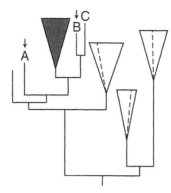

**FIGURE 5.4** Choosing new taxa for optimal supertree construction. Overlap might be maximised by sequencing a few model organisms (indicated by dashed lines within large radiations) when sequencing a particular marker for a clade of interest (indicated by grey triangle), as many other markers will also be sequenced for these organisms. If some idea of the relationships of the sequenced organisms is known, more effective overlap might be obtained by sequencing closely related outgroups that form sequential sister groups to the clade of interest (perhaps taxa A and B would be the best choices here). This relates to existing taxonomic practice, in which closely related outgroups are chosen to root phylogenies, but care should be taken that the outgroups do not form a monophyletic group to the exclusion of the clade of interest (as taxa B and C would), as this would result in no effective overlap.

## 5.4 FAST QUARTET-BASED SUPERTREE CONSTRUCTION

Perhaps the best known and most widely used divide-and-conquer approach in phylogenetics is the QP method of Strimmer and von Haeseler[12]. This method has three steps. First, trees are inferred for all quartets of leaves using some objective function. Second, 'puzzling' is used to combine the quartet inferences to produce a tree for all the leaves. In puzzling, an initial quartet is selected and additional leaves are added sequentially, with the position at which they are grafted to the growing tree determined as a function of the votes cast by the relevant quartets, those that include the new leaf and any three leaves already in the tree and thus convey information on the position of the new leaf on the growing tree. The result may be contingent on the choice of starting quartet and the order of addition of leaves. Consequently, the puzzling step is usually repeated many times, and the third step of QP is the construction of a majority rule consensus of the resulting trees. The frequencies of relationships found in these trees can be taken as an indication of their relative support.

### 5.4.1 VOTING SYSTEMS

The puzzling step is a supertree method, and it can take also, as input, any set of weighted quartets, including those displayed by a set of input trees[39]. However, QP differs from other supertree methods in having been designed for the analysis of a single data set from which inferences about all or most quartets can be made. In contrast, supertree construction is more typically based on input trees that display relationships amongst only a fraction of the quartets. It has been suggested that to be well suited to the latter, the voting method used in puzzling requires some modification[40,41]. Consider the twelve quartets in Figure 5.5a, which are all compatible and jointly entail a single tree (Figure 5.5b). Taking these quartets as input, we would expect a supertree method to yield just this tree, as does, for example, parsimony analysis of character encodings of the quartets. In contrast, applying the puzzling step (1,000 times) with these quartets as input does not yield the expected tree, and all support values are low (Figure 5.5c), a quite unsatisfactory result.

In the original puzzling voting procedure of Strimmer and von Haeseler[12], each relevant quartet is taken in turn. In each of these, the new leaf is paired with one of the three leaves already in the

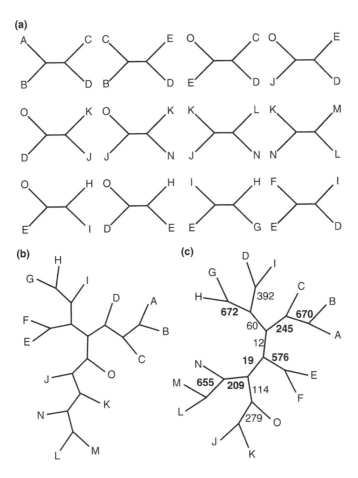

**FIGURE 5.5** Performance of QP in the supertree setting. (a) Twelve quartets; (b) the unique tree displaying all 12 quartets; (c) the majority-rule component consensus of trees constructed from the 12 quartets using 1,000 replicates of the voting method of QP. Numbers indicate frequencies of occurrence of splits in the QP trees, with those for splits entailed by the 12 quartets shown in bold.

growing tree. We find the path between the other two leaves in the growing tree and give a score of +1 to every edge on that path (Figure 5.6a). This is a vote against the grafting of the new leaf to the tree on any of those edges. The votes of all relevant quartets are counted, and the new leaf is attached to an edge with the smallest vote against, ties being broken randomly (Figure 5.6c). Strimmer et al. subsequently developed an approximate system for weighting the votes of quartets according to their posterior probabilities that is employed in TREE-PUZZLE[42,43].

The inadequacy of the QP voting system in the more general supertree case, that is, where we do not have the luxury of votes from all possible relevant quartets and have to rely upon a subset of them, is readily demonstrated and diagnosed. Consider in our example (Figure 5.6) that we have only the single quartet AE/BC to vote on the position of the new leaf E. The puzzling voting system leaves a tie between two branches, which are thus equally likely inferred placements of E. However, only one of these placements (with A) is consistent with what the quartets actually entail about the relationships of E. The other (with D) actually contradicts the information in the relevant quartet because it entails AB/CE. That the puzzling voting system does not provide a vote against this illogical position may not matter when all or nearly all quartets are available, because other quartets (for example AE/BD) may vote directly against this position, but it is expected to compromise its performance, as we have already seen, when not all relevant quartets are available.

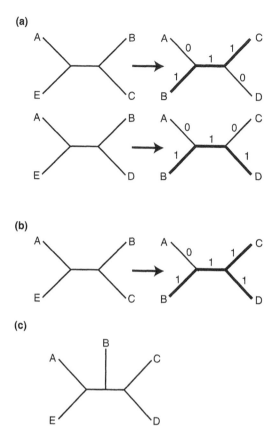

**FIGURE 5.6** Quartet voting systems. (a) The QP voting system showing the votes cast by two quartets relevant to the placement of E (on the left) on the quartet AB/CD; (b) the Vinh and von Haeseler[44] voting system showing votes cast by a single quartet relevant to the addition of E to AB/CD; (c) the fully resolved tree these quartets entail.

An alternative voting procedure, used by Vinh and von Haeseler[44] in the somewhat different context of an algorithm which efficiently elucidates the landscape of possible optimal trees, seems much better suited for the supertree context. For any three leaves, A, B and C, in a tree there is a unique node or vertex where the paths connecting each pair of these leaves intersect and which is subtended by three subtrees (one containing A, one containing B and one containing C), which we shall call the subtrees of the node. The resolution of a quartet on A, B, C and a new leaf tells us to which subtree the new leaf must be grafted in order for the quartet to be displayed by the tree. Other positions contradict the quartet. Thus, instead of voting against branches lying on a particular path, we can vote either for all branches in the subtree in which any grafting of the new taxon would display the quartet, or against all branches in the subtrees in which grafting would contradict the quartet. What is entailed by a pair of quartets is governed by dyadic inference rules of which there are just two[45–47]. Vinh and von Haeseler's[44] voting system reflects these simple inference rules extended to the case where one of the quartets being compared is embedded in a larger tree.

Although not suggested by Vinh and von Haeseler[44], we could use their voting system in place of the original puzzling step of QP. Taking the twelve quartets in Figure 5.5a as input, a QP-type analysis using this alternative voting system would return the unique tree defined by the quartets (Figure 5.5b and Figure 5.7) with maximum support for all splits, a far more satisfactory result than unmodified QP (Figure 5.5c). The comparative performance of this alternative voting system in more typical QP-type analysis merits further investigation. Certainly we would expect it to offer

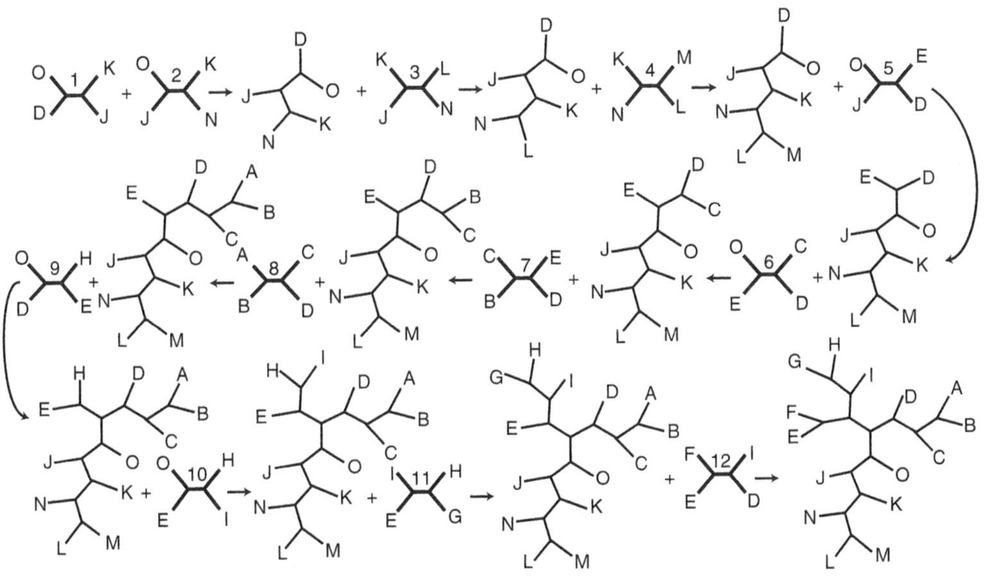

**FIGURE 5.7** Quartet joining of the quartet trees shown in Figure 5.5. Choosing a starting quartet at random, a supertree is built up by sequentially adding a single taxon using the information from relevant quartets. For the minimal set of quartets used here, there is always only one relevant quartet, and the order in which leaves are selected does not matter, but it will matter in general.

improvements if QP is used to analyse very incomplete or patchy supermatrices which do not support resolutions of many quartets and might therefore be expected to expose the limitations of the current QP voting system.

### 5.4.2 Using Fewer Quartets

One potential problem with quartet methods is that the number of possible quartets increases substantially (c. $n^4$) with the number of leaves ($n$). For example, for 500 taxa there are more than $2.5 \times 10^9$ quartets. While the number of quartets is far fewer than the number of possible trees, solving all quartets seriously limits the efficiency that may be obtained from breaking a large problem into more tractable quartet problems. In addition, the number of quartets whose votes must be counted to determine the placement of each new taxon in the puzzling step also increases polynomially (c. $n^3$) with the number of leaves in the growing tree, giving a complexity for the puzzling step of $O(n^4)$.

In contrast, the minimum number of quartets needed to uniquely specify a tree increases only linearly with the number of leaves[48]. This suggests that considerable improvements in speed might be obtained by focussing only upon privileged subsets of quartets that are sufficient to specify a tree. Knowing the tree, it is easy to find minimal sets of quartets that fully specify the tree (for example, the quartets in Figure 5.5a are one such set). More typically in phylogenetics we are trying to infer an unknown tree, which would make the selection of appropriate quartets rather more difficult. Fortunately, our precise problem is to find a privileged set of quartets sufficient to efficiently place a single leaf on an otherwise known tree, which is considerably more tractable.

Vinh and von Haeseler[44] defined a natural ranking of the leaves of the subtrees of a node in terms of their 'distance' (number of edges) from that node and used this to define subsets of leaves called $k$-representative sets comprising the $k$ leaves closest to the node (with random breaking of ties). Motivated by the desire to speed puzzling by relying upon the votes of fewer quartets (hence their unexplained modified voting system), while using quartets likely to provide the most accurate placements, they defined the $k^3$ important quartets with respect to a node as those including a new leaf and one leaf from each of the $k$-representative sets of each of the three subtrees of the node. The important quartets of a tree are all those that are important quartets of any node in the tree, and time can be saved by permitting only these quartets to vote.

Important quartets were used by Vinh and von Haeseler[44] to vote on the reattachment of leaves that have been deleted from a tree, as part of a method for exploring tree space. They note that using only important quartets in QP would yield a decrease in the complexity of puzzling, from $O(n^4)$ to $O(n^2)$, but because of poor performance in simulations they did not pursue this further. Better performance might be expected using their alternative voting procedure, given the frailty of unmodified QP when not all relevant quartets are available. In their study, Vinh and von Haeseler set $k$ to four, but in the extreme $k$ could be set to one, so that there are just $N - 2$ important quartets for the tree. This is the minimum number needed to uniquely specify a tree for $N + 1$ leaves, but there is no guarantee that minimal sets of important quartets will specify a tree: there may be conflict.

### 5.4.3 QUARTET JOINING

Concentrating on speed, we suggest an alternative approach which carries the divide-and-conquer approach a step further. Each relevant quartet provides information on the placement of a new leaf with respect to one internal node, by indicating to which subtree the leaf must be added so as to display the quartet. Thus, if we accept the subtree placement implied by one or more relevant quartets, the problem then becomes that of placement of the leaf within that subtree. This can be further addressed with one or more quartets relevant to its position with respect to a node in the subtree, and so on, until the position is uniquely specified or there are no more relevant quartets (in which case a placement in the remaining subtree is chosen at random or its addition delayed in favour of another leaf). The number of relevant quartets that need to be consulted is thus bounded by the number of nodes and increases linearly with the number of leaves giving a complexity of $O(n^2)$. The one or more quartets used in each step of this iterative divide-and-conquer procedure could be important quartets in the sense of Vinh and von Haeseler[44], but need not be.

The basic idea of this approach, which we call 'quartet joining' is to grow a tree through a series of refinements of the problem of adding leaves that make use of dyadic inference rules. Importantly, the order in which refinements of the problem are sought can further affect the speed of tree construction. We can rank the nodes in a growing tree or any subtree by the number of possible relevant quartets (which is equal to the product of the numbers of leaves in the three subtrees of the node). If we always resolve first the position of the new leaf with respect to the nodes with the highest number of possible relevant quartets, this (at least) halves the problem at each step, leading to a complexity of only $O(n \log n)$.

Quartet joining requires only a single relevant quartet to resolve the placement of a leaf with respect to a particular node (Figure 5.7). With the 12 quartets of Figure 5.5a as input, quartet joining returns the single tree jointly entailed by the quartets with maximal support. More generally, for any compatible set of quartets the method will return one or more trees that display or extend the dyadic closure of the quartets. Thus to maximise speed, quartet joining would consult a single quartet when considering the placement of a new leaf with respect to a node. Compared to the democracy of QP, and the oligarchy of Vinh and von Haeseler's important QP, each quartet consulted in this extreme form of quartet joining dictates the placement of the new leaf, its vote is decisive, and the consulted quartets never conflict. However, the method can also accommodate the votes of multiple relevant quartets at each step (for example, those of important quartets) in order to improve accuracy with only linear increase in complexity. Further speedups could be obtained by using the placement of a new leaf as an opportunity to graft a larger piece of an input tree onto the supertree. For example, if the position of leaf X is finalised by quartet XA/BC drawn from input tree T, the subtree of T at the node defined by X, A and either of B or C that includes leaf X can be grafted to the growing supertree. Note that the method does not demand the starting trees be quartets; they could be the trees inferred using any designed decomposition, and this obviates any concern that quartet trees are difficult to infer accurately because of poor taxon sampling. As with QP, this approach may be sensitive to the starting quartet and to the order in which new leaves are added, with the extent of any variation reflecting conflict and/or ineffective overlap.

## 5.5 CONCLUSION

Supertree methods provide ways of combining phylogenetic information in diverse trees. As such they can be used to produce large-scale phylogenetic trees from sets of trees that are culled from the literature or produced anew through mining of genomic data, and they are essential to any more formal divide-and-conquer analysis of single data sets. To date, supertree methods have been used mostly to produce composite phylogenetic trees from previously published trees, but there has been a recent increase in their application to the phylogenetic analysis of genomic data[49,50]. Molecular data is still available for relatively few genes from relatively few taxa[51], but this is rapidly improving as more complete genomes are sequenced and as 'shallow genomics' projects such as expressed sequence tag (EST) surveys[52] and organelle genome sequences are completed[53], and we expect the use of supertree methods to increase along with the available genomic data.

Much has been made of the potential for supertree methods to combine 'data' that are otherwise difficult to combine in a single phylogenetic analysis[54]. Increasingly, however, most phylogenetic work will be based on molecular data that could, in principle, be combined so that this justification for supertree methods will become less important[55,56]. While there has been a debate between advocates of supertree methods and those who prefer simultaneous analysis of data, we agree with others in not seeing a stark choice between mutually exclusive alternatives[57,58]. This is perhaps most clear in the use of supertree methods as part of divide-and-conquer approaches to finding best fitting trees for a given set of data, that is to efficiently and accurately perform simultaneous analysis of large datasets.

Whereas we do not know whether supertrees constructed from published trees are particularly accurate, we do know that supertree methods embedded in a rationally designed divide-and-conquer strategy can improve heuristic searches, producing better trees faster[19,20,59]. We also know that some sort of supertree analysis will be needed to join together disparate parts of the tree of life inferred using different markers. We consider the question of how best to achieve effective overlap to be an extremely important one, because good answers have the potential to help us target our future research efforts to build the tree of life as efficiently as possible. Efficient supertree construction also requires polynomial time algorithms. The quartet joining method we have outlined is a very fast method of supertree construction that should work well in the absence of conflict. However, its accuracy when confronted with real inference problems is unknown. A priori, one might anticipate some trade-off between speed and accuracy, given that speed is achieved partly by considering less evidence. Hence, accuracy might be improved by considering the evidence from multiple relevant quartets, should they be available. We are currently developing an implementation of quartet joining that will allow the performance of the method to be investigated when input trees conflict.

## ACKNOWLEDGEMENTS

This work was supported by BBSRC grant 40/G18385. We thank Melissa Pentony for running the QP. supertree analysis.

## REFERENCES

1. Haldane, J.B.S., *Possible Worlds and Other Essays,* Chatto and Windus, London, 1927.
2. Cracraft, J. and Donoghue, M.J., *Assembling the Tree of Life,* Oxford University Press, New York, 2004.
3. Soltis, P.S. and Soltis, D.E., Molecular systematics: assembling and using the tree of life, *Taxon,* 50, 663, 2004.

4. Hillis, D.M., Inferring complex phylogenies, *Nature,* 383, 130, 1996.

5. Kim, J., General inconsistency conditions for maximum parsimony: effects of branch lengths and increasing numbers of taxa, *Syst. Biol.,* 45, 363, 1996.

6. Purvis, A. and Quicke, D.L.J., Building phylogenies: are the big easy? *Trends Ecol. Evol.,* 12, 49, 1997.

7. Pollock, D.D. et al., Increased taxon sampling is advantageous for phylogenetic inference, *Syst. Biol.,* 51, 664, 2002.

8. Rosenberg, M.S. and Kumar, S., Taxon sampling, bioinformatics and phylogenomics, *Syst. Biol.,* 52, 119, 2003.

9. Wilkinson, M., Missing data and multiple trees: stability and support, *J. Vertebr. Paleontol.,* 23, 311, 2003.

10. Nylander, J.A.A. et al., Bayesian phylogenetic analysis of combined data, *Syst. Biol.,* 53, 47, 2004.

11. Pagel, M. and Meade, A., A phylogenetic mixture model for detecting pattern-heterogeneity in gene sequence or character-state data, *Syst. Biol.,* 53, 571, 2004.

12. Strimmer, K. and von Haeseler, A., Quartet puzzling: a quartet maximum likelihood method for reconstructing tree topologies, *Mol. Biol. Evol.,* 13, 964, 1996.

13. Cormen, T.H. et al., *Introduction to Algorithms,* MIT Press, Cambridge, MA, 2001.

14. Garvey, M.R. and Johnson, D.S., *Computers and Intractability: A Guide to the Theory of NP-Completeness,* Freeman, NY, 1979.

15. Graham, R.L. and Foulds, L.R., Unlikelihood that minimal phylogenies for a realistic biological study can be constructed in reasonable computation time, *Math. Biosci.,* 60, 133, 1982.

16. Day, W.H.E. and Sankoff, D., Computational complexity of inferring phylogenies from dissimilarity matrices, *Syst. Zool.,* 35, 224, 1986.

17. Day, W.H.E., Computational complexity of inferring phylogenies from dissimilarity matrices, *B. Math. Biol.,* 49, 461, 1987.

18. Addario-Berry, L. et al., Ancestral maximum likelihood of evolutionary trees is hard, *J. Bioinformatics Comput. Biol.,* 2, 257, 2004.

19. Nakleh, L. et al., Designing fast converging phylogenetic methods, *Bioinformatics,* 17, S190, 2001.

20. Roshan, U. et al., Performance of supertree methods on various data set decompositions, in *Phylogenetic Supertrees: Combining Information to Reveal the Tree of Life,* Bininda-Emonds, O.R.P., Ed., Kluwer Academic, Dordrecht, The Netherlands, 2004, chap. 15.

21. Semple, C. and Steel, M., A supertree method for rooted trees, *Discrete Appl. Math.,* 105, 147, 2000.

22. Page, R.D.M., Modified mincut supertrees, in *Proceedings of WABI 2002,* Gusfield, D. and Guigó, R., Eds., 2002, 537.

23. Huson, D.H., Nettles, S., and Warnow, T., Disk-covering, a fast-converging method for phylogenetic tree reconstruction, *J. Comp. Biol.,* 6, 369, 1999.

24. Eulenstein, O. et al., Performance of flip supertree construction with a heuristic algorithm, *Syst. Biol.,* 53, 299, 2004.

25. Wilkinson, M. et al., The shape of supertrees to come: tree shape related properties of fourteen supertree methods, *Syst. Biol.,* 54, 419, 2005.

26. Gordon, A.D., Consensus supertrees: the synthesis of rooted trees containing overlapping sets of labelled leaves, *J. Classif.,* 3, 335, 1986.

27. Price, S.A., Bininda-Emonds, O.R.P., and Gittleman, J.L., A complete phylogeny of the whales, dolphins and even-toed hoofed mammals (Cetartiodactyla), *Biol. Rev.,* 80, 445, 2005.

28. Sanderson, M.J. et al., Obtaining maximal concatenated phylogenetic data sets from large sequence databases, *Mol. Biol. Evol.,* 20, 1036, 2003.

29. Yan, C., Burleigh, J.G., and Sanderson, M.J., Identifying optimal incomplete phylogenetic data sets from sequence databases, *Mol. Phylogenet. Evol.,* 35, 528-535, 2005.

30. Wilkinson, M. et al., Towards a phylogenetic supertree of Platyhelminthes? in *Interrelationships of the Platyhelminthes,* Littlewood, D.T.J. and Bray, R.A., Eds., Taylor and Francis, London, 2001, 292.

31. Faith, D.P., Conservation evaluation and phylogenetic diversity, *Biol. Conserv.,* 61, 1, 1992.

32. Kennedy, M. and Page, R.D.M., Seabird supertrees: combining partial estimates of procellariiform phylogeny, *Auk,* 119, 88, 2002.

33. Pisani, D. et al., A genus-level supertree of the Dinosauria, *Proc. Royal Soc. B,* 269, 915, 2002.

34. Cardillo, M. et al., A species-level phylogenetic supertree of marsupials, *J. Zool.,* 264, 11, 2004.

35. Huson, D.H., Vawter, L., and Warnow, T., Solving large scale phylogenetic problems using DCM2, in *Proceedings of the Seventh International Conference on Intelligent Systems for Molecular Biology,* Lengauer, T. et al., Eds., AAAI Press, Menlo Park, CA, 1999, 118.

36. Roshan, U.W. et al., Rec-I-DCM3: a fast algorithmic technique for reconstructing large phylogenetic trees, in *Proceedings of the 2004 IEEE Computational Systems Bioinformatics Conference,* 2004, 98.

37. Wilkinson, M. and Thorley, J.L., Efficiency of strict consensus trees, *Syst. Biol.,* 50, 610, 2001.

38. Thorley, J.L., Wilkinson, M., and Charleston, M., The information content of consensus trees, in *Advances in Data Science and Classification,* Rizzi, A., Vichi, M., and Bock, H.H., Eds., Springer-Verlag, Berlin, 1998, 91.

39. Pentony, M.M., *Quartet Puzzling Supertrees,* Ph.D. thesis, National University of Ireland, Maynooth, 2004.

40. Pisani, D. and Wilkinson, M., Matrix representation with parsimony, taxonomic congruence, and total evidence, *Syst. Biol.,* 51, 151, 2002.

41. Wilkinson, M. et al., Some desiderata for liberal supertrees, in *Phylogenetic Supertrees: Combining Information to Reveal the Tree of Life,* Bininda-Emonds, O.R.P., Ed., Kluwer Academic, Dordrecht, The Netherlands, 2004, chap. 11.

42. Strimmer, K., Goldman, N., and von Haeseler, A., Bayesian probabilities and quartet puzzling, *Mol. Biol. Evol.,* 14, 210, 1997.

43. Schmidt, H.A. et al., TREE-PUZZLE: maximum likelihood phylogenetic analysis using quartets and parallel computing, *Bioinformatics,* 18, 502, 2002.

44. Vinh, L.S. and von Haeseler, A., IQPNNI: moving fast through tree space and stopping in time, *Mol. Biol. Evol.,* 21, 1565, 2004.

45. Bryant, D., *Building Trees, Hunting for Trees and Comparing Trees,* Ph.D. thesis, University of Canterbury, 1997.

46. Dekker, M.C.H., *Reconstruction Methods for Derivation Trees,* Masters thesis, Vrije Universiteit, 1986.

47. Wilkinson, M., Cotton, J.A., and Thorley, J.L., The information content of trees and their matrix representations, *Syst. Biol.,* 53, 989, 2004.

48. Steel, M., The complexity of reconstructing trees from qualitative characters and subtrees, *J. Classif.,* 9, 91, 1992.

49. Creevey, C.J. et al., Does a tree-like phylogeny only exist at the tips in the prokaryotes? *Proc. Royal Soc. Lond. B,* 271, 2551, 2004.

50. Beiko, R.G., Harlow, T.J., and Ragan, M.A., Highways of gene sharing in prokaryotes, *Proc. Natl. Acad. Sci. USA,* 102, 14332, 2005.

51. Sanderson, M.J. and Driskell, A.C., The challenge of constructing large phylogenetic trees, *Trends Plant Sci.,* 8, 374, 2003.

52. Theodorides, K. et al., Comparison of EST libraries from seven beetle species: towards a framework for phylogenomics of the Coleoptera, *Insect Mol. Biol.,* 11, 467, 2002.

53. Miya, M., Kawaguchi, A., and Nishida, M., Mitogenomic exploration of higher teleostean phylogenies: a case study for moderate-scale evolutionary genomics with 38 newly determined complete mitochondrial DNA sequences, *Mol. Biol. Evol.,* 18, 1993, 2001.

54. Sanderson, M.J., Purvis, A., and Henze, C., Phylogenetic supertrees: assembling the trees of life, *Trends Ecol. Evol.,* 13, 105, 1998.

55. Rokas, A. et al., Genome-scale approaches to resolving incongruence in molecular phylogenies, *Nature,* 42, 798, 2003.

56. Scotland, R.W., Olmstead, R.G., and Bennett, J.R., Phylogeny reconstruction: the role of morphology, *Syst. Biol.,* 52, 539, 2003.

57. Levausser, C. and Lapointe, F.-J., War and peace in phylogenetics: a rejoinder on total evidence and consensus, *Syst. Biol.,* 50, 881, 2001.

58. Holmes, S., Statistics for phylogenetic trees, *Theor. Popul. Biol.,* 63, 17, 2003.

59. Fuellen, G., Wagele, J.W., and Giegerich, R., Minimum conflict: a divide-and-conquer approach to phylogeny estimation, *Bioinformatics,* 17, 1168, 2001.

# 6 Taxon Sampling versus Computational Complexity and Their Impact on Obtaining the Tree of Life

## O. R. P. Bininda-Emonds

Institut für Spezielle Zoologie und Evolutionsbiologie mit Phyletischem Museum, Friedrich-Schiller-Universität Jena, Germany

## A. Stamatakis

Swiss Federal Institute of Technology, School of Computer and Communication Sciences, Lausanne, Switzerland

## CONTENTS

## ABSTRACT

The scope of phylogenetic analysis has increased greatly in the last decade, with analyses of hundreds, if not thousands, of taxa becoming increasingly common in our efforts to reconstruct the tree of life and study large and species rich taxa. Through simulation, we investigated the potential to reconstruct ever larger portions of the tree of life using a variety of different methods (maximum parsimony, neighbour joining, maximum likelihood and maximum likelihood with a divide-and-conquer search algorithm). For problem sizes of 4, 8, 16 … 1,024, 2,048 and 4,096 taxa sampled from a model tree of 4,096 taxa, we examined the ability of the different methods to reconstruct the model tree and the running times of the different analyses. Accuracy was generally good, with all methods returning a tree sharing more than 85% of its clades with the model tree on average, regardless of the size of the problem. Unsurprisingly, analysis times increased greatly with tree size. Only neighbour joining, by far the fastest of the methods examined, was able to solve the largest problems in under 12 hours. However, the trees produced by this method were the least accurate of all methods (at all tree sizes). Instead, the strategy used to sample the taxa had a larger impact on both accuracy and, somewhat unexpectedly, analysis times. Except for the largest problem sizes, analyses using taxa that formed a clade generally both were more accurate and took less time than those using taxa selected at random. As such, these results support recent suggestions that taxon number in and of itself might not be the primary factor constraining phylogenetic accuracy and also provide important clues for the further development of divide-and-conquer strategies for solving very large phylogenetic problems.

## 6.1  INTRODUCTION

Reconstructing the tree of life accurately and precisely represents the holy grail of phylogenetics and systematics. However, the impact of obtaining the tree goes well beyond these research fields to include all of the life sciences because, as it was nicely put recently by Rokas and Carroll[1], the conclusions we make as phylogeneticists form part of the assumptions underlying the analyses of the other biologists. Evolutionary information is now becoming increasingly included in fields as diverse as comparative biology, genomics and pharmaceutics. In the past decade, the increasing accumulation of phylogenetic data, made possible by the molecular revolution, has brought the dream of realising a highly comprehensive tree of life tantalisingly close.

Currently, however, the continued lack of suitable phylogenetic data represents a proximate hindrance in our efforts to reconstruct the tree of life. Although whole genomic data is becoming available at an increasing rate (but more so for prokaryotic organisms with their smaller genomes), molecular sampling has generally been sparse and restricted largely to model organisms and model genes[2,3]. However, even with the prospect of abundant whole-genome sequence data, the ultimate hindrance is the sheer size of the tree of life itself, which has been estimated to comprise anywhere from 3.6 million to 100+ million species (but most commonly 10–15 million)[4].

It has long been appreciated that the number of possible phylogenetic trees increases superexponentially with the number of taxa[5]. For example, there are three distinct rooted phylogenetic trees for three species, 15 for four species, 105 for five species, and so on. For only 67 species, the number of possible trees is on the order of 10 to the power of 111 trees, a number that just exceeds the volume of the universe in cubic Ångstroms (a comparison first heard by the first author from David Hillis). Phylogenetic analyses are now routinely conducted on data sets of this size and larger (up to hundreds of taxa). Albeit comparatively rare, analyses of thousands of taxa have also been performed, mostly as proof of concepts for new algorithmic implementations. These include a neighbour joining (NJ) analysis of nearly 8,000 sequences[6], a maximum likelihood (ML) analysis of 10,000 taxa[7], and a maximum parsimony (MP) analysis of 13,921 taxa[8,9]. However, we are unsure of the prospects of achieving a correct or nearly correct answer for studies of these size, given the literally astronomical size of 'tree space'.

Compounding this limitation is the fact that the general problem of reconstructing a tree (or a network, given that the tree of life is not always tree-like) from a given data set is one of a set of non-deterministic polynomial time (NP) problems for which no efficient solution is known or, more pessimistically, one for which no such solution potentially exists (NP-complete)[10]. Thus, the analysis of larger data sets requires a disproportionately longer time (or disproportionately more computer resources) and/or the use of increasingly less efficient heuristic search strategies, with both factors impacting negatively on our ability to recover the best solution for that given data set. Fortunately, several studies using empirical and/or simulated data have shown that even phylogenetic analyses at the high end of the scale currently examined are both tractable and show acceptable, if not surprising, accuracy with shorter sequence lengths than might be expected[11–13], thereby reinforcing some theoretical work in the latter area[14,15]. Additionally, advances in computer technology and architecture such as parallel and distributed computing and programs that exploit them efficiently in combination with the continual development of faster search strategies promise to make even larger phylogenetic problems increasingly tractable. However, the NP-completeness of the phylogeny problem represents a fundamental limitation in our efforts to unearth the tree of life.

As such, we face a dilemma in attempting to reconstruct the tree of life (or even major portions thereof). Smaller problems are computationally easier to solve, but at the extreme, have been demonstrated to be susceptible to the adverse effects of taxon sampling and, for parsimony in particular, long branch attraction[16] (for a review of the latter, see Bergsten[17]). In these cases, the fact that DNA has only four character states can lead to a high number of convergent changes (noise) along two long branches leading to unrelated taxa. These convergent changes can pull the two branches together, thereby leading the phylogenetic analysis astray. Thus, the general consensus is that, given a suitable sampling strategy[18], the addition of species to a phylogenetic analysis is usually beneficial in terms of accuracy because it ameliorates the effects of these two problems[19,20] (see Rosenberg and Kumar[21] for a contrary view). At some point, however, the computational complexity of the phylogeny problem must begin to outweigh the benefits of adding taxa. Although it is not stated explicitly in the literature, it seems that the general expectation is that phylogenetic accuracy shows a convex distribution with respect to the number of taxa in the analysis, with taxon sampling and computational complexity limiting accuracy when species numbers are low and high, respectively.

It remains to be demonstrated whether or not this expectation is true and, if so, at what point accuracy is maximised, while simultaneously considering the running time of the analysis. Establishing the latter could be especially important to the further development of the so-called 'divide-and-conquer' search strategies such as quartet puzzling[22] and disk-covering[9,23,24]. These strategies generally seek to solve large phylogenetic problems by breaking them down into numerous smaller subproblems that are computationally easier to solve precisely because they are smaller with respect to both the number of taxa and the evolutionary distance between those taxa. The results from the subproblems are then combined to provide an answer for the initial, global problem. As such, divide-and-conquer strategies essentially attempt to bridge the gap between the problems of taxon sampling and computational complexity. However, it is unknown what the optimal sizes of the subproblems should be in order to achieve the greatest accuracy in the shortest time possible. To date, subproblem sizes have usually been determined empirically on a case-by-case basis.

Thus, the goal of this chapter is to extend on previous analyses examining the scalability of phylogenetic accuracy with respect to the number of species in the analysis (the 'size' of the analysis). Specifically, we use simulation to investigate the changes in various parameters (accuracy, resolution and running time) related to the analysis of increasingly larger phylogenetic problems under different optimisation criteria (NJ, MP and ML) and methods of data set selection (random or clade sampling). Our results elucidate the prospects for phylogenetic analyses of very large phylogenetic problems, as might be needed to infer the tree of life or study large and species rich taxa, and provide additional insights into the potential of divide-and-conquer search strategies within this context.

## 6.2 MATERIALS AND METHODS

### 6.2.1 SIMULATION PROTOCOL

The simulation protocol used was modelled on that followed by Bininda-Emonds et al.[13] to examine the scaling of accuracy in very large phylogenetic trees. For each run, a model tree of 4,096 taxa was generated according to a stochastic Yule birth process using the default parameters of the

YULE_C procedure in the program r8s v1.60[25]. Branch lengths on the tree were modelled assuming a model of substitution that departs from a molecular clock. Specifically, branch-specific rates of evolution were determined by drawing random normal variates (mean of 1.0 and standard deviation of 0.5, truncated outside of [0.1, 2.0]) and multiplying by an overall tree-wide rate of substitution. Branch lengths were determined by multiplying branch-specific rates with branch durations obtained from the Yule process model.

A model data set was then created by evolving a nucleotide sequence down the model tree using a standard Markov process model as implemented in Seq-Gen v.1.2.7[26]. The sequence length was 2,000 bp, which is of sufficient length for simulated data with its stronger signal to achieve good accuracy for even the largest tree examined herein[13], but is also short enough to keep running times within acceptable limits. Sequences were generated under a Kimura 2-parameter model[27] with a transition/transversion ratio (ti:tv) of 2.0, site-to-site rate heterogeneity (that is, Gamma model) with shape parameter of 0.5, and an overall average rate of evolution of 0.1 substitutions/site, measured along a path from the root to a tip of the tree. No invariant sites were explicitly modelled.

The model data set was then sampled to create test data sets where the number of taxa varied on a $\log_2$ scale from 4 to 2,048. No sampling of characters was performed so that the sequence length was always 2,000 bp. Taxon sampling was accomplished by either selecting taxa at random (random sampling) or by selecting a single clade from the model tree of the same size as the number of taxa to be retained (clade sampling); all other taxa were pruned from the test data sets. The expectation is that clade sampling should result in improved accuracy, given that it minimises the evolutionary diameter of the problem; this is the logic underlying the disk-covering family of divide-and-conquer methods[28]. By contrast, random sampling will tend to result in an increased number of long branches and/or extend the diameter of the problem, especially when the proportion of taxa sampled is very low. Both factors have been demonstrated to reduce the accuracy of phylogenetic inference.

Clade sampling requires the model tree to possess at least one clade for all the test sizes. Because this situation was difficult to achieve, clades that were within ±2.5% of the desired size were used when there was no clade of exactly the size desired. When multiple clades for a given size existed, one was chosen at random. If the model tree did not contain clades of all the desired sizes, it was discarded, and a new model tree was generated.

Each subsampled data set (for both random and clade sampling) as well as the full data set were analysed using three optimisation criteria, each of which accounted for the model of evolution Kimura 2-parameter + Gamma (K2P + G) as far as possible: MP, NJ, and ML. For the four largest matrices (512; 1,024; 2,048; and 4,096 taxa), a ML analysis in conjunction with a disk-covering divide-and-conquer framework (ML-DCM3) was also used. Bayesian analysis was not examined due to time and memory constraints[29]. Because Bayesian analysis samples from the posterior distribution of trees, it is necessarily significantly slower than the other methods examined here, especially if a high number of generations is employed to ensure reliable results. Even without Bayesian analysis, each replicate required just over five days to complete.

Thus, the results for each individual run were based on data matrices all derived from the same model set of molecular data evolved along the same model tree. This procedure differs substantially from that used by Bininda-Emonds et al.[13], in which model trees of the desired problem size were generated (that is, there was no sampling performed). Additionally, for each subproblem size and sampling strategy, the same alignment was analysed by each of MP, NJ, ML and where appropriate ML-DCM3. In total, 50 runs were conducted, comprising nearly eight CPU months of analysis time.

## 6.2.2 Phylogenetic Analysis

MP analyses used PAUP* v4.0b10[30] with transversions weighted twice as much as transitions. Different search strategies were employed depending on the size of the alignment. Below 16 taxa, a branch-and-bound search was used, thereby guaranteeing that all optimal trees were found. For matrices with ≥16 taxa, various heuristic searches were used depending on the size of the problem:

a thorough heuristic (<256 taxa), the parsimony ratchet (<1,024 taxa[31]), and finally a greedy heuristic (Σ 4,096 taxa). The thorough heuristic consisted of 100 random addition sequences with TBR branch-swapping, with a maximum of 10,000 trees being retained at any time during the analysis. The parsimony ratchet consisted of 10 batches of 100 iterative weighting steps, with 25% of the characters receiving a weight of two at each step. Thereafter, all equally most parsimonious trees were used as starting trees for a heuristic search using TBR branch swapping and limited to one hour of CPU time. Each replicate used the same command file for the ratchet, which was created using the Perl script PerlRat v1.0.9a. However, for the largest matrices, even the parsimony ratchet proved to be too slow during the test phase, especially because of the use of a step matrix to account for the ti:tv ratio. Therefore, a greedy heuristic was used, consisting of a single simple stepwise addition sequence followed by TBR branch-swapping with a maximum of 10 trees being retained at any time.

NJ analyses used QuickTree[32] using a Kimura translation to determine the pairwise distances.

ML analyses used RAxML-V (Randomized Axelerated Maximum Likelihood)[33], which is one of the fastest and most accurate programs for ML-based phylogenetic inference. A key feature of RAxML is its comparatively low memory consumption[29], which in combination with its advanced search algorithms and accelerated likelihood function[33,34] makes it uniquely suitable for ML analyses of large numbers of taxa. All RAxML analyses used the default hill-climbing search option (–f c) using an HKY85 substitution model with an estimate of 50 distinct per site evolutionary rate categories (CAT). This HKY + CAT model is essentially empirically equivalent to the better known HKY + I + G model, but requires fewer floating point operations and memory.

Finally, we also performed ML analyses using a divide-and-conquer search algorithm at the largest problem sizes (512 or more taxa) using RAxML in concert with the Recursive Iterative Disk Covering Method (Rec-I-DCM3)[9]. This combination of methods has been more formally referred to as Rec-I-DCM3(RAxML); however, we use the simpler ML-DCM3 throughout this chapter. Based on an initial 'guide tree' containing all taxa (here, the starting tree for the ML analyses as computed by RAxML), Rec-I-DCM3 intelligently decomposes the data set into smaller subproblems that overlap in their taxon sets. These subproblems are then solved using RAxML (using the same parameters as above), with the respective subtrees merged into a comprehensive tree with the Strict Consensus Merger[23]. This global tree was then further improved using RAxML (using the fast hill climbing heuristic; option –f f) to construct the new guide tree. The processes of decomposition, subproblem inference, subtree merging and global refinement were repeated for three iterations. The maximum size of the subproblems was 25% of the size of the full data set, as suggested in the user notes to Rec-I-DCM3.

A time limit of 12 hours was imposed on each individual analysis. This limit was never invoked for the NJ analyses and only for the largest matrices for MP (4,096 only), ML (2,048 and 4,096, but not always for both sizes) and ML-DCM3 (2,048 and 4,096). The use of a time limit will obviously impact accuracy negatively and potentially penalise the more computationally intensive ML analyses to a greater extent. However, the reality is that shortcuts of various types (for example, time limits or less thorough search strategies) must be employed when analysing very large matrices, so this constraint might represent a reasonable one. To judge the effects of imposing a time limit, one additional run was performed for the full data set of 4,096 taxa with all methods being allowed to run to completion.

In all cases, the inferred tree was held to be the strict consensus of all equally optimal solutions. All analyses were conducted on a cluster of unloaded 2.4-GHz Opteron 850 processors, each with 8 GB of RAM, located at the Department of Informatics at the Technical University of Munich. All programs used (including those used to simulate the data) were compiled as needed for this platform.

### 6.2.3 VARIABLES EXAMINED

Results were analysed with respect to three variables that are particularly relevant to the phyloge-netic analysis of very large data sets: resolution, accuracy and running time. Resolution is the number of clades on the inferred tree relative to the total number of clades on a fully bifurcating tree of the same size ($n - 2$ for an unrooted tree, where $n$ = number of taxa). Resolution varies between 0 and 1, with the former value indicating a completely unresolved bush and the latter indicating a fully resolved tree. This parameter reflects the decisiveness of the analysis and is most relevant for the MP analyses. NJ always returns a single, fully resolved tree, and ML analyses invariably do so as well.

Accuracy was measured as the ability to reconstruct the model tree. In computer science, the optimality score of an analysis (either in isolation or in relation to that of the model tree) is also often used as a proxy for accuracy. However, the use of three different optimality criteria in this study prevents such an approach, and the comparison to a known 'true' tree is perhaps more intuitive to biologists. Accuracy was quantified using both the consensus fork index (CFI[35,36]) of the strict consensus of the inferred and model trees and the symmetric difference (or partition metric) between these two trees ($d_S$[37]). The CFI indicates the proportion of clades shared between the two trees, whereas $d_S$ indicates the number of clades found on one tree or the other, but not both. To make these values comparable, $d_S$ was normalised according to the number of taxa on the trees (by dividing by $2n - 6$, where $n$ = number of taxa[38]) and subtracted from one to derive a similarity measure equivalent to CFI. Although it is not strictly accurate, we continue to refer to this metric as $d_S$ for convenience.

CFI and $d_S$ differ most importantly in how they treat polytomies in the inferred tree (the model tree is always fully bifurcating). CFI treats all polytomies as errors, whereas $d_S$ essentially ignores them because they do not specify any unique clades. Thus, in comparing a fully resolved tree with a fully unresolved one, CFI = 0 and $d_S = 0.5$. As such, the difference between CFI and $d_S$ is again most relevant for the MP analyses, which are the only ones expected to produce trees that are not fully resolved. For the comparison of two fully resolved trees, CFI = $d_S$.

Finally, the running time for each analysis was recorded in seconds. Again, an upper limit of 12 hours (43,200 seconds) was imposed on all analyses. However, analysis times could still substantially exceed this limit in some cases due to the discrete nature of the stopping mechanisms. For instance, a search can be terminated only after the completion of an iteration or calculation of an optimality score, both of which can represent long-running operations at the largest tree sizes.

For each variable, results were compared using a multivariate analysis of variance (ANOVA), with the method of analysis and sampling strategy as factors, and the size of the data set as a covariate. The level of significance was $\alpha = 0.05$. Fisher's protected least significant difference (PLSD) test was used to determine significant differences between categories within a factor.

### 6.2.4 SOFTWARE AVAILABILITY

The following software and/or source code used in this study are freely available at the following URLs:

- PerlRat.pl: www.uni-jena.de/~b6biol2/ProgramsMain.html
- RAxML: diwww.epfl.ch/~stamatak (under 'software')
- Rec-I-DCM3: www.cs.njit.edu/usman/RecIDCM3.html

## 6.3 RESULTS

### 6.3.1 RESOLUTION

Resolution was always one for each individual NJ, ML, and ML-DCM3 analysis. MP produced trees that were significantly less resolved ($P < 0.0001$ for all pairwise comparisons) and, except for a tree size of four with random sampling, were never fully resolved on average (Figure 6.1A). Nevertheless, the MP trees were generally well resolved at all tree sizes, with the average resolution being always greater than 0.90. Resolution for MP differs significantly with tree size ($P < 0.0001$),

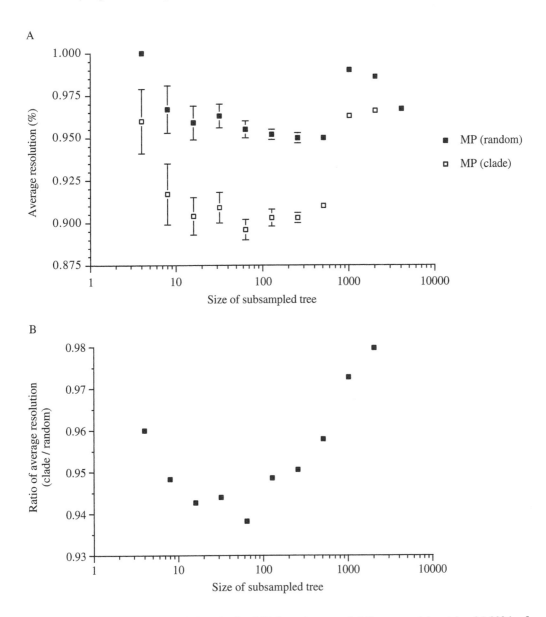

**FIGURE 6.1** Resolution of trees inferred using MP from data sampled from a model matrix of 2,000 bp for 4,096 taxa. (A) Average resolution over 50 individual runs; error bars represent standard errors. (B) Ratio of average resolutions from clade sampling as compared to random sampling. Resolution for all other optimisation criteria was always 1.

showing a concave pattern that is noticeably higher at extremely small and extremely large tree sizes. In the latter case, however, this is an artefact of only 10 trees being retained in analyses of 1,024 or more taxa. Otherwise, it appears that resolution reaches a plateau of about 0.90 for clade sampling and 0.95 for random sampling. The average resolution for the MP analyses using clade sampling was always significantly less than that for random sampling ($P < 0.0001$); the ratio of the values for clade versus random sampling fell between 0.94 and 0.98 at all tree sizes (Figure 6.1B). All methods yielded fully resolved trees, or nearly so for MP, in the time-unlimited analyses (Table 6.1).

**TABLE 6.1**

**Statistics Relating to a Time-Unlimited Analysis of the Full Dataset of 4,096 Taxa**

| Optimisation Criterion/Method | Resolution | Accuracy | | Time (seconds) |
| | | CFI | $(1 - d_S)$ | |
|---|---|---|---|---|
| MP | 1.000 | 0.903 | 0.917 | 69,392 |
| NJ | 1.000 | 0.857 | 0.857 | 193 |
| ML (fast hill climbing) | 1.000 | 0.912 | 0.912 | 38,737 |
| ML (standard hill climbing) | 1.000 | 0.923 | 0.923 | 303,450 |
| ML-DCM3 | 1.000 | 0.921 | 0.921 | 195,371 |

### 6.3.2 ACCURACY

Accuracy, whether measured by CFI or $d_S$ was generally good at all tree sizes and for all methods (Figure 6.2A and Figure 6.3A). In all cases, accuracy was greater than 80% on average and often better than 90%. Tree size had a variable impact on accuracy. It did not influence accuracy for either ML-DCM3 ($P = 0.4812$; although only four sizes were tested for this method), MP as measured by $d_S$ ($P = 0.4132$), or ML ($P = 0.1995$), but had a significant effect for both NJ ($P = 0.0244$) and MP as measured by CFI ($P = 0.0087$). However, the only clear trend is for NJ under clade sampling where accuracy decreases with the size of the problem. In all the remaining cases, the curves are reasonably flat and/or sigmoidal. Except for ML-DCM3, allowing all methods to run to completion in the time-unlimited analyses produced significantly more accurate results when compared to the 12-hour limited analyses ($P < 0.0001$ according to a one sample $t$-test).

The different optimisation criteria/methods used also had an impact on the accuracy of the solutions. When CFI was used to measure accuracy (Figure 6.2A), ML and ML-DCM3 were not significantly different ($P = 0.0763$), and neither were MP and NJ ($P = 0.6982$). However, the trees derived using the former methods were significantly more accurate than those from the latter ($P < 0.0001$). When $d_S$ was used (Figure 6.3A), ML trees were statistically indistinguishable from those from either MP ($P = 0.7037$) or ML-DCM3 ($P = 0.0618$), although the latter two were significantly different from one another ($P = 0.0340$). NJ yielded significantly worse trees in all cases ($P < 0.0001$).

Only the MP analyses showed a difference in accuracy as measured by the two metrics (compare Figure 6.2A and Figure 6.3A), with the analogous values for $d_S$ being either equal to, or more commonly, greater than those for CFI. The effect was the most pronounced for clade sampling, which also produced solutions that were less resolved than were those from random sampling (Figure 6.2B and Figure 6.3B).

For both NJ and MP ($d_S$ only), the sampling strategy had a significant effect on accuracy ($P < 0.0001$), with clade sampling generally leading to increasingly accurate solutions as the size of the problem decreased. However, the two sampling strategies showed similar performance with respect to accuracy for trees of 512 or more taxa. No effect was present for MP when accuracy was measured using CFI ($P = 0.1248$). Likewise, there was no significant trend for ML with respect to the sampling strategy ($P = 0.0698$). Random sampling produced slightly, but significantly more accurate trees with ML-DCM3 at the three relevant problem sizes examined for it (512; 1,024; and 2,048 taxa; $P < 0.0001$).

### 6.3.3 RUNNING TIME

Except for NJ, no method obtained a solution for the full model data set (4,096 taxa) within the 12-hour time limit. The running times for the unlimited MP, ML and the three iteration ML-DCM3 analyses of 4,096 taxa (see Table 6.1) were 1.6, 7.0 and 4.5 times longer than the limit of 12 hours.

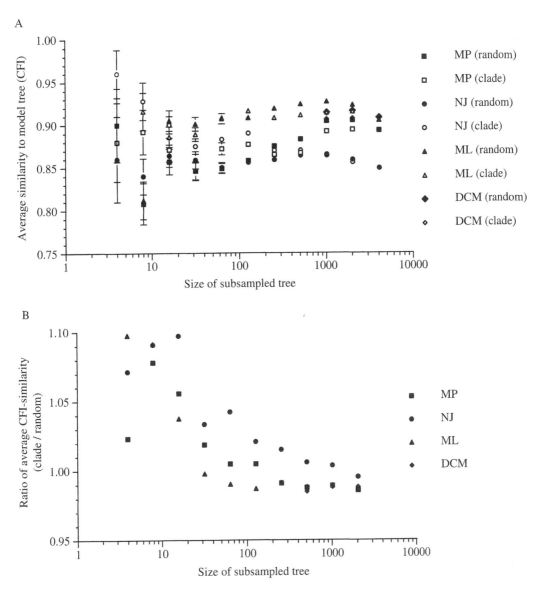

**FIGURE 6.2** Phylogenetic accuracy of trees inferred using different methods from data sampled from a model matrix of 2,000 bp for 4,096 taxa. Accuracy was measured as the value of the CFI between the inferred tree and the model tree upon which the data were simulated; both trees were pruned so as to have identical taxon sets. (A) Average accuracy over 50 individual runs; error bars represent standard errors. (B) Ratio of average accuracy from clade sampling as compared to random sampling.

Running times were significantly influenced by all three factors and covariates examined, either in isolation or in combination (all $P < 0.0001$). Fisher's PLSD tests also revealed highly significant differences (all $P < 0.0001$) between all pairs of categories within the factors of sampling strategy and method of analysis.

For all optimisation criteria, running times increased approximately linearly with tree size on a log-log scale (Figure 6.4A and Figure 6.4C). For each doubling in tree size, the running time of NJ increased by a factor of about three on average (random sampling: $3.22 \pm 0.60$ (mean $\pm$ SE); clade sampling: $3.33 \pm 0.61$). MP showed both the largest and most variable increases in running

A

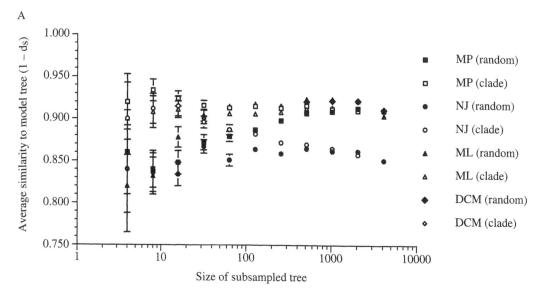

B

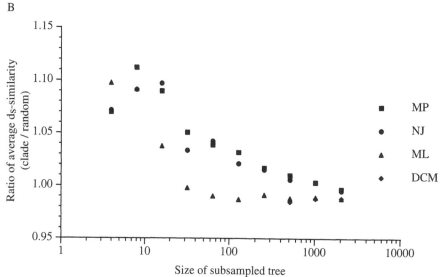

**FIGURE 6.3** Phylogenetic accuracy of trees inferred using different methods from data sampled from a model matrix of 2,000 bp for 4,096 taxa. Accuracy was measured as one minus the normalised value of the partition metric between the inferred tree and the model tree upon which the data were simulated; both trees were pruned so as to have identical taxon sets. (A) Average accuracy over 50 individual runs; error bars represent standard errors. (B) Ratio of average accuracy from clade sampling as compared to random sampling.

time (random sampling: $6.54 \pm 2.31$; clade sampling: $12.26 \pm 5.84$). The largest increases for MP occurred for the comparisons 8–16 and 64–128 taxa (random sampling) and 16–32, 32–64 and 64–128 taxa (clade sampling). Many of these high rates corresponded with either the adoption of a new, less thorough search strategy or when a given search strategy was apparently becoming 'overloaded' for a given problem size. Finally, the rate increases for ML were intermediate between NJ and MP and with low variation (random sampling: $4.34 \pm 0.69$; clade sampling: $5.03 \pm 0.87$).

Compared to the other methods (Figure 6.4A), NJ always produced the shortest running times ($P < 0.0001$), with the differences becoming the most marked at tree sizes of 16 taxa or greater.

A

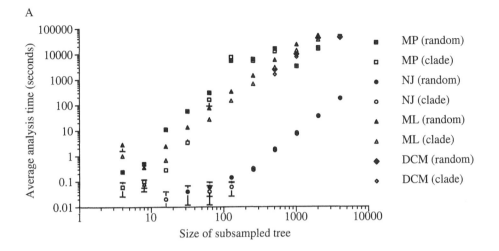

B

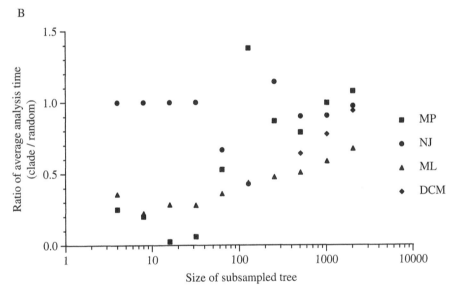

**FIGURE 6.4** Analysis times for trees inferred using different methods from data sampled from a model matrix of 2,000 bp for 4,096 taxa. The time used for 4,096 taxa derives from the single time-unlimited analysis for MP, ML and ML-DCM3. (A) Average analysis time over 50 individual runs; error bars represent standard errors. (B) Ratio of average analysis time from clade sampling as compared to random sampling. (C) Ratio of average analysis time for a given sample size as compared to previous sample size.

The running times of MP and ML were roughly comparable, although MP was significantly faster ($P < 0.0001$) on the whole. MP tended to run faster for the smallest and largest tree sizes with clade sampling, whereas ML generally ran faster for all tree sizes under random sampling. In those cases where the time limit was not exceeded, ML-DCM3 was faster on average than ML, but slower than MP (in both cases by a factor of two to three).

Analysis times under clade sampling were almost always less than those for random sampling, with the differences becoming smaller with increasing tree size (Figure 6.4B). For ML, running times under clade sampling were faster by a factor of at least two for all tree sizes except four and 2,048. The most marked differences for MP were limited to tree sizes of 64 or fewer taxa, where the differences were the largest for all the methods examined (including a factor of nearly 40 with

C

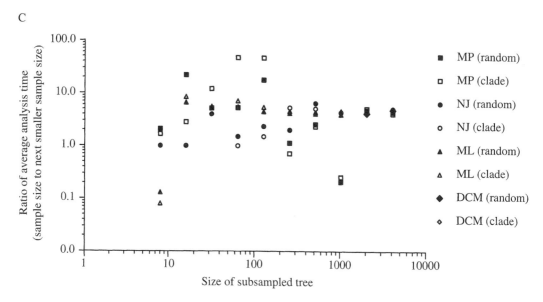

**FIGURE 6.4** (Continued).

16 taxa). Beyond 128 taxa, the respective running times for the different sampling strategies were approximately equal for MP. For 256 and 512 taxa, at least, this result reflects the more deterministic nature of the parsimony ratchet, which must perform a set number of iterations.

## 6.4  DISCUSSION

Our simulation study produced three key findings, some unexpected:

- That accuracy is at best only weakly influenced by the size of the problem
- That the methods of inference examined produce solutions of comparable, and good, accuracy
- That the sampling strategy employed has a significant effect on both accuracy (to a point) and, more strongly, on the running time of the analysis

Naturally, caveats abound. Our study used simulated data, which tend to be 'cleaner' and contain more phylogenetic signal than real sequence data. Moreover, the model of evolution used (HKY + G) is less complicated, and therefore less computationally intensive, than those more commonly used. Finally, the methods of inference were refined to match the known model as precisely as possible. Thus, our results might, in isolation, represent a best case scenario. However, the comparative aspects of our study should be accurate.

### 6.4.1  ACCURACY AND SPEED

Although it is widely accepted that larger phylogenetic problems are more difficult to solve and should therefore show decreased accuracy, there is now accumulating evidence to suggest that the performance dropoff is not as severe as many would believe, at least up to problem sizes of about 10,000 taxa[11,13]. Our finding that accuracy is essentially flat with respect to tree size (Figure 6.2A and Figure 6.2B) lends further support to these latter findings, indicating that good accuracy is achievable even for very large phylogenetic problems and within a reasonable timeframe. Moreover, our work indirectly supports recent work by Rokas, Carroll and colleagues[1,39] that strongly argues that the amount of sequence data (number of genes), and not the number of taxa, is the more critical factor influencing phylogenetic accuracy. In our case, the alignment length of 2,000 bp was specifically chosen because it apparently contained sufficient phylogenetic signal for all problem sizes examined here[13], thereby minimising the effects of sequence length. In real terms, 2,000 bp would represent perhaps one or two genes of lengths that are typically used in phylogenetic analysis. Although this might be an insufficient number of genes to achieve good accuracy[1,39] it must again be remembered that the simulated sequence data often contain much more signal than real data due to the absence of gaps and a lack of noise that would arise because of alignment errors.

Similarly, good evidence exists that the performance of NJ with respect to accuracy, although acceptable, is generally inferior to that produced by ML and weighted MP[40]. This difference in performance, as well as the vastly shorter running times of NJ, derives from the absence of branch-swapping in NJ to correct for suboptimal topologies created during the tree construction process. As such, NJ by itself seems ideally suited as a method to very quickly generate a relatively accurate starting tree for subsequent and more computationally intensive branch-swapping. Exactly such an approach is implemented in PHYML[41] and could equally well be applied to MP or within a distance framework (minimum evolution, ME).

Our results also attest to the recent advances in heuristic search strategies, particularly in a ML framework. Despite the increased complexity of ML as compared to MP (which has long been viewed as an obstacle to ML analyses), both accuracy and running times were comparable between the two optimisation criteria. Moreover, the more complex nature of the likelihood surface means that, at most, only a few equally optimal solutions are usually found (and typically only a single solution)[42], thus providing a more resolved solution than is usually the case for the analogous MP analyses (Figure 6.1A). Many would see this as being a desirable feature in that the ML estimate of the phylogeny of a large, species-rich group could be argued to be more decisive and definitive than the MP estimate.

In addition, whereas MP running times were kept in check by applying increasingly faster heuristic search algorithms, all ML searches used the same standard hill climbing searches in RAxML. A less thorough, but faster hill climbing strategy also exists, which was used to optimise the global tree in the ML-DCM3 analyses based on previous empirical work showing it to work the best of all methods in this context. As revealed by the single analysis of the full data set, the use of this option makes the ML analyses faster (by a factor of 7.8) with virtually no loss in accuracy (see Table 6.1). In fact, the 'fast' ML analyses were both faster and more accurate than the MP analyses performed. However, it should also be realised that faster implementations of MP searches than those used here, such as those implemented in TNT (available from http://www.zmuc.dk/public/phylogeny/TNT)[43] also exist. Rec-I-DCM3 has also been used in conjunction with TNT[9], boosting performance even further. A faster implementation of NJ than that used here was also recently published[6]. At the same time, it should be pointed out that the computational 'arms race' is still ongoing, with the latest version of RAxML, RAxML-VI-HPC (v2.1), showing significant speed improvements over the version used here, particularly for very large data sets.

Altogether, these findings bode well, not only for reconstructing very large phylogenies, but also for estimating support for the groups present in those phylogenies[44]. Analogous to our finding that accuracy is relatively flat with respect to the size of the problem and, in the case of MP, to the use of greedier heuristic search strategies, Salamin et al.[44] found that estimated bootstrap frequencies are apparently robust to the use of less effective branch-swapping methods during the tree searching operations (for example, in decreasing order of searching thoroughness, TBR versus SPR versus NNI; see Swofford et al.[27] for descriptions). Again, the use of NJ offers a means to quickly generate a reasonably accurate starting tree for further branch-swapping operations. Finally, an additional option for quickly determining support values in a ML framework is the use of the resampling of estimated log-likelihood (RELL) approximation[45,46], which apparently can estimate bootstrap proportions for a given tree more accurately than a true bootstrap analysis that uses fast heuristics to search through tree space[47].

## 6.4.2 The Importance of Sampling Strategy

Instead of sample size, it was the form of the sampling strategy that had the greatest impact on both phylogenetic accuracy (Figure 6.2A and Figure 6.2B) and, perhaps more unexpectedly, the running time of the analyses (Figure 6.4A and Figure 6.4B). Given that we will always be working with samples from the entire tree of life, the sampling strategy used, therefore, becomes a crucial consideration in phylogenetic systematics.

The influence of sampling strategy on accuracy has long been discussed, but more in the context of sampling density and the diameter of the problem. The former variable relates to how many taxa for a given group are included in the analysis, whereas the latter roughly corresponds to how much evolutionary history the sampled group contains. An especially stimulating paper was that of Kim[48], which showed that adding taxa to an analysis usually decreased phylogenetic accuracy. However, the protocol used by Kim added taxa outside the reference group, thereby expanding the phylogenetic diameter of the problem and decreasing the overall sampling density. Subsequent studies designed to address Kim's findings instead added the taxa within the reference group. As such, the phylogenetic diameter of the problem was unchanged, and the sampling density was increased. These studies instead demonstrated the general benefit of adding taxa to the analysis (see [18] and references therein). Moreover, taxa that were added specifically to break up any long branches (and therefore making the greatest increase in local density) were shown to improve accuracy to the greatest extent. Thus, a general, long-standing recommendation for phylogenetic analyses is to add taxa in such a way as to best represent the overall diversity of the group and/or to potentially break up any long branches[18].

The two sampling strategies used herein likewise can be viewed with respect to sampling density and evolutionary diameter. For a given problem size, clade sampling always yields the maximum density (there are no unsampled taxa for that group) and minimises the diameter. By comparison, random sampling will usually yield samples of greater diameter and less density (and therefore contain more long branches) for the same problem size, especially when the sample size is only a small percentage of the overall problem size. As the sampling size increases, the difference between clade and random sampling decreases, with the two strategies obviously converging when the sampling size equals the overall problem size. This explains why the performance differences between the strategies decreased as the sampling size increased.

More important, however, was the generally beneficial effect of clade sampling on running time (Figure 6.4A and Figure 6.4B), an effect that was only absent at the largest subproblem sizes and for ratchet-based MP searches (which will have a more rigid running time due to their more deterministic nature). To our knowledge, although this general finding might have been suggested informally, our results are the first to document it. The speed advantages to performing an analysis in a divide-and-conquer framework have usually been ascribed to the smaller problem sizes, with clade sampling being held to improve accuracy[28].

As such, our findings in this regard provide another important reason (in addition to increased accuracy) to ensure that the sampling is as complete as possible for a phylogenetic problem of a given size. Moreover, this recommendation applies equally to conventional phylogenetic analyses and to those performed in a divide-and-conquer framework. With respect to the latter, methods such as the disk-covering family[9,23,24] or IQPNNI[49] that intelligently subsample the data matrix are therefore to be preferred for reasons of both speed and accuracy over those such as classic quartet puzzling[22] that employ random sampling.

That said, technical considerations can occasionally override this general recommendation. A primary example here includes the parallel implementation of a divide-and-conquer algorithm for large scale phylogeny reconstruction, where the selected strategy has important practical implications. Recent work on a parallel version of Rec-I-DCM3(RAxML) revealed significant problems of processor load imbalance due to the great variation in subproblem sizes yielded by the intelligent decomposition method of Rec-I-DCM3[50]. However, such load imbalance problems could be resolved based on our findings that the sampling strategy has a decreasing effect on accuracy and inference times for proportionately larger sampling sizes, and that the choice of sampling strategy does not significantly affect the accuracy of the ML analyses, which are known to be more immune to the adverse effects of taxon sampling and long-branch attraction. As such, it should be possible in a parallel implementation of Rec-I-DCM3(RAxML) to initially split the alignment naïvely into relatively large subsets of approximately equal size (comprising approximately 12.5–50% of the original dataset) based on the guide tree. This strategy should improve load balance without any undue loss of performance. In turn, these large initial subproblems would then be optimised using the more intelligent subdivision method employed by Rec-I-DCM3, where the benefits of clade sampling take on greater importance. The potential utility of a similar naïve division method has been observed with a proprietary divide-and-conquer algorithm implemented in RAxML (Stamatakis unpublished data).

### 6.4.3 Implications for the Divide-and-Conquer Framework

Although the ML-DCM3 strategy employed here showed slightly less accuracy (~1–2% less) than a ML search using RAxML alone, it showed tremendous savings in terms of running time, running faster by a factor of about 1.5 or greater on average. Admittedly, the fast ML heuristic was even faster than the ML-DCM3 strategy for the time-unlimited analyses of 4,096 taxa; however, the latter could be easily adapted to use the fast heuristic throughout and so also benefit from the speed improvement. Similar, if not even greater, performance gains with respect to both running time and especially accuracy for a MP analysis performed within a divide-and-conquer framework have also been reported[9,28].

Thus, it seems clear that a divide-and-conquer based strategy will form a key component for studying very large phylogenetic problems. In this sense, it is instructive to compare the cumulative running times for multiple analyses of a given subproblem size such that the total number of taxa examined equals the global problem size of 4,096 taxa (Figure 6.5). For instance, for a subproblem

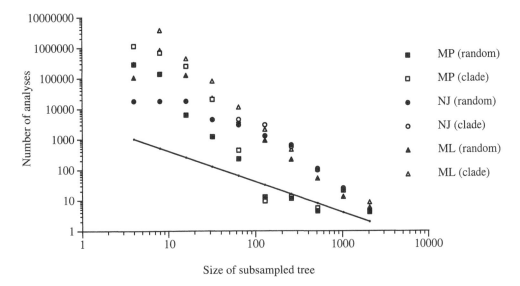

**FIGURE 6.5** Number of analyses for a given tree size that could be completed in the same time needed for an analysis of 4,096 taxa. For each tree size, the average running time over the 50 runs (see Figure 6.4) was compared against the time required to analyse 4,096 taxa from the time-unlimited analyses (see Table 6.1). The black line represents the number of analyses required such that 4,096 taxa are analysed in total.

size of 32 taxa, this amounts to 128 individual analyses (although a global tree of all 4,096 taxa could not be derived from these analyses because the trees do not overlap). However, Figure 6.5 reveals that over 20,000 MP analyses of 32 taxa selected using clade sampling could be conducted in the time taken for a single MP analysis of 4,096 taxa, or 176 sets of 128 analyses. Note also that these particular numbers are underestimates, given that the MP search strategy used for 4,096 taxa was considerably less robust, and therefore comparatively faster, than that used for 32 taxa. For all optimisation criteria (including NJ) and at virtually all subproblem sizes, the time savings are similarly enormous; only MP analyses at tree sizes of 128 to 512 taxa show a decrease in time (Figure 6.5). Thus, there is tremendous scope in a divide-and-conquer framework for many individual analyses to ensure high overlap between the trees, a factor that has been shown to improve the accuracy of the merged supertree[51].

Although it is tempting to try and derive the optimal subproblem size from Figure 6.5 under the assumption that accuracy is approximately flat with respect to the size of the subproblem, it must be remembered that these numbers do not account for the initial accuracy of the merged solution and, therefore, the time required for any global optimisations of it. Because such optimisations are computationally expensive (which accounts for the proportionately longer analysis times of larger solutions), they represent an important performance bottleneck. For example, the global optimisation step, even with the use of the fast ML algorithm, consumed the most execution time for the ML-DCM3 analyses. Moreover, there is a general consensus among researchers involved in the development of divide-and-conquer algorithms that global optimisations must be applied at some point to obtain the most accurate trees possible[52]. As such, the role for divide-and-conquer strategies will be, as for NJ, to yield as good a starting tree in as little time as possible. Research should now focus, therefore, on determining the optimal subproblem size and merger method that maximise the accuracy of the merged tree (so as to minimise the global optimisation time) in as short a time as possible. To our knowledge, there has been little work in this area (although the Rec-I-DCM3 user guide suggests a maximum subproblem size of 25% of the global size), nor in examining the accuracy of the merged tree without any subsequent global optimisation. Additional benefits would derive from pursuing this course of action with an eye toward the development of efficient parallel optimisation methods, particularly for the computationally intensive global optimisation step.

## 6.5  CONCLUSIONS

Together with other similar findings[11–13], the results we present here are encouraging for the prospects of building ever larger phylogenetic trees in our efforts to reconstruct the tree of life. Continued developments in computer technology and algorithm development can only increase our feeling of optimism. Even so, it must be remembered that even 10,000 taxa, the approximate limit for all simulations performed to date, represent only a minute fraction of the entire tree of life. Larger problems have been analysed successfully, but without any real knowledge of how accurate the answers might be. We simply do not know at this point how far the scalability of acceptable accuracy extends. As such, it seems clear that a divide-and-conquer approach, whereby we can break the problem down into pieces where we are confident of achieving good accuracy (and in less time), must form a necessary part of our efforts to obtain the tree of life.

## ACKNOWLEDGEMENTS

We thank Trevor Hodkinson and John Parnell for the invitation to contribute to this volume. We are also grateful to the Department of Informatics at the Technical University of Munich for providing access to their Infiniband computer cluster and to Usman Roshan for allowing us to include the (at the time) unpublished Rec-I-DCM3(RAxML) procedure in our simulations. This work was funded as part of the NGFN-funded project Bioinformatics for the Functional Analysis of Mammalian Genomes (BFAM) (Olaf Bininda-Emonds).

## REFERENCES

1. Rokas, A. and Carroll, S.B., More genes or more taxa? The relative contribution of gene number and taxon number to phylogenetic accuracy, *Mol. Biol. Evol.,* 22, 1337, 2005.
2. Bininda-Emonds, O.R.P., Supertree construction in the genomic age, in *Molecular Evolution: Producing the Biochemical Data, part B,* Zimmer, E.A. and Roalson, E., Eds., *Methods in Enzymology,* Vol. 395, Elsevier, Amsterdam, 2005, 745.
3. Sanderson, M.J. and Driskell, A.C., The challenge of constructing large phylogenetic trees, *Trends Pl. Sci.,* 8, 374, 2003.
4. Wilson, E.O., Taxonomy as a fundamental discipline, *Philos. Trans. R. Soc. Lond. B,* 359, 739, 2004.
5. Felsenstein, J., The number of evolutionary trees, *Syst. Zool.,* 27, 27, 1978.
6. Mailund, T. and Pedersen, C.N., QuickJoin: fast neighbour-joining tree reconstruction, *Bioinformatics,* 20, 3261, 2004.
7. Stamatakis, A., Parallel inference of a 10,000-taxon phylogeny with maximum likelihood, in *Proceedings of 10th International Euro-Par Conference (Euro-Par 2004),* Springer Verlag, 2004, 997.
8. Coarfa, C. et al., PRec-I-DCM3: a parallel framework for fast and accurate large scale phylogeny reconstruction, in *The First IEEE Workshop on High Performance Computing in Medicine and Biology (HiPCoMP 2005),* Fukuoka, Japan, 2005.
9. Roshan, U. et al., Rec-I-DCM3: a fast algorithmic technique for reconstructing large phylogenetic trees, in *Proceedings of the IEEE Computational Systems Bioinformatics conference (CSB),* IEEE Computer Society Press, Stanford, California, 2004.
10. Garey, M.R. and Johnson, D.S., *Computers and Intractability: a Guide to the Theory of NP-Completeness,* W.H. Freeman, San Francisco, 1979.
11. Salamin, N., Hodkinson, T.R., and Savolainen, V., Towards building the tree of life: a simulation study for all angiosperm taxa, *Syst. Biol.,* 54, 183, 2005.
12. Hillis, D.M., Inferring complex phylogenies, *Nature,* 383, 130, 1996.
13. Bininda-Emonds, O.R.P. et al., Scaling of accuracy in extremely large phylogenetic trees, in *Pacific Symposium on Biocomputing 2001,* Altman, R.B. et al., Eds., World Scientific Publishing Company, River Edge, NJ, 2000, 547.
14. Erdös, P.L. et al., A few logs suffice to build (almost) all trees (I), *Random Struc. Alg.,* 14, 153, 1999.
15. Erdös, P.L. et al., A few logs suffice to build (almost) all trees: part II, *Theoret. Comput. Sci.,* 221, 77, 1999.
16. Huelsenbeck, J.P. and Hillis, D.M., Success of phylogenetic methods in the four-taxon case, *Syst. Biol.,* 42, 247, 1993.
17. Bergsten, J., A review of long-branch attraction, *Cladistics,* 21, 163, 2005.
18. Hillis, D.M., Taxonomic sampling, phylogenetic accuracy, and investigator bias, *Syst. Biol.,* 47, 3, 1998.
19. Pollock, D.D. et al., Increased taxon sampling is advantageous for phylogenetic inference, *Syst. Biol.,* 51, 664, 2002.
20. Graybeal, A., Is it better to add taxa or characters to a difficult phylogenetic problem? *Syst. Biol.,* 47, 9, 1998.
21. Rosenberg, M.S. and Kumar, S., Incomplete taxon sampling is not a problem for phylogenetic inference, *Proc. Natl. Acad. Sci. USA,* 98, 10751, 2001.
22. Strimmer, K. and von Haeseler, A., Quartet puzzling: a quartet maximum-likelihood method for reconstructing tree topologies, *Mol. Biol. Evol.,* 13, 964, 1996.
23. Huson, D.H., Nettles, S.M., and Warnow, T.J., Disk-covering, a fast-converging method for phylogenetic tree reconstruction, *J. Comput. Biol.,* 6, 369, 1999.
24. Huson, D.H., Vawter, L., and Warnow, T.J., Solving large scale phylogenetic problems using DCM2, in *Proceedings of the Seventh International Conference on Intelligent Systems for Molecular Biology,* Lengauer, T. et al., Eds., AAAI Press, Menlo Park, California, 1999, 118.

25. Sanderson, M.J., r8s: inferring absolute rates of molecular evolution and divergence times in the absence of a molecular clock, *Bioinformatics,* 19, 301, 2003.

26. Rambaut, A. and Grassly, N.C., Seq-Gen: an application for the Monte Carlo simulation of DNA sequence evolution along phylogenetic trees, *Comput. Appl. Biosci.,* 13, 235, 1997.

27. Swofford, D.L. et al., Phylogenetic inference, in *Molecular Systematics,* Hillis, D.M., Moritz, C., and Mable, B.K., Eds., Sinauer Associates, Sunderland, Massachusetts, 1996, 407.

28. Roshan, U. et al., Performance of supertree methods on various data set decompositions, in *Phylogenetic Supertrees: Combining Information to Reveal the Tree of Life,* Bininda-Emonds, O.R.P., Ed., Kluwer Academic, Dordrecht, Netherlands, 2004, 301.

29. Stamatakis, A., Ludwig, T., and Meier, H., New, fast and accurate heuristics for inference of large phylogenetic trees, in *Proceedings of 18th IEEE/ACM International Parallel and Distributed Processing Symposium (IPDPS2004), High Performance Computational Biology Workshop,* IEEE Computer Society, Santa Fe, New Mexico, 2004.

30. Swofford, D.L., *PAUP*. Phylogenetic Analysis Using Parsimony (*and Other Methods). Version 4,* Sinauer Associates, Sunderland, MA, 2002.

31. Nixon, K.C., The parsimony ratchet, a new method for rapid parsimony analysis, *Cladistics,* 15, 407, 1999.

32. Howe, K., Bateman, A., and Durbin, R., QuickTree: building huge neighbour-joining trees of protein sequences, *Bioinformatics,* 18, 1546, 2002.

33. Stamatakis, A., Ludwig, T., and Meier, H., RAxML-III: a fast program for maximum likelihood-based inference of large phylogenetic trees, *Bioinformatics,* 21, 456, 2005.

34. Stamatakis, A.P. et al., AxML: a fast program for sequential and parallel phylogenetic tree calculations based on the maximum likelihood method, *Proc. IEEE Comput. Soc. Bioinform. Conf.,* 1, 21, 2002.

35. Colless, D.H., Congruence between morphometric and allozyme data for *Menidia* species: a reappraisal, *Syst. Zool.,* 29, 288, 1980.

36. Colless, D.H., Predictivity and stability in classifications: some comments on recent studies, *Syst. Zool.,* 30, 325, 1981.

37. Robinson, D.F. and Foulds, L.R., Comparison of phylogenetic trees, *Math. Biosci.,* 53, 131, 1981.

38. Steel, M., The complexity of reconstructing trees from qualitative characters and subtrees, *J. Classif.,* 9, 91, 1992.

39. Rokas, A. et al., Genome-scale approaches to resolving incongruence in molecular phylogenies, *Nature,* 425, 798, 2003.

40. Hillis, D.M., Huelsenbeck, J.P., and Cunningham, C.W., Application and accuracy of molecular phylogenies, *Science,* 264, 671, 1994.

41. Guindon, S. and Gascuel, O., A simple, fast, and accurate algorithm to estimate large phylogenies by maximum likelihood, *Syst. Biol.,* 52, 696, 2003.

42. Rogers, J.S. and Swofford, D.L., Multiple local maxima for likelihoods of phylogenetic trees: a simulation study, *Mol. Biol. Evol.,* 16, 1079, 1999.

43. Goloboff, P.A., Analyzing large data sets in reasonable times: solutions for composite optima, *Cladistics,* 15, 415, 1999.

44. Salamin, N. et al., Assessing internal support with large phylogenetic DNA matrices, *Mol. Phylogenet. Evol.,* 27, 528, 2003.

45. Hasegawa, M. and Kishino, H., Accuracies of the simple methods for estimating the bootstrap probability of a maximum-likelihood tree, *Mol. Biol. Evol.,* 11, 142, 1994.

46. Kishino, H., Miyata, T., and Hasegawa, M., Maximum likelihood inference of protein phylogeny and the origin of chloroplasts, *J. Mol. Evol.,* 31, 151, 1990.

47. Waddell, P.J., Kishino, H., and Ota, R., Very fast algorithms for evaluating the stability of ML and Bayesian phylogenetic trees from sequence data, in *Genome Informatics 2002,* Lathrop, R. et al., Eds., Universal Academy Press, Tokyo, 2002, 82.

48. Kim, J., General inconsistency conditions for maximum parsimony: effects of branch lengths and increasing numbers of taxa, *Syst. Biol.,* 45, 363, 1996.

49. le Vinh, S. and Von Haeseler, A., IQPNNI: moving fast through tree space and stopping in time, *Mol. Biol. Evol.,* 21, 1565, 2004.

50. Du, Z. et al., Parallel divide-and-conquer phylogeny reconstruction by maximum likelihood, in *High Performance Computing and Communications: First International Conference, HPCC 2005, Sorrento, Italy, September, 21–23, 2005, Proceedings,* Dongarra, J. et al., Eds., Springer Verlag, Berlin, 2005, 776.

51. Bininda-Emonds, O.R.P. and Sanderson, M.J., Assessment of the accuracy of matrix representation with parsimony supertree construction, *Syst. Biol.,* 50, 565, 2001.

52. Roshan, U., *Fast Algorithmic Techniques for Large Scale Phylogenetic Reconstruction,* Ph.D. thesis, University of Texas at Austin, 2004.

# 7 Tools to Construct and Study Big Trees: A Mathematical Perspective

*M. Steel*
Biomathematics Research Centre, University of Canterbury,
Christchurch, New Zealand

## CONTENTS

## ABSTRACT

This chapter describes some of the ways in which mathematical techniques provide insights and useful tools for reconstructing and analysing large trees and networks that will be required for species rich groups. Classically, 'mathematical biology' conjures up images of complex systems of differential equations; however, in phylogenetics quite different approaches are appropriate. Discrete mathematics, particularly algorithmic methods, graph theory and combinatorics, along with probability theory and statistics are the tools of choice. We describe how they can be used to address a range of topical issues relevant to constructing a tree of life. Why might large trees be useful? Does it even make sense to talk about a tree in the presence of reticulate evolution (and how does tree incongruence allow one to quantify the extent of reticulation?). How can one best combine trees on overlapping sets of taxa into supertrees or supernetworks? And how is biodiversity lost as species go extinct? At present most of these questions have only partial answers that are undergoing constant revision. Their full solution is a challenge for the future that will involve a close interplay between (at least) four disciplines: biology, mathematics, statistics and computer science.

## 7.1 TREES (AND NETWORKS) OF LIFE

### 7.1.1 Introduction and Terminology

It is forty years since the first phylogenetic trees of vertebrates were constructed from amino acid sequence data for cytochrome c[1]. Since then our understanding of life has undergone continual revision thanks to a stream of technical advances, the discovery of new data types (DNA sequences, SINEs, gene order, AFLPs, etc.), better computing resources, new methodology and the enthusiasm of many researchers. Today with whole genome sequences, and gene databases with many billions of base pairs, the pace set by the early pioneers seems likely to continue. In short, we are living in a golden age for molecular phylogenetics.

So what role can mathematics and its associated fields, statistics and computer science, play? First it can help design faster and more accurate algorithms for reconstructing, analysing and comparing phylogenetic trees and networks. Many of the problems biologists would like solved are computationally intractable on large data sets ('NP-hard' in computer science jargon). However this often suggests variations on the problem that can be solved exactly and quickly. Mathematical techniques can also help explain why existing methods can be misleading for data that evolves under certain processes, and provide techniques to correct for this[2-4]; or help answer more basic questions, such as how much sequence data is needed to accurately reconstruct a large tree, or some deep divergence within it[5]? Perhaps the most tantalising questions are those that involve determining whether different processes necessarily lead to different signals in the data (and thereby allowing the data to test between these models), or whether some processes are effectively indistinguishable from each other. A further use of mathematics is in formalising ideas so that the assumptions required are explicit and unambiguous; often this leads to insights into the limits of what is possible with any method of tree building (as in Steel et al.[6]).

In this chapter we first outline why large trees might (or might not) be needed. In particular we give some statistical arguments in support of large phylogenies as a tool to better understand evolutionary processes and detail. We then consider the more basic question of whether the notion of a 'tree of life' can be rigorously defended, particularly in the face of reticulate evolution. We present the first of two new results in this chapter, namely a simple argument showing there is a well defined notion of a tree of life, even though reticulation and other processes may also play an important role. We then review some of the recent work that has helped set out a mathematical foundation for representing and studying reticulate evolution, and we describe some of the approaches that have been developed for constructing both supertrees and supernetworks. In Section 7.3 we consider one of the uses of large trees, namely the quantification of phylogenetic diversity, and its loss as species go extinct. Here we describe our second new result, namely we derive equations that show that the concave relationship between phylogenetic diversity and taxon sampling that have been reported in the literature is generic, and not particular to certain trees or branch lengths. The chapter ends with some brief concluding comments regarding progress and challenges that lie ahead in mathematical phylogenetics. Before going further, we introduce some terminology that will be used throughout the chapter.

**Definitions.** Mostly we follow the notation used in Semple and Steel[7]. We let $\mathcal{T}$ denote a *phylogenetic X–tree*, that is, a tree whose leaves comprise the set $X$ of taxa (generally species or populations) under study, and whose remaining vertices (nodes) are of degree at least 3 (the degree of a vertex is the number of edges (branches) that are incident with it). The vertices at the tips are called *leaves*. If all the non-leaf vertices in a tree have three incident edges the tree is said to be *fully resolved* (sometimes called 'binary', these are the trees without polytomies, and so are maximally informative). We also deal with *rooted trees* which have some vertex (often the mid-point of an edge) distinguished as a root vertex; if we direct all the edges of the tree away from the root (so they are consistent with a time direction if the root is the ancestral taxon) then we can talk about the *clusters* of the tree, which are the subsets of $X$ that lie below the different vertices of the tree (it is a classical result that any rooted phylogenetic tree can be uniquely reconstructed from its set of clusters). Often the edges of the trees (rooted or unrooted) will have a *(branch) length*, corresponding perhaps to the expected amount of evolutionary change on that edge. For a rooted tree, if the sum of these branch lengths from the root to any leaf is the same (for each leaf) we say the branch lengths satisfy a *molecular clock*. Of course, for any tree (including ones constructed from non-molecular data), if the vertices all have (temporal) dates and the leaves represent contemporaneous taxa, then assigning each branch the difference between the dates of its two vertices also gives branch lengths that satisfy a molecular clock.

### 7.1.2 WHY BIG TREES?

A variety of impressive tree of life projects such as the National Science Foundation funded CIPRES initiative[8] have set forth the challenge of reconstructing phylogenies of unprecedented size and scope. To build and analyse trees on thousands of taxa demands clever computational tools, including new supertree methods (these combine existing phylogenies into larger parent phylogenies) and more slick techniques for quickly and accurately reconstructing trees from primary genetic data. A further issue is how to cope with processes that can 'mess up' the tree such as lineage sorting, horizontal gene transfer, recombination and the formation of hybrid taxa (Rønsted et al., *Chapter 9*).

Given the substantial effort (and funding) being expended in this task, it is timely to discuss why such large trees are needed in the first place, beyond the obvious challenge ('because it's there').

After all, many of the central evolutionary questions in systematic evolutionary biology have come down to the relationship between three or four taxa or groups; the human-chimp-gorilla debate of the 1990s is one of many such examples. Four other arguments against the wholesale reconstruction of large trees can be summarised by the following viewpoints:

- **The traditionalist.** If one regards life as organised according to a Linnean hierarchy (species, genus, family, order etc.) then one need only build (smallish) trees *within* the taxonomic level of interest.
- **The logician.** If one knows the (rooted) tree for every set of three taxa, then this uniquely specifies the entire tree, so we need only ever build trees on three taxa.
- **The pessimist.** Large trees are only useful if they are accurate, yet to build very large accurate trees surely needs huge data, since the number of possible trees grows so quickly (superexponentially with the number of taxa). Moreover, a large tree would be so complex and difficult to visualise that it would obscure rather than illuminate interesting biological relationships and processes.
- **Life is a mess.** Although locally much of life may be described by a tree, a global tree of life does not exist; it is rather a tangled network comprising both vertical (tree-like) and horizontal (reticulation, horizontal gene transfer) evolution. Attempts to reconstruct a large tree of life are therefore misplaced.

Each viewpoint conveys some truth, but none is compelling in itself. For example, the traditionalist imposes a hierarchy a priori, whilst a more objective approach would allow the data to reveal how well it fits this scheme; experience with genetic data has shown that numerous relationships have turned out to violate a traditional classification (for example, see Maley and Marshall[9] and the references therein). The logician is technically correct (it is a mathematical theorem that the collection of rooted trees on triples of species determines uniquely the global tree), but it ignores a statistical reality: namely, one simply cannot infer all 3-taxon trees accurately due to sampling effects, model violation and site saturation on long branches (more about this below). Similarly, the view concerning life as a tangled network championed by Ford Doolittle and colleagues[10,11] is no doubt correct up to a point; however, it may still make sense to talk about a tree of life, and we describe below two ways this can be done.

In answer to the pessimist, simulations[12,13] (see also Bininda-Emonds and Stamatakis, *Chapter 6*) and analytical results[14] have suggested that trees for large numbers of taxa can be reconstructed with reasonable accuracy from moderate data. Furthermore, the accuracy of inferring historical relationships for a given set of taxa is often improved by sampling more taxa[15]. Additional taxa can break up long edges that give rise to misleading signals due to 'long branch attraction', and addition of taxa can also help address another problem in molecular phylogenetics, namely site saturation. This latter property of sequences results from a combination of high rates of nucleotide substitution and long time scales. It leads to the character state of a taxon at a leaf of the tree being essentially independent of the states in the rest of the tree (and therefore inferring the placement of this leaf in the tree is unreliable). Site saturation is a problem for any tree reconstruction method, it is not particular to methods such as maximum parsimony. Mathematical formulae based on information theory can provide a useful way to quantify the effect of site saturation, and the extent to which it can be ameliorated by including sequences for additional taxa[15,16]. These mathematical bounds typically require $n$ (the number of taxa) to grow exponentially with the amount of evolutionary change, suggesting that the resolution of some deep divergences in a tree may require looking at very many taxa.

Regarding the pessimist's second point, rather than obscuring biological processes, large trees may instead be essential for testing them for two reasons. First, insufficient or biased sampling can mislead inferences; second, large trees may be required to perform meaningful statistical tests. One field of study where both these factors are important involves the analysis of tree shape (see Davies and Barraclough, *Chapter 10*; Hodkinson et al., *Chapter 17*). A number of studies[17–19] have considered various measures of 'tree balance' as a way of testing between different models of speciation (such as the Yule-Harding process, or the uniform (PDA) model). However if taxon sampling is incomplete and highly skewed by the availability of sequences or the interests of the particular investigator, then this may be the main influence on tree shape. Even if these problems can be overcome, it is difficult to reject one model in favour of another with small numbers of taxa unless the trees are extremely unbalanced; one frequently needs 50 or 100 taxa to decide for a given tree which model generated the data[20]. As models become more refined and the questions more delicate, it is clear that much larger trees, involving many hundreds or perhaps thousands of taxa, will be needed to tease these processes apart.

Another area where large trees are invaluable is in the study of models of DNA site substitution. Whilst some small trees (with just a few leaves) can reject certain models (for example, if some taxa are highly GC rich and others highly AT rich, then one can reject a stationary reversible process); for more delicate studies larger trees are needed. This is apparent, for example, in attempting to distinguish between covarion drift models of sequence evolution and rates-across-sites models. The two models 'look' very similar, in the sense that they produce similar site patterns, but trees involving many taxa are needed to tell them apart[21]. As models become more refined, it will require even more taxa to test between them; large trees may also provide a way to estimate underlying parameters in more standard models of sequence evolution (Elchanan Mossel, personal communication).

Finally, large trees are convenient from a statistical perspective, since various limit theorems exist for certain probability distributions. For example, the parsimony score of a random character on any fixed large fully resolved tree is normally distributed[22], and the parsimony score of a character evolved under a standard Markov process at low rates on a large tree is Poisson distributed[23]. For small trees, these convenient, familiar distributions must be replaced by a tedious ad hoc analysis or by simulations.

### 7.1.3 Is There a Tree of Life?

The notion of a 'tree of life' is central to evolutionary biology, yet problems arise when one tries to precisely formalise the notion. One issue is the thorny and long-standing question of what constitutes a 'species' (there are myriad definitions; see for example Wheeler and Meier[24]); another is that evolution has involved reticulate processes such as horizontal gene transfer and the formation of hybrid species (for example in certain plant, insect and animal species[25]), gene fusion and endosymbiosis. Furthermore, a species is not a single entity, but rather a population of individuals, and under sexual reproduction recombination can further complicate a tree-like description of ancestry. These and other details throw into doubt the plausibility of constructing any well defined notion of a tree of life, as noted by several authors (for example, Bapteste[10]). Wayne Maddison[26] has also explored the related question of what really constitutes a 'phylogeny'.

In this section we describe a simple mathematical result that shows how an underlying tree of life always can be defined (and exists) even in the presence of these various complications. To explain this result, first recall that a *hierarchy C* on *X* is a collection of subsets of *X*, containing *X*, and satisfying the property

$$A, B \in C \Rightarrow A \cap B \in \{\phi, A, B\},$$

that is, the sets in *C* are *nested*; if they have one or more species in common then one set is a subset of the other. It is a classical result that a hierarchy on *X* forms a tree whose leaves are labelled with subsets of *X* that partition X.

One can define a tree of life and avoid problematic notions concerning the definition of species by working at the level of individual organisms. Furthermore, all one needs to assume about evolution for this definition is that each organism on earth (now or in the past, excluding the first organisms) had at least one parent who originated before that organism. We show now how this assumption alone allows one to define an underlying tree. This tree does not represent the detailed history of ancestry of individual organisms (after all, sexual reproduction is inherently reticulate and so is represented by a pedigree graph rather than a tree). Rather, we describe a coarser structure, based on subsets of extant organisms that form nested clusters (and hence a tree) according to a property of their ancestry.

Of course this definition of a tree of life should not be taken too literally; the purpose here is much more modest, namely to show that one can define such a tree even in the face of the many complications of evolutionary biology mentioned above. Also, although the tree we discuss can in principle be computed (and in polynomial time), it requires knowing some detailed information about ancestry, and is unlikely to be feasible, at least at present. Nevertheless it is interesting to speculate what the tree we describe here looks like, and the approach may provide some enticing questions and fruitful approaches for future work.

Let *X* be the set of *all* extant living taxa, that is, all living organisms currently on Earth. Note that we are not regarding here *X* as a set of species or populations, but of individuals. Let $\Omega$ denote the (large but finite) collection of all living organisms throughout the history of life on Earth, and for any real number $t > 0$, let $\Omega_t$ denote all organisms that were alive anytime up to *t* years ago. Thus $X = \Omega_0$. For $x \in \Omega$, let $t(x)$ denote the time when organism *x* first arose (i.e., was born), measured, say, in years. Thus:

$$t(x) = \max\{t \geq 0 : x \in \Omega_t\}.$$

We suppose that each organism that has ever existed arose from one or more parent organism(s) either:

- By haploid reproduction, or
- By diploid (sexual) reproduction, or
- By some higher-level process involving two or more parent organisms (for example a complex endosymbiosis event), or
- By being part of the initial population $P_0$ that constituted an origin of life.

Stated formally, for each organism $x \in \Omega - P_0$ there is a subset $p(x)$ of $\Omega$ (of size 1, 2 or possibly higher) and with the following property (for some fixed $\varepsilon > 0$):

(P1)    For $x, y \in \Omega$,    if $y \in p(x)$    then    $t(y) > t(x) + \varepsilon$,

which merely formalises a familiar fact: *parents originate before their offspring*.

We refer to the triple $\mathcal{L} = ((\Omega_t, t \geq 0), P_0, p)$ as a *history of life*; it is essentially a pedigree on a grand scale showing all organisms and their parents back through time to the origin of life. There are two ways to define a natural system of clusters from $\mathcal{L}$, and we will see below (Proposition 7.1.1) that they are actually equivalent and form a tree.

For $a \in X$, let $P_t(a)$ denote the set of organisms that lived up to $t$ years ago and which have $a$ as an descendant. Formally, $P_t(a)$ is the subset of $\Omega_t$ consisting of those $x$ in $\Omega_t$ for which there is a sequence of organisms, $x = x_0, x_1, x_2, \ldots, x_k = a$ (for some $k$) and with $x_i \in p(x_{i+1})$ for each $i$. We say that a set of extant organisms $A \subseteq X$ is an $\mathcal{L}$-cluster if it satisfies the following property: there is some time $t$ for which the following holds for all $a, a' \in A$ and all $x \in X - A$:

$$P_t(a) \cap P_t(a') \neq \phi \quad \text{and} \quad P_t(a) \cap P_t(x) = \phi. \tag{7.1}$$

In words, this property states there was some time $t$ (measured, say, in years) for which any two organisms in $A$ shared an ancestor that lived at most $t$ years ago, and such that any organism that was an ancestor of both an organism in $A$ and an organism not in $A$ lived more than $t$ years ago. Let $\mathcal{C}_{\mathcal{L}}$ denote the set of $\mathcal{L}$-clusters of $X$.

The second way to define a collection of subsets of $X$ uses distances. Define a 'distance' $d_{\mathcal{L}}$ on $X$ as follows: let $d_{\mathcal{L}}(x, y)$ be the first time before the present when $x$ and $y$ shared an ancestral organism. Formally:

$$d_{\mathcal{L}}(x, y) := \min\{t \geq 0 : P_t(x) \cap P_t(y) \neq \phi\}.$$

Note that $d$ is symmetric $(d_{\mathcal{L}}(x, y) = d_{\mathcal{L}}(y, x))$ and finite $(d_{\mathcal{L}}(x, y) < \infty)$, though $d_{\mathcal{L}}$ may fail to satisfy the triangle inequality (that is, $d_{\mathcal{L}}(x, y)$ may be larger than $d_{\mathcal{L}}(x, z) + d_{\mathcal{L}}(z, y)$). Given any distance function $d$ on $X$, the *Apresjan clusters* of $d$ are those subsets $A$ of $X$ for which

$$\max\{d(a, a') : a, a' \in A\} < \min\{d(a, x) : a \in A, x \in X - A\}.$$

It is well known, and easily shown, that for any distance function $d$ (even if it fails the triangle inequality) the set of associated Apresjan clusters forms a hierarchy (see for example Bryant and Berry[27]; Devauchelle et al.[28]).

**Proposition 7.1.1** *For any life history, $\mathcal{L}$ satisfying (P1), the set $\mathcal{C}_{\mathcal{L}}$ is precisely the set of Apresjan clusters of $d_{\mathcal{L}}$ and so forms a hierarchy (or equivalently a rooted tree). Furthermore, $\mathcal{C}_{\mathcal{L}}$ can be reconstructed from $d_{\mathcal{L}}$ in $O(|X|^2)$ time.*

**Proof.** Suppose that $A$ is a $\mathcal{L}$-cluster of $X$. Let $t_A$ denote a value of $t$ for which (7.1) holds for all $a, a' \in A$ and $x \in X - A$. Then for all $a, a' \in A, x \in X - A$, we have $\max\{d_{\mathcal{L}}(a, a') : a, a' \in A\} \leq t_A$ and, by (P1), $\min\{d_{\mathcal{L}}(a, x) : a \in A, x \in X - A\} > t_A$ so that $A$ is an Apresjan cluster of $d_{\mathcal{L}}$. Conversely, suppose that $A$ is an Apresjan cluster of $d_{\mathcal{L}}$. Let $t = \max\{d_{\mathcal{L}}(a, a') : a, a' \in A\}$. Then $t$ satisfies (7.1) for all $a, a' \in A$ and $x \in X - A$, and so $A$ is an $\mathcal{L}$-cluster. The last part of Proposition 7.1.1, follows from Corollary 2.1 of Bryant and Berry[27] that provides an explicit algorithm to reconstruct the Apresjan clusters from any distance function.

Although it may be reassuring to know that a tree of life can still be defined in the presence of complications such as reticulation, such a tree will inevitably miss much of the detail and richness of evolutionary history, and may be largely unresolved in places. To describe a second type of tree that underlies evolution, though at a higher species level, we first need to talk about the sorts of networks that have been proposed to represent reticulate evolution.

### 7.1.4 MODELLING RETICULATE EVOLUTION

To represent evolutionary reticulation an appropriate tool is a directed acyclic graph or DAG. As with a rooted tree, this graph has an orientation (time direction); the 'acyclic' condition simply ensures that we cannot follow a path forward in time and arrive back at the same vertex (time travel).

Until recently approaches for representing reticulate evolution by DAGs were somewhat ad hoc. For example, a biologist might construct a tree and then insert some reticulate branches, perhaps guided by a minimization principle, as in Legendre and Makarenkov[29]. In the last few years intensive work by mathematicians and computer scientists has provided a more sound basis for representing and analysing reticulate evolution[30-35]. Formally, a hybrid phylogeny is a connected, directed graph $\mathcal{H} = (V, A)$ consisting of a set $V$ of vertices, and a set $A$ of arcs linking pairs of vertices, and with no directed cycles. There is generally a single 'root' vertex $\rho$ of in-degree 0, and the set of vertices of out-degree 0 is the set $X$ of extant taxa being classified (the *in-degree* of a vertex $v$ is the number of arcs that end at $v$; the *out-degree* of $v$ is the number of arcs that start at $v$). Usually some other minor conditions are imposed in the definition to avoid trivialities.

Let $\mathcal{H} = (V, E)$ be a hybrid phylogeny on $X$ with root vertex $\rho$ (Figure 7.1). Let $V_T$ be the set of vertices of $\mathcal{H}$ consisting of $\rho$ together with those vertices that do not lie on any undirected cycle. For a vertex $v$ of $V$, let $c(v)$ denote the set of species on $X$ for which there is a directed path from $v$ to $x$ (that is, the extant species for which $v$ is an ancestor).

The proof of the next result is given in Baroni et al.[36] (related results appear in Gusfield and Bansal[32] and Huson et al.[37]). It essentially (and informally) says that if we 'squash down' all the parts of a hybrid network that are involved in reticulation, we always end up with a tree. Or put another way, any hybrid phylogeny can be thought of as a tree, with certain vertices expanded out to reveal reticulation.

**Proposition 7.1.2**   *Let $\mathcal{H} = (V, E)$ be a hybrid phylogeny on X. Then the collection $\{c(v) : v \in V_T\}$ forms a hierarchy on X, and so forms a tree.*

An interesting optimization question is to quantify the extent of reticulate evolution. A natural measure of how much reticulation occurs in a hybrid phylogeny $\mathcal{H}$ is

$$h(\mathcal{H}) = \sum_{v \in V - \rho} (d^-(v) - 1)$$

where $d^-(v)$ is the in-degree of vertex $v$. It is easily seen that $h(\mathcal{H}) = 0$ precisely if $\mathcal{H}$ is a tree. For the hybrid phylogeny $\mathcal{H}$ in Figure 7.1, $h(\mathcal{H}) = 2$.

Now, suppose we have a collection of phylogenetic trees constructed from different genes. Assuming each tree correctly represents the history of the corresponding gene, then any incompatibility between the trees must be due to other processes such as reticulate evolution or lineage sorting. One can then ask for the fewest reticulate events required to explain the incompatibility. This question can be phrased more precisely as follows: Given rooted phylogenetic $X$–trees $T_1, T_2, \ldots, T_k$ find a hybrid phylogeny $\mathcal{H}$ that minimises $h(\mathcal{H})$ and displays $T_1, \ldots, T_k$. For example, consider Figure 7.1. This hybrid phylogeny displays both of the trees on the right, and, and this is the minimum value possible.

It was recently shown that, even for two rooted binary trees, this optimisation problem is computationally intractable[38]; nevertheless there are useful mathematical theorems that allow for lower bounds on $h(\mathcal{H})$ to be established, and these are often strong enough to pin down its exact

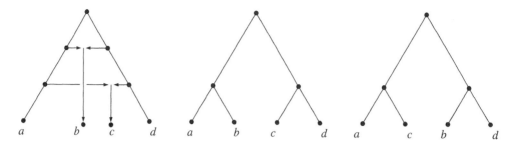

**FIGURE 7.1** A hybrid phylogeny (left) that displays the two trees on the right.

value if the degree of reticulation is not too extreme[36,39]. A different, information-based approach to quantify reticulation, based on a notion of phylogenetic 'compression', has also been described recently by Ané and Sanderson[30].

## 7.2 CONSTRUCTING SUPERTREES AND SUPERNETWORKS

Large-scale trees and networks can be built from two types of input, either from existing phylogenetic trees (or networks), or directly from primary data, possibly combined from different sources. These two strategies have been referred to as a supertree (and supernetwork) versus a total evidence or supermatrix approach[40]. The latter viewpoint attracts considerable support from those who widely advocate maximum parsimony, although supertree approaches enjoy certain advantages, in particular, the availability of large databases of trees (such as TreeBASE[41]) to combine. Supertree approaches can also provide a useful 'divide-and-conquer' approach to tree reconstruction, where more complex models of sequence and genome evolution may preclude an analysis directly on large trees. In that case it may be more feasible to build many small trees, assigning them confidence values, and perhaps branch lengths, and then to combine these trees into a supertree or supernetwork. A further feature of supertree methods is that they provide a way of testing (and measuring) the extent to which the input trees may be incorrect, since if the taxa do indeed have a tree-like history, then the input trees should be consistent with some global tree. In practice, the nature of this inconsistency can be informative; for example, it may be due to one or two 'rogue' taxa that appear in different places in several input trees, or it may be that one tree has been built from poorly aligned sequence data.

Many methods for constructing supertrees have been developed recently. However, by far the most widely used supertree method is a traditional approach called matrix recoding with parsimony (MRP). This method recodes each tree as a set of partial binary characters, then combines them and applies standard maximum parsimony tree reconstruction to the resulting set of characters. The method's popularity is due more to the historical precedent and the availability of existing software than any compelling mathematical justification. However, it appears to produce reasonable trees and some basic desirable properties have been mathematically established. For example, if the input trees have the same leaf set, then any MRP tree will refine the strict consensus tree of the input trees (Theorem 2.16 of Bryant[42]).

Despite the widespread use of MRP, there has been some intensive development of alternative methods by computer scientists and mathematicians over the last six years or so. One of the problems with MRP is that it is computationally intensive; the point at which a large parsimony search is concluded is arbitrary, and there will generally be many (thousands) of output trees, so one typically then applies some consensus method to those trees to produce a single output. More direct algorithmic approaches to supertree methods include variations on the original (1981) BUILD algorithm of Aho et al.[43] including MinCutSupertree and its variants (see for example Bininda-Emonds[44]); these methods are provably fast (polynomial-time) and may be useful for building very large trees in realistic time.

### 7.2.1 Methods for Constructing Consensus Networks and Supernetworks

Methods to build phylogenetic networks (rather than trees) serve two related but distinct purposes, although the distinction has often become blurred in applications. These are:

- To exhibit conflicting signals in data, and uncertainty as to an underlying tree by providing support for alternative resolutions
- To explicitly model reticulate evolution due to processes such as the formation of hybrid species, hybridisation, recombination and so forth

Methods of the first type include Splits Graphs[45], NeighborNet[46], Median Neworks[47], Consensus Networks[48] and Z-closure networks[49]; all of which were developed by mathematicians. They provide useful representations of data. Methods of the second type include the supernetwork approach of Huson et al.[37] based on modifying split decomposition, and several other approaches[34,50,51]. A particularly simple and general approach to network construction is to construct the 'cover digraph' of a set of clusters (subsets of $X$); in some cases this can be an effective strategy for reconstructing a reticulate network when sufficient phylogenetic signals 'accumulate' in an evolutionary process[50]. A mathematically elegant technique to construct a network that displays a tree with multiple labels (arising from polyploidy) has also recently been developed[52].

### 7.2.2 Direct Methods for Analysing Genomic Data

A number of promising new techniques have also recently been developed for using genomic data to directly infer phylogeny and to better understand evolutionary processes. This area, often called phylogenomics, has caught the attention of many computer scientists. Methods range from those that deal just with the gene content of the taxa to more elaborate techniques for analysing differences in gene order (for recent surveys, see Moret et al.[53] and the comprehensive overview by Delsuc et al.[54]). Rare insertion events (such as SINEs) have also provided useful phylogenetic markers, and phylogenetic methods are currently being developed to use raw (non-aligned) genomic sequence based on relative information and compression measures (see for example Burstein et al.[55] and Otu and Sayood[56] and the references therein). Because of the potentially large state space (depending on how characters are coded) some genomic data, such as SINEs and gene order, often exhibit little or no homoplasy (parallel or convergent evolution), and so multistate maximum parsimony and related character-based methods are often quite efficient. The information content of such data is also generally high in the sense that the number of characters required to reconstruct a tree accurately can be shown mathematically to be relatively modest[5].

## 7.3 AN APPLICATION FOR LARGE TREES: PHYLOGENETIC DIVERSITY

A large tree, with branch lengths, provides a way to measure how much of the total evolutionary history is spanned by various subsets of the taxa. This measure, called 'phylogenetic diversity' (PD) and defined more precisely below, has been used as a comparative measure in biodiversity conservation, following its introduction by Dan Faith in 1992[57]. Subsequent authors[58–61] (see also the references therein) have explored its application further. In this section we describe some mathematical properties of this measure that are useful when applying it to large trees. In particular we describe how certain optimisation problems can be solved quickly by a greedy approach, and we also study the statistical properties of the process whereby PD is lost due to random extinction of taxa.

To define PD precisely, consider first an unrooted phylogenetic tree (one can think of such a tree as that obtained from any rooted phylogenetic tree by ignoring the root vertex; for precise definitions see Semple and Steel[7]). Let $\lambda$ be an assignment of branch lengths to the edges of $\mathcal{T}$.

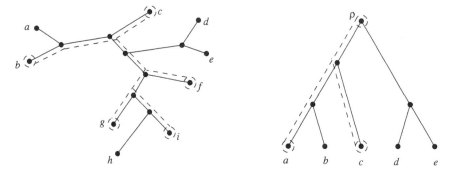

**FIGURE 7.2** Left: For an unrooted tree $PD(W)$ for $W = \{b, c, f, g, i\}$ is the sum of the lengths of the dashed edges. Right: For a rooted tree $PD(W)$ for $W = \{b, c\}$ is the sum of the lengths of the dashed edges.

Given a subset $W$ of $X$, consider the induced phylogenetic $W$-tree, denoted $\mathcal{T}/W$ that connects just those species in $W$ and its associated edge weighting $\lambda_W$ which assigns to each edge $e$ of $\mathcal{T}/W$ the sum of the $\lambda(e)$values over those edges of $\mathcal{T}$ in the path that corresponds to $e$. The PD value of $W$, denoted $PD(W)$, is defined as

$$PD(W) := \sum_e \lambda_W(e)$$

where the summation is over all edges $e$ in the tree $\mathcal{T}/W$. An example is illustrated in Figure 7.2 for $W = \{b, c, f, g, i\}$. Note that $PD(W)$ also depends on $(\mathcal{T}, \lambda)$, but we will think of these as fixed. Also, when $|W| = 1$ we set $PD(W) = 0$. In the case of a rooted phylogenetic tree, with root vertex $\rho$, we can regard the root as a leaf of an unrooted tree (with associated edge length 0), and then it is usual to define the phylogenetic diversity of a set $W$ as $PD(W \cup \{\rho\})$ as illustrated in Figure 7.2 (this quantity for rooted trees has also been referred to as 'evolutionary history'[62]).

The PD score provides some indication of how much genetic variation each possible subset $W$ contains in relation to the entire variation in the tree (by comparing $PD(W)$ to the total length of the tree $PD(X) = \sum_e \lambda(e)$). The PD score also turns out to have some interesting mathematical properties. In particular, it is possible to quickly find subsets of $X$ of a given size that maximise PD by using a simple greedy approach. This was established for trees whose branch lengths satisfy a molecular clock[62] and extended to arbitrary trees[63]. The latter extension also allows for a subset $Y$ of $X$ of given size to be found that maximises $PD(Y) + a \cdot \sum_{y \in Y} f(y)$, where $f(y)$ is some value (or cost) of species $y$, and $a$ is any (scale conversion) constant. In particular, one can ensure that certain species (including the root of a rooted tree) are always in the set $Y$ (by giving them a large enough $f$ value), and one can also find the taxa that are in all (or in none) of the maximal PD sets of given size. The fact that a fast (in this case greedy) approach works is vital for applications to large trees, since if one has a tree with (say) 1,000 taxa, and one wishes to find a subset of (say) 100 taxa that maximises the PD, then it is impossible for any computer to search all subsets of size 100 from the 1,000.

The combinatorial properties of PD have also been investigated[64], although for a different purpose, namely to show that the PD values of subsets of given size $m$ suffice to uniquely determine the underlying tree (provided $m$ is less than half the number of leaves of the tree). This approach has been developed further[65] to extend the popular neighbor joining tree reconstruction method so that it uses the PD values of taxa of given size (estimated, for example, by maximum likelihood) rather than just pairwise distance data.

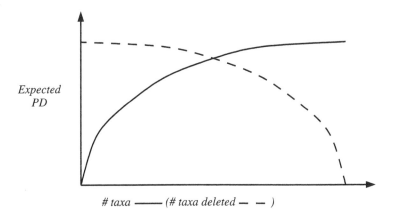

**Expected PD**

*# taxa ——— (# taxa deleted — —  )*

**FIGURE 7.3** Concave relationships between PD gain/loss in a tree with addition/deletion of taxa.

We turn now to the statistical properties of PD as species go extinct. Nee and May[62] investigated the loss of PD as taxa are randomly deleted from random trees under a simple model in which each taxon is equally likely to be the next to go extinct (the 'field of bullets' model). The trees were generated by a random birth model with branch lengths that satisfy a molecular clock. They found a characteristic concave shape in the relationship between expected PD and the proportion of taxa deleted. This relationship was further investigated recently[66] on random deletion of taxa from certain biological trees. Once again the relationship between taxa deleted and PD was concave, as illustrated schematically in Figure 7.3. Recall that a sequence $x = (x_1, x_2, \ldots, x_n)$ of real numbers is *concave* if, when we let $\Delta x_r = x_r - x_{r-1}$, the following inequality holds for all $r$:

$$\Delta x_r - \Delta x_{r+1} \geq 0$$

and the sequence is strictly concave if the inequality is strict for all $r$; geometrically this means that the slope of the line joining adjacent points in the graph of $x_r$ versus $r$ is decreasing. Note that $x_r$ is concave precisely if the complementary (reverse) sequence $y_r = x_{n-r}$ is. The significance of (strict) concavity for PD is that it says (informally) that most of the loss of PD comes near the end of an extinction process, as illustrated in Figure 7.3.

In this section we investigate the following question: is the concave relationship observed between the average PD and the number of taxa deleted particular to the trees (and the data or processes that generated them), or is it a generic property that applies to any tree with any set of branch lengths? We will see that the latter is true for any fully resolved tree with positive branch lengths. This makes intuitive sense because each interior branch survives until the point where there is no taxon that lies below that branch (which is likely to occur towards the end of a random extinction process). However, one could suspect that some trees with a certain assemblage of branch lengths might still lead to a violation of the concavity relationship, but the argument below rules this out. Perhaps the most satisfying aspect of the argument, however, is that we obtain exact expressions to describe the degree of concavity, in terms of the topology and branch lengths of the trees.

### 7.3.1 CONCAVITY OF EXPECTED PD IS GENERIC

Consider a rooted phylogenetic tree having a leaf set $X$ of size $n$. Let $E(\mathcal{T})$ denote the set of edges of $\mathcal{T}$, and let $W$ be a random subset of taxa of size $r$ sampled uniformly from $X$ (for example, by selecting uniformly at random a set $S$ of $n - r \geq 0$ elements of $X$ and deleting them, in which case

$W = X - S$). For $r \in \{1, \ldots, n\}$ let $\mu_r = \mathbb{E}[PD(W)]$, the expected value of $PD(W)$ over all such choices of $W$. Equivalently,

$$\mu_r = \binom{n}{r}^{-1} \sum_{W \subseteq X : |W| = r} PD(W).$$

where $\binom{n}{r}$ is the binomial coefficient ($= \frac{n!}{r!(n-r)!}$), the number of ways of selecting $r$ elements from a set of size $n$. Clearly $\mu_n = PD(X)$. For an edge $e$ of $\mathcal{T}$ and a positive integer $r$ let $\theta(e,r) = \frac{\binom{n-n_e}{r}}{\binom{n}{r}}$, where $n_e$ denotes the number of leaves of $\mathcal{T}$ that lie 'below' (i.e., separated from the root by) $e$.

**Proposition 7.3.1** *Consider a rooted phylogenetic tree $\mathcal{T}$ with an assignment $\lambda$ of positive branch lengths. Then, for all $r \in \{0, \ldots, n\}$,*

$$\mu_r = PD(X) - \sum_{e \in E(\mathcal{T})} \lambda(e)\theta(e,r).$$

---

**Proof.** For each $e \in E(\mathcal{T})$, and $W$ selected uniformly at random from all subsets of $X$ of size $r$, consider the random variable $X_W(e)$ defined by setting

$$X_W(e) = \begin{cases} 1, & \text{if } W \text{ contains an element that lies below } e; \\ 0, & \text{otherwise.} \end{cases}$$

Then $PD(W) = \sum_{e \in E(\mathcal{T})} \lambda(e) X_W(e)$, and so

$$\mu_r = \mathbb{E}[PD(W)] = \sum_{e \in E(\mathcal{T})} \lambda(e)\mathbb{E}[X_W(e)]. \tag{7.2}$$

Now, $\mathbb{E}[X_W(e)] = 1 - \mathbb{P}[X_W(e) = 0]$, and the event $X_W(e) = 0$ occurs precisely if all the $r$ elements of $W$ are selected from amongst the leaves that are not below $e$. The probability of this occurring, when these $r$ leaves are chosen randomly without replacement, is $\frac{\binom{n-n_e}{r}}{\binom{n}{r}}$, which is $\theta(e,r)$. Thus, $\mathbb{E}[X_W(e)] = 1 - \theta(e,r)$, which, combined with (7.2), establishes the Proposition.

---

To illustrate Proposition 7.3.1,

$$\mu_{n-1} = PD(X) - \frac{1}{n} \sum_{e \in E_{\text{ext}}(\mathcal{T})} \lambda(e),$$

where $E_{\text{ext}}(\mathcal{T})$ denotes the set of $n$ (exterior) edges of $\mathcal{T}$ (leaves incident with a leaf).

For $r \in \{1, \ldots, n\}$, let $\Delta\mu_r = \mu_r - \mu_{r-1}$. Note that, since $\mu_0 = 0$, we have $\Delta\mu_1 = \mu_1$. For an edge $e$ of $\mathcal{T}$, and $r \in \{1, \ldots, n-1\}$ let

$$\psi(e,r) := \frac{n_e(n_e - 1)}{r(r+1)} \cdot \frac{\binom{n-n_e}{r-1}}{\binom{n}{r+1}}.$$

We now describe the main consequence of Proposition 7.3.1. It shows that for any fully resolved tree PD decays in a strictly concave fashion as taxa are randomly deleted, and the only trees for which the decay of PD is linear are fully unresolved 'star' trees.

**Corollary 7.3.2** *Consider a rooted phylogenetic tree $\mathcal{T}$ with an assignment $\lambda$ of positive branch lengths. Then,*

1. For each $r \in \{1,...,n-1\}$,

$$\Delta\mu_r - \Delta\mu_{r+1} = \sum_{e \in E(\mathcal{T})} \lambda(e)\psi(e,r).$$

   In particular, $\mu$ is concave over this domain.
2. $\mu$ is strictly concave if and only if $\mathcal{T}$ has a cherry (i.e., there is an cluster of $\mathcal{T}$ that has precisely two leaves).
3. $\mu$ is linear if and only if $\mathcal{T}$ has no interior edges (i.e., is an unresolved 'star' tree).

---

**Proof.** From Proposition 7.3.1,

$$\Delta\mu_r - \Delta\mu_{r+1} = 2\mu_r - \mu_{r-1} - \mu_{r+1} = -\sum_{e \in E(\mathcal{T})} \lambda(e)[2\theta(e,r) - \theta(e,r-1) - \theta(e,r+1)]$$

and using a straightforward though tedious manipulation of (ratios of) binomial coefficients leads to the formula in the corollary.

For part (ii), if $\mathcal{T}$ has a cherry, let $e$ be an edge with two leaves below it. Then $\psi(e,r) > 0$ for all $r \in \{1,...,n-1\}$. Conversely, if $\Delta\mu_{n-1} - \Delta\mu_n > 0$, then there exists an edge $e$ for which $\psi(e,n-1) > 0$, in which case $n_e = 2$, and so $\mathcal{T}$ has a cherry.

For part (iii), note that $\mathcal{T}$ is a star tree, if and only if $(n_e - 1) = 0$ for all edges $e$ of $\mathcal{T}$, and this holds precisely if $\psi(e,r) = 0$ for all edges $e$ of $\mathcal{T}$ and all values of $r$.

---

## 7.4  CONCLUDING COMMENTS

Mathematics has a long and successful history of application in the 'hard sciences', and Eugine Wigner[67] once talked of the "unreasonable effectiveness of mathematics" in physics. By contrast, the mathematician Gian Carlo-Rota wrote in 1986, "the lack of real contact between mathematics and biology is either a tragedy, a scandal or a challenge, it is hard to decide which"[68]. However, much has changed over the last two decades, and it seems that, in evolutionary biology, mathematics and related fields are starting to play a central role, and many of the techniques (such as NeighborNet, Split Decomposition, Median Networks and Hadamard transformation) have sprung from some elegant mathematical theory. In this chapter we have listed only some of these, and there are many others.

Despite these successes, many challenges lie ahead. For example, although many supertree methods have been proposed, very few also incorporate branch length information in the input trees and provide corresponding branch length estimates in the output tree (or network). Recent work by Stephen Willson[69,70] has provided a useful lead as to how this can be done, but much more work is needed. Similarly, supertree and supernetwork methods that take account of confidence estimates (or Bayesian posterior probabilities) on branches of the tree would seem to be desirable, and the Bayesian techniques described in Rønquist et al.[71] may point the way forward. A further challenge for the future will be the analysis of more complex models in molecular systematics, particularly since even comparatively simple 'mixture' processes of DNA site evolution can be problematic for standard computational approaches in statistics such as MCMC[72].

## ACKNOWLEDGEMENTS

We thank the New Zealand Marsden Fund and the Allan Wilson Centre for Molecular Ecology and Evolution for supporting this research. I also thank Arne Mooers, Peter Lockhart, Klaas Hartmann, an anonymous referee and the editors for some helpful comments on an earlier version of this chapter.

## REFERENCES

1. Fitch, W.M. and Margoliash, E., Construction of phylogenetic trees, *Science*, 155, 279, 1967.
2. Hendy, M.D. and Penny, D., A framework for the quantitative study of evolutionary trees, *Syst. Zool.,* 38, 297, 1989.
3. Lockhart, P.J. et al., Recovering evolutionary trees under a more realistic model of sequence evolution, *Mol. Biol. Evol.,* 11, 605, 1994.
4. Susko, E., Inagaki, Y., and Rogers, A.J., On inconsistency of the neighbor-joining, least squares, and minimum evolution estimation when substitution processes are incorrectly modeled, *Mol. Biol. Evol.*, 21, 1629, 2004.
5. Mossel, E. and Steel, M., How much can evolved characters tell us about the tree that generated them? in *Mathematics of Evolution and Phylogeny*, Gascuel, O., Ed., Oxford University Press, 2005, chap. 14.
6. Steel, M., Böcker, S., and Dress, A.W.M., Simple but fundamental limits for supertree and consensus tree methods, *Syst. Biol.,* 49, 363, 2000.
7. Semple, C. and Steel, M., *Phylogenetics,* Oxford University Press, 2003.
8. CIPRES, Building the Tree of Life: A national resource for phyloinformatics and computational phylogenetics (http://www.phylo.org).
9. Maley, L.E. and Marshall, C.R., The coming of age of molecular systematics, *Science*, 279(5350), 505, 1998.
10. Bapteste, E. et al., Do orthologous gene phylogenies really support tree-thinking? *BMC Evol. Biol.,* 5, 33, 2005.
11. Doolittle, W.F., Phylogenetic classification and the universal tree, *Science*, 284, 2124, 1999.
12. Salamin, N., Hodkinson, T.R., and Savolainen, V., Towards building the tree of life: a simulation study for all angiosperm genera, *Syst. Biol.,* 54, 183, 2005.
13. Zwickl, D.J. and Hillis, D.M., Increased taxon sampling greatly reduces phylogenetic error, *Syst. Biol.,* 51, 588, 2002.
14. Erdös, P.L. et al., A few logs suffice to build (almost) all trees (Part 1), *Rand. Struct. Algor.,* 14(2), 153, 1999.
15. Pollock, D.D. et al., Increased taxon sampling is advantageous for phylogenetic inference, *Syst. Biol.*, 51, 664, 2002.
16. Sober, E. and Steel, M., Testing the hypothesis of common ancestry, *J. Theor. Biol.,* 218, 395, 2002.
17. Aldous, D., Stochastic models and descriptive statistics for phylogenetic trees, from Yule to today, *Stat. Sci.,* 16, 23, 2001.
18. Chan, K.M.A. and Moore, B.R., Whole-tree methods for detecting differential diversification rates, *Syst. Biol.,* 51, 855, 2002.
19. Heard, S.B. and Mooers. A.O., The signatures of random and selective mass extinctions in phylogenetic tree balance, *Syst. Biol.,* 51, 889, 2002.
20. McKenzie A. and Steel, M., Distributions of cherries for two models of trees, *Math. Biosci.* 164, 81, 2000.
21. Lockhart, P.J. et al., How molecules evolve in Eubacteria, *Mol. Biol. Evol.,* 17, 835, 2000.
22. Steel, M.A., Goldstein, L., and Waterman, M., A central limit theorem for parsimony length of trees, *Adv. Appl. Prob.,* 28, 1051, 1996.
23. Steel, M., and Penny, D., Maximum parsimony and the phylogenetic information in multi-state characters, in *Parsimony, Phylogeny and Genomics*, Albert, V., Ed., Oxford University Press, 2005, chap. 9.
24. Wheeler, Q. and Meier, R., *Species Concepts and Phylogenetic Theory*, Columbia University Press, New York, 2000.
25. Mallet, J., Hybridization as an invasion of the genome, *Trends. Ecol. Evol.,* 20, 229, 2005.
26. Maddison, W., Gene trees in species trees, *Syst. Biol.,* 46, 523, 1997.

27. Bryant, D. and Berry, V., A structured family of clustering and tree construction methods, *Adv. Appl. Math.*, 27, 705, 2001.

28. Devauchelle, C., et al., Constructing hierarchical set systems, *Ann. Combin.*, 8, 441, 2004.

29. Legendre, P. and Makarenkov, V., Reconstruction of biogeographic and evolutionary networks using reticulograms, *Syst. Biol.*, 51, 199, 2002.

30. Ané, C. and Sanderson, M.J., Missing the forest for the trees: phylogenetic compression and its implications for inferring complex evolutionary histories, *Syst. Biol.*, 54(1), 146, 2005.

31. Baroni, M., Semple, C., and Steel. M., A framework for representing reticulate evolution, *Ann. Combin.*, 8, 391, 2004.

32. Gusfield, D. and Bansal, V., A fundamental decomposition theory for phylogenetic networks and incompatible characters, in *Proc. RECOMB 2005*, Miyato, S. et al. Eds., LNBI 3500, Springer-Verlag, Berlin Heidelberg, 2005, 217.

33. Huynh, T.N.D., Jansson, J., Nguyen, N.B. and Sung, W.-K., Constructing a smallest refining galled phylogenetic network, in *Proc. RECOMB 2005*, Miyato, S. et al. Eds., LNBI 3500, Springer-Verlag, Berlin Heidelberg, 2005, 265.

34. Moret, B. M. E. et al., Phylogenetic networks: modeling, reconstructibility, and accuracy, *IEEE/ACM Trans. Comput. Biol. Bioinf.*, 1, 1, 2004.

35. Song, Y. and Hein, J., On the minimum number of recombination events in the evolutionary history of DNA sequences, *J. Math. Biol.*, 48, 160, 2003.

36. Baroni, M., Semple, C. and Steel, M., Hybrids in real time, *Syst. Biol.*, 55, 46, 2006.

37. Huson, D.H. et al., Reconstruction of reticulate networks from gene trees, in *Proc. RECOMB 2005*, LNBI 3500 Miyano S. et al. Eds., Springer-Verlag, Berlin Heidelberg, 2005, 233.

38. Bordewich, M. and Semple, C., Computing the minimum number of hybridisation events for a consistent evolutionary history, Research Report (UCDMS2004/21), Department of Mathematics and Statistics, University of Canterbury, Christchurch, New Zealand, 2005.

39. Baroni, M. et al., Bounding the number of hybridisation events for a consistent evolutionary history, *J. Math. Biol.*, 51, 171, 2005.

40. Faith, D. P., From species to supertrees: Popperian corroboration and some current controversies in systematics, *Austr. Syst. Bot.*, 17, 1, 2004.

41. Sanderson, M.J. et al., TreeBASE: A prototype database of phylogenetic analyses and an interactive tool for browsing the phylogeny of life, *Am. J. Bot.*, 81, 183, 1994, (http://www.treebase.org/treebase).

42. Bryant, D., A classification of consensus methods for phylogenies, in *BioConsensus*, Janowitz, M., Lapointe, F.-J., McMorris, F.R., Mirkin, B., and Roberts, F.S. Eds., American Mathematical Society, 2003, 163.

43. Aho, A. V., Sagiv, Y., Szymanski, T. G., and Ullman, J. D., Inferring a tree from lowest common ancestors with an application to the optimization of relational expressions. *SIAM Journal on Computing*, 10, 405, 1981.

44. Bininda-Emonds, O.R.P., *Phylogenetic Supertrees: Combining Information to Reveal the Tree of Life*, Kluwer Academic Publishers, Dordrecht, 2004.

45. Dress, A.W.M. and Huson, D.H., Constructing splits graphs, *IEEE/ACM Trans. Comput. Biol. and Bioinf.*, 1, 109, 2004.

46. Bryant, D. and Moulton, V., NeighborNet: an agglomerative algorithm for the construction of phylogenetic networks, *Mol. Biol. Evol.*, 21, 255, 2004.

47. Bandelt, H.-J., Forster, P., and Röhl, A., Median-joining networks for inferring intraspecific phylogenies, *Mol. Biol. Evol.*, 16, 37, 1999.

48. Holland, B. et al., Using consensus networks to visualize contradictory evidence for species phylogeny, *Mol. Biol. Evol.*, 21, 1459, 2004.

49. Huson, D. H. et al., Phylogenetic super-networks from partial trees, *IEEE/ACM Trans. Comput. Biol. Bioinf.*, 1, 151, 2004.

50. Baroni, M. and Steel, M., Accumulation phylogenies, *Ann. Combin.*, 10, 19, 2006.

51. Nakhleh, L., Warnow, T., and Linder, C.R., Reconstructing reticulate evolution in species—theory and practice. In *Proc. RECOMB 2004*, ACM, 2004, 337.

52. Huber, K.T. and Moulton, V., Phylogenetic networks from multi-labelled trees, *J. Math. Biol.* 52, 613, 2006.

53. Moret, B.M.E., Tang, J., and Warnow, T., Reconstructing phylogenies from gene-content and gene order data, in *Mathematics of Evolution and Phylogeny* Gascuel, O. Ed., Oxford University Press, chap. 12.

54. Delsuc, F., Brinkmann, H., and Philippe, H., Phylogenomics and the reconstruction of the tree of life, *Nature Rev. Genet.* , 6, 361, 2005.

55. Burstein, D. et al. Information theoretic approaches to whole genome phylogenies, in *Proc. RECOMB 2005*, Miyato, S. et al., Eds., LNBI 3500, Springer-Verlag, Berlin Heidelberg, 2005, 283.

56. Otu, H.H. and Sayood, K., A new sequence distance measure for phylogenetic tree construction, *Bioinf.*, 19, 2122, 2003.

57. Faith, D.P., Conservation evaluation and phylogenetic diversity, *Biol. Conserv.,* 61, 1, 1992.

58. Barker, G. M., Phylogenetic diversity: a quantitative framework for measurement of priority and achievement in biodiversity conservation, *Biol. J. Linn. Soc.,* 76, 165, 2002.

59. Crozier, R.H., Dunnet, L.J., and Agapow, P.-M., Phylogenetic biodiversity assessment based on systematic nomenclature, *Evol. Bioinf. Online* , 1, 11, 2005.

60. Mooers, A.O., Heard, S.B., and Chrostowski, E., Evolutionary heritage as a metric for conservation, in *Phylogeny and Conservation*, Purvis, A., Brooks, T.L. and Gittleman, J.L. Eds., Cambridge University Press, Cambridge, 2005, 120.

61. Pavoine, S., Ollier, S., and Dufour, A.-B., Is the originality of species measurable? *Ecol. Lett.*, 8, 579, 2005.

62. Nee, S. and May, R.M., Extinction and the loss of evolutionary history, *Science*, 278, 692, 1997.

63. Steel, M., Phylogenetic diversity and the greedy algorithm, *Syst. Biol.,* 54, 527, 2005.

64. Pachter, L. and Speyer, D., Reconstructing trees from subtree weights, *Appl. Math. Lett.,* 17, 615, 2004.

65. Levy, D., Yoshida, R., and Pachter, L., Beyond pairwise distances: neighbor joining with phylogenetic diversity estimates, *Mol. Biol. Evol.* 23, 491, 2006.

66. Soutullo, A. et al., Distribution and correlates of Carnivore phylogenetic diversity across the Americas, *Animal Conserv.* , 8, 249, 2005.

67. Wigner, E.P., The unreasonable effectiveness of mathematics in the natural sciences, *Comm. Pure Appl. Math.,* 13, 1, 1960.

68. Kac, M., Rota, G.C., and Schwartz, J., *Discrete Thoughts*, Birkhauser, 1993.

69. Willson, S.J., Constructing rooted supertrees using distances, *Bull. Math. Biol.,* 66, 1755, 2004.

70. Willson, S.J., Unique solvability of certain hybrid networks from their distances, *Ann. Combin.,* 10, 165, 2005.

71. Ronquist, F., Huelsenbeck, J.P., and Britton, T., Bayesian supertrees, in *Phylogenetic Supertrees: Combining Information to Reveal the Tree of Life* , Bininda-Emonds, O.R.P. Ed., Kluwer Academic Publishers, Dordrecht, 2004, 193.

72. Mossel, E. and Vigoda, E., Phylogenetic MCMC algorithms are misleading on mixtures of trees, *Science*, 309, 2207, 2005.

# 8 The Analysis of Molecular Sequences in Large Data Sets: Where Should We Put Our Effort?

*W. C. Wheeler*
Division of Invertebrate Zoology, American Museum
of Natural History, New York City, USA

## CONTENTS

## ABSTRACT

The problems of nucleotide homology determination and tree search are intertwined and complex issues for phylogenetic reconstruction. Both present NP-hard optimisations. One step and two step heuristic procedures are reviewed and compared through the analysis of example data sets using multiple sequence alignment plus tree search and direct optimisation techniques. The examples here show that extraordinary effort on the tree search side cannot overcome the shortcomings of poor sequence homology heuristics. Direct optimisation using the most simple heuristics can offer solutions with 30% better optimality scores in larger data sets.

## 8.1  THE PROBLEM PRESENTED BY UNALIGNED SEQUENCE DATA

DNA sequences are rich sources of data for systematic analysis. Three problems rise above all others today in the use of such data: cladogram search, homology determination and optimality choice. All three of these are particularly relevant to the analysis of large collections of sequences from species rich taxa as well as smaller data sets. Cladogram search, given a set of homologies and an optimality criterion, has long been understood to be a computational challenge; homology determination less so. These are the topics of this discussion. The notion of whether to employ parsimony or likelihood as a criterion to identify 'best' hypotheses is beyond the scope of this discussion. Systematists expend a great deal of computational effort in optimal topology search, that is searching tree space for the best solution, supported by algorithmic research, especially in large (currently defined as >500 taxa) data sets. The result of this is clear; we are able to generate better (that is, more optimal) solutions than ever before. What of homology determination of unaligned DNA sequences? Phylogenetic trees of DNA sequences are based on assumptions of character homology (usually alignment of nucleotides), and although this process of homology determination is critical to the analysis, less attention has been paid to this component of the problem than cladogram search or choice of optimality criterion. We would like to know how homology determination compares to cladogram search in influencing the final result.

### 8.1.1  TWO STEP VERSUS ONE STEP ANALYSIS

Phylogenetic analysis of a sequence data set often begins with a process of multiple alignment, where unaligned sequence data are arrayed in a series of columns via the insertion of gaps denoting insertion–deletion (indel) events. These aligned data now have an equal number of positions forming a putative or primary homology statement scheme sensu DePinna[1]. The aligned data are then subjected to a cladogram search (for example, using PAUP[2] or TNT[3]). This is referred to as 'two step' phylogenetics[4] in opposition to the 'one step' optimisation approach[5]. In this methodology, there is no separation of a homology determination phase from that of cladogram diagnosis and evaluation[6–8]. Such methods attempt to create an optimal set of homologies specific to each cladogram topology. As such, they do not create multiple alignments for processing through standard phylogeny reconstruction programs. The transformation series specific to each of these topologies can be arrayed and presented in the form of an implied alignment[9]. The use of an implied alignment from an optimisation analysis as a multiple alignment is problematic, since it was constructed with a particular phylogenetic hypothesis (cladogram) as a foundation; hence its use to evaluate competing topologies may seem 'unfair'. This does, however, provide a framework for comparison of the effectiveness of analysis, since optimality values can be directly compared.

### 8.1.2  NP-COMPLETENESS

The complexities of the problem of converting observations into cladograms come from the difficulty in optimising its two components. The joint problem is composed of two NP-complete (nondeterministic polynomial complete) problems. Both cladogram search and the assignment of optimal ancestral sequences, such that any overall tree is optimal, are NP-hard optimisations[10]. Hence, heuristic solutions are required to find usable solutions to both of these problems. Their joint nature makes the challenge even greater. Two step multiple alignment analysis seeks to simplify the problem by separating homology and cladogram search into separate, tractable operations. One step optimisation methods attack the issue as a nested problem, dealing directly with the complexity of both operations.

### 8.1.3 WHAT TO DO?

Given a world of finite computational resources, how do we apportion our effort? Much has been achieved in the realm of cladogram search heuristics, increasing the manageable size of data sets from tens to thousands in the past few years[3,11]. Are these advances paying off in the homology problem of unaligned sequences?

## 8.2 CLADOGRAM SEARCH HEURISTICS

Ever larger data sets have required increasingly powerful heuristic search procedures to search tree space (the set of possible trees given the number of taxa). Briefly reviewed here are several major approaches.

### 8.2.1 INITIAL BUILD

Often called the Wagner procedure or Wagner tree[12], initial cladogram construction is usually a straightforward procedure where taxa are added in turn to each possible location on a cladogram (Figure 8.1). The computational complexity of this operation is proportional to the square of the number of taxa [$O(n^2)$]. This operation is repeated until all taxa are added and an initial, complete cladogram is produced.

### 8.2.2 LOCAL SEARCH REFINEMENT

The initial Wagner tree may be complete, but it is quite unlikely to be very near a usefully optimal value. To improve on this solution, tree rearrangements can be performed where the basic Wagner tree is progressively rearranged and improved (Figure 8.2). These refinement operations yield a local search within the optimality neighbourhood defined by the initial tree (and include such methods as nearest neighbour interchange (NNI), subtree pruning and reconnection (SPR) and tree bisection and reconnection (TBR), depending on the level of rearrangement[13]).

### 8.2.3 RANDOM STARTING POINTS

Both the initial build and rearrangement are trajectory methods that follow a predefined and completely repeatable sequence to find a solution. An early method to attempt to break out of the local minima found in such searches was the randomisation of the taxon addition sequence used in the initial Wagner build. Such randomisations greatly improved search results and have been components in phylogenetic software for some time[14,15].

### 8.2.4 SIMULATED ANNEALING

As with many complex optimisation problems, simple heuristics can get stuck in local optima. Randomisation can help find more global solutions, but local search techniques such as NNI, SPR and TBR are prone to wallowing in local minima. This is due to the fact that the path to a more

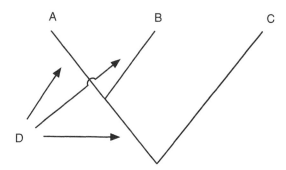

a.

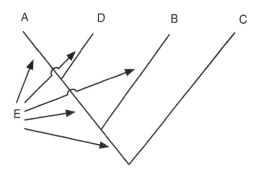

b.

**FIGURE 8.1** Basic Wagner build procedure of Farris[12] showing the addition of each taxon in turn to each possible edge (branch) on the tree. (a) The fourth taxon D is added to each of three places on the rooted tree. (b) The fifth taxon E is added at each of five positions.

globally satisfying solution may require traveling through suboptimal, intermediate states (Figure 8.3). This is the situation presented by the annealing of metals and applied to computational problems by Metropolis et al.[16].

**Ratcheting.** Nixon[17] brought simulated annealing to phylogenetic analysis. In Nixon's approach, called the 'ratchet', characters are randomly reweighted and searches performed on the newly weighted data. The weights are then set back to their initial values, and a search is performed with the reweighted tree as a starting point. The method has been extremely effective in finding lower-cost solutions in data sets thought to be refractory to further analysis, such as a large angiosperm matrix[18].

**Drifting.** A phylogenetic method much closer to the original description of simulated annealing was proposed by Goloboff[11] and termed 'drifting'. Unlike the ratchet, where suboptimal solutions were arrived at via weighting, drifting explicitly creates a probability of topology acceptance based on the extent of its suboptimality. This is implemented in TNT[3].

**Monte Carlo Markov Chain.** A probabilistic form of local search uses the relative probability of successive tree rearrangements as a criterion for the acceptance of a rearrangement. As with drifting and other simulated annealing techniques, suboptimal solutions can be accepted as intermediate solutions on the way to more globally optimal scenarios. As a search strategy, Monte Carlo Markov Chains have had their greatest impact on Bayesian estimates of clade probabilities[19].

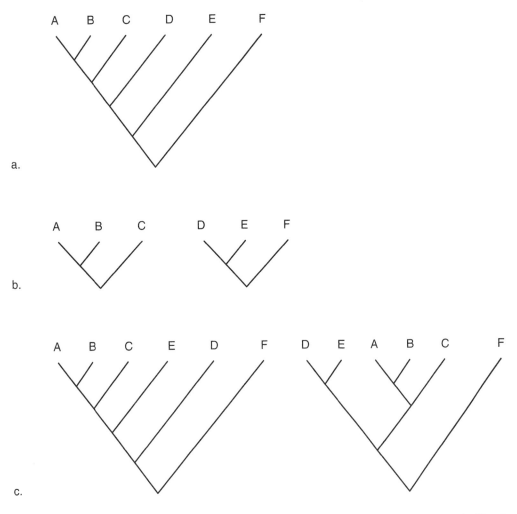

**FIGURE 8.2** Simple tree rearrangement showing SPR branch swapping. Clade of tree (a) is pruned off leaving two subtrees (b). The subtree is then added back to each possible place on the subtree (c), avoiding its original position and yielding new trees closely related topologically to the first.

### 8.2.5 GENETICAL ALGORITHM

An extremely productive class of algorithms mimic the process of natural selection and go collectively under the name 'genetical algorithms' (GA). Unlike trajectory searches and simulated annealing perturbations, GA methods operate on collections or populations of trees. This was first introduced to systematics by Moilanen[20]. There are usually three steps to such a procedure, namely: generation of variation, recombination and selection. Beginning with a collection of locally optimal or near optimal trees, variation can be generated variously, drifting and ratcheting being example strategies. After this, a recombination stage occurs where, in this case, trees exchange compatible branches (Figure 8.4; this has been called 'tree fusing' by Goloboff[11,13]). Selection then takes place based on the optimality values at hand, retaining optimal and perhaps suboptimal trees at an intermediate stage to preserve variation. Combined search strategies implemented in TNT[3] have reduced search times for some iconic data sets (for example, 'Zilla'[21]) from months to minutes on commodity hardware.

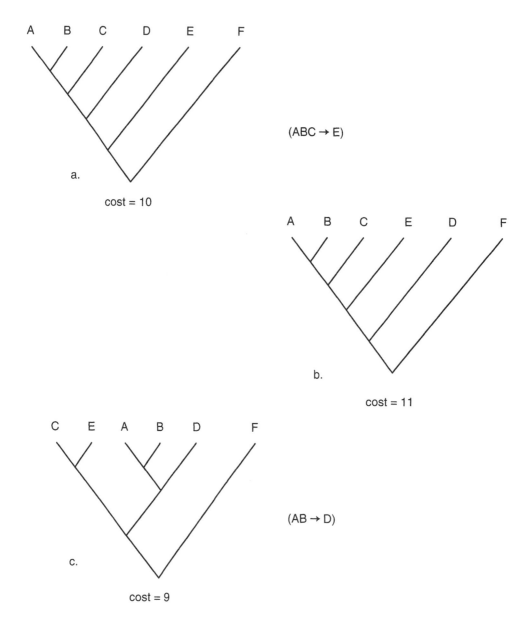

**FIGURE 8.3** A simulated annealing trajectory. Globally optimal topology (c) can only be reached from locally optimal (a) by passing through the suboptimal topology (b).

## 8.2.6 SECTORIAL SEARCHES

Goloboff[11] introduced the notion of effective taxon reduction as a means of focusing the search on the relationships among segments or sectors of a large tree. The central idea is that there may be subsets of taxa in optimal or near optimal arrangements that are in suboptimal arrangements with respect to each other. Goloboff[11] showed that random addition of taxa and simple TBR branch swapping can be very effective at creating the well ordered sectors of a cladogram, but when the number of taxa approaches hundreds or thousands, this strategy is much less effective at determining the overall structure of the tree. Basically, the topology of a candidate tree is divided into a series of

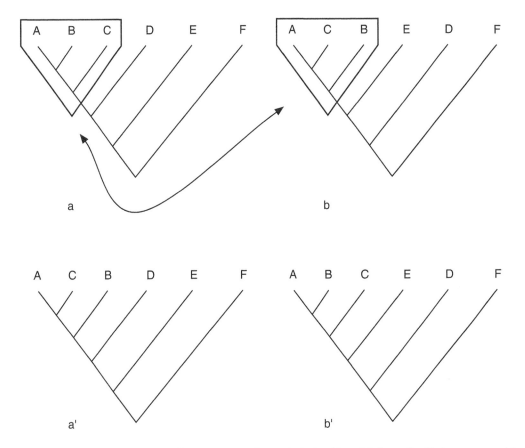

**FIGURE 8.4** Tree fusing (Goloboff[11]) component of genetical algorithm (Moilanen[20]). Cladograms a and b exchange the (ABC) groups yielding two new arrangements a′ and b′.

sectors of 35–50 taxa that are then treated as a single terminal (Figure 8.5). Branch swapping is then performed on this reduced data set. Sectors and searches are dynamically defined and alternated as the topology evolves until a stable solution is found. This approach has been further explored as 'disk covering methods'[22,23] (see also Wilkinson and Cotton, *Chapter 5;* Bininda-Emonds and Stamatakis, *Chapter 6*), yielding improvements in many areas of phylogenetic tree searching.

### 8.2.7 COMPLEX SEARCH STRATEGIES

The search methods described above are most profitably used in concert. This approach was advocated by Goloboff[11] when he first described tree fusing, drifting and sectorial searches. Goloboff described a variety of combinatorial strategies for different size data sets with different analytical properties. This sort of flexible approach is implemented in TNT[3], allowing the user to explore these procedures and their combinations to solve very large (thousands of taxa) data sets.

## 8.3 HOMOLOGY DETERMINATION HEURISTICS

As pointed out in Wheeler et al.[5], multiple sequence alignment and optimisation techniques can be viewed as alternate modes of homology heuristic. Both strive to create optimal (that is, lowest cost) cladograms (but see Simmons[24] for a different view), yet the approaches differ significantly.

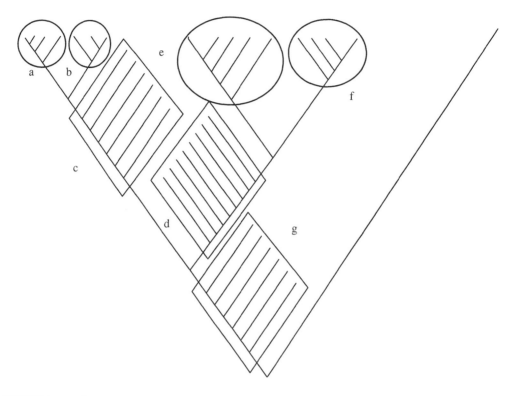

**FIGURE 8.5** Definition of cladogram sectors for use in a sectorial search (Goloboff[11]); (a–g) usually 35–50 taxa.

Multiple alignment methods seek, in general, a single set of homologies, upon which all cladograms are evaluated, whereas optimisation methods create potentially unique schemes for each cladogram. The fundamental methods are briefly reviewed below.

### 8.3.1 MULTIPLE ALIGNMENT

As described by Sankoff and Cedergren[25], an exact multiple alignment can be constructed via a recursive method, if the tree of relationships is known. This will have a time complexity of $O(n^k 2^k)$ for $k$ sequences of length $n$; an impossibly large number for real data sets. This represents the enumeration of all possible multiple alignments for a given tree, described by Wang and Jiang[10] as a 'tree alignment'. Two problems are presented by this approach: time complexity and the lack of a 'known' tree. Sankoff and Cedergren[25] suggest an approximation in $O(n^3)$ time, but even this can be extremely daunting. In general, multiple alignment methods reduce the problem to a series of $k–1$ pairwise [$O(n^2)$] alignments using a guide tree. Since the alignments are performed two at a time, no phylogenetic tree is required to determine alignment costs (Figure 8.6). This type of approach is commonly used, for example in CLUSTALW[26] reviewed by Phillips et al.[27].

One of the key issues is the generation of the guide tree. CLUSTALW generates a neighbour joining distance tree[28] based on pairwise alignment distances. Other methods used an interactive tree alignment building procedure (TREEALIGN[29,30]) or an explicit search on multiple guide trees attempting to find the alignment of lowest cost (MALIGN[31,32]). In general, pairwise, guide tree–based multiple alignment procedures are fast, but tend to become coarse as the number of sequences increases[27].

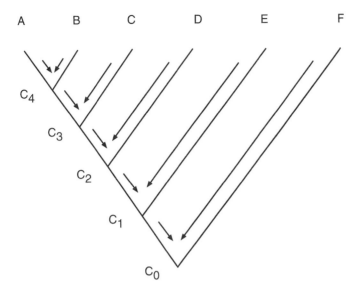

A    B    C    D    E    F

$C_4$

$C_3$

$C_2$

$C_1$

$C_0$

**FIGURE 8.6** Guide tree-based multiple alignment. Sequences are accreted in turn as the procedure moves from the tips of the tree (A–E) to the root. Intermediate vertices ($C_i$) may be consensus sequences as in CLUSTALW (Thompson et al.[26]) or partial alignments as in MALIGN[32].

### 8.3.2 OPTIMISATION APPROACHES

As opposed to multiple alignment methods, optimisation approaches seek to deal with the cladogram directly by determining the sequence states of internal vertices without the use of a pre-existing alignment. Each topology, in essence, would be granted its own set of ancestral sequences and transformations, hence homology relationships. Sankoff[33] pioneered this approach with an $O(n^k)$ for $k$ sequences of length $n$ (as above). As with multiple alignment, this time complexity placed the procedure well beyond the reach of real data sets. A series of heuristic solutions to this problem have been proposed for parsimony[6,34–36] and likelihood[7,8,37]. The simplifications can be divided into approaches where medians are calculated from two and three sequences to reduce the complexity to a manageable level, and those where no medians are calculated, but the set of possible vertex sequences specified a priori.

**Median approaches.** Median-based heuristics calculate vertex sequences from adjacent nodes (Figure 8.7). This is done with the same objective as the exact case[33], so that the total cladogram cost be minimised. An $O(n^2)$ method was described by Wheeler (1996) that used string matching to create a median sequence with the constraint that there are no sequence gaps in the median sequences. This was improved[35] in the sense of a better (that is, lower cost) median, but with additional complexity [$O(n^3)$] based on Sankoff and Blanchette[38] and Gladstein[39], again without sequence gaps in the medians. Likelihood medians have been derived for complex models[7,8] and simple ones[37]. Given the greater numerical complexity of likelihood calculations, these procedures are far more time consuming than parsimony based methods.

**Search-based approaches.** Unlike median based heuristics, search-based approaches[36,40] make no effort to calculate sequences de novo, but choose them from a prespecified set. This has the benefit of being relatively rapid when the set is not too large. After determining the edit cost between each pair of candidate sequences (a series of pairwise alignments), dynamic programming[41] is used to choose the optimal set of sequences for each vertex (Figure 8.8). For a single cladogram, with $n$ sequences of length $m$ and a set of $k$ additional candidate sequences, such an optimisation would require

$$O(n \cdot (n+k)^2)$$

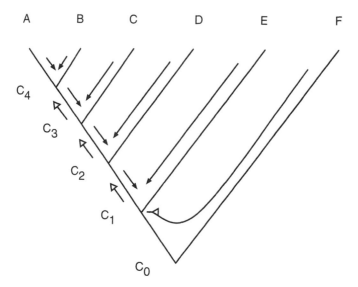

**FIGURE 8.7** $O(n^2)$ and $O(n^3)$ sequence optimisation medians. $O(n^2)$ (closed arrows) and $O(n^3)$ (closed and open arrows). The median sequences ($C_i$) are calculated based on either their two descendants, or their descendants and immediate ancestor. $C_5$ would not be calculated in the $O(n^3)$ case. Multiple passes may be performed on the cladograms to update the vertex sequences and improve median quality.

whereas a simple median approach would require $O(n \cdot m^2)$. As long as $n + k < m$, search-based methods should win out as far as time is concerned. The set size $k$, however, will determine the quality (cost) of the result, with $k$ varying from 0 in fixed state optimisation[40] to the set of all sequences resulting in an exact solution through explicit enumeration[36]. Little work has been done to examine how large $k$ should be, or how best to choose the sequences to be included in the heuristic set.

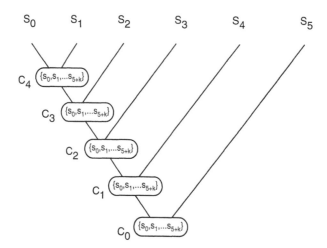

**FIGURE 8.8** Search-based optimisation of a set of observed sequences ($S_i$) to determine vertex sequences ($C_j$) using a set of candidate sequences ($S_0,...,S_{5+k}$).

## 8.4  EXAMPLE DATA

### 8.4.1  DATA SETS

As a demonstration of the effects of these procedures on real data, four collections of unaligned sequences were analysed. The data sets were the 62-taxon set of mantid (Mantodea) 18S rDNA from Svenson and Whiting[42], a 208-sequence collection (18S rRNA) of Metozoa from G. Giribet (personal communication), a 585-taxon archaeal small subunit sequence data set from the European Ribosomal RNA Database (http://www.psb.ugent.be), and a 1,040 mitochondrial small subunit data set from the same source.

### 8.4.2  CLUSTALW ALIGNMENT

CLUSTALW[26] was used to align the four data sets under two sets of conditions. The first was with the default pairwise and multiple alignment parameters (initial indel cost of 10 and extension indel cost of 0.2). A second run was performed where all events were set to 1. Alignments were generated and then analysed using TNT[3] to assay the optimality (tree length) of the homology statements in terms of equal weighted parsimony. This was done under basic simple (10 random addition sequences and TBR swapping) and more aggressive searches (options 'mxram 1,000 hold 10,000 xmult = replications 10 ratchet 50 drift 20 fuse 5') (Table 8.1).

### 8.4.3  POY OPTIMISATION

POY[43] was used to perform optimisation-based analyses. Searches were done under cost parameters matching CLUSTALW default and equal weighting scenarios. Searches were performed using direct optimization [ $O(n^2)$ medians] and simple single addition sequence Wagner build

---

**TABLE 8.1**
**Summary of CLUSTALW Multiple Alignment and POY Optimisation Analyses**

| Data Set | Parameters | Taxa | Aligned Length | Variable Positions | POY Cost | TNT Simple | TNT Aggressive |
|---|---|---|---|---|---|---|---|
| Mantodea | CLUS Default | 62 | 1,873 | 510 | — | 1,052 | 1,052 |
| | Equal | | 1,828 | 479 | — | 1,037 | 1,037 |
| Metazoa | CLUS Default | 208 | 2,868 | 2,521 | — | 30,980 | 30,980 |
| | Equal | | 4,263 | 3,888 | — | 29,089 | 29,089 |
| Archaea | CLUS Default | 585 | 2,722 | 2,516 | — | 39,084 | 39,062 |
| | Equal | | 1,768 | 1,679 | — | 39,533 | 39,499 |
| Mitochondria | CLUS Default | 1,040 | 3,054 | 3,002 | — | 90,640 | 90,619 |
| | Equal | | 7,662 | 7,162 | — | 115,686 | 115,649 |
| Mantodea | CLUS Default | 62 | 1,900 | 847 | 4,582 | 1,292 | 1,292 |
| | Equal | | 1,990 | 1,900 | 941 | 981 | 981 |
| Metazoa | CLUS Default | 208 | 21,430 | 21,238 | 171,949 | 48,675 | 48,675 |
| | Equal | | 6,190 | 5,838 | 27,146 | 26,783 | 26,783 |
| Archaea | CLUS Default | 585 | 20,831 | 20,765 | 208,060 | 54,733 | 54,717 |
| | Equal | | 6,725 | 6,682 | 37,931 | 37,751 | 37,750 |
| Mitochondria | CLUS Default | 1,040 | 30,648 | 30,638 | 486,950 | 132,586 | 132,529 |
| | Equal | | 13,978 | 13,978 | 79,769 | 79,594 | 79,593 |

*Note:* Results from CLUSTALW are shown above the line (upper part of table) and those of POY shown below the line (lower part of table). 'CLUS Default' denotes CLUSTALW default parameters, 'Equal' all events = 1. The POY costs for CLUS Default runs are high due to the parameter setting ('-gap 50 -extensiongap 1 -change 5) in those runs. 'TNT Simple' denotes the TNT command 'mult', whilst 'TNT Aggressive' signifies 'xmult=replications 10 ratchet 50 drift 20 fuse 5'. The rightmost three column values are cladogram costs.

*Source:* Data from CLUSTALW Thompson et al.[26] and POY Wheeler et al.[43]

---

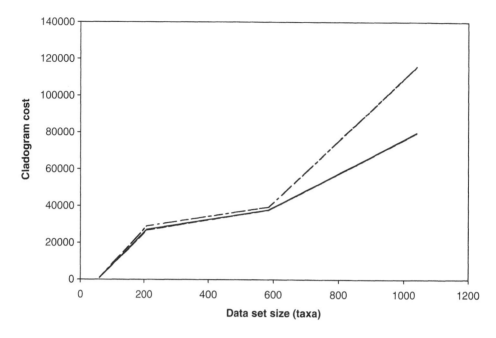

**FIGURE 8.9** Cladogram optimality as a function of data set size based on the 1:1:1 values of Table 8.1. Dashed line represents POY, POY-TNT 'Simple' and POY-TNT 'Aggressive'. Solid line represents CLUST-ALW-TNT 'Simple' and CLUSTALW-TNT 'Aggressive' (full values can be seen in Table 8.1).

without further refinement. Implied alignments[9] were created to allow for direct comparison with CLUSTALW results. These were subjected to the same TNT analysis conditions as the CLUSTALW alignments to find the tree lengths.

### 8.4.4 RESULTS

The results of the CLUSTALW multiple alignment and POY optimisation analyses are shown in Table 8.1 and Figure 8.9.

## 8.5 COMPARISONS

Several patterns are immediately apparent. The CLUSTALW alignments do not differ much in optimality value (tree length) from the default to equal weighting scenarios. Given that the relative indel costs differ by a factor of 50, this is striking. POY analyses contrast sharply with this. The Mantodea data show a difference in equally weighted TNT cost (tree length) of 1.4%, whereas the POY runs are 32% different; Metazoa data show 6.5% (CLUSTALW) and 81% (POY), and for Archaea data the difference was 1.1% (CLUSTALW) and 44% (POY). The mitochondrial data set has a 22% difference for CLUSTALW, the highest of the alignment-based runs, but still lower than all the optimisation comparisons. This difference is so great that the CLUSTALW alignments were superior to the POY optimisations in every case where the homology and cladogram cost parameters differed (CLUSTALW default settings). Whilst the POY optimisation analyses are very responsive to cost parameters, the CLUSTALW runs are not. Whilst each case where equal weighted optimi-sation was used yielded superior (that is, lower cost) optimality values for POY optimisation, only half of the alignment cases showed this pattern.

The two methods also differed in their response to increased severity of cladogram search heuristics. Neither CLUSTALW nor POY implied alignments displayed any better solutions under more exhaustive cladogram searching for the smaller Mantodea or Metazoa data sets. The 585-taxon archaeal and 1,040-taxon mitochondrial data sets did, with the CLUSTALW showing an average improvement factor of $4.93 \times 10^{-4}$ and the POY a factor of $1.09 \times 10^{-4}$. The results based on these simple POY homology searches were from 4.7% (Metazoa) to 45% (mitochondrial) less costly than those based on the CLUSTALW alignments. This is especially pointed in concert with the search severity improvement being 20% as great for the POY runs. Cladogram search effort is much more productive for the CLUSTALW alignments. Even with the very aggressive phylogenetic search options of TNT, in no case where the homology and search parameters were the same did the CLUSTALW alignment match the cost of the rudimentary Wagner build procedure used in the POY optimisation.

## 8.6 WHAT IS HAPPENING IN LARGE DATA SETS?

As data sets grow (more taxa are added), however, the relative importance of homology and cladogram search heuristics change. For small data sets, homology heuristics seem to be relatively capable, and cladogram searching is more important. As the data sets grow, however, the importance of homology determination increases, eventually dominating the result. Multiple sequence alignment, at least as implemented in CLUSTALW, is not effective enough. The comparisons here are based on only the most simple homology heuristics. Methods such as more aggressive median calculation[35] and search[36] coupled with better cladogram search, will likely yield results 10% lower in cost for optimisation techniques, making the comparison even more stark.

Whilst tree search space is relatively well studied and understood, we are only at the beginning stages of understanding the space of homology optimisation. Homology problems are at least as computationally complex and certainly less well known. Increased understanding in this area will no doubt yield much greater improvements in the optimality of our results in the future. Cladogram searching has undergone a huge change in the last ten years; homology assessment needs the same amount of attention. A good cladogram search can never make up for a poor homology search. Systematics requires that more attention be paid, in both methodological innovation and computer time, to homology determination.

## ACKNOWLEDGEMENTS

The US National Science Foundation and NASA for research support. Louise Crowley, Gonzalo Giribet, Megan Harrison, Camilo Mattoni, Kurt Pickett and Andrés Varón for data sets, discussion and commentary on this manuscript. The temporal forbearance of Trevor Hodkinson and John Parnell while reviewing this paper.

## REFERENCES

1. DePinna, M.C.C., Concepts and tests of homology in the cladistic paradigm, *Cladistics,* 7, 367, 1991.
2. Swofford, D.L., *PAUP*: Phylogenetic Analysis Using Parsimony (* and Other Methods),* version 4.0b 10, Sinauer Associates, Sunderland, MA, 2002.
3. Goloboff, P.A., Farris, J.S., and Nixon, K., TNT (Tree analysis using New Technology) version 1.0 ver. beta test v. 0.2., 2003, Tucumán, Argentina (http://www.zmuc.dk/public/phylogeny/tnt).
4. Giribet, G., Generating implied alignments under direct optimization using POY, *Cladistics,* 21, 396, 2005.
5. Wheeler, W.C. et al., *Dynamic Homology and Phylogenetic Systematics: A Unified Approach Using POY,* American Museum of Natural History, 2005.
6. Wheeler, W.C., Optimization alignment: the end of multiple sequence alignment in phylogenetics? *Cladistics,* 12, 1, 1996.
7. Hein, J. et al., Statistical alignment: computational properties, homology testing, and goodness-of-fit, *J. Mol. Biol.,* 302, 265, 2000.

8.  Hein, J., Jensen, C.J.L., and Pedersen C.N.S., Recursions for statistical multiple alignment, *Proc. Natl. Acad. Sci. USA,* 100, 14960, 2003.

9.  Wheeler, W.C., Implied alignment, *Cladistics,* 19, 261, 2003.

10. Wang, L. and Jiang, T., On the complexity of multiple sequence alignment, *J. Comput. Biol.,* 1, 337, 1994.

11. Goloboff, P.A., Analyzing large data sets in reasonable times: solutions for composite optima, *Cladistics,* 15, 415, 1999.

12. Farris, J.S., A method for computing Wagner trees. *Syst. Zool.,* 19, 83, 1970.

13. Goloboff, P.A., Techniques for analysing large data sets, in *Techniques in Molecular Systematics and Evolution,* DeSalle, R., Giribet, G. and Wheeler, W., Eds., Birkhäuser Verlag, Basel, 2002, 7.

14. Felsenstein, J., *PHYLIP,* 1980 (http://evolution.genetics.washington.edu/phylip.html)

15. Mickevich, M.F. and Farris, J.S., *PHYSYS: Phylogenetic Analysis System,* 1980.

16. Metropolis, N.A. et al., Equation of state calculations by fast computing machine, *J. Chem. Phys,* 21, 1087, 1953.

17. Nixon, K.C., The parsimony ratchet, a new method for rapid parsimony analysis, *Cladistics,* 15, 407, 1999.

18. Rice, K.A., Donoghue, M.J., and Olmstead. R.G., Analyzing large data sets: rbcl 500 revisited, *Syst. Biol.,* 46, 554, 1997.

19. Huelsenbeck, J.P. and Ronquist, F., *MrBayes: Bayesian inference of phylogeny,* 3.0 edition, 2003. (http://mrbayes.csit.fsu.edu).

20. Moilanen, A., Searching for most parsimonious trees with simulated evolutionary optimization, *Cladistics,* 15, 39, 1999.

21. Chase, M.W. et al., Phylogenetics of seed plants: an analysis of nucleotide sequences from the plastid gene rbcl, *Ann. Mol. Bot. Gard.,* 80, 528, 1993.

22. Huson, D., Nettles, S., and Warnow, T., Disk-covering, a fast converging method for phylogenetic tree reconstruction, *J. Comput. Biol.,* 6, 368, 1999.

23. Roshan, U., et al., Rec-i-dcm3: A fast algorithmic technique for reconstructing large phylogenetic tree, in *Proc. IEEE Computer Society Bioinformatics Conference CSB 2004,* Stanford University, 2004.

24. Simmons, M.P., Independence of alignment and tree search, *Mol. Phylogenet. Evol.,* 31, 874, 2004.

25. Sankoff, D.M. and Cedergren, R.J., Simultaneous comparison of three or more sequences related by a tree, in *Time Warps, String Edits, and Macromolecules: The Theory and Practice of Sequence Comparison,* Sankoff, D.M. and Kruskall, J.B., Eds., Addison Wesley, Reading, MA, 1983, chap. 9.

26. Thompson, J.D., Higgins, D.G., and Gibson, T.J., CLUSTAL W: improving the sensitivity of progressive multiple sequence alignment through sequence weighting, position specific gap penalties and weight matrix choice, *Nucleic Acids Res.,* 22, 4673, 1994.

27. Phillips, A., Janies, D., and Wheeler. W., Multiple sequence alignment in phylogenetic analysis, *Mol. Phylogenet. Evol.,* 16, 317, 2000.

28. Saitou, N. and Nei. M., The neighbor-joining method: a new method for reconstructing phylogenetic trees, *Mol. Biol. Evol.,* 4, 406, 1987.

29. Hein, J., A new method that simultaneously aligns and reconstruct ancestral sequences for any number of homologous sequences, when the phylogeny is given, *Mol. Biol. Evol.,* 6, 649, 1989.

30. Hein, J., A tree reconstruction method that is economical in the number of pairwise comparisons used, *Mol. Biol. Evol.,* 6, 669, 1989.

31. Wheeler, W.C., Sources of ambiguity in nucleic acid sequence alignment, in *Molecular Ecology and Evolution: Approaches and Applications,* Schierwater, G.W.B., Streit, B., and DeSalle, R., Eds., Birkhäuser Verlag, Basel Switzerland, 1994, 323.

32. Wheeler, W.C. and Gladstein, D.S., (documentation by Janies, D. and Wheeler, W.C.), *MALIGN,* New York, NY, 1991–1998 (http://research.amnh.org/scicomp/projects/malign.php).

33. Sankoff, D.M., Minimal mutation trees of sequences, *SIAM J. Appl. Math.,* 28, 35, 1975.

34. Wheeler, W.C., Fixed character states and the optimization of molecular sequence data, *Cladistics,* 15, 379, 1999.

35. Wheeler, W.C., Iterative pass optimization, *Cladistics,* 19, 254, 2003.

36. Wheeler, W.C., Search-based character optimization, *Cladistics,* 19, 348, 2003.

37. Wheeler, W.C., Dynamic homology and the likelihood criterion, *Cladistics,* 2005.

38. Sankoff, D.M. and Blanchette, M., The median problem for breakpoints in comparative genomics, *Computing and Combinatorics 3rd Annual Int. Conf. COCOON 97,* 1276, 251, 1997.
39. Gladstein, D.S., Efficient incremental character optimization, *Cladistics,* 13, 21, 1997.
40. Wheeler, W.C., Measuring topological congruence by extending character techniques, *Cladistics,* 15, 131, 1999.
41. Sankoff, D.M. and Rousseau, P., Locating the vertices of a Steiner tree in arbitrary space, *Math. Program.,* 9, 240, 1975.
42. Svenson, G.J., and Whiting, M.F., Phylogeny of Mantodea based on molecular data: evolution of a charismatic predator, *Syst. Ent.,* 29, 359, 2004.
43. Wheeler, W.C., Gladstein, D.S., and De Laet, J.D., (documentation by Janies, D. and Wheeler, W.C.—commandline documentation by De Laet, J.D. and Wheeler W.C.), *POY* version 3.0.11, American Museum of Natural History, New York, 1996–2005 (http://research.amnh.org/scicomp/projects/poy.php).

# 9 Species-Level Phylogenetics of Large Genera: Prospects of Studying Coevolution and Polyploidy

*N. Rønsted, E. Yektaei-Karin, K. Turk,*
*J. J. Clarkson and M. W. Chase*
Jodrell Laboratory, Royal Botanic Gardens, Kew,
Richmond, UK

## CONTENTS

## ABSTRACT

The problems facing workers trying to produce phylogenetic hypotheses of large genera are usually surmountable if the group in question is well studied previously. The major problems faced in plants are caused by hybridisation between species and low levels of variability in the standard phylogenetic markers. Many researchers use plastid DNA and the internal transcribed spacers of nuclear ribosomal DNA (nrDNA), neither of which alone is suitable for the detection of hybrids, the former because it is inherited through the maternal lineage and the latter because it is subject to concerted evolution via gene conversion. Sequencing low copy, protein coding, regions is a good alternative, but these are often neither easily amplified nor suitable for other reasons. Low levels of variability in the standard markers can alternatively be dealt with by using markers such as amplified fragment length polymorphisms (AFLPs). Some prospects, problems and solutions will be discussed and exemplified with work on figs (*Ficus*, Moraceae) and tobacco (*Nicotiana*, Solanaceae).

## 9.1  INTRODUCTION: PROSPECTS OF STUDYING LARGE GENERA

There are many reasons for basing a book around the theme of species rich taxa and more specifically focusing attention on large genera (Hodkinson and Parnell, *Chapter 1*). Due to their size alone, they constitute a large proportion of the biodiversity that we seek to describe and gain knowledge of. For example, more than 50 seed plant genera have over 500 species each, and roughly 20 of these have over 1,000 species each[1]. Large genera have often been neglected for taxonomic studies, and many suffer from superficial or outdated classifications as well as uncertain species numbers and delimitations. In addition to these obvious reasons for studying large genera, they also offer unique opportunities for investigations of comparative biology. Much work on biological patterns and processes focuses on a single case or a small set of related taxa, which can easily be studied in detail. Such case stories are exciting, but are not as potentially useful for generating more general hypotheses about ecological and evolutionary processes behind the observed cases. Large genera offer a much better chance for evaluating patterns and processes based on comparison between several taxa. Hypotheses of evolutionary processes posed in studies of one group can be evaluated by examining other lineages. The possibility of extensive replication is rare in comparative biology. For instance, the enormous diversity of large genera offers the possibility of separating features of a coevolved interaction that are highly constrained from those that are more plastic[2].

Large genera are ideal for studying broader biological questions. Large genera may be used to get a time scale for important evolutionary events; we can determine rates of speciation and evaluate which factors may have played an important role in speciation or the evolution of pollination syndromes[3]. We can also examine biogeographic patterns at various scales[4–6]. Finally, it is important to ask why large genera have become large. What explains explosive speciation, and how can diversification be linked to adaptive innovations, ecological specialisations or accelerated rates of morphological or molecular change? For instance, Davies and coworkers[7–9] used a phylogenetic tree of the angiosperms to study the correlation between environmental factors and shifts in diversification rates (see also Davies and Barraclough, *Chapter 10*).

The precondition for all these exciting prospects of studying large genera is a comprehensive and robust phylogenetic hypothesis, but in most cases phylogenetic work on large genera is still in its early days, with limited sampling and an array of potential problems. Sampling is important because incomplete sampling may affect several of the analyses that can be used to interpret broader biological questions from phylogenetic trees. Linder et al.[10] used a dataset including 95% of the nearly 300 South African *Restio* (Restionaceae) species to study the effect of incomplete taxon sampling on obtaining molecular age estimates by various methods. All methods were sensitive to incomplete taxon sampling to some degree, resulting in underestimation of ages. Likewise, ancestral area analysis[11] is affected by incomplete sampling at the basal nodes of a phylogenetic tree.

Phylogenetic hypotheses are currently typically based on molecular data, which surpasses morphological and other types of data in various ways, particularly in the ability to obtain sufficient amounts of information for species-level comparison and the need for phylogenetic hypotheses that are independent of the biological traits that one may wish to evaluate. However, large genera often display low levels of variation in the standard markers used and problems with interspecific hybridisation. In addition, the mere size of the genera poses problems with the time required for phylogenetic analyses. However, various strategies can be used to reduce analysis time. For instance, Chase et al.[12] and Rønsted et al.[13] employed maximum parsimony analyses in two steps. Trees from a limited search were used as starting trees for a second search with no limit or a preset limit, thereby avoiding swapping on large numbers of trees.

This chapter will discuss problems and potential prospects in species-level phylogenetics of large genera exemplified by ongoing work on figs (*Ficus*, Moraceae) and their coevolution with hymenopteran wasps. It also discusses issues relating to phylogenetic incongruence, polyploidy and hybridisation, using tobacco (*Nicotiana*, Solanaceae) as a case study.

## 9.2 COEVOLUTION OF FIGS AND THEIR POLLINATING WASPS

*Ficus*, the figs, is the most successful genus in the mulberry family (Moraceae) and one of the largest genera of angiosperms with over 750 species in tropical and subtropical regions worldwide. Frodin cited them as the thirty-first largest genus of seed plants[1]. The Asian-Australasian region has the richest and most diverse fig flora, with over 500 species. Figs play a big role in the maintenance of the local diversity in the rainforest by setting fruit and providing food resources throughout the year. The most important characteristic of *Ficus* to humans is the fig, as we consume *F. carica* and other edible figs. The fig is not a fruit, but a closed inflorescence called a syconium.

*Ficus* is one of the most diverse genera with regard to habit and life form, with both deciduous and evergreen free-standing trees, small shrubs, climbers and creepers. It is especially diverse in Asia and Australasia, where it includes hemi-epiphytes that establish themselves in the canopy and send aerial roots down to reach the soil (many of them are potential stranglers). *Ficus* also contains rheophytes adapted to life in running water and lithophytes growing on rocks or rocky soil. The diversity of *Ficus* in Africa is less, and much less in the Neotropics, than in Asia and Australasia.

One noteworthy feature of the figs is that they can only be pollinated by female hymenopteran wasps of the family Agonidae. The wasps can only lay their eggs within *Ficus* inflorescences, where their offspring feed on some of the developing seeds (Figure 9.1). Roughly half of the *Ficus* species are monoecious, and the other half are functionally dioecious. In the monoecious fig system, pollen-carrying female wasps are attracted by chemical signals. The wasps enter syconia via the ostiole, a bract-covered pore. Once inside, they deposit pollen on stigmas, then oviposit directly into some ovaries via their styles. However, because style length is highly variable, ovaries vary in accessibility to wasps, which guarantees that a mixture of seeds and seed-eating wasp larvae matures within each syconium. Weeks later, the new generation of wasps emerge. After mating, female wasps actively collect pollen from anthers, which they deposit in small thoracic pollen pockets. In other species pollination is passive. In the meantime, males make a tunnel through the wall of the syconium. Females then depart through the tunnel in search of other syconia in which to oviposit. Males are wingless and die in their native fig.

Pollination biology of the dioecious species is less well understood. Dioecious fig trees either bear syconia with both female and male flowers or with female flowers only. When wasps enter hermaphroditic syconia, they can oviposit in all flowers because the styles are uniformly short. These trees support wasp larvae and function only as pollen donors. Syconia on female trees offer only long-styled flowers. Wasps deposit their pollen load but cannot oviposit, and these trees produce only seeds[14-15].

This mutualism, which has persisted over 60 million years[16-19], is thought to be highly species specific and has become one of the best model systems for studying the comparative biology of mutualisms or coevolution[2]. Coevolution can be defined as reciprocal changes in traits of interacting

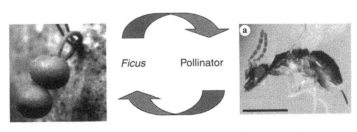

Ficus sect. *Sycomorus* fig                                   *Ceratosolen* fig wasp

**FIGURE 9.1** Coevolution in the fig-wasp system. Figs can only be pollinated by female agonid wasps. The wasps can only lay their eggs inside fig inflorescences, where their larvae feed on some of the developing seeds. (Reproduced with permission from Weiblen and Bush[81].)

species and is a widespread phenomenon[20]. In addition to the fig–wasp model system, a range of other examples of coevolution between plants and their pollinators have been reported, such as the well studied interaction between *Yucca* (Asparagaceae) and yucca moths (Lepidoptera)[21]. A diversity of insect pollination mutualisms have been described for palms[22], for instance an obligate weevil pollination mutualism of the dwarf palm, *Chamaerops humilis*[23]. Another recent case is provided by *Epicephala* moths (Gracillariidae) pollinating trees of the genus *Glochidion* (Phyllanthaceae)[24]. Coevolution of plants and their pollinators is widespread, and the need for improving our understanding of the principles of coevolution continues to grow. Persistence of mutualisms over long periods of time is noteworthy, considering that they are essentially unstable systems due to the underlying conflict between partners[25].

However, our knowledge of the nature and extent of coevolution in systems such as the fig–wasp mutualism, with over 750 pairs of interacting species, has been limited because an accurate evaluation of patterns and processes of species diversification in a coevolution system can only be performed if phylogenetic trees of both partners are analysed and can therefore be compared[26]. Classifications of both *Ficus* and Agaonidae have previously been based on morphology and reproductive traits closely linked to their interaction, but these are potentially misleading due to convergence/parallelism. We therefore risk circularity if we base our study of coevolution on such classifications, and evolutionary relationships are perhaps more appropriately revealed by DNA sequence analyses[27].

Previous DNA sequence–based phylogenetic studies of *Ficus* have shown that taxonomic categories are not natural and revealed several parallel transitions in growth habit and breeding system[14,27,28]. Correlation between inflorescence characters with head shapes of pollinating wasps and pollination behaviour indicates that reciprocal adaptations of morphological characters in these mutualistic partners have occurred[28,29]. However, previous studies have only included limited sampling (less than 50 species, or about 6%) of this large genus and have detected insufficient genetic variation to allow a detailed estimation of relationships of fig species, especially at species level.

Ongoing work by Rønsted and collaborators (for example, Rønsted et al.[18]) aims first to use molecular techniques to produce comprehensive phylogenetic trees for *Ficus* and second to combine fig and wasp phylogenetic data to explore the causes for the extraordinary diversification at this plant–animal interface. The rest of this chapter will present some problems and prospects of phylogenetic work with *Ficus* and other similarly large genera.

## 9.3   LOW LEVELS OF VARIATION IN STANDARD MARKERS

The first decades of molecular phylogenetic work have focused on plastid regions, which are easily amplified and have provided sufficient resolution in many studies. A review by Shaw et al.[30] discussed the relative utility of 21 noncoding plastid regions for phylogenetic analyses. However, in an increasing number of cases, plastid regions provide too little variation to be of infrageneric use, and this problem is especially pronounced when dealing with large genera.

The first phylogenetic analysis of *Ficus* was based on plastid *rbcL* and included only 15 species[27]. Before beginning our work on *Ficus*, we screened a number of plastid regions (the *matK* gene; *rpl16* intron; *rps16* intron; *trnL* intron, *trnL-F* spacer; *trnS-trnG* spacer; *accD-psaA* spacer; *trnH-psbA* spacer and *psbB-psbF* spacer), for variation within a small subset of figs (six species), but phylogenetic analyses using maximum parsimony as implemented in PAUP* 4.0b10[31] showed little variation (1–2% variable characters for each region). Four of the plastid regions (*rps16* intron, *trnL* intron, *trnL-F* spacer and *psbB-psbF* spacer) were combined for a set of 18 fig species and six other genera of Moraceae as an outgroup[32]. Phylogenetic analyses showed 100% bootstrap support for the ingroup, but there was no support for groupings within the genus (Figure 9.2). Plastid DNA sequences are also known to evolve slowly in palms,

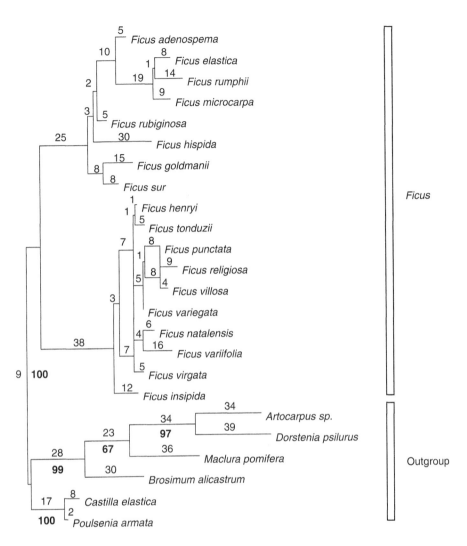

**FIGURE 9.2** One of over 2,000 most parsimonious trees obtained from the combined analysis of sequences from four plastid regions (*rpl16* intron, *trnL* intron, *trnL-F* spacer, and *psbB-psbF* spacer) of *Ficus*. Tree length 511 steps, consistency index (CI) = 0.88, and retention index (RI) = 0.72. Branch lengths and bootstrap percentages (>50%) are shown above and below the branches, respectively.

although palms are resolved on a long branch relative to other major clades of monocots[33–35]. Such patterns (high bootstrap support for a genus and little variation within) are common among plants[36,37].

In addition to using plastid regions, the nuclear ribosomal (nrDNA) internal transcribed spacer regions (ITS)[38] have proven useful in many species-level phylogenetic studies of a wide range of taxa[39–41]. ITS occurs in high copy number, which makes it easy to amplify, and the region has been widely employed for systematic studies. Furthermore, the whole ribosomal complex undergoes rapid concerted evolution, meaning that sequence similarity between individual copies is extremely high in most taxa. Divergent copies are detected in some cases[37,42–46]. Paralogs often require cloning of individual copies to separate them from orthologs (see section on hybrids and polyploids later).

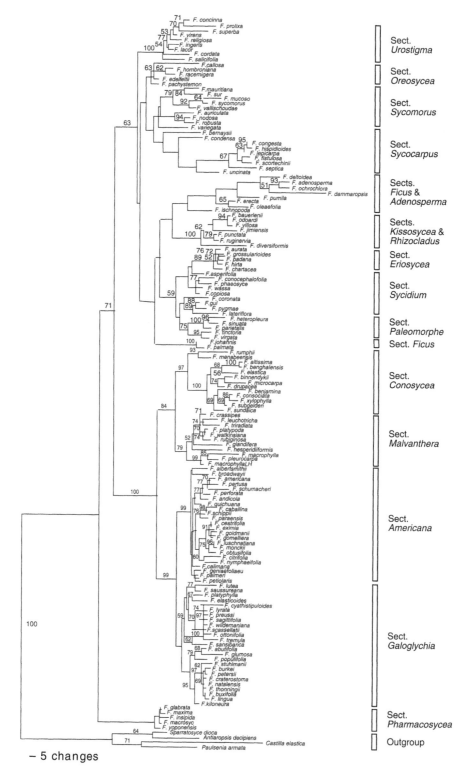

**FIGURE 9.3** One of 74 most parsimonious trees obtained from combined analysis of ITS and ETS rDNA sequences of *Ficus*. Tree length 2,010 steps, CI = 0.52, and RI = 0.83. Bootstrap percentages (>50%) are shown above the branches.

The ITS region has been used in a phylogenetic analysis of 46 dioecious *Ficus* species[14]. Amplification gave single bands, and cloning of four species showed no problems with heterogeneity among ITS copies. However, interspecific variability among ITS sequences within *Ficus* was limited, and the matrix was combined with a morphological character set. Later on, Jousselin and coworkers[28] combined ITS sequences with sequences of the external transcribed spacer (ETS)[47] in a study including 41 fig species representing most of the sections. ETS sequences evolve more rapidly than ITS sequences and can be a useful complement to ITS[47-50]. However, in comparison with plastid regions and ITS, the ETS region is notoriously difficult to amplify and necessitates high template quality. The combined analyses of ITS and ETS sequences of *Ficus* produced six trees, which were better resolved and supported down to sectional level than trees obtained with either of the separate ITS and ETS datasets[28]. Four genera of Moraceae (*Artocarpus*, *Brossemum*, *Broussonetia* and *Morus*) were included in the study as outgroups, but *Ficus* ITS and ETS sequences were too divergent to be aligned with these other genera, and the trees were rooted internally based on concepts of morphological change in the genus.

Sequencing of ITS and ETS was continued by Rønsted and coworkers[18] on a much larger dataset including 146 fig taxa. Based on other studies[27,51] including a recent phylogenetic analysis of Moraceae[17], sequences of four putatively closely related Moraceae genera (*Antiaropsis*, *Castilla*, *Paulsenia* and *Sparratosyce*) were successfully aligned and included as an outgroup. With this large dataset, sectional relationships of monoecious figs were clarified, but still only limited support was obtained for the dioecious groups and other groups within sections in general (Figure 9.3).

## 9.4 LOW COPY NUCLEAR MARKERS: THE IDEAL TOOLS

Nuclear genes evolve four- to five-fold faster than the plastid genes and should provide good alternatives to using the standard DNA regions (plastid and nuclear ribosomal spacers) for resolving relationships between closely related species[52]. Ideal candidate regions should occur in single or low copy numbers and include one or more sizeable introns flanked by conserved regions for which primers can be designed. Unfortunately, nuclear genes often occur in large multigene families, which can lead to problems with unidentified paralogy and partial concerted evolution[53].

Over the last decade, several low-copy nuclear genes have proven useful for examining divergence among closely related species in which nuclear ribosomal spacers and plastid spacers do not provide sufficient variation for phylogenetic reconstruction. Examples include: the structural nuclear gene for granule bound starch synthase (GBSSI or waxy gene)[54,55]; the MADS box genes *pistillata*[56] and cycloidea and homologs[57]; LEAFY/FLORICAULA[50,58]; the phytochrome family[59-61]; plastid expressed glutamine synthetase (*ncpGS*)[3,62]; alcohol dehydrogenase (*adh*)[49,53,63,64]; malate synthase[65]; and the RNA polymerase family (*rpb2*)[66,67]. Although these regions often provide more informative variation than the standard commonly used regions, there will typically still be a need for combining several regions to obtain a substantial increase in phylogenetic resolution. In some cases resolution and support for groupings can also be slightly improved by including gaps in analyses (for example, Jousselin et al.[28]).

Low-copy nuclear regions are often difficult to amplify, and generation of phylogenetic trees may be confounded by dynamic evolutionary processes of the gene families, such as gene duplication/deletion, gene conversion and recombination. Such notoriously difficult amplification compared to plastid or nuclear ribosomal regions necessitates high template quality, thereby limiting use of herbarium samples and other material with highly degraded DNA. Amplification difficulties can be overcome to some degree by varying polymerase chain reaction (PCR) conditions, for instance by allowing more time for denaturation and annealing.

However, such less specific PCR conditions may allow for nonspecific annealing of primers and increase the number of PCR products obtained. Designing specific primers can often improve amplification and reduce the number of bands obtained during amplification as well as improve specificity in general, but in most cases cloning will still be required for a portion of the samples, often due to heterozygosity in some individuals.

Gene duplication is also common and can result in erroneous phylogenetic trees if undetected paralogs are included in the analysis. For example, gene duplication of *rpb*2 has been reported in two major groups of asterid plants[67] and *Hibiscus* and related Malvaceae[68]. Duplication of *adh* genes has occurred in grasses, palms and other monocot clades and may reflect multiple duplication events rather than a single ancestral duplication[63]. In one tribe of legumes, LEAFY has been duplicated[69].

The problems discussed do not occur consistently for all regions and taxa, and often the best approach for finding a new useful region for a specific phylogenetic study is to try out an array of regions on a subset of the taxa (five to eight species spanning the range of expected variation, that is, two closely related species, plus one to two more distant species plus one to two outgroups).

Looking for alternative regions for phylogenetic analyses of figs, Rønsted and coworkers[32] screened various nuclear regions. Nuclear plastid-expressed glutamine synthetase gene (*ncpGS*)[62] yielded strong amplification of one and occasionally two bands in a small set of taxa and was chosen as an additional region; *ncpGS* is a nuclear gene responsible for assimilation of ammonia from photorespiration. It is a member of a multigene family but diverged long ago from the cytosolic expressed members; it contains several introns and is expected to diverge at a higher rate than ITS. The primers designed by Emshwiller and Doyle[62] for a range of dicotyledoneous plants amplify a region with four introns, and the size of the amplified product varies between 500 and 1,600 base pairs (bp). The region varies between 1,050 and 1,350 bp in figs with one to two indels missing in some taxa[32]. To obtain specific and strong amplification, a new set of primers was designed based on fig sequences.

In a provisional dataset containing 60 aligned *ncpGS* sequences of *Ficus* and 1,695 characters, 211 characters were potentially parsimony informative (12%) and 492 (29%) were variable[32]. This is somewhat less variation than in Sinningieae (Gesneriaceae), where *ncpGS* provided 24% potentially parsimony informative characters[3]. A preliminary analysis of a combined dataset of ITS, ETS and *ncpGS* sequences of 33 figs and two outgroup taxa was performed using 500 replicates of random stepwise addition with tree bisection-reconnection (TBR), equal weights, and the maximum parsimony criterion as implemented in PAUP* 4.0b10[31]. This analysis produced seven trees[32]. One of the trees is shown in Figure 9.4. A total of 500 bootstrap replicates with simple stepwise addition and TBR swapping was performed.

Compared with the previous combined analysis of ITS and ETS sequences (Figure 9.3)[18] both resolution and support is improved in the three-region set. For instance, in the ITS/ETS analysis the relationship of the African *F.* section *Galoglychia* and the New World *F.* section *Americana* was uncertain, with some trees showing *Galoglychia* and *Americana* as sisters and other trees showing *Galoglychia* paraphyletic to *Americana*. In the three-region analysis, section *Galoglychia* (63 bootstrap percentage, BP) is sister (100 BP) to section *Americana* (100 BP).

This dataset is limited and includes more monoecious than dioecious figs, but when many more taxa are sequenced for *ncpGS* and included, the three regions combined are expected to provide a "well supported" phylogenetic hypothesis for sectional relationships within figs. However, to obtain a well resolved and supported species-level tree, additional nuclear regions will be needed. However, direct sequencing of currently known nuclear regions may not provide sufficient resolution, and an alternative could be to use other molecular techniques, such as amplified fragment length polymorphism (AFLP), which is discussed in the next section.

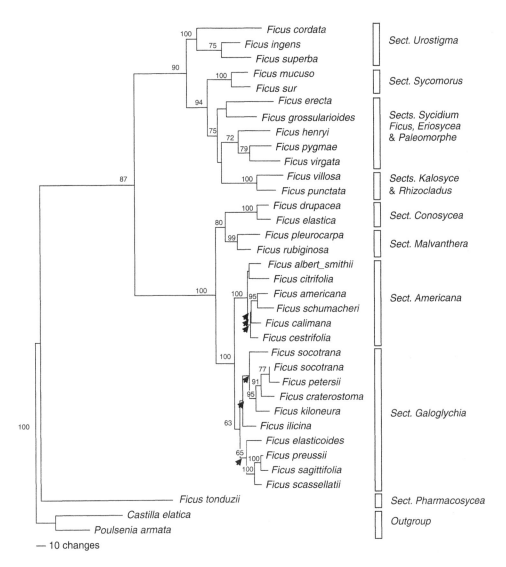

**FIGURE 9.4** One of seven most parsimonious trees obtained from combined analysis of ITS, ETS and *ncpGS* sequences of *Ficus*. Tree length 3,061 steps, CI = 0.75, and RI = 0.80. Bootstrap percentages (>50%) are shown above the branches. Arrowheads indicate nodes that collapse in the strict consensus tree.

## 9.5   USING AFLP AND OTHER FINGERPRINTING TECHNIQUES

Another way of dealing with low levels of variability in the standard markers is to use fingerprinting techniques such as AFLP, which generates genetic markers using selective PCR amplification of restriction fragments[70]. The AFLP technique was originally designed for crop genetics, but it is now routinely used in many systematics laboratories for phylogenetic and population studies, but sequencing approaches are usually preferred when comparing species and higher taxonomic levels. A major obstacle is that generating AFLPs includes several steps, making the technique laborious compared to sequencing. Moreover, the ability to isolate specific AFLP bands for sequencing can be complicated by band overlap and the need to clone bands into vectors in most cases. Li and Quiros[71] developed a simpler technique called sequence-related amplified polymorphism (SRAP) to overcome this problem. SRAPs were easily amplified in crops such as lettuce (*Lactuca*), potato

(*Solanum*) and rice (*Oryza*), but more experience is needed if the technique is to become more widely used.

Many researchers also feel that AFLP data are not suitable for parsimony and other types of phylogenetic analyses. According to Koopman and coworkers[72], critics raise two main points of concern. First, because AFLP fragments are identified by their length and not by their base composition, nonidentical fragments of equal length will mistakenly be scored as homologous. Second, AFLPs are usually scored as dominant characters, that is, with only the character states present (1), and absent (0), whereas in reality, at least some of the bands may represent codominant markers. Both sources of error introduce homoplasies and redundancies into the data and could lead to erroneous tree topologies in phylogenetic analyses. However, Koopman and coworkers[72] concluded that the impact of these homoplasies on conclusions regarding species relationships will be minor.

Several other empirical studies have likewise demonstrated that AFLP data are suitable for analyses with parsimony. In a study by Hodkinson et al.[73] AFLPs were used to investigate phylogenetic relationships of *Phyllostachys,* a large economically important genus of woody bamboos. DNA bands ranging from 50 to 500 bp in size were scored as presence/absence characters, and weak bands were removed from the matrix, which was then subjected to parsimony analyses. Hodkinson and coworkers also used AFLP on *Miscanthus,* a close relative of sugarcane (*Saccharum*), and found that major groupings were consistent with those determined from DNA sequences[74]. Other examples include cultivated lettuce (*Lactuca*)[72], the orchid genus *Dactylorhiza*[75], *Phylica*[76] and wild potato (*Solanum*)[77].

Summarising, AFLP is an effective technique for systematic studies in groups for which DNA sequence analyses have provided insufficient variation. The biggest problem with AFLP markers is that nonhomologous but similarly sized fragments may be scored as homologous, but this problem is overcome by not working with distantly related taxa[75]. Another problem with these markers is that they are not amenable to use with models of molecular evolution, which are needed in many studies, such as in molecular clock approaches. A preliminary study of the utility of AFLPs for figs indicated that the levels of variation between a subset of species produced results comparable to the DNA sequence results, but with greater levels of variation, and band homology could be easily assigned[78,79].

## 9.6  DOUBLE DATING OF FIG AND WASP LINEAGES: EVIDENCE FOR CODIVERGENCE

The interaction between figs and their pollinating wasps has become a model system for studies of coevolution as outlined previously. Interspecific coevolution involves reciprocal, selected changes in traits of interacting species, whereas codivergence can arise purely from maintenance of a specialised association between two lineages[20,26,80]. Patterns of codivergence are expected in fig–pollinator relationships owing to extreme host fidelity, and comparisons of phylogenetic trees for both figs and wasps with taxonomy of the other partner indicate that this might be the case[14,16,27,28]. Molecular phylogenetic trees of figs and their pollinators are compatible with the hypothesis of cospeciation[27,81]. However, due to lack of robust and comprehensive phylogenetic hypotheses for both partners, there was until recently no critical test of temporal congruence, that is, whether dates of divergence are correlated between interacting lineages[20,80].

Rønsted et al.[18] produced a comprehensive analysis of 146 diverse *Ficus* species based on nuclear ribosomal sequences (ITS and ETS) using maximum parsimony and Bayesian[82] tree construction methods (Figure 9.3). They also obtained an independent estimate of fig–wasp relationships[16]. Divergence times were calculated for both figs and wasps using independent fossil calibrations for each partner[16,83] and both nonparametric rate smoothing (NPRS)[84] and penalised likelihood[85] to account for deviations from the assumption of a molecular clock. Confidence intervals for ages were calculated by reapplying NPRS to 100 bootstrapped matrices. Ten interacting lineages of figs and pollinators were identified and their ages compared in a plot of the age of figs versus the corresponding pollinating wasps (Figure 9.5). Linear regression through the origin gave $r = 0.968$, $p = 0.386$, which is not

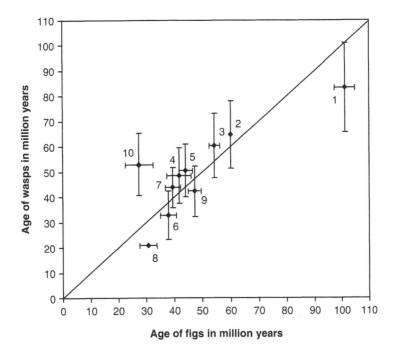

**FIGURE 9.5** Temporal congruence of fig lineages and their associated pollinator wasp lineages based on independent, fossil-calibrated molecular phylogenetic trees. Horizontal and vertical bars indicate standard errors of ages inferred from fig and wasp phylogenetic trees, respectively. (Reproduced with permission from Rønsted et al.[18])

significantly different from $r = 1$. To evaluate whether the relationship could be due to chance alone, the sum of squares of perpendicular offsets from a perfect linear regression (slope = 1) were compared to 10,000 randomised sets of 10 pairs of ages drawn from both phylogenetic trees. This analysis showed that the pattern observed in Figure 9.5 was highly significant, and the correlation between interacting fig and wasp lineages could therefore not be due to chance alone.

The strength of relationship between independently inferred ages of closely associated fig and wasp lineages provided the most compelling published evidence for long-term codivergence in this mutualism during at least the last 60 million years.

## 9.7 INCONGRUENCE IN PHYLOGENETIC TREES: EFFECTS OF POLYPLOIDS AND HYBRIDS

### 9.7.1 ALLOPOLYPLOID HYBRIDS

Allopolyploid taxa (having three or more sets of chromosomes from different diploid progenitor species) can make the interpretation of phylogenetic trees extremely difficult, particularly if the cytological constitution of the taxa under study is poorly assessed or unknown. Progenitors of an allopolyploid are typically determined by comparing one or more target DNA loci with the same loci in possible diploid progenitors; high sequence similarity indicates a genome donor. Plastids are generally maternally inherited organelles, and therefore in complex allopolyploids, trees based on plastid loci are often a good starting point because they indicate the maternal genome donor[86–88]. If allopolyploids have been recently synthesized and parental species are extant, then sequences from the polyploid will be almost exact matches with those from its parents, regardless of whether they are from the nucleus or plastid genome.

This is the situation with recently formed allotetraploids in *Nicotiana,* such as *N. tabacum*[86,89], which is less than 200,000 years old[90]. However, if the polyploid is older, then the match with parental loci will be less exact (due to subsequent divergence in both parent and hybrid progeny), and it is possible that parental species may have become extinct, both of which may obscure the origins of allopolyploids. Parentage of older allotetraploid groups in *Nicotiana,* such as *N.* sect. *Suaveolentes* (25 species found in Australia and southwestern Africa, about 8 million years old[91]), was much more difficult for Goodspeed[92] to sort out on the basis of chromosome studies and has been more difficult to determine on the basis of plastid DNA, nuclear ITS DNA and low-copy nuclear genes[86,89,91]. DNA sequences in such taxa are so divergent from all extant species that assessments of phylogenetic relationships are problematic.

ITS sequences of nuclear ribosomal DNA in allopolyploids are subject to concerted evolution and/or gene conversion and, given enough time, are usually converted to one of the parental types, most often that of the maternal parent[86,89,93,94]. However these loci can be unpredictable within some genera, and conversion can occur in both directions in closely related species[89]. For example, in *N. rustica* conversion favoured the maternal parent, so both plastid loci and ITS data indicated the same phylogenetic placement; thus if it had not been known that *N. rustica* was an allotetraploid, sequencing a plastid and nuclear locus like ITS would not have revealed its hybrid origin. In *N. tabacum,* ITS conversion to the paternal copy occurred, so a comparison of the plastid and ITS trees clearly revealed discordant relationships for this species[89]. Therefore, ITS and plastid data are difficult to correctly interpret without other sources of independent information.

ITS and ETS loci occupy the same 35S rDNA array and have been combined in phylogenetic analyses for diploid species (for example,[18,28]). However, few comparisons between these two loci have been made in allopolyploids, and it cannot be assumed that they will trace the same evolutionary history in all cases. Conversion of both loci to the same parental copies cannot be assumed, although we know of no documented cases of such in the angiosperms. The 5S ribosomal gene is typically located on a different chromosome from that of the 35S array, and thus it is more likely that different conversion patterns will occur, which would generate incongruent results in studies that use both loci. A striking example of this is the allopolyploid *N.* section *Repandae* (Figure 9.6), in which the 5S copy is inherited solely from the paternal parent, *N. obtusifolia,* whereas the ITS type is inherited from only the maternal parent, *N. sylvestris*[90]. Single-copy nuclear genes do not generally undergo concerted evolution, and copies of each progenitor type can usually be sequenced from allopolyploids. In *Nicotiana* single-copy nuclear genes (for example, plastid expressed glutamine synthetase[62]) have provided important information about parentage for some allopolyploid species[91]. These data have proven to be particularly useful in allopolyploids in which the ITS region is converted to the maternal copy, "making plastid and ITS based trees agree".

Allotetraploids (polyploids with two distinct diploid genomes) obviously contain twice as many copies of the genes a plant needs to function. There are four main ways in which homeologous nuclear genes can interact, and most produce a specific signature in phylogenetic studies:

- **Both homeologous copies are expressed and therefore have intact reading frames.** Here both types have usually diverged from their respective progenitor types at an approximately equal rate. This appears to be the most common fate of duplicated genes, and the literature is filled with examples, including studies of floral gene sequences in Hawaiian silverswords (Asteraceae)[95] and 16 loci surveyed in cotton (*Gossypium*)[96]. In fact, polyploid cotton is probably a special case, as both copies have been shown to be expressed but not always in the same tissue[97]. This form of specialisation could result in each copy being under different selection pressures; hence each type may evolve at a different rate.
- **Only one of the duplicated genes is expressed, and therefore the other becomes redundant.** A relaxation in the constraints on the redundant gene results in a build up of deleterious mutations, eventually making it a pseudogene (reviewed in Wendel[98]).

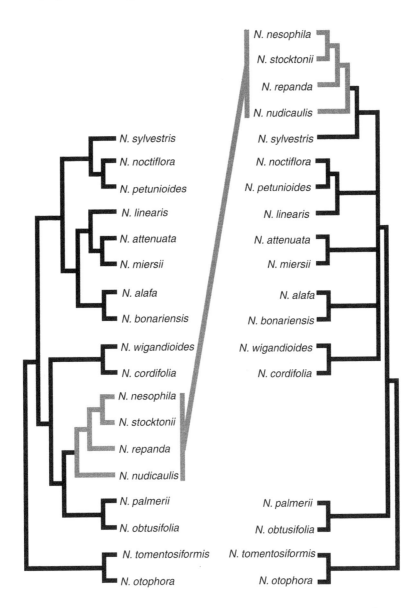

**FIGURE 9.6** The single most parsimonious NTS tree (left) versus the strict consensus of all most parsimonious ITS trees (right) for 18 species of *Nicotiana*. Both resulted from analysis using identical parameters[90]. The striped clade indicates the allopolyploid section *Repandae,* and all other species are diploid. (Reproduced with permission from Clarkson et al.[91])

- **Both homeologous copies of the gene are expressed, but there is relaxed selection on one, which, given the right selection pressure, can evolve another function.** This results in accelerated nonsynonymous rates of substitution. This situation appears to be rare, although it has been shown to be a possible fate of duplicated genes in *Petunia* and *Ipomoea*[99,100].
- **There is 'cross talk' between the two copies, which results in chimeric 'hybrid' sequences.** This situation appears to be rare and, as far as we are aware, has only been reported once in homeologous gene copies, in *Nicotiana tabacum*[101]. Here members of the glucan endo-1,3 β-glucosidase gene family were reported to be hybrid sequences between the two progenitor types. Crucially, however, the two progenitors of this allotetraploid were not sequenced for the study.

### 9.7.2 HOMOPLOID HYBRIDS

Homoploid (usually diploid) hybrids are hybrids with no change in ploidy relative to their parents. Here there are no genome duplications, but rather some segments of the nuclear genome come from one progenitor and others come from the other, creating linkage disequilibrium, which is the best way of detecting homoploid hybrids[102]. Under the model of hybrid (recombinational) speciation[103], phylogenetic reconstructions of individual DNA regions or loci that are part of a tightly linked set of loci should exhibit phylogenetic relationships of only one parent of the hybrid, and these trees would be discordant with trees based upon clusters of regions or loci from the other parent of the hybrid.

An example of this is the diploid *Nicotiana glutinosa,* which proved difficult to place on morphological grounds in sectional classifications of the genus because it displays floral traits of *N.* section *Tomentosae*[92] and vegetative traits of *N.* section *Undulatae*[86]. Recent phylogenetic studies have revealed that *N. glutinosa* is embedded in *N.* section *Undulatae* in trees based on plastid loci[86] and ITS[89]. However when nuclear genes have been sequenced, it is a member of *N.* section *Tomentosae* in trees based on glutamine synthetase[91] or *N.* section *Undulatae* in alcohol dehydrogenase trees[104], showing the expected differences in trees under the recombinational model. Sequencing single nuclear loci will not aid in the detection of homoploid hybrids unless, by chance, the results of plastid loci show one parental pattern and the nuclear region selected exhibits the other. Chimeric ITS sequences have been reported in hybrids in which the ITS1 spacer came from one progenitor and the ITS2 spacer from another[105]. This situation is caused by melting of the partially synthesised new strand when the polymerase has progressed only as far as the conserved 5.8S gene; reannealing to a copy from the other parent then permits extension of new strand, resulting in a chimeric copy. Such sequences should only be recovered by cloning from hybrid individuals; the highly conserved 5.8S sequence in the middle of the two ITS spacers makes this chimeric pattern possible.

## 9.8 CONCLUSIONS

Large genera have often been neglected for taxonomic studies, and many suffer from superficial or outdated classifications as well as uncertain species numbers and delimitations. However, large genera offer unique opportunities for studies of comparative biology. The precondition is a comprehensive and robust phylogenetic hypothesis, but in most cases phylogenetic work on large genera is still in its early days, with limited sampling and an array of potential problems. Phylogenetic hypotheses are currently typically based on molecular data, which surpasses morphological and other types of data in various ways, particularly the ability to obtain sufficient amounts of information for species-level comparisons and the need for phylogenetic hypotheses that are independent of biological traits that one may wish to evaluate.

The first decades of molecular phylogenetic work have focused on plastid regions, which are easily amplified and have provided sufficient resolution in many studies. However, in an increasing number of cases, a limited number of relatively small plastid regions provide too little variation to be of use for infrageneric studies, and this problem is especially pronounced when dealing with large genera. Several low-copy nuclear genes have proven useful for examining divergence among closely related species in which nuclear ribosomal spacers and plastid spacers do not provide sufficient variation for phylogenetic reconstruction. However, low-copy nuclear regions are often difficult to amplify, and generation of phylogenetic trees may be confounded by dynamic evolutionary processes of the gene families, such as gene duplication/deletion, gene conversion and recombination. Although these regions often provide more informative variation than commonly used standard regions, there will typically still be a need for combining several regions to obtain a substantial increase in resolution. In addition, mere size of the genera poses problems with phylogenetic analysis time, which can be reduced by various strategies.

In cases for which direct sequencing of nuclear regions may not provide sufficient resolution, an alternative could be using other molecular marker techniques, such as AFLP. The biggest problem with AFLP markers is that nonhomologous but similarly sized fragments may be scored as homologous, but this problem is overcome by not working with distantly related taxa. Another problem with these markers is that they are not amenable to use with models of molecular evolution, which are needed in many studies (for example, molecular clock approaches).

In general, there is no easy way to detect hybrids using DNA sequences in phylogenetic studies unless one has at hand multiple trees from independent loci. Sequencing of just plastid and nuclear ribosomal DNA will in many cases lead to the conclusion that hybrids are not present due to conversion to the maternal copy type in the ribosomal DNA. Inclusion of several single or low-copy regions is the most likely route to discovery of hybrids, particularly allopolyploids, but detection of a homoploid hybrid requires a much greater number of loci and some luck with choice of regions such that different linkage groups have been selected.

## ACKNOWLEDGEMENTS

This work was supported by the Danish Carlsberg Foundation (Nina Rønsted), a Marie Curie Outgoing International Fellowship within the Sixth European Community Framework Program (Nina Rønsted), the Natural Environment Research Council (NERC) (James Clarkson) and the Jeff Metcalf Fellows Programme of the University of Chicago (Kathrine Turk).

## REFERENCES

1. Frodin, D.G., History and concepts of big plant genera, *Taxon,* 53, 753, 2004.
2. Bronstein, J.L. and McKey, D., The fig-pollinator mutualism: a model system for comparative biology, *Experientia,* 45, 601, 1989.
3. Perret, M. et al., Systematics and evolution of tribe Sinningieae (Gesneriaceae): evidence from phylogenetic analyses of six plastid DNA regions and nuclear *ncpGS, Amer. J. Bot.,* 90, 445, 2003.
4. Albach, D.C., Martinez-Ortega, M.M., and Chase, M.W., *Veronica:* parallel morphological evolution and phylogeography in the Mediterranean, *Pl. Syst. Evol.,* 246, 177, 2004.
5. Pennington, R.T., Cronk, Q.C.B. and Richardson, J.A., Introduction and synthesis: plant phylogeny and the origin of major biomes, *Phil. Trans. R. Soc. Lond. B,* 359, 1455, 2004.
6. Rutschmann, F. et al., Did Crypteroniaceae really disperse out of India? Molecular dating evidence from *rbcL, ndhF,* and *rpl16* intron sequences, *Int. J. Plant Sci.,* 165, S69, 2004.
7. Davies, T.J. et al., Darwin's abominable mystery: insights from a supertree of the angiosperms, *Proc. Natl. Acad. Sci. USA,* 101, 1904, 2004.
8. Davies, T.J. et al., Environmental causes for plant biodiversity gradients, *Phil. Trans. R. Soc. Lond. B,* 359, 1645, 2004.
9. Davies, T.J. et al., Environmental energy and evolutionary rates in flowering plants, *Phil. Trans. R. Soc. Lond. B.,* 271, 2195, 2004.
10. Linder, H.P., Hardy, C.R., and Rutschmann, F., Taxon sampling effects in molecular clock dating: an example from the African Restionaceae, *Mol. Phyl. Evol.,* 35, 569, 2005.
11. Bremer, K., Ancestral areas: a cladistic reinterpretation of the center of origin concept, *Syst. Biol.,* 41, 436, 1992.
12. Chase, M.W. et al., When in doubt, put it in Flacourtiaceae: a molecular phylogenetic analysis based on plastid *rbcL* DNA sequences, *Kew Bull.,* 57, 141, 2002.
13. Rønsted et al., Molecular phylogenetic evidence for the monophyly of *Fritillaria* and *Lilium* (Liliaceae; Liliales) and the infrageneric classification of *Fritillaria, Mol. Phyl. Evol.,* 35, 509, 2005.
14. Weiblen, G.D., Phylogenetic relationships of functionally dioecious *Ficus* (Moraceae) based on ribosomal DNA sequences and morphology, *Amer. J. Bot.,* 87, 1342, 2000.
15. Cook, J.M. and Rasplus, J.-Y., Mutualists with attitude: coevolving fig wasps and figs, *Trends Ecol. Evol.,* 18, 241, 2003.
16. Machado, C.A. et al., Phylogenetic relationships, historical biogeography and character evolution of fig-pollinating wasps, *Proc. R. Soc. Lond. B,* 268, 685, 2001.
17. Datwyler, S.L. and Weiblen, G.D., On the origin of the fig: phylogenetic relationships of Moraceae from *ndhF* sequences, *Amer. J. Bot.,* 91, 767, 2004.

18. Rønsted, N. et al., 60 million years of co-divergence in the fig-wasp symbiosis, *Proc. R. Soc. Lond. B,* 272, 2593, 2005.

19. Zerega, N.J.C. et al., Biogeography and divergence times in the mulberry family (Moraceae), *Mol. Phyl. Evol.,* 37, 402, 2005.

20. Page, R.D.M., Introduction, in *Tangled Trees: Phylogeny, Cospeciation and Coevolution,* Page, R.D.M., Ed., The University of Chicago Press, Chicago, 2003, 1.

21. Pellmyr, O., Yuccas, yucca moths, and coevolution: a review, *Ann. Missouri Bot. Gard.,* 90, 35, 2003.

22. Henderson, A., A review of pollination studies in the Palmae, *Bot. Rev.,* 52, 221, 1986.

23. Meekijjaroenroj, A. and Anstett, M.C., A weevil pollinating the Canary Islands date palm: between parasitism and mutualism, *Naturwissenschaften,* 90, 452, 2003.

24. Kato, M., Takimura, A., and Kawakita, A., An obligate pollination mutualism and reciprocal diversification in the tree genus *Glochidion* (Euphorbiaceae), *Proc. Natl. Acad. Sci. USA,* 100, 5264, 2003.

25. Bronstein, J.L., The costs of mutualism, *Amer. Zool.,* 41, 825, 2001.

26. Page, R.D.M., Clayton, D.H., and Patterson, A.M., Lice and cospeciation: a response to Barker, *Int. J. Parasit.,* 26, 213, 1996.

27. Herre, E.A. et al., Molecular phylogenies of figs and their pollinator wasps, *J. Biogeogr.,* 23, 521, 1996.

28. Jousselin, E., Rasplus, J.-Y., and Kjellberg, F., Convergence and coevolution in a mutualism: evidence from a molecular phylogeny of *Ficus, Evolution,* 57, 1255, 2003.

29. Weiblen, G.D., Correlated evolution in fig pollination, *Syst. Biol.,* 53, 1, 2004.

30. Shaw, J. et al., The tortoise and the hare II: relative utility of 21 noncoding chloroplast DNA sequences for phylogenetic analyses, *Amer. J. Bot.,* 92, 142, 2005.

31. Swofford, D.L., *PAUP*: Phylogenetic Methods Using Parsimony (*and Other Methods),* version 4, Sinauer, Sunderland, MA, 2002.

32. Rønsted, N. et al., unpublished data, 2005.

33. Chase, M.W., Fay, M.F., and Savolainen, V., Higher-level classification in the angiosperms: new insights from the perspective of DNA sequence data, *Taxon,* 49, 685, 2000.

34. Asmussen, C.B. and Chase, M.W., Coding and noncoding plastid DNA in palm systematics, *Amer. J. Bot.,* 88, 1103, 2001.

35. Chase, M.W. et al., Multi-gene analyses of monocot relationships: a summary, in *Monocots: Comparative Biology and Evolution,* Columbus, J.T. et al., Eds., Rancho Santa Ana Botanic Garden, Claremont, CA, USA, in press.

36. Reeves, G. et al., Molecular systematics of Iridaceae: evidence from four plastid DNA regions, *Amer. J. Bot.,* 88, 2074, 2001.

37. Goldblatt, P. et al., Radiation in the Cape flora and the phylogeny of peacock irises *Moraea* (Iridaceae) based on four plastid DNA regions, *Mol. Phyl. Evol.,* 25, 341, 2002.

38. Baldwin, B.G. et al., The ITS region of nuclear ribosomal DNA: a valuable source of evidence on angiosperm phylogeny, *Ann. Missouri Bot. Gard.,* 82, 247, 1995.

39. Rønsted, N. et al., Phylogenetic relationships within *Plantago* (Plantaginaceae): evidence from nuclear ribosomal ITS and plastid *trnL-F* sequence data, *Bot. J. Linn. Soc.,* 139, 323, 2002.

40. Park, S.J. and Kim, K.J., Molecular phylogeny of the genus *Hypericum* (Hypericaceae) from Korea and Japan: evidence from nuclear rDNA ITS sequence data, *J. Plant Biol.,* 47, 366, 2004.

41. Huang, J.L., Giannasi, D.E., and Huang, J., Phylogenetic relationships in *Ephedra* (Ephedraceae) inferred from chloroplast and nuclear DNA sequences, *Mol. Phyl. Evol.,* 35, 48, 2005.

42. Buckler, E.S., Ippolito, A., and Holtsford, T.P., The evolution of ribosomal DNA: divergent paralogues and phylogenetic implications, *Genetics,* 145, 821, 1997.

43. Campbell, C.S. et al., Persistent nuclear ribosomal DNA sequence polymorphism in the *Amelanchier* agamic complex (Rosaceae), *Mol. Biol. Evol.,* 14, 81, 1997.

44. Denduangboripant, J. and Cronk, Q.C.B., High intraindividual variation in internal transcibed spacer sequences in *Aeschynanthus* (Gesneriaceae): implications for phylogenetics, *Proc. R. Soc. Lond. B,* 267, 1407, 2000.

45. Kita, Y. and Ito, M., Nuclear ribosomal ITS sequences and phylogeny in East Asian *Aconitum* subgenus *Aconitum* (Ranunculaceae), with special reference to extensive polymorphism in individual plants, *Pl. Syst. Evol.,* 225, 1, 2000.

46. Rapini, A., Chase, M.W., and Konno, T.U.P., Phylogenetics of the South American Asclepiadoideae (Apocynaceae), *Taxon,* 55, 119, 2006.

47. Baldwin, B.G. and Markos, S., Phylogenetic utility of the external transcribed spacer (ETS) of 18S-26S rDNA: congruence of ETS and ITS trees of *Calycadenia* (Compositae), *Mol. Phyl. Evol.,* 10, 449, 1998.

48. Béna, G. et al., Ribosomal external and internal transcribed spacers: combined use in the phylogenetic analyses of *Medicago* (Leguminosae), *J. Mol. Evol.,* 46, 299, 1998.

49. Roalson, E.H. and Friar, E.A., Phylogenetic analysis of the nuclear alcohol dehydrogenase (*Adh*) gene family in *Carex* section *Acrocystis* (Cyperaceae) and combined analyses of *Adh* and nuclear ribosomal ITS and ETS sequences for inferring species relationships, *Mol. Phyl. Evol.,* 33, 671, 2004.

50. Oh, S.-H. and Potter, D., Molecular phylogenetic systematics and biogeography of tribe Neillieae (Rosaceae) using DNA sequences of cpDNA, rDNA and LEAFY, *Amer. J. Bot.,* 92, 179, 2005.

51. Sytsma, K.J., et al., Urticalean rosids: circumscription, rosid ancestry, and phylogenetics based on *rbcL, trnL-F* and *ndhF* sequences, *Amer. J. Bot.,* 89, 1531, 2002.

52. Soltis, D.E. and Soltis, P.S., Choosing an approach and an appropriate gene for phylogenetic analysis, in *Molecular Systematics of Plants II: DNA Sequencing,* Soltis, D.E., Soltis, P.S., and Doyle, J.J., Eds., Kluwer Academic, Dordrecht, 1998, 1.

53. Clegg, M.T., Cummings, M.P., and Durbin, M.L., The evolution of plant nuclear genes, *Proc. Natl. Acad. Sci. USA,* 94, 7791, 1997.

54. Mason-Gamer, R.J. and Kellogg, E.A., Potential utility of the nuclear gene waxy for plant phylogenetic analyses, *Amer. J. Bot.* (suppl.), 83, 178, 1996.

55. Mason-Gamer, R.J., Weil, C.F., and Kellogg, E.A., Granule-bound starch synthase: structure, function, and phylogenetic utility, *Mol. Biol. Evol.,* 15, 1658, 1998.

56. Bailey, C.D. and Doyle, J.J., Potential phylogenetic utility of the low-copy nuclear gene *pistillata* in dicotyledonous plants: comparison to nrDNA ITS and *trnL* intron in *Sphaerocardamum* and other Brassicaceae, *Mol. Phyl. Evol.,* 13, 20, 1999.

57. Wang, C.N., Moller, M., and Cronk, Q.C.B., Phylogenetic position of *Titanotrichum oldhamii* (Gesneriaceae) inferred from four different gene regions, *Syst. Bot.,* 29, 407, 2004.

58. Grob, G.B.J., Gravendeel, B., and Eurlings, M.C.M., Potential phylogenetic utility of the nuclear FLORICAULA/LEAFY second intron: comparison with three chloroplast DNA regions in *Amorphophallus* (Araceae), *Mol. Phyl. Evol.,* 30, 13, 2004.

59. Mathews, S., Lavin, M., and Sharrock, R.A., Evolution of the phytochrome gene family and its utility for phylogenetic analyses of angiosperms, *Ann. Missouri Bot. Gard.,* 82, 296, 1995.

60. Davis, C.C. et al., Laurasian migration explains Gondwanan disjunctions: evidence from Malpighiaceae, *Proc. Natl. Acad. Sci. USA,* 99, 6833, 2002.

61. Samuel, R. et al., Molecular phylogenetics of Phyllanthaceae: evidence from plastid *matK* and nuclear *PHYC* sequences, *Amer. J. Bot.,* 92, 132, 2005.

62. Emshwiller, E. and Doyle, J.J., Chloroplast-expressed glutamine synthetase (*ncpGS*): potential utility for phylogenetic studies with an example from *Oxalis* (Oxalidaceae), *Mol. Phyl. Evol.,* 12, 310, 1999.

63. Morton, B.R., Gaut, B., and Clegg, M.T., Evolution of alcohol dehydrogenase gene in the palm and grass families, *Proc. Natl. Acad. Sci. USA,* 93, 11735, 1996.

64. Small, R.L. et al., The tortoise and the hare: choosing between noncoding plastome and nuclear ADH sequences for phylogeny reconstruction in a recently diverged plant group, *Amer. J. Bot.,* 85, 1301, 1998.

65. Lewis, C.E. and Doyle, J.J., Phylogenetic utility of the nuclear gene malate synthase in the palm family (Arecaceae), *Mol. Phyl. Evol.,* 19, 409, 2001.

66. Denton, A.L., Hall, B.D., and McConaughy, B.L., *RPB2,* a nuclear gene for tracing angiosperm phylogeny, *Amer. J. Bot.,* 83 (suppl.), 150, 1996.

67. Oxelman, B. et al., *RPB2* gene phylogeny in flowering plants, with particular emphasis on asterids, *Mol. Phyl. Evol.,* 32, 462, 2004.

68. Pfeil, B.E. et al., Paralogy and orthology in the Malvaceae *RPB2* gene family: investigation of gene duplication in *Hibiscus, Mol. Biol. Evol.,* 21, 1428, 2004.

69. Archambault, A. and Bruneau, A., Phylogenetic utility of the LEAFY/FLORICAULA gene in the Caesalpinioideae (Leguminosae): gene duplication and a novel insertion, *Syst. Bot.,* 29, 609, 2004.

70. Vos, P. et al., AFLP: a new technique for DNA fingerprinting, *Nucleic Acids Res.,* 23, 4407, 1995.

71. Li, G. and Quiros, C.F., Sequence-related amplified polymorphism (SRAP), a new marker system based on a simple PCR reaction: its application to mapping and gene tagging in *Brassica, Theor. Appl. Genet.,* 103, 455, 2001.

72. Koopman, W.J.M., Zevenbergen, M.J., and van den Berg, R.G., Species relationships in *Lactuca* s.l. (Lactuceae, Asteraceae) inferred from AFLP fingerprints, *Amer. J. Bot.*, 88, 1881, 2001.

73. Hodkinson, T.R. et al., A comparison of ITS nuclear rDNA sequence data and AFLP markers for phylogenetic studies in *Phyllostachys* (Bambusoideae, Poaceae), *J. Pl. Res.*, 113, 259, 2000.

74. Hodkinson, T.R. et al., Phylogenetics of *Miscanthus, Saccharum* and related genera (Saccharinae, Andropogoneae, Poaceae) based on DNA sequences from ITS nuclear ribosomal DNA and plastid *trnL* intron and *trnL-F* intergenic spacers, *J. Pl. Res.*, 115, 381, 2002.

75. Hedrén, M., Fay, M.F., and Chase, M.W., Amplified fragment length polymorphisms (AFLP) reveal details of polyploid evolution in *Dactylorhiza* (Orchidaceae), *Amer. J. Bot.*, 88, 1868, 2001.

76. Richardson, J.E. et al., Species delimitation and the origin of populations in island representatives of *Phylica* (Rhamnaceae), *Evolution*, 57, 816, 2003.

77. Lara-Cabrera, S.I. and Spooner, D.M., Taxonomy of North and Central American diploid wild potato (*Solanum* sect. *Petota*) species: AFLP data, *Pl. Syst. Evol.*, 248, 129, 2004.

78. Parrish, T., Krakatau: genetic consequences of island colonization, Ph.D. thesis, University of Utrecht, Utrecht, 2002.

79. Parrish, T., personal communication, 2002.

80. Percy, D.M., Page, R.D.M., and Cronk, Q.C.B., Plant-insect interactions: double-dating associated insect and plant lineages reveals asynchronous radiations, *Syst. Biol.*, 53, 120, 2004.

81. Weiblen, G.D. and Bush, G.L., Speciation in fig pollinators and parasites, *Mol. Ecol.*, 11, 1573, 2002.

82. Hulsenbeck, J.P. and Ronquist, F., Mr Bayes: Bayesian inference of phylogeny, *Bioinformatics*, 17, 754, 2001.

83. Collinson, M.E., The fossil history of the Moraceae, Urticaceae (including Cecropiaceae), and Cannabaceae, in *Evolution, Systematics, and Fossil History of the Hamamelidae Vol. 2: Higher Hamamelidae*, Crane, P. and Blackmore, S., Eds., Clarendon Press, Oxford, The Systematics Association, 1989, 319.

84. Sanderson, M.J., A nonparametric approach to estimating divergence times in the absence of rate constancy, *Mol. Biol. Evol.*, 14, 1218, 1997.

85. Sanderson, M.J., Estimating absolute rates of molecular evolution and divergence times: a penalized likelihood approach, *Mol. Biol. Evol.*, 19, 101, 2002.

86. Clarkson, J. et al., Phylogenetic relationships in *Nicotiana* based on multiple plastid loci, *Mol. Phyl. Evol.*, 33, 75, 2004.

87. Doyle, J.J., Doyle, J.L., and Brown, A.H.D., Incongruence in the diploid B-genome species complex of *Glycine* (Leguminosae) revisited: histome H3-D alleles vs. chloroplast haplotypes, *Mol. Biol. Evol.*, 16, 354, 1999.

88. Wendel, J.F., New world tetraploid cottons contain old world cytoplasms, *Proc. Natl. Acad. Sci. USA*, 86, 4132, 1989.

89. Chase, M.W. et al., Molecular systematics, GISH and the origin of hybrid taxa in *Nicotiana* (Solanaceae), *Ann. Bot.*, 92, 107, 2003.

90. Clarkson, J. et al., Long-term genome diploidization in allopolyploid *Nicotiana* section *Repandae* (Solanaceae), *New Phytol*, 168, 241, 2005.

91. Clarkson, J. et al., unpublished data, 2005.

92. Goodspeed, T.H., The genus *Nicotiana*, Chronica Botanica Company, MA, USA, 1954.

93. Aoki, S, and Ito, M., Molecular phylogeny of *Nicotiana* (Solanaceae) based on the nucleotide sequence of the *matK* gene, *Pl. Biol.*, 2, 316, 2000.

94. Pillon, Y. et al., Insights into the evolution and biogeography of western European species complexes in *Dactylorhiza* (Orchidaceae), *Taxon*, in press.

95. Barrier, M. et al., Interspecific hybrid ancestry of a plant adaptive radiation: allopolyploidy of the Hawaiian silversword alliance (Asteraceae) inferred from floral homeotic gene duplications, *Mol. Biol. Evol.*, 16, 1105, 1999.

96. Cronn, R.C., Small, R.L., and Wendel, J.F., Duplicated genes evolve independently after polyploid formation in cotton, *Proc. Natl. Acad. Sci. USA*, 96, 14406, 1999.

97. Adams, K.L. et al., Genes duplicated by polyploidy show unequal contributions to the transcriptome and organ-specific reciprocal silencing, *Proc. Natl. Acad. Sci. USA*, 100, 4649, 2003.

98. Wendel, J.F., Genome evolution in polyploids, *Pl. Mol. Biol.*, 42, 225, 2000.

99. Durbin, M.L. et al., Evolution of the chalcone synthetase gene family in the genus *Ipomoea*, *Proc. Natl. Acad. Sci. USA*, 92, 3338, 1995.

100. Huttley, G.A. et al., Nucleotide polmorphism in the chalcone synthase: a locus and evolution of the chalcone synthase multigene family of common morning glory, *Ipomoea purpurea, Mol. Ecol.,* 6, 549, 1997.

101. Sperisen, C., Ryals, J., and Meins, F., Comparison of cloned genes provides evidence for intergenomic exchange of DNA in the evolution of a tobacco glucan endo-1,3-β-glucosidase gene family, *Proc. Natl. Acad. Sci. USA,* 88, 1820, 1991.

102. Linder, C.R. and Rieseberg, L.H., Reconstructing patterns of reticulate evolution in plants, *Amer. J. Bot.* 91, 1700, 2004.

103. Müntzing, A., Outlines to a genetic monograph of the genus *Galeopsis, Hereditas,* 13, 185, 1930.

104. Timberlake, J., unpublished data, 2005.

105. Barkman, T.J. and Simpson, B.B., Hybrid origin and parentage of *Dendrochilum acuiferum* (Orchidaceae) inferred in a phylogenetic context using nuclear and plastid DNA sequence data, *Syst. Bot.,* 27, 209, 2002.

# 10 The Diversification of Flowering Plants through Time and Space: Key Innovations, Climate and Chance

*T. J. Davies*
Department of Biology, University of Virginia,
Charlottesville, USA

*T. G. Barraclough*
Division of Biology and NERC Centre for Population Biology,
Imperial College London, Silwood Park Campus, Ascot, Berkshire

## CONTENTS

## ABSTRACT

The flowering plants represent one of the largest terrestrial evolutionary radiations within recent geological times. Current estimates indicate there may be as many as half a million extant species, yet within the angiosperms species richness can vary over several orders of magnitude between closely related clades and between geographical regions. Understanding why some regions and some lineages contain more species than others has been a major challenge in biology. To date, approaches for studying these two patterns have been mostly separate. Traditional explanations for taxonomic imbalance have focused upon key biological traits, whilst regional variation in species richness has been ascribed largely to environmental factors. Using a tree of life for flowering plants, we demonstrate that environment can explain much of the taxonomic imbalance evident within phylogenetic trees not explained by key traits, and unequal rates of diversification, a product of the interaction between traits and environment, may contribute to regional patterns in species richness.

## 10.1  INTRODUCTION

One of the principal goals of ecology and evolutionary biology is to understand the diversity and distribution of life on Earth. The expansion of molecular approaches to phylogenetics has provided a wealth of data for reconstructing the evolutionary events behind why some groups have flourished whilst others have floundered. Flowering plants (angiosperms) have been one focus for such studies. Flowering plants represent a highly species rich group with an estimated 500,000 extant species[1-3] and were the subject of early coordinated efforts to reconstruct a complete family level phylogenetic tree of a higher taxonomic group[4,5]. Flowering plant species richness varies greatly among taxonomic groups and geographic regions; traditionally such patterns have been treated as largely separate phenomena. Here we outline recent efforts to explore patterns and processes of diversification in flowering plants using large-scale phylogenetic trees.

The phylogenetic distribution of species richness can vary over several orders of magnitude, even between closely related families, indicating considerable variation in net diversification rates between clades. Fossil evidence suggests a first appearance of angiosperms in the Cretaceous[6]; however, early diverging lineages, such as Amborellaceae and Nymphaceae, tend to be relatively species poor. Furthermore, a number of more recently derived families are unexpectedly species rich, notably within Euasterids I[7] sensu APG II[8]. There are two broad explanations for low species richness in older clades, either extinction rates have been higher, or speciation rates lower. Although the fossil record is insufficient to provide accurate estimates of extinction rates, there is little evidence that species poor clades were previously more diverse. It is therefore most likely that it was not until after the initial branching events of the clade that significant shifts in speciation rate arose[9]. Hence, high species richness is a result of elevated diversification rates within a subset of angiosperm lineages and therefore is of uneven distribution within flowering plants. However, the frequency, magnitude and location of shifts in diversification rates across the group have been poorly documented.

The geographical distribution of flowering plant species richness varies at a magnitude similar to that observed between lineages. Pollen records indicate ecological dominance was first attained at low latitudes between 20°N and 20°S[10]. Subsequent latitudinal expansion of the clade coincided with significant changes in the diversity of other plant groups, for example a decline in bryophytes and pteridophytes[10]. By the late Cretaceous flowering plants were the dominant flora of low latitudes, but comprised only 30–50% of diversity at higher latitudes. Over the past 65 million years before present (mybp) flowering plants have become the predominant vegetation type across all latitudes, but perhaps the most striking, and certainly the most frequently cited, spatial pattern in species richness remains the latitudinal gradient in diversity[11]. Tropical regions, for example Brazil's Atlantic forest, the Eastern Arc and coastal forests of Tanzania/Kenya and Sundaland, are recognised hotspots of flowering plant species richness[12]. Species richness tends to decrease at higher latitudes, although there are a number of exceptions, notably within Mediterranean climates such as the Cape of South Africa, the Mediterranean basin and the Californian chaparral.

Numerous studies have sought explanation for why some lineages are more diverse than others, concentrating on the role of key biological traits, such as pollination syndrome, but in flowering plants (as in other groups) such traits apparently explain relatively little of the variation in species numbers. At the same time, ecological studies have explored the effects of environment on floristic richness within regions, but have not traditionally addressed evolutionary explanations as to why some lineages or regions have more species. Phylogenetics provides a means to combine these approaches. Here we review how information on diversification rates inferred from phylogenetic trees can offer insights into the processes shaping both taxonomic and geographic patterns of species richness. Specifically, we consider whether differences in environment experienced by lineages can explain the extreme imbalance in species richness among clades.

## 10.2  MEASURING DIVERSIFICATION RATES

Among most organisms studied, it is commonly observed that relatively few taxonomic groups are species rich, the majority being species poor, and the shape of this frequency distribution has been termed the hollow curve (see Hilu, *Chapter 11*). Null models, such as the broken stick[13], typically fail to explain the extremes of the distribution. The lack of fit between empirical data and null expectations has been interpreted as revealing differences in speciation and extinction rates[14]. However, contrasting species richness between higher taxa can be misleading and may be confounded by taxonomic artefacts[15]. An alternative approach uses information from phylogenetic trees to infer diversification rates.

First, branching pattern of phylogenetic trees can provide information on the processes that shaped them[16–19]. By comparing phylogenetic tree shape against an appropriate null model, it is possible to estimate whether diversification rates have varied significantly among lineages. Second, contrasts in species richness between sister clades provide a means to identify where large shifts in net diversification rates have occurred on the tree[20]. Third, calibration of the branches on the tree allows estimation of absolute diversification rates[21]. Conclusions drawn from such studies are, however, critically dependent upon the accuracy of the underlying estimate of phylogeny[22–24]. Until recently, insufficient phylogenetic information had been a limiting factor in our understanding and interpretation of the evolutionary history of flowering plants[25,26].

The revolution in molecular techniques and phylogenetic methods in the 1990s saw an explosive growth in both the production and analysis of phylogenetic data, including the publication of a first draft of a phylogenetic tree for all flowering plant families[5,27]. Molecular studies of more than 100 taxa are now commonplace[28]. Davies et al.[29] used a supertree approach to summarise this wealth of phylogenetic data in a single tree, providing the first complete representation of evolutionary relationships within flowering plants above the family level. Unlike traditional consensus methods, supertrees can deal with source trees that do not share the same terminal taxa. They are therefore able to provide a more comprehensive phylogenetic tree than any represented in the individual source trees on which they are based. Although supertree construction, particularly that based upon matrix representation with parsimony, has attracted some criticism[30,31] empirical results suggest it performs well[32–35].

Using the mean tree imbalance measure of Fusco and Cronk[36], as modified by Purvis et al.[37], Davies et al.[29] demonstrated that the uneven distribution of species richness among higher clades in the supertree of flowering plants was much greater than that predicted from a purely stochastic process, in which the propensity to diversify is equal across all lineages. It is possible that bias in supertree construction led to preferential resolution of imbalanced nodes[38]; however, comparisons with nodal support metrics used in the source trees indicate that imbalanced nodes are as strongly supported as more balanced nodes. Furthermore, the supertree reflected the highly imbalanced topology of previous estimates of phylogeny based on incomplete sampling[39].

High tree imbalance suggests that speciation rates have been higher, or extinction rates lower, in some lineages over others. Sister clade comparisons using the method of Slowinski and Guyer[20] revealed numerous significant shifts in net diversification rates across all major clades (Table 10.1). Whilst there was some evidence for phylogenetic clustering of nodes subtending exceptionally imbalanced clades, for example, nodes falling within Lamiales, Asparagales and Caryophyllales, indicating a potentially heritable component to rate shifts, large shifts were evident throughout the tree topology.

The maximum likelihood estimate of diversification rate may be estimated as

$$LogN/t$$

**TABLE 10.1**
**Taxonomic Distribution of Imbalanced Nodes Using the Imbalance Measure of Slowinski and Guyer[20] on the Phylogenetic Tree of Flowering Plant Families from Davies et al.[29]**

| Higher Clade | Order | Number of Nodes | % Imbalanced Nodes |
|---|---|---|---|
| N/A | Austrobaileyales | 2 | 50 |
| Magnolids | Canellales | 2 | 0 |
| Magnolids | Laurales | 6 | 33 |
| Magnolids | Magnoliales | 5 | 40 |
| Magnolids | Piperales | 2 | 50 |
| Monocots | Alismatales | 13 | 23 |
| Monocots | Asparagales | 20 | 30 |
| Monocots | Dioscoreales | 2 | 50 |
| Monocots | Liliales | 6 | 17 |
| Monocots | Pandanales | 4 | 0 |
| Commelinids | Commelinales | 4 | 25 |
| Commelinids | Poales | 14 | 36 |
| Commelinids | Zingiberales | 7 | 14 |
| Eudicots | Proteales | 1 | 100 |
| Eudicots | Ranunculales | 6 | 33 |
| Core Eudicots | Caryophyllales | 19 | 21 |
| Core Eudicots | Saxifragales | 11 | 36 |
| Asterids | Cornales | 4 | 25 |
| Asterids | Ericales | 24 | 21 |
| Euasterids I | Gentianales | 5 | 20 |
| Euasterids I | Lamiales | 20 | 35 |
| Euasterids I | Solanales | 6 | 33 |
| Euasterids II | Apiales | 9 | 22 |
| Euasterids II | Aquifoliales | 4 | 25 |
| Euasterids II | Asterales | 11 | 27 |
| Euasterids II | Dipsacales | 1 | 0 |
| Rosids | Crossosomatales | 2 | 0 |
| Rosids | Geraniales | 2 | 0 |
| Rosids | Myrtales | 10 | 10 |
| Eurosids I | Celastrales | 2 | 50 |
| Eurosids I | Cucurbitales | 6 | 17 |
| Eurosids I | Fabales | 3 | 100 |
| Eurosids I | Fagales | 6 | 17 |
| Eurosids I | Malpighiales | 27 | 22 |
| Eurosids I | Oxalidales | 4 | 0 |
| Eurosids I | Rosales | 7 | 29 |
| Eurosids II | Brassicales | 15 | 20 |
| Eurosids II | Malvales | 6 | 50 |
| Eurosids II | Sapindales | 9 | 22 |

under the assumption that the diversification rate during time $t$ has been approximately exponential[40,41], where $N$ is the number of species in the clade, and $t$ is the time since the clade diverged from its sister clade on the dated tree. Shifts in net diversification rates were therefore calculated as:

$$\left(\frac{LogN(des)}{t(des)}\right) - \left(\frac{LogN(anc)}{t(anc)}\right)$$

where *des* is the descendent clade and *anc* is the ancestral clade. A positive shift in net diversification rate indicates an increase in rates from the ancestral to the descendent clade. Mapping the magnitude of rate shifts on the topology of the tree confirms the impression of frequent large shifts in diversification rate, indicating that the propensity to diversify is a highly labile trait. However, the direction and magnitude of shifts in net rates appeared to vary nonrandomly across the phylogenetic tree.

The 10 greatest shifts in net diversification rates were negative, from high ancestral rates to low descendent rates, for example, the nodes subtending Ecdeiocoleaceae (tussocky cord rush; one species) sister to Poaceae (grasses; c. 12,000 species), Stegnospermaceae (Cuban tangle; three species), sister to a number of families within Caryophyllales (for example, cacti, carpetweeds and fig-marigolds; c. 4,300 species), and Calyceraceae (calycera family; 40 species) sister to Asteraceae [daisies; c. 13,000 species). The two clades with the greatest positive shift in rates were identified as the sister family pair Moraceae (figs and mulberrys; 1,675 species) and Urticaceae (nettles; 825 species). The mean age for the top 10 greatest positive shifts was significantly younger than that for the negative shifts (mean age 38.5 mybp versus 53.4 mybp for the positive and negative shifts respectively; $P < 0.05$, Mann-Whitney test).

In general, older nodes tended to exhibit greater taxonomic imbalance, associated with a negative shift in net diversification rates, and more recent nodes tended to be more balanced than expected, with several sister family pairs displaying correlated positive shifts in rates. One possible explanation for this would be a general increase in diversification rates within recent time periods, and the imbalance of older nodes might reflect the accumulated effect of past shifts in diversification rate. However, an alternative explanation is that this pattern reflects a bias due to the use of families as terminal taxa; shifts occurring within families can only be reconstructed as occurring in the entire family in the analyses. Furthermore, extinction will have had less time to operate within more recently derived clades, thereby inflating diversification rate estimates[7]. The overriding impression is of a history littered with tales of evolutionary successes and failures. We explore what might explain this chequered past in the following sections.

## 10.3 KEY INNOVATIONS

Much emphasis continues to be placed upon the possession of a few key traits that might have influenced rates of diversification. By opening up new adaptive zones, such traits may have enabled those lineages that possess them to proliferate at an increased rate[42]. Salamin and Davies[43] employed phylogenetically independent contrasts[22] from the supertree of flowering plants to evaluate a number of putative key traits: generation time (herbaceous versus woody and annual versus perennial), dispersal (biotic versus abiotic), pollination (biotic versus abiotic) and sex (dioecy versus monoecy). Generation time might be negatively correlated to evolutionary rates[44–46], resulting in greater evolutionary time over equivalent absolute time periods for fast-lived species. Biotic dispersal may enhance the probability of long-distance dispersal events[44]; further, both biotic pollination and biotic dispersal might reduce the frequency of outcrossing between geographically isolated populations, thereby providing the reproductive isolation necessary for allopatric speciation[47,48]. Among biotically pollinated taxa, monoecious species are more likely to have specialist pollinators[49,50], further reinforcing reproductive isolation. Within monoecious taxa, selfing species may be more likely to form new species following hybridisation[51], thereby enabling the establishment of isolated populations founded by rare dispersal events[52].

The utility of phylogenetic trees in the comparative method in controlling for nonindependence and confounding variables is well recognised[22,53]. Independent contrasts provide a statistically powerful approach for identifying correlates of diversification. Any changes between sister taxa must have occurred since the time of divergence and thus represent independent evolutionary events. As sister taxa are the same age, comparisons of species richness directly reflect variation in net rates of diversification. Finally, the effects of confounding variables, for example additional traits also affecting speciation rates, are minimised[54]. However, Salamin and Davies[43] found no significant association between the traits studied and species richness among higher clades.

There are several possible explanations as to why no support was found for the key innovation hypothesis:

- **Poor phylogenetic data**; phylogenetic error will tend to reduce signal and thereby increase the probability of type II errors when independent contrasts are employed[24].
- **Poor trait data**; if clades were miscoded in terms of trait value, we would also predict an associated increase in type II error rates.
- **The wrong traits were examined**; Gorelick[55] lists 20 hypotheses that have variously been proposed to explain the evolutionary success of the flowering plants, including many key traits. Doubtlessly a comprehensive survey of the literature would reveal many more putative key innovations, and a number of significant results have been reported within a subset of clades[43,56–60]. It is possible that, if sufficient data were to become available to test these hypotheses in the future, significant associations may be found across the flowering plants.
- **Contingency upon other traits and the environment**; whether a certain trait influences diversification rates is likely to depend on a number of factors, including the abiotic environment, other biological traits, and other taxa[61].
- **Contrasts at higher taxonomic levels may be too insensitive**; the majority of flowering plant diversity is encompassed within, rather than between, families, hence a stronger association between traits and species richness might be observed for more fine-scale analyses.

Although there were a possible 378 contrasts (nodes in the supertree), the sample size of unambiguous state changes was small; life form and mode of pollination were both limited to two comparisons, the maximum being 15 comparisons (mode of dispersal). The limited number of contrasts was partly a product of lack of variation in the traits under examination; for example, abiotic pollination characterises both clades subtending the most imbalanced node identified above, the grasses and their sister group. However, the predominant limiting factor was within family variation, resulting in many clades being classified as polymorphic for the majority of traits, indicating that, for the traits examined, the taxonomic scale of this analysis was inappropriate.

Where strong associations between species richness and biological traits have been found, they are often environment or clade specific, for example, annual life form in grasses[43], floral nectar spurs in columbines[57], climbing habit in predominantly tropical taxa[58] and fleshy fruit in the tropical understorey[59]. As we look further back in time, at nodes deeper in the phylogenetic tree, we would expect a proportional increase in the impact of other factors, such as mass extinctions, biogeography and other traits on diversification rates[61,62]. The difference between the findings of Smith[59] and those of Salamin and Davies[43] on the importance of biotic dispersal, for which fleshy fruit is an indicator, is likely a result of the former study restricting comparisons to taxa found only within a narrow environmental niche, the tropical understorey. The significant association between annual life form and species richness in grasses and the absence of significance in contrasts between families of flowering plants is also a likely product of scale, but taxonomic rather than environmental. It is therefore unsurprising that key traits do not always generalise across disparate taxa. For example, biogeography appears to have left a greater imprint on patterns of current species richness than presence or absence of nectar spurs in the genus *Halenia*[63], yet nectar spurs may remain important when only young, geographically restricted, clades are considered.

In summary, there is a growing appreciation that explanations based upon one or a few traits are too simplistic to explain patterns of flowering plant species richness[7,8,64]. Where significant correlations between biological traits and species richness have been found, they tend to be in comparisons between recently radiated taxa sharing similar ecological conditions. Whether a biological trait influences net diversification rates is therefore likely to depend on a number of other factors, including abiotic environment. If the efficacy of a trait in influencing speciation rates were environment dependent we might also predict that different traits would have been advantageous at different geological times, with those taxa that happened to be pre-adapted to changes in environmental conditions radiating rapidly. Such a scenario has been suggested as explaining the rapid radiation of the grasses (previously restricted to marginal habitats) coinciding with the late Tertiary change towards a drier climate, which enabled the exploitation of new niches and a dramatic increase in their ecological dominance[65,66], and might explain the apparent lag between the origin of particular traits and the increase in the proportion of taxa possessing them in the fossil record[67].

Environment clearly has the potential to greatly enhance our understanding of the evolutionary history of flowering plant diversity; in the following section we explore the effects of one aspect of environment, namely latitudinal gradients.

## 10.4 EVOLUTIONARY RATES AND THE LATITUDINAL GRADIENT IN SPECIES RICHNESS

One of the most pervasive patterns in ecology is the latitudinal gradient in species richness. In most taxa, species richness tends to be greatest at the equator and declines towards the poles[68]. Despite the wealth of literature on this phenomenon, the underlying causes remain unclear[69,70]. One possible explanation is that high levels of environmental energy promote higher species richness nearer the equator[71–73]. This hypothesis is supported by observations that energy-rich regions tend to support more species than energy-poor regions[74–78]. Energy gradients can explain c.70–80% of the variation in species richness between regions[70]. The existence of alternative gradients in species richness, for example those with altitude and depth, provide additional support for a species–energy relationship[72]. Furthermore, when energy is controlled for, the latitudinal gradient in species richness can disappear[79], suggesting that in many cases the latitudinal gradient in species richness may be more accurately described as an energy gradient. The reasons why species richness might vary with energy are examined below.

### 10.4.1 BIOMASS THEORY

Higher productivity at lower latitudes might allow a greater biomass and hence more species to be supported[80,81]. Wright[71] formally stated the species–energy theory as an extension of the species–area theory. Wright argued that area was a surrogate measure of available resources; a direct indicator of resource availability, such as energy, would therefore provide a more accurate predictor of species richness. This assumes that energy-rich environments support more populations rather than simply more individuals per population. The lack of a consistent relationship between productivity and species richness has led some to largely dismiss this theory[69], but a better understanding of the effects of spatial scale and resource partitioning upon these relationships may provide a clearer picture[82,83].

### 10.4.2 FASTER EVOLUTION HYPOTHESIS

Increased environmental energy speeds up evolutionary rates and species production. Rohde[72] argued that the higher diversity of the tropics could be explained by greater effective evolutionary time. The higher temperature of tropical regions may increase metabolic rates and decrease development times, leading to shorter generation times, faster mutation rates, and as a consequence, a faster response to selection pressures (Figure 10.1a). The faster evolution hypothesis assumes that environment-mediated variation in diversification rates is sufficient to produce geographical gradients in species richness. If correct, we would expect variation in diversification rates between clades encompassing areas with different energy loads. The faster-evolution hypothesis is consistent with the tropics as a cradle of diversity[84–87]. Previous work established a link between evolutionary rates, estimated from sequence divergence data, and species richness in flowering plants[88,89], one step in the faster evolution theory. However, a study of evolutionary rates in birds by Bromham and

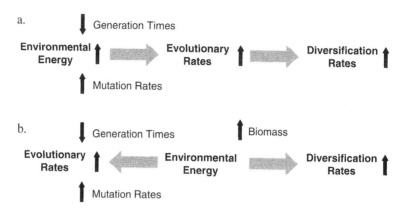

**FIGURE 10.1** Alternate pathways depicting the potential relationships among species richness, mutation rates and environmental energy. (a) Higher temperature of tropical regions may increase evolutionary rates, via shortening generation times and elevating mutation rates, thereby driving net diversification rates; the faster evolution hypothesis. (b) Alternatively, environmental energy may influence both evolutionary rates and diversification rates independently.

Cardillo[90] failed to find any association between molecular rates and latitude; the second step in the theory. Therefore, despite widespread interest, support for the faster evolution hypothesis has been equivocal, and the direction of causality unclear (compare Figure 10.1a and Figure 10.1b).

### 10.4.3 EVALUATING THE FASTER EVOLUTION HYPOTHESIS

Davies et al.[91] reviewed evidence for the faster evolution hypothesis using estimates of energy load across families of flowering plants derived from GIS data, based upon contemporary distributions. As different aspects of the environment might influence evolutionary rates versus biomass more strongly in plant taxa, three energy measures were employed: ultraviolet (UV) radiation, actual evapotranspiration (AET) and temperature. UV radiation might influence mutation rates via the formation of harmful photoproducts and has been described as a driving force in evolutionary rates[92]. AET represents water-energy dynamics and reflects the amount of biomass an area can support; it is often strongly correlated with regional plant species richness[75,79,93,94]. Finally, temperature might have an effect on both biomass and rates of molecular evolution[72,73]. Therefore, if the faster evolution hypothesis were correct, we would predict that UV or temperature would display the stronger relationship with species richness, via an intermediate link with molecular evolutionary rates.

Present day distributions and climate might provide a poor estimate of conditions experienced over evolutionary time. Families of flowering plants range in age from c. 25 mybp (for example, Moraceae, Urticaceae and Asteropeiaceae) to >100 mybp (for example, Amborellaceae, Nymphaeaceae and Chloranthaceae), while climatic shifts occur in the order of every 10,000–100,000 years[95]. It may therefore seem unrealistic to expect contemporary measures to provide a useful index of past environment. However, correlated range dynamics amongst related but disjunct taxa indicate ecological niche conservatism over time periods spanning tens of millions of years[96]. Environmental tracking by plant lineages can result in a migration like response to climatic change[97]; hence mean environment for a plant lineage might be relatively constant over time. Although geographical barriers are likely to prohibit exact environmental matching, if contemporary environment was independent from the environmental conditions experienced over evolutionary time, it would most likely confound attempts to detect true relationships, particularly those with evolutionary variables such as rate of molecular change.

Sister family comparisons of Davies et al.[91] supported the broad predictions of the species–energy theory, revealing a strong correlation between environmental energy and species richness. Temperature was the best predictor of the alternate energy measures, explaining 19% of the variation in species numbers between families, once area had been accounted for. Energy was also found to be a good predictor of molecular evolutionary rates, with faster rates in high-energy environments, confirming the second step in the faster evolution theory. Although a relationship between molecular evolutionary rates and environmental energy had not previously been reported[90], there was no evidence that energy increased diversification rates by this pathway. Instead the effects of energy on both molecular rates and species richness was direct (Figure 10.1b), leading Davies et al.[91] to reject the faster evolution hypothesis.

Diversification rates co-vary with environment; lineages occupying higher energy regions tend to have higher net speciation rates. The direct link between energy and species richness is compatible with the biomass hypothesis. More productive environments might reduce extinction through supporting higher population densities, thereby elevating net diversification rates, although the precise relationship between productivity and density remains controversial[70,98–100]. Bonn et al.[101] recently proposed an alternative explanation, in which more productive environments were likely to contain a greater sample of taxon-specific critical resources. Therefore, high energy environments could sustain a greater number of viable populations, not by increasing population density, but rather increasing the probability of the occurrence of a limiting resource, which may vary between taxa. This hypothesis remains to be evaluated in plants.

Both environment and biological traits may explain a proportion of the variation in net diversification rates among lineages within flowering plants. Key traits are difficult to evaluate for older nodes; similarly, we might expect causal relationships between environment and species richness to be more difficult to detect for more ancient splits due to post speciation range movement and historical climate change. In the final section of this chapter, we explore how combining information on biology and environment can be mutually informative, using the iris family (Iridaceae) as a test case. By using younger and more narrowly distributed taxa, it may be possible to more accurately discriminate environment and the effect of species-specific traits on geographical and taxonomic patterns of species richness.

## 10.5 TRAITS × ENVIRONMENT: DIVERSIFICATION OF IRISES IN THE CAPE OF SOUTH AFRICA

We contend that species richness of a clade depends on both biological traits and the environment. In similar environments biological traits might therefore be predicted to explain much of the variation in species richness between taxa. For taxa sharing similar traits but occupying different habitats, environment will be the major determinant of species richness. However, it may more often be the case that the interaction between traits and environment will predominate. The interaction term might be positive, neutral or negative[61].

The Cape Region of South Africa is renowned for its high levels of plant species richness and endemism but is an outlier from global trends relating environment to flowering plant species richness[102–106]. However, it is possible that clades containing lineages that have radiated extensively in the Cape, such as the iris family, are characterised by biological traits that have resulted in a different functional response to the environment that can still explain geographical variation in species richness. If the influence of the biological traits on diversification rates within these clades were environment dependent, we would first predict that Cape clades lacking those traits would be species poor and, second, that non-Cape clades would be species poor regardless of the traits that define them.

Irises are a highly diverse family of perennial herbs with around 1,800 species in 65 genera[107], including several familiar cultivars of well known genera, for example *Crocus*, *Gladiolus* and *Iris*. Adapted to neither the intense competition for light nor the rapid growth required for gap colonisation

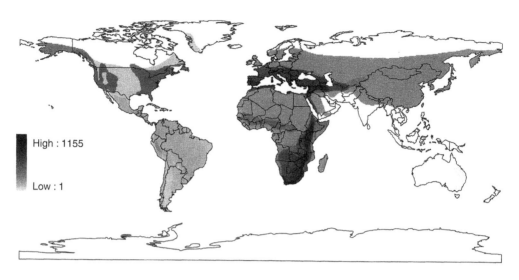

**FIGURE 10.2** Geographical distribution of iris family species richness. (Generated by summing species counts from overlaid generic distribution maps from Davies et al.[110,111])

within dense vegetation typical of tropical regions, species numbers tend to be highest in seasonally dry environments (Figure 10.2). Irises conform to typical patterns of Cape diversity: the family contains 677 species in the region, of which 80% are endemic[105], and many genera including *Gladiolus* (260 species) and the peacock irises (*Moraea*) (196 species) have radiated extensively within the Cape[108]. Although a few genera such as *Neomarica* (eight species) and *Eleutherine* (two species) are found in the Neotropics, they are relatively species poor. The family is characterised by an isobilateral leaf held vertically, perhaps the single most important morphological innovation of the family, and underground storage organs, such as a corms (for example, *Crocus* and *Gladiolus*), rhizomes (for example, *Isophysis* and *Sisyrinchium*) or bulbs (for example, *Cypella* and *Tigridia*). Floral morphology is highly variable and matches several different pollination syndromes[109]. These traits have been thought important in the group's evolution within the Cape[105].

Phylogenetically independent contrasts[22] of species richness from a generic level phylogenetic tree of irises revealed several significantly imbalanced nodes[110,111]. Although the node subtending *Isophysis* (one species) and the remainder of Iridaceae is the most imbalanced, the next most basal nodes are also highly imbalanced, suggesting that that the ancestral state may have been a low net diversification rate. The early diverging lineages, *Diplarrhena* (two species), *Patersonia* (21 species) and *Geosiris* (one species), tend to be species poor, rhizomatous, of limited geographical distribution and suggest an Australasian origin for the family. Repeating the family-level analysis of environment and species richness at the generic level showed that abiotic environment plus area could explain up to 85% of the variation in species richness between sister clades, including the highly imbalanced nodes identified above.

Environmental factors associated with warm, dry and topologically diverse habitats were the best predictors of species richness, reflecting the specific preferences of the family and confirming its departure from the more general trend towards higher species richness in tropical environments. However, environment alone was insufficient to explain the high diversity of Cape clades. By using contrasts between sister clades, lineages within the Cape were revealed to have speciated at faster rates than those found elsewhere, even than in regions with similar Mediterranean type climates. Molecular evidence suggests that the Cape may have undergone a period of rapid diversification coinciding with a change in oceanic currents leading to the aridification of the region 8–7 mybp[104,112]. While the majority of branching events in the generic phylogenetic tree predate this shift in climate, there is evidence in at least one genus, *Moraea*, that this geological background provided the setting for the diversification leading to high species richness of irises in the Cape[108.]

It is possible that the high net diversification rates observed in Cape clades are a product of abiotic factors not included in the model parameters or that the spatial scale of the analysis was too insensitive to accurately characterise the great physical diversity of the Cape Region. For example, the Cape may have been more climatically stable during the Pleistocene radiations, allowing greater time for gradual speciation, elevating net diversification rates[95]. However, of the genera with the greatest deviation from the model, all those that contain more species than predicted by environment, *Geissorhiza, Hesperantha, Ixia* and *Therianthus,* fall within a single clade, Crocoideae. Although no biological traits changed state frequently enough on the tree to allow tests for a general correlation with diversity, a number of traits are characteristic of this clade and may have been instrumental in its diversification in the Cape. The evolution of the perianth tube and zygomorphic flowers was likely important in allowing floral plasticity in Crocoideae, and subsequent pollinator specialisation[105,113,114]. A cormous rootstock, again typical of Crocoideae, may have enabled rapid regrowth in fire-dominated landscapes and also promoted establishment by vegetative reproduction following rare long-distance dispersal events.

Even if sufficient data were available to evaluate these putative key traits more rigorously, it seems unlikely that they would be identified as key innovations from simple analyses of taxonomic imbalance, as several species poor genera share many of these traits with their species rich counterparts. Ancestral state reconstructions for both flower symmetry and rootstock type reveal that shifts in character states are correlated with neither significantly imbalanced nodes nor large deviance from the expected values derived from the climate variables[110,111]. Biological traits associated with high species richness in genera of irises may therefore have only had those effects in particular places, most notably the highly heterogeneous environment of the Cape. This is consistent with our second prediction: clades outside the Cape have not diversified, irrespective of their intrinsic biological attributes. The limited number of comparisons meant that it was not possible to examine our first prediction: whether the absence of particular traits results in lower diversification rates in the region.

## 10.6 CONCLUSIONS

Phylogenetic trees of the flowering plants are too imbalanced to be a product of an equal rates Markov process, in which all lineages have an equal probability of diversifying, but shifts in diversification rate appear to be too frequent for it to be explained by the inheritance of a few key traits. The interaction between traits and the environment may offer a resolution to this apparent paradox. If the influence of heritable biological traits upon the likelihood of diversifying was dependent on environmental conditions and environmental change was frequent, repeated shifts in diversification rate would be expected. Within any set of conditions some lineages would be favoured over others. However, the identity of these lineages would fluctuate with a changing environment, as conditions favourable to speciation within one lineage may not be so in another. A strong relationship between environment and species richness would, however, be evident at any single point in the evolutionary history of flowering plants.

Extant species richness may be best explained with reference to the contemporary environment. However, lineages characterised by different suites of traits might be expected to display different functional responses to their physical surroundings. Irises represent a family that departs from global trends in species richness, yet the environment can explain a large proportion of the variation in species richness among lineages. Even within this clade, it is likely that particular biological traits have favoured rapid diversification within the Cape of South Africa. A wider sample of Cape clades may provide sufficient data to evaluate the interaction between traits and environment in this unique and species rich region.

## ACKNOWLEDGEMENTS

We thank our collaborators on the published work described here, Vincent Savolainen, Nicolas Salomin, Mark Chase, Justin Moat, Peter Goldblatt, Pam Soltis and Doug Soltis. We also thank Camille Barr for helpful comments on an earlier draft of this manuscript. The work was supported by a NERC Ph.D. studentship and by the Royal Society.

## REFERENCES

1. Govaerts, R., How many species of seed plants are there? *Taxon,* 50, 1085, 2001.
2. Bramwell, D., How many plants species are there? *Plant Talk,* 28, 32, 2002.
3. Willis, K.J. and McElwain, J.G., *The Evolution of Plants,* Oxford University Press, Oxford, 2002.
4. APG, An ordinal classification for the families of flowering plants, *Ann. Mo. Bot. Gard.,* 85, 531, 1998.
5. Chase, M.W. and Albert, V.A., A perspective on the contribution of plastid *rbc*L DNA sequences to angiosperm phylogenetics, in *Molecular Systematics of Plants II: DNA Sequences,* Soltis, D.E., Soltis, P.S. and Doyle, J.A., Eds., Kluwer Academic Publishers, Boston, 1998, 488.
6. Niklas, K.J., Tiffney, B.H., and Knoll, A.H., Patterns in vascular land plant diversification, *Nature,* 303, 614, 1983.
7. Magallón, S. and Sanderson, M.J., Absolute diversification rates in angiosperm clades, *Evolution,* 55, 1762, 2001.
8. APG II, An update of the Angiosperm Phylogeny Group classification for the orders and families of flowering plants: APG II, *Bot. J. Linn. Soc.,* 141, 399, 2003.
9. Sanderson, M.J. and Donoghue, M.J., Shifts in diversification rate with the origin of angiosperms, *Science,* 264, 1590 1994.
10. Crane, P.R. and Lidgard, S., Angiosperm diversification and paleolatitudinal gradients in cretaceous floristic diversity, *Science,* 246, 675, 1989.
11. Gaston, K.J. and Williams, P.H., Spatial patterns in taxonomic diversity, in *Biodiversity: a Biology of Numbers and Differences,* Gaston, K.J., Ed., Blackwell Science, Oxford, 1996, 202.
12. Myers, N. et al., Biodiversity hotspots for conservation priorities, *Nature,* 403, 853, 2000.
13. MacArthur, R., On the relative abundance of bird species, *Proc. Natl. Acad. Sci. USA,* 43, 293, 1957.
14. Dial, K.P. and Marzluff, J.M., Nonrandom diversification within taxonomic assemblages, *Syst. Zool.,* 38, 26, 1989.
15. Scotland, R.A. and Sanderson, M.J., The significance of few versus many in the tree of life, *Science,* 303, 643, 2004.
16. Slowinski, J.B. and Guyer, C., Testing the stochasticity of patterns of organismal diversity: an improved null model, *Am. Nat.,* 134, 907, 1989.
17. Purvis, A., Using Interspecies phylogenies to test macroevolutionary hypotheses, in *New Uses for New Phylogenies,* Harvey, P.H. et al., Eds., Oxford University Press, Oxford, 1996, 153.
18. Mooers, A.Ø. and Heard, S.B., Inferring evolutionary process from phylogenetic tree shape, *Q. Rev. Biol.,* 72, 31, 1997.
19. Mooers, A.Ø. and Heard, S.B., Using tree shape, *Syst. Biol.,* 51, 833, 2002.
20. Slowinski, J.B. and Guyer, C., Testing whether certain traits have caused amplified diversification: an improved method based on a model of random speciation and extinction, *Am. Nat.,* 142, 1019, 1993.
21. Nee, S., Mooers, A.Ø., and Harvey, P.H., Tempo and mode of evolution revealed from molecular phylogenies, *Proc. Natl. Acad. Sci. USA,* 89, 8322, 1992.
22. Harvey, P.H. and Pagel, M.D., *The Comparative Method in Evolutionary Biology,* Oxford University Press, Oxford, 1991.
23. Symonds, M.R.E., Life histories of the Insectivora: the role of phylogeny, metabolism and sex differences, *J. Zool. Soc. Lond.,* 249, 315, 1999.
24. Symonds, M.R.E., The effect of topological inaccuracy in evolutionary trees on the phylogenetic comparative method of independent contrasts, *Syst. Biol.,* 51, 541, 2002.
25. Doyle, J.A. and Donoghue, M.J., Phylogenies and angiosperm diversification, *Paleobiology,* 19, 141, 1993.
26. Friedman, W.E. and Floyd, S.K., Perspective: the origin of flowering plants and their reproductive biology — a tale of two phylogenies, *Evolution,* 55, 217, 2001.

27. Soltis, D.E. et al., Inferring complex phylogenies using parsimony: an empirical approach using three large DNA data sets for angiosperms, *Syst. Biol.,* 47, 32, 1998.
28. Savolainen, V. and Chase, M.W., A decade of progress in plant molecular phylogenetics, *Trends Genet.,* 19, 717, 2003.
29. Davies, T.J. et al., Darwin's abominable mystery: insights from a supertree of the angiosperms, *Proc. Natl. Acad. Sci. USA,* 101, 1904, 2004.
30. Pisani, D. and Wilkinson, M., Matrix representation with parsimony, taxonomic congruence, and total evidence, *Syst. Biol.,* 51, 151, 2002.
31. Gatesy, J. et al., Resolution of a supertree/supermatrix paradox, *Syst. Biol.,* 51, 652, 2002.
32. Bininda-Emonds, O.R.P. et al., Supertrees are a necessary not-so-evil: a response to Gatesy et al., *Syst. Biol.,* 52, 724, 2003.
33. Bininda-Emonds, O.R.P. and Bryant, H.N., Properties of matrix representation with parsimony analyses, *Syst. Biol.,* 47, 497, 1998.
34. Bininda-Emonds, O.R.P. and Sanderson, M.J., Assessment of the accuracy of matrix representation with parsimony analysis supertree construction, *Syst. Biol.,* 50, 565, 2001.
35. Salamin, S., Hodkinson, T.R., and Savolainen, V., Building supertrees: an empirical assessment using the grass family (Poaceae), *Syst. Biol.,* 51, 136, 2002.
36. Fusco, G. and Cronk, Q.C.B., A new method for evaluating the shape of large phylogenies, *J. Theor. Biol.,* 175, 235, 1995.
37. Purvis, A., Katzourakis, A., and Agapow, P.M., Evaluating phylogenetic tree shape: two modifications to Fusco & Cronk's method, *J. Theor. Biol.,* 214, 99, 2001.
38. Wilkinson, M. et al., The shape of supertrees to come: tree shape related properties of fourteen supertree methods, *Syst. Biol.,* 54, 419, 2005.
39. Guyer, C. and Slowinski, J.B., Comparisons of observed phylogenetic topologies with null expectations among three monophyletic lineages, *Evolution,* 45, 340, 1991.
40. Eriksson, O. and Bremer, B., Pollination systems dispersal modes, life forms, and diversification rates in angiosperm families, *Evolution,* 46, 258, 1992.
41. Stanley, S.M., *Macroevolution: Pattern and Process,* Freeman, San Francisco, 1979.
42. van Valen, L., Adaptive zones and the orders of mammals, *Evolution,* 25, 420, 1971.
43. Salamin, N. and Davies, T.J., Using supertrees to investigate species richness in grasses and flowering plants, in *Phylogenetic Supertrees: Combining Information to Reveal the Tree of Life, Computational Biology Vol. 4,* Bininda-Emonds, O.R.P., Ed., Kluwer Academic, Dordrecht, 2004, 461.
44. Eriksson, O. and Bremer, B., Pollination systems, dispersal modes, life forms, and diversification rates in angiosperm families, *Evolution,* 46, 258, 1992.
45. Gaut, B.S. et al., Substitution rate comparisons between grasses and palms: synonymous rate differences at the nuclear gene *Adh* parallel rate differences at the plastid gene *rbcL, Proc. Natl. Acad. Sci. USA,* 93, 10274, 1996.
46. Gaut, B.S. et al., Relative rates of nucleotide substitution at the *rbcL* locus of monocotyledonous plants, *J. Mol. Evol.,* 35, 292, 1992.
47. Dodd, M.E., Silvertown, J., and Chase, M.W., Phylogenetic analysis of trait evolution and species diversity variation among angiosperm families, *Evolution,* 53, 732, 1999.
48. Ricklefs, R.E. and Renner, S.S., Species richness within families of flowering plants, *Evolution,* 48, 1619, 1994.
49. Bawa, K.S. and Opler, P.A., Dioecism in tropical forest trees, *Evolution,* 29, 167, 1975.
50. Bawa, K.S., Pollinators of tropical dioecious angiosperms: a reassessment? No, not yet, *Am. J. Bot.,* 81, 456, 1994.
51. Rieseberg, L.H., Hybrid origins of plant species, *Annu. Rev. Ecol. Syst.,* 28, 359, 1997.
52. Baker, H.G., Self-compatability and establishment after 'long-distance' dispersal, *Evolution,* 9, 347, 1955.
53. Felsenstein, J., Phylogenies and the comparative method, *Am. Nat.,* 125, 1, 1985.
54. Barraclough, T.G., Nee, S., and Harvey, P.H., Sister-group analysis in identifying correlates of diversification: comment, *Evol. Ecol.,* 12, 751, 1998.
55. Gorelick, R., Did insect pollination cause increased seed plant diversity? *Biol. J. Linn. Soc.,* 74, 407, 2001.
56. Farrell, B.D., Dussourd, D.E., and Mitter, C., Escalation of plant defence — do latex and resin canals spur plant diversification? *Am. Nat.,* 138, 881, 1991.

57. Hodges, S.A. and Arnold, M.L., Spurring plant diversification: are floral nectar spurs a key innovation? *Proc. R. Soc. Lond. B,* 262, 343, 1995.
58. Gianoli, E., Evolution of a climbing habit promotes diversification in flowering plants, *Proc. R. Soc. Lond. B,* 271, 2011, 2004.
59. Smith, J.F., High species diversity in fleshy-fruited tropical understory plants, *Am. Nat.,* 157, 646, 2001.
60. Sargent, R.D., Floral symmetry affects speciation rates in angiosperms, *Proc. R. Soc. Lond. B,* 271, 603, 2003.
61. de Queiroz, A., Contingent predictability in evolution: key traits and diversification, *Syst. Biol.,* 51, 917, 2002.
62. Ree, R.H., Detecting the historical signature of key innovations using stochastic models of character evolution and cladogenesis, *Evolution,* 59, 257, 2005.
63. Von Hagen, K.B. and Kadereit, J.W., The diversification of *Halenia* (Gentianaceae): ecological opportunity versus key innovation, *Evolution,* 57, 2507, 2003.
64. Sims, H.J. and McConway, K.J., Nonstochastic variation of species-level diversification rates within angiosperms, *Evolution,* 57, 460, 2003.
65. Axelrod, D.I., A theory of angiosperm evolution, *Evolution,* 6, 29, 1952.
66. Chapman, G.P., *The Biology of Grasses,* CAB International, Oxon, 1996.
67. Crane, P.R., Friis, E.M., and Pedersen, K.J., The origin and early diversification of angiosperms, *Nature,* 374, 27, 1995.
68. Hillebrand, H., On the generality of the latitudinal diversity gradient, *Am. Nat.,* 163, 192, 2004.
69. Willig, M.R., Kaufman, D.M., and Stevens, R.D., Latitudinal gradients of biodiversity: pattern, process, scale, and synthesis, *Annu. Rev. Ecol. Evol. Syst.,* 34, 273, 2003.
70. Currie, D.J. et al., Predictions and tests of climate-based hypotheses of broad-scale variation in taxonomic richness, *Ecol. Lett.,* 7, 1121, 2004.
71. Wright, D.H., Species-energy theory: an extension of species-area theory, *Oikos,* 41, 496, 1983.
72. Rohde, K., Latitudinal gradients in species-diversity: the search for the primary cause, *Oikos,* 65, 514, 1992.
73. Allen, A.P., Brown, J.H., and Gillooly, J.F., Global biodiversity, biochemical kinetics, and the energetic-equivalence rule, *Science,* 297, 1545, 2002.
74. Turner, J.R.G., Lennon, J.J., and Lawrenson, J.A., British bird species distributions and the energy theory, *Nature,* 335, 539, 1988.
75. Currie, D.J., Energy and large-scale patterns of animal-species and plant-species richness, *Am. Nat.,* 137, 27, 1991.
76. Wright, D.H., Currie, D.J., and Maurer, B.A., Energy supply and patterns of species richness on local and regional scales, in *Species Diversity in Ecological Communities,* Ricklefs, R.E. and Schluter, D.S., Eds., Chicago Press, Chicago, 1993, 66.
77. Roy, K. et al., Marine latitudinal diversity gradients: tests of causal hypotheses, *Proc. Natl. Acad. Sci. USA,* 95, 3699, 1998.
78. Francis, A.P. and Currie, D.J., A globally consistent richness-climate relationship for angiosperms, *Am. Nat.,* 161, 523, 2003.
79. Wylie, J.L. and Currie, D.J., Species-energy theory and patterns of species richness: I. Patterns of bird, angiosperm, and mammal species richness on islands, *Biol. Cons.,* 63, 137, 1993.
80. Hutchinson, G.E., Homage to Santa Rosalina or why are there so many kinds of animals? *Am. Nat.,* 93, 145, 1959.
81. Pianka, E.R., Latitudinal gradients in species diversity: a review of the concepts, *Am. Nat.,* 100, 33, 1966.
82. Gaston, K.J., Global patterns in biodiversity, *Nature,* 405, 220, 2000.
83. Willis, K.J. and Whittaker, R.J., Species diversity: scale matters, *Science,* 295, 1245, 2002.
84. Stehli, F.G., Douglas, R.D., and Newell, N.D., Generation and maintenance of gradients in taxonomic diversity, *Science,* 164, 947, 1969.
85. Jablonski, D., The tropics as a source of evolutionary novelty through geological time, *Nature,* 364, 142, 1993.
86. Cardillo, M., Latitude and rates of diversification in birds and butterflies, *Proc. R. Soc. Lond. B,* 266, 1221, 1999.
87. Buzas, M.A., Collins, L.S., and Culver, S.J., Latitudinal difference in biodiversity caused by higher tropical rate of increase, *Proc. Natl. Acad. Sci. USA,* 99, 7841, 2002.

88. Barraclough, T.G. and Savolainen, V., Evolutionary rates and species diversity in flowering plants, *Evolution,* 55, 677, 2001.
89. Webster, A.J., Payne, R.J.H., and Pagel, M., Molecular phylogenies link rates of evolution and speciation, *Science,* 301, 478, 2003.
90. Bromham, L. and Cardillo, M., Testing the link between the latitudinal gradient in species richness and rates of molecular evolution, *J. Evol. Biol.,* 16, 200, 2003.
91. Davies, T.J. et al., Environmental energy and evolutionary rates in flowering plants, *Proc. R. Soc. Lond. B,* 271, 2195, 2004.
92. Rothschild, L.J., The influence of UV radiation on protistan evolution, *J. Eukaryotic Microbiol.,* 46, 548, 1999.
93. Currie, D.J. and Paquin, V., Large-scale biogeographical patterns of species richness of trees, *Nature,* 329, 326, 1987.
94. O'Brien, E.M., Climatic gradients in woody plant species richness: towards an explanation based on an analysis of southern Africa's woody flora, *J. Biogeog.,* 20, 181, 1993.
95. Dynesius, M. and Jansson, R., Evolutionary consequences of changes in species' geographical distributions driven by Milankovitch climate oscillations, *Proc. Natl. Acad. Sci. USA,* 97, 9115, 2000.
96. Qian, H. and Ricklefs, R.E., Geographical distribution and ecological conservatism of disjunct genera of vascular plants in eastern Asia and eastern North America, *J. Ecol.,* 92, 253, 2004.
97. Huntley, B. and Webb, T.I., Migration: species' response to climate variations caused by changes in the Earth's orbit, *J. Biogeog.,* 16, 5, 1989.
98. Currie, D.J. and Fritz, J.T., Global patterns of animal abundance and species energy use, *Oikos,* 67, 56, 1993.
99. Kaspari, M., O'Donnell, S., and Kercher, J.R., Energy, density, and constraints to species richness: ant assemblages along a productivity gradient, *Am. Nat.,* 155, 280, 2000.
100. Enquist, B.J. and Niklas, K.J., Invariant scaling relations across tree-dominated communities, *Nature,* 410, 655, 2001.
101. Bonn, A., Storch, D., and Gaston, K., Structure of the species-energy relationship, *Proc. R. Soc. Lond. B,* 271, 1685, 2004.
102. Cowling, R.M., Holmes, P.M., and Rebelo, A.G., Plant diversity and endemism, in *The Ecology of Fynbos,* Cowling, R.M., Ed., Oxford University Press, Cape Town, 1992, 62.
103. Simmons, M.T. and Cowling, R.M., Why is the Cape Peninsular so rich in plant species? An analysis of the independent diversity components, *Biodiv. Cons.,* 5, 551, 1996.
104. Richardson, J.E. et al., Rapid and recent origin of species richness in the Cape flora of South Africa, *Nature,* 412, 181, 2001.
105. Goldblatt, P. and Manning, J.C., Plant diversity of the Cape region of southern Africa, *Ann. Mo. Bot. Gard.,* 89, 281, 2002.
106. Linder, H.P., Radiation of the Cape flora, southern Africa, *Biol. Rev.,* 78, 597, 2003.
107. Goldblatt, P., Phylogeny and classification of the Iridaceae and the relationship of *Iris, Annali di Botanica, Nouva Serie,* 1, 13, 2001.
108. Goldblatt, P. et al., Radiation in the Cape flora and the phylogeny of peacock irises *Moraea* (Iridaceae) based on four plastid DNA regions, *Mol. Phylog. Evol.,* 25, 341, 2002.
109. Goldblatt, P., Phylogeny and classification of Iridaceae, *Ann. Mo. Bot. Gard.,* 77, 607, 1990.
110. Davies, T.J. et al., Environmental causes for plant biodiversity gradients, *Phil. Trans R. Soc. Lond. B,* 359, 1645, 2004.
111. Davies, T.J. et al., Environment, area and diversification in the species-rich flowering plant family Iridaceae, *Am. Nat.,* 166, 1537, 2005.
112. Klak, C., Reeves, G., and Hedderson, T., Unmatched tempo of evolution in Southern African semi-desert ice plants, *Nature,* 427, 63, 2004.
113. Goldblatt, P., An overview of the systematics, phylogeny and biology of the African Iridaceae, *Contributions from the Bolus Herbarium,* 13, 1, 1991.
114. Bernhardt, P. and Goldblatt, P., The diversity of pollination mechanisms in the Iridaceae of southern Africa, in *Monocots: Systematics and Evolution,* Wilson, K.L. and Morrison, D.A., Eds., CSIRO, Melbourne, 2000, 301.

# 11 Skewed Distribution of Species Number in Grass Genera: Is It a Taxonomic Artefact?

_K. W. Hilu_

Department of Biological Sciences, Virginia Tech, Blacksburg, USA

## CONTENTS

## ABSTRACT

The grass family (Poaceae) comprises about 10,000 species distributed in some 785 genera, seven large subfamilies and a few small ones. The distribution of species in genera appears skewed toward monotypic genera and those with few species. This pattern follows the hollow curve distribution documented by Willis[1]. Explanations of the pattern have been attributed to statistical, biological and taxonomic factors. This study explores potential biological and statistical explanations for species distribution in Poaceae. Patterns of species distribution in the family and its major subfamilies were investigated, and the influence of age, habit and habitat on these patterns was assessed. Results showed that species distribution is not only skewed for the number of small genera but also for the total number of species in larger genera. Phylogenetic position does not appear to explain species distribution in the family and in fact refutes the age and area theory proposed by Willis. Genus size appears to be correlated with habit where larger genera are predominantly perennial. Genera with mixed annual and perennial species do not reflect the hollow curve pattern. These patterns of species distribution may be explained by polyploidy and hybridisation, two prominent features in the evolution of the family.

## 11.1 INTRODUCTION

A striking pattern for distribution of taxa in their respective higher categories points to a skewed distribution towards monotypic and small groups. This phenomenon was first documented by Willis[1] and Willis and Yule[2] in a study of the flora of Ceylon (Sri Lanka). They dubbed this pattern the hollow curve distribution (HCD) and indicated that such a pattern exists at all taxonomic levels. Willis and Yule[2] asserted that the longer the group has existed, the more area it will occupy. They further stated that monotypic genera are in general 'beginners' and are descendents of larger ones. The HCD was later demonstrated in other organisms, such as arthropods, birds and mammals[3–7]. Although this skewed pattern is evident across a broad range of biological diversity and at all taxonomic levels, explanations of its causes vary, and different hypotheses and models have been proposed (see Hodkinson and Parnell, *Chapter 1*; Davies and Barraclough, *Chapter 10*; Parnell et al., *Chapter 16*).

Willis[1] cited biological, historical, mathematical, psychological and statistical elements as potential causes of the HCD. Willis and Yule[2] stressed age and area as the principal factors behind the biological patterns of diversification and the emergence of the HCD. Dial and Marzluff[5] indicated that early authors favoured deterministic explanations, whereas more recent authors incline toward stochastic models. In a study aimed at determining patterns and causes of species diversification, Dial and Marzluff[5] compared species distributions in 85 taxonomic units from six groups of animals and one group of plants to those predicted by five null models. Their sample comprised 53 taxonomic assemblages based on traditional classifications and 32 based on phylogenetic schemes. They found that real assemblages were dominated to a significantly higher extent by one unit than predicted by all five models. They also noted that the pattern is evident in both traditional and phylogenetic schemes and concluded that such skewed distributions reflect real differences in the evolutionary successes of the groups. Dial and Marzluff concluded that overdominance of an assemblage by one unit is a common and nonrandom feature of taxonomic diversity distribution and proposed that such a pattern might be the consequence of differences in life history traits such as fecundity, age of first reproduction, longevity and mobility. Cardillo et al.[6] studied the pattern of diversification in 76 genera (210 species) of Australian mammals and contrasted the observed distribution with the Poisson and geometric models. They observed that species distribution based on real data is significantly different from those predicted by Poisson and geometric null distribution, with the observed distribution having more species poor and species rich genera than predicted by the models. Scotland and Sanderson[7] tested the HCD in birds and the three angiosperm families Fabaceae (legumes), Orchidaceae (orchids) and Asteraceae (asters or daisies). They compared their new model, simultaneous broken tree (SBT), with distributions based on real data, the simultaneous broken stick (SBS) and the geometric distribution. Their study showed that the SBT model overestimated the monotypes and dominance (large genera), whereas the SBS underestimated them. Consequently, they suggested that lack of fit between real data and the SBT model is taxonomic and not evolutionary, contending that taxonomists are averse to studying genera that are too large or too small.

In this study, the grass family (Poaceae) is chosen to assess the potential influence of some biological traits on species distribution in genera. Biological traits examined here are habit, eco-geographic preferences, habitat and polyploidy. Genus size is also considered in a phylogenetic context using a consensus tree for the grass phylogeny. The grass family is chosen because of its large size (approximately 10,000 species and 785 genera), wide distribution over diverse habitats, variation in habit that provides a sizeable sample and wealth of data on chromosome number and polyploidy. This study is based on real (observed) data and does not include null model assessments.

## 11.2 THE GRASS FAMILY (POACEAE)

Poaceae is the fifth most species rich flowering plant family. It spans the globe in distribution and uniquely forms large stretches of grass-dominated communities, the grasslands. Grasses are found at almost all altitudes and latitudes that allow plant life to exist, and its species are among the pioneers in primary and secondary succession communities. Various species and lineages of grasses have evolved ecophysiological, anatomical and morphological adaptations that have allowed them to flourish and radiate in a variety of habitats ranging from humid tropics to seasonal tropics to temperate regions and from xeric to aquatic[8–10]. The major grass subfamilies also tend to display geographic/ecological preferences; for instance, Bambusoideae are found predominantly in tropical regions, Pooideae generally occupy temperate areas, Chloridoideae tend to flourish in drier habitats and Ehrhartoideae (Oryzoideae) exist in wet environments[11]. Certain grass lineages, such as pooids and bambusoids, employ a $C_3$ type of photosynthetic system, whereas others, such as Chloridoideae, have $C_4$ types. Grass species range in life span from annuals to biennials and from short-lived perennials to long-lived perennials. In the latter type, some woody bamboo plants may live for over 100 years before they flower, set seed and die.

Grass reproduction is intriguing. The flowers lack a showy perianth and are reduced to mostly three or six stamens and an ovary enclosed in bracts, features that promote wind pollination. They have adopted various forms of sexual reproduction, like outcrossing, inbreeding and cleistogamy, and added asexual means of reproduction such as apomixis, vivipary and vegetative propagation. Hybridisation and polyploidy are common in grasses[12]. These features are evident in the estimated 80% polyploidy and the predominance of allopolyploidy[13,14]. Basic chromosome numbers in grasses[13–15] are remarkably variable, covering a range of $x = 2$–14 and 18, having somatic numbers[13] that range between $2n = 4$ and $2n = 263$–265, and with a 2C DNA content that varies from 0.7 in *Chloris gayana* Kunth to 27.6 in *Lygeum spartum* L.[16], an impressive 40-fold range. Chromosome number and structure and DNA content have undertaken a number of evolutionary pathways that include numerous reversals[15].

To place the pattern of variation in genus size in a phylogenetic context, a robust phylogeny for the grass family is required. Our understanding of grass systematics has recently made major leaps with the application of molecular approaches to the family. General consensus exists on the delimitation of the major grass lineages and the patterns of their divergence (see GPWG[17]; Hilu[15]; and Hodkinson et al., *Chapter 17*). These phylogenetic hypotheses are based on sequence information from different genomic regions and sometimes incorporate structural characters. The consensus grass phylogenetic tree depicts a basal grade of *Anomochloa, Streptochaeta,* Pharoideae (*Pharus*) and Puelioideae (*Puelia* and *Guaduella*)[17,18]. Following this grade, the remaining nine recognised subfamilies fall into the strongly supported PACCAD (Panicoideae, Arundinoideae, Chloridoideae, Centothecoideae, Aristidoideae and Danthonioideae) and the less substantiated BEP (Bambusoideae, Ehrhartoideae and Pooideae) clades (reviewed in Hilu[15]). Therefore, our deep understanding of grass phylogenetics and systematics can provide a reliable guideline for evolutionary trends at various taxonomic levels, including patterns of variation in genus size.

## 11.3  MATERIALS AND METHODS

Information on genus size, habit, habitats, geographic distribution and chromosome numbers was obtained from two major publications on the grasses: Genera Graminum[11] and The Grass Genera of the World[19]. In total, 651 genera and 9,654 species were assessed from all grass subfamilies and tribes recognised in these publications. Sources for additional information are referenced in the text. The Poaceae consensus tree is derived from two recent comprehensive studies on grass phylogeny[17,18]. Statistical analysis of means and Chi square test for frequencies were conducted in JMP version 3[20].

## 11.4  RESULTS AND DISCUSSION

Considering the whole grass family, it is evident that species distribution is strongly skewed towards monotypic or small genera (Figure 11.1A), following the HCD depicted by Willis[1] and also discussed by Clayton and Renvoize[11] and references therein. In Poaceae, 35% (230) of the genera are monotypic, and those with one or two species represent almost half (315) of the total number of genera. In fact, 78% (506) of the grass genera are relatively small, containing 10 or less species (Table 11.1). Due to their small size and large proportion, these genera encompass only 15% (1,416) of the grass species. In contrast, only 3% (22) of the grass genera contain 100 or more species. Those 3%, however, encompass an astonishing 50% (4,820) of the species in the family. Therefore, although

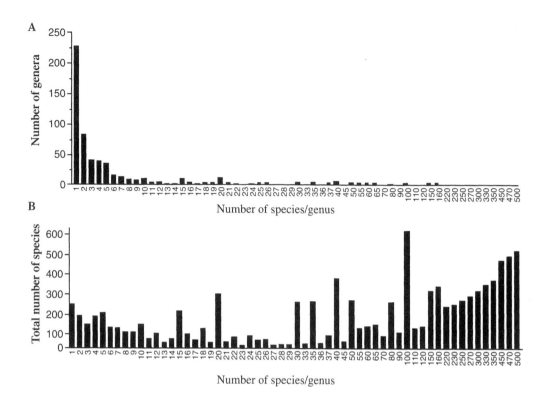

**FIGURE 11.1** The distribution of genus size in the grass family. (A) The pattern follows the HCD with predominance of monotypic and small genera. (B) Species concentration (dominance) occurs at the opposite end of the curve. (Data from Clayton and Renvoize[11] and Watson and Dallwitz[19].)

the distribution curve is strongly skewed toward small genera, the overwhelming majority of species (dominance) is found at the other end of the curve in large genera (Figure 11.1B). Excluding the two ends of the spectrum in terms of genus size and species distribution leaves genera with 11–99 species that make up 19% of the genera (123) with 35% (3,418) of the species, and an average of

**TABLE 11.1**
**Distribution of Species in Grass Genera**

|  | Number of Genera | Percentage of Genera | Number of Species | Percentage of Species |
|---|---|---|---|---|
| Poaceae family | 651 | not applicable | 9,654 | not applicable |
| Monotypic genera | 230 | 35 | 230 | 2 |
| Genera of 1–10 sp. | 506 | 78 | 1,416 | 15 |
| Genera of 100 and more sp. | 22 | 3 | 4,820 | 50 |
| Genera of 11–99 sp. | 123 | 19 | 3,418 | 35 |
| Annuals | 134 | 21 | 384 | 4 |
| Perennials | 371 | 57 | 3,824 | 40 |
| Mixed genera | 146 | 22 | 5,446 | 56 |

*Note:* Statistics calculated for the whole family, annuals, perennials and genera with mixed annual and perennial species (noted as mixed genera). Also noted is the distribution of species in genera as grouped into three arbitrary categories: 1–10 species, 11–99 species and 100 and more species. Data obtained from Hilu et al.[18] and Watson and Dallwitz[20].

28 species per genus. The species distribution in this group is in sharp contrast with the two extremes, where an average of 2.8 species per genus for genera containing 10 or less species and 219 species per genus for genera with 100 or more species is found (Table 11.1).

### 11.4.1  AGE AND DIVERSIFICATION

Although the skewed distribution towards monotypic or very small genera is evident in Poaceae and elsewhere, explanation of its causes has varied. Willis and Yule[2] proposed the concept of age and geographic distribution to explain the HCD originally documented by Willis[1]. They asserted that large genus size is a manifestation of age and wide distribution and that those genera have given rise to small ones with restricted distribution. Considering this scenario, one may expect genera at the base of the grass family tree (taxa sister to the rest of the grasses), as well as those at the base of major lineages (PACCAD) and the individual subfamilies, to be large in size while genera in terminal lineages would be small in size. However, this is not evident among extant grasses (Figure 11.2); the four basal lineages in the grass phylogeny are represented by monotypic to very small genera: *Streptochaeta* (two to three), *Anomochloa* (one), *Pharus* (five to six), *Puelia* (six), and *Guaduella* (eight). The early diverging lineages are not the ancestors of the other grasses; it is not known how diverse the stem lineages (ancestors) leading to the split of these basal lineages and their sister group were. However, it is clear that the early diverging lineages are not diverse in all cases examined here.

When geographic distribution is considered, *Streptochaeta* is restricted to shady places from Mexico to Argentina, *Anomochloa* to the forest of tropical America, *Pharus* to shady places of tropical America and the West Indies, *Puelia* to shady places of Sierra Leone and Angola and *Guaduella* to tropical rainforest of Africa. The pattern of small basal taxa is also evident in individual major grass lineages. *Micraira* appears as the sister genus to the large PACCAD clade (not shown);

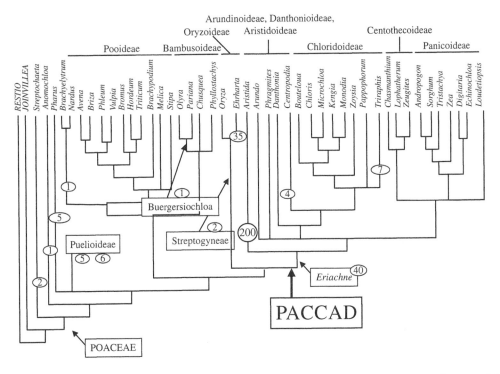

**FIGURE 11.2** A consensus tree for the grass family on which genus size is mapped for basal lineages of the grass family and its major subfamilies. (The consensus tree is based on trees obtained from GPWG[17] and Hilu et al.[18]; information on genus size is from Clayton and Renvoize[11] and Watson and Dallwitz[19].)

the genus comprises eight species confined to Australia (not shown). Considering the base of individual subfamilies for instance (Figure 11.2), in Pooideae, both *Brachyelytrum* (shady places of woodlands in North America, Japan and Korea) and *Nardus* (Europe and Western Asia) are monotypic; in Chloridoideae, *Triraphis* includes seven species found in Africa and Arabia and one species in Australia; and in Ehrhartoideae, *Streptogyna* contains only two species distributed in the forest shade from West Africa to Southern India and Sri Lanka, and Mexico to Brazil. Therefore, genus size at the base of individual subfamilies is small, but geographic distribution varies. In contrast, a wide range of genus size exists at the terminal branches that include monotypics and large genera. Thus, the size of early diverging genera in the family and the subfamilies is small rather than large. The small size and mostly restricted distribution may either represent the prehistorical geographic pattern or could be the outcome of species extinction and endemism. Cardillo et al.[6] indicated that diversification rate is an outcome of a differential rate of speciation and extinction. In conclusion, age alone cannot be used to explain genus size or the predominance of small-sized genera.

### 11.4.2 Habitat and Ecophysiologically Related Traits

Next to be examined is the potential impact of ecophysiological factors and habitat on genus size. To address the potential underlying impact of these traits on species distribution, three subfamilies with different preferences have been chosen: Bambusoideae with tropical and forest habitats and $C_3$ photosynthesis; Chloridoideae with drier habitats and $C_4$ photosynthesis; and Ehrhartoideae with primarily aquatic or wet habitats and $C_3$ photosynthesis. Deviation from the HCD would be construed as a potential influence of one or more of these variables on mode of diversification.

Comparing species distribution in these three subfamilies, skewed distribution towards small genera (HCD) is evident (Figure 11.3; Ehrhartoideae is not shown). Similarly, concentration of the majority of the species in a few large genera (dominance) matches what has been documented for Poaceae as a whole (distributions not shown). Consequently, these results do not point to differences in habitats as potential factors that impact species distribution. Bambusoideae is of special interest

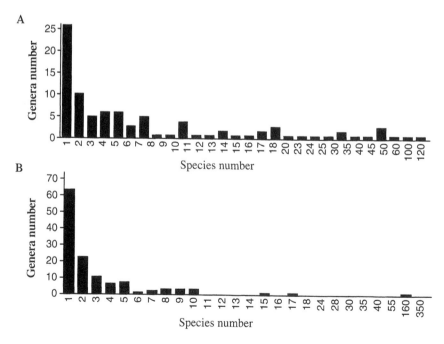

**FIGURE 11.3** Genus size distribution in Bambusoideae (A) and Chloridoideae (B). Genus size distribution follows the HCD. The two subfamilies differ in their ecophysiological preferences and photosynthetic systems.

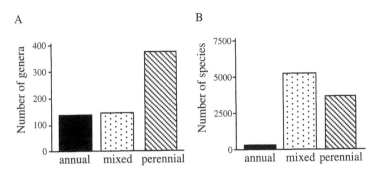

**FIGURE 11.4** The distribution of genera and species in annuals, perennials and mixed genera (genera with both types of life histories). (A) The distribution of genera. (B) The distribution of species. (Information on genus size is from Clayton and Renvoize[11] and Watson and Dallwitz[19].)

here as, due to infrequent flowering, taxonomic decisions are in some cases based on vegetative characters, a situation that may favour a trend toward splitting but not lumping of species in a genus. Consequently, one would expect species distribution to be skewed away from monotypic or very small genera and to favour relatively larger ones; this assumption is based on potentially higher variation amongst populations in vegetative characters relative to reproductive traits. In this case, taxonomic factors would become more pronounced in assessments of species distribution. This, obviously, does not seem to be the case (Figure 11.3A).

Thus, these contrasting subfamilies all show the HCD despite having differing geographic distribution, photosynthetic pathways and other adaptive traits correlating to ecological factors.

### 11.4.3 Impact of Life History on Genus Size

The next question to ask is whether life history has an impact on genus size across the Poaceae. Different life forms exist in the grass family as a response to adaptation to scores of environmental conditions. This chapter focuses on plant longevity and groups species into annuals and perennials. Perennialism in the grass family has a very wide scale, as it can span a life history period from a few years to over 100 years. To address life history as a factor, grass genera are labeled as annuals, perennials or mixed. In the latter case, a genus is labeled as such when it encompasses both annual and perennial species regardless of their proportions.

The number of genera composed of perennials in Poaceae is almost three times that of the number of genera composed of annuals (371 versus 134), leaving 146 genera with a mixture of annual and perennial species (Figure 11.4A, Table 11.1). The difference between genera composed of annual species and those composed of perennial species becomes more acute when the number of species in each group is considered. Genera composed of perennials contain 10 times as many species as genera composed of annuals (3,824 versus 384). The number of species (5,446) in mixed genera is high (Figure 11.4B, Table 11.1). On average, the number of species in genera composed of annuals is 2.9, in genera composed of perennials 10.3, and in mixed genera 37.3. These differences are statistically significant at the 5% level ($\chi^2 = 39$, $p < 0.05$).

More striking differences in genus size in relation to different life histories can be seen in the plots of the distribution of species across these three classes (Figure 11.5 and Figure 11.6). Plotting the number of genera against genus size, the HCD was evident for perennial genera (Figure 11.5A), paralleling the pattern observed for the whole grass family and for individual subfamilies (Figure 11.1–11.3). Similar patterns are also observed in the annuals (Figure 11.5B). However, when the mixed genera were examined, the pattern of distribution of genus size failed to fit the HCD (Figure 11.6A). For instance, the number of genera with five species is larger than the number of monotypic genera, and genera with four species are as frequent as monotypic genera. This pattern

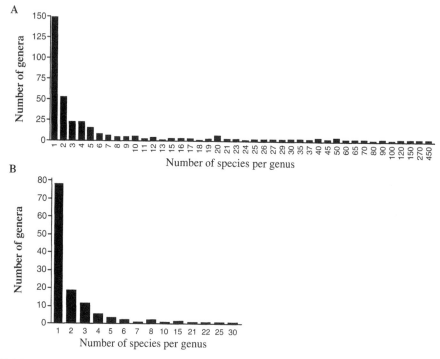

**FIGURE 11.5** The distribution of genus size in perennial (A) and annual (B) genera. Genus size distribution follows the HCD. (Information on genus size is from Clayton and Renvoize[11] and Watson and Dallwitz[19].)

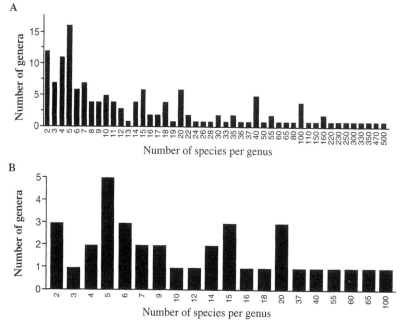

**FIGURE 11.6** Deviations from the HCD. Genera containing both annual and perennial species deviate from the HCD at the whole grass family level (A) and in individual tribes such as Andropogoneae (B). A similar pattern is found in other grass tribes such as the Paniceae and Poeae/Aveneae. (Information on genus size is from Clayton and Renvoize[11] and Watson and Dallwitz[19].)

of distribution was also apparent in mixed genera of the tribes Andropogoneae (Figure 11.6B), Paniceae and Poeae/Aveneae (not shown).

The small number of annual genera and their small average size and the substantially larger number of perennial genera and their larger size in grasses are unexpected findings at first glance, as an annual but not a perennial habit is generally considered to favour higher diversification (see Hodkinson et al., *Chapter 17*). Annuals reach flowering stage in the same season, and variation in length of vegetative (juvenile) period is a matter of days. In contrast, perennials go through juvenile vegetative stages which may extend from months to years[21]. Furthermore, Harper[21] noted that in perennial herbs, such as grasses, years of flowering and seed production are often interrupted by years of purely vegetative growth. Thus regeneration of offspring in perennials, or at least herbaceous ones, is at a lower rate than with annuals. This gives annuals a definite advantage over perennials in terms of rapid generation turnover, increased variability due to more frequent sexual recombination, and potential for higher rate of fixation of adaptive mutations. Considering this scenario, annuals are expected to have better options for enhanced speciation compared with perennials. It has been shown that fast life history is conducive to higher diversification and may increase probability of speciation[6,22,23]. Evidently this does not seem to be the case in grasses, as annual genera are by far less species rich than perennial ones. The causes of this pattern of diversification require explanation.

Considering these observed patterns of species distribution in relation to habit, three questions can be posed:

- Why do annual genera tend to be smaller in size, whereas perennial genera are more species rich?
- Why does the HCD theory break down in mixed annual/perennial genera?
- What causes genera with mixed species to be larger in size on the average than either annual or perennial genera?

### 11.4.4 Polyploidy and Diversification in Poaceae

To partly address the above three questions, two of the prominent features of Poaceae, namely polyploidy and hybridisation, were examined to see if they could provide possible answers. Polyploidy and hybridisation are considered as two important biological forces in grass diversity and evolution[13–15,24]. Estimates of polyploidy in grasses are as high as 80%, basic chromosome numbers[24] range from x = 2 to 18, and somatic numbers vary from 2n = 6 to 265. Hybridisation is common in Poaceae, and wide crosses in nature and under artificial environments are well known[13,14,19,24]. Polyploidy often confers immediate reproductive isolation from parental species and may lead to speciation. This isolation can be achieved whether polyploidy has arisen via hybridisation and chromosome doubling (allopolyploidy) or through simple doubling of the chromosome complement of an individual plant (autopolyploidy). In both cases, chromosome doubling can result in unbalanced meiotic behaviour and subsequent seed sterility (or lowered viability), at least in the first few generations. Mutations to correct chromosome pairing are thus crucial for the success of those polyploid genotypes regardless of their autoploid or alloploid nature.

The role of mutations in meiotic chromosome pairing is well documented, and simple mutations can result in bivalent associations and subsequently gametic balance (see Hilu[25]). In this case, perennial species would have by far the better chance to accommodate polyploidy, as they can survive beyond a single generation without the need for seed set, increasing the chance of acquiring such mutations. As an annual, correcting chromosome pairing in the first generation is a prerequisite for survival. Should polyploidy be a factor in speciation and larger genus size in Poaceae, then one could expect annuals to be predominantly diploids or low polyploids and perennials to accommodate polyploidy of different magnitude in proportion to genus size. Following this scenario, the size of mixed genera should relate to the proportion of perennial species in the genus and the degree of polyploidy.

To assess the potential impact of polyploidy in grass biodiversity, samples of annual, perennial and mixed genera of various sizes were examined for the presence and degree of polyploidy. Large perennial genera examined are highly polyploid, displaying a series of aneuploid and/or euploid gametic chromosome numbers. Standing out among these genera are *Festuca* (450 species), *Calamagrostis* (270 species), *Bambusa* (120 species) and *Rytidosperma* (90 species). In *Festuca*, somatic chromosome numbers reported include 2n = 14, 28, 35, 42, 56 and 70; in *Calamagrostis* 2n = 28, 42 and 56, or 56-91 (apomicts); *Bambusa* 2n = 24, 46, 48, 70 and 72; and *Rytidosperma* 2n = 24, 48, 72, 96 and 120. With the exception of *Festuca*, diploid types are not found in these genera. In contrast, perennial genera of small size tend to occupy the other end of the spectrum in terms of polyploidy. For instance, the monotypic perennial *Calderonella* is a diploid with x = 12, *Thysanolaena* is a diploid based on x = 11, and *Asthenochloa* and *Cleistachne* are tetraploids based on x = 9. The perennial small genus *Sartidia* (4) is also a diploid based on x = 11. This genus is of special interest, as other members of its tribe, Aristideae, are much larger, with *Stipagrostis* containing 50 species and *Aristida* 250. However, *Stipagrostis* contains primarily perennial species with a few annuals, and species are diploid or tetraploid based on x = 11. *Aristida* on the other hand includes a mixture of annuals and perennials but displays a more extensive polyploid series based on x = 11 or 12 (2n = 22, 24, 36, 44, 48 and 66). These three Aristideae genera display a grade of genus size that parallels their polyploidy levels and habit, suggesting a correlation between species number and polyploidy levels in perennials. It is to be noted that *Bambusa* belongs to the woody bamboo subfamily (Bambusoideae), a group containing long-lived perennials. Polyploidy is the norm in these, and diploid genotypes/species have for the most part gone extinct[26].

Looking at genera with different habits, it appears from examining representative genera that there is a tendency towards lower frequency of polyploidy in annuals than perennials. *Asthenochloa* and *Cleistachne* are monotypic annuals, both are tetraploids based on x = 9, most likely rediploidised tetraploids that lack other chromosomal variation. The annual genera *Gastridium*, *Mibora* and *Elytrophorus* each contain two species; they are all diploids. *Gaudinia* contains four annual species that sometimes behave like biennials; the species are diploids based on x = 7, but occasionally possess a somatic number of x = 15. Larger-size annual genera seem to have been able to accommodate some degree of polyploidy. Examples of those are *Sacciolepis* (30) with x = 9 and 2n = 16, 18, 36 and 45, and *Avena* (25) with x = 7 and 2n = 14, 28, 42, 48 and 63. The presence of 5x ploidy level and aneuploid numbers in species of these genera may suggest the presence of some degree of apomictic reproduction, again a system that can accommodate and promote polyploidy by producing seeds without sexual reproduction. In this case, the apparent large genus size may be an artefact of taxonomic splitting caused by apomictic perpetuation of morphologically distinct biotypes.

The second question to be addressed is why mixed genera display a pattern that does not fit the HCD. The answer to this question may relate to the proportion of each life history type in any given genus. If polyploid perennials have an accelerated rate of speciation and the converse is true for the mostly diploid or low polyploid annuals, then the size of a genus with mixed annuals and perennials would depend on the proportion of the two types of species. A mixed genus with proportionally more perennial species would tend to be larger in size than that with higher proportion of annuals. This scenario fits well the large mixed genera *Sporobolus* (160), *Muhlenbergia* (160), *Isachne* (100) and *Pennisetum* (80), where both annual and perennial habits are present but the annual species are rare. Polyploidy is extensive in all four genera[19]. In *Sporobolus*, chromosome numbers are based on x = 9 and 10, and somatic numbers are 2n = 18, 24, 36, 38, 54, 72, 80, 88, 90, 198 and 124. Similarly, *Pennisetum* species display basic chromosome numbers of x = 9, and somatic numbers are 2n = 14, 18, 22, 34, 35, 36, 45, 52 and 54.

The large genus *Paspalum* (330) is described as 'usually' perennial, indicating more annuals than in the above discussed genera. However, extensive series of polyploidy based on x = 10 and 12 augmented by aneuploidy have been established in this genus. *Digitaria* contains 230 annual and perennial species. Although it is not described as primarily perennial, it has evolved a wide range of polyploidy numbers based on x = 9, 15 and 17; such basic chromosome series appear to have originated via aneuploidy, most likely coupled with hybridisation. *Panicum* (470) comprises both annuals and perennials with no predominance of either habit, but diploidy (2n = 18) is seemingly rare, while polyploidy is extensive and based on x = 7, 9 and 10, and 2n = 36, 37, 54 and 72. In these cases, it appears that species diversification is augmented by extensive patterns of polyploidisation.

At the other end of the spectrum of mixed genera, namely small-sized ones, one would expect either annuals to be found in higher proportion or polyploidy to be rare, or both. *Amphicarpum* (2) is described as an annual and perennial genus; only diploid chromosome numbers have been reported (x = 9, 2n = 18). *Tricholaena* contains four species, mostly perennial, that are all tetraploids based on x = 9. *Echinolaena* is slightly larger with eight annual and perennial species, but only tetraploidy is found (2n = 60). Only diploid genotypes of 2n = 20 have been reported for *Chionachne* (7), although it is composed primarily of perennial species. Lack of polyploidy may account for the small size of these genera. *Loudetia* (26) is rarely annual, but the larger size is associated with a more elaborate polyploid series based on x = 6, 12, and 2n = 20, 24, 40 and 60. Moving up in genus size, *Echinochloa* contains 35 annual and perennial species. Polyploidy has progressed considerably in this genus where, based on x = 9, odd and even euploidy as well as aneuploidy have been established (2n = 27, 36, 42, 48, 54, 72 and 108). Therefore, it appears that small genus size is correlated with lack of polyploidy, and an increase in genus size tends to be associated with progressive emergence of polyploid genotypes.

The third question is why, on average, genera with mixed annual and perennial species tend to be more species rich. Part of the answer may be found in the combined accelerated rates of speciation of the two life forms through different biological routes. Although perennials take advantage of polyploidy, annuals are also well suited for speciation at the diploid level. In these cases, a combination of rapid generation turnover inherent to annuals and extensive polyploidy in perennials, possibly with apomictic mode of reproduction superimposed, could play a role in determining genus size. This question, however, remains intriguing and in need of further investigation.

## 11.5  CONCLUSION

It is evident from this survey that small genera either have more annual species or less polyploidy, or a combination of both. With increase in genus size, the proportion of perennial species and the incidence of polyploidy increases. This phenomenon could have contributed to the punctuated genus size pattern observed for the mixed genera. It also may explain the correlation between annuals and small genus size and perennials and explosive speciation. Although perenniality combined with polyploidy sets the stage for increased diversity at the species level in Poaceae, other factors such as breeding systems (outbreeding, inbreeding, cleistogamy and apomixis) impose yet additional parameters that could influence speciation. All these factors together should be considered to act in promoting adaptive radiation to exploit a diverse variety of habitats. As such, ecological parameters represent major factors that work in accordance with the genetic factors when degree and patterns of speciation are to be explained. Thus, the nonrandom distribution of species appears to have contributed to the skewed distribution favouring monotypic and small genera and towards that skewed distribution of dominance in terms of species richness. Biological factors, such as habit and polyploidy, appear to have had some impact on these patterns in the grass family. These biological factors may be group specific and should not be overly generalised.

## ACKNOWLEDGEMENTS

I thank Scott Parker for discussion and assistance with statistical tests and Kieran Hilu for valuable comments on a draft of the manuscript.

## REFERENCES

1. Willis, J.C., *Age and Area*, Cambridge University Press, Cambridge, England, 1922.
2. Willis, J.C. and Yule, G.U., Some statistics of evolution and geographic distribution in plants and animals and their significance, *Nature*, 109, 177, 1922.
3. Williams, C.B., *Patterns in the Balance of Nature and Related Problems in Quantitative Ecology* Academic Press, New York, 1964.
4. Anderson, S., Patterns of faunal evolution, *Rev. Biol.* 49, 1, 1974.
5. Dial, K.P. and Marzluff, J.M., Nonrandom diversification within taxonomic assemblages, *Syst. Zool.*, 38, 26, 1989.
6. Cardillo, M., Huxtable, J.S., and Bromham, L., Geographic range size, life history and diversification of Australian mammals, *J. Evolution. Biol.*, 16, 282, 2003.
7. Scotland, R.W. and Sanderson, M.J., The significance of few versus many in the tree of life, *Science*, 303, 643, 2004.
8. Clayton, W.D., Evolution and distribution of grasses, *Ann. Missouri Bot. Gard.*, 68, 5, 1981.
9. Hilu, K.H. and Soderstrom, T.R., Biological basis of adaptation in grasses: an introduction, *Ann. Missouri Bot. Gard.*, 72, 823, 1985.
10. Redman, R.E., Adaptation of grasses to water stress-leaf rolling and stomata distribution, *Ann. Missouri Bot. Gard.*, 72, 833, 1985.
11. Clayton, W.D. and Renvoize, S.A., *Genera Graminum*, HMSO Publications, London, 1986.
12. Connor, H.E., Evolution of reproductive systems in Gramineae, *Ann. Missouri Bot. Gard.*, 72, 48, 1981.
13. de Wet, J.M.J., *Hybridization and Polyploidy in the Poaceae*, Smithsonian Institution Press, Washington, DC, 1987.
14. Hunziker, J.H. and Stebbins, G.L., *Chromosomal Evolution in the Gramineae*, Smithsonian Institution Press, Washington, DC, 1987.
15. Hilu, K.H., Phylogenetics and chromosomal evolution in the Poaceae (grasses), *Aust. J. Bot.*, 52, 13, 2005.
16. Bennett, M.D. and Leitch, I.J., *Angiosperm DNA C-value Database*, http://www.rbgkew.org.uk/cval/homepage/html, 2001.
17. GPWG, Phylogeny and subfamilial classification of the grasses (Poaceae), *Ann. Missouri Bot. Gard.*, 88, 373, 2001.
18. Hilu, K.W., Alice, L.A., and Liang, H., Phylogeny of Poaceae inferred from *matK* sequences, *Ann. Missouri Bot. Gard.*, 86, 835, 1999.
19. Watson, L. and Dallwitz, M.J., *The Grass Genera of the World*, CAB International, Wallingford, Australia, 1992.
20. JMP Version 3, SAS Institute Inc., Cary, NC, 1996.
21. Harper, J.L., *Population Biology of Plants*, Academic Press, London, 1977.
22. Marzluff, J.M. and Dial, K.P., Life-history correlates of taxonomic diversity, *Ecology*, 72, 428, 1991.
23. Rosenheim, J.A. and Tabashnik, B.E., Generation time and evolution, *Nature*, 365, 791, 1993.
24. Stebbins, G.L., Polyploidy, hybridization and the invasion of new habitats, *Ann. Missouri Bot. Gard.*, 72, 824, 1985.
25. Hilu, K.W., Identification of the 'A' genome of finger millet using chloroplast DNA, *Genetics*, 118, 163, 1988.
26. Pohl, R.W. and Clark, L.G., New chromosome counts for *Chusquea* and *Aulonemia* (Poaceae: Bambusoideae), *Am. J. Bot.*, 79, 478, 1992.

# 12 Reconstructing Animal Phylogeny in the Light of Evolutionary Developmental Biology

*A. Minelli*
Department of Biology, University of Padova, Italy

*E. Negrisolo*
Department of Public Health, Comparative Pathology and Veterinary Hygiene, University of Padova, Italy

*G. Fusco*
Department of Biology, University of Padova, Italy

## CONTENTS

## ABSTRACT

The relevance of evolutionary developmental biology (evo-devo) to our effort to reconstruct the tree of life has until recently been very poorly explored. However, the contribution of an evo-devo approach to the main steps of phylogenetic analysis, such as evaluation of homology, selection of characters and assessment of character polarity can be critically important, especially in species rich groups.

As independence of traits is a prerequisite for the use of coded information in the reconstruction of phylogeny, the identification of developmentally independent units is one of the areas where evo-devo may offer an especially important contribution. The way in which characters originate and change in evolution has fundamental consequences on the patterns of evolutionary change we can reconstruct from character distribution. The remoulding of pre-existing features, genetic networks or developmental trajectories, can operate at any level of biological organisation. Comparative developmental biology supports a view that homology cannot be a relationship of the all-or-nothing kind.

## 12.1  DEVELOPMENT, PHYLOGENY AND THE HISTORICAL ROOTS OF EVO-DEVO

Early in the nineteenth century, the importance of developmental information in assessing relationships among different kinds of animals was already clear to some of the brightest zoologists of the time. For example, Geoffroy Saint-Hilaire[1] used congruence in ossification centres as a criterion to assess what we now call the homology between the bones of different vertebrate species, while von Baer's law[2] encapsulated the principle according to which traits diagnostic for classes appear first during development, followed by those diagnostic for orders, families, genera and species. Later criticisms to the suggestions of these pioneers, such as de Beer's magisterial analysis of heterochrony[3], did not subtract from the increasing faith in the relevance of developmental evidence in the assessment of phylogenetic relationships, but the field became increasingly complex and in need of in-depth analysis and reformulation.

Assessing homologies, however, is not the only step in phylogenetic reconstruction where a sensible evaluation of developmental evidence may be of value. As we will briefly illustrate in this article, information about development can also be critically important in evaluating the phylogenetic signal we may expect to retrieve from a character, in assessing the degree of independence between traits, and in formulating hypotheses of character polarity. Use of developmental information, however, is still uncommon in systematic literature, despite recent advances in evo-devo, a newly emerging field in which the previously independent research traditions in evolutionary biology and developmental biology are merging around a broad and still imperfectly circumscribed array of problems[4,5].

During the last two decades or so of evo-devo, the most productive field of inquiry has been the search for genes involved in body patterning and the comparative analysis of their spatial and temporal expression patterns in selected developmental stages of different organisms; for example, during segmentation or the specification of anteroposterior (AP) body patterning in metazoan embryos, or in the specification of the identity of the individual floral whorls during early flowering stages of angiosperms. The comparative framework, progressively broadening around the handful of model species to which experimental research in this field was initially confined, has revealed many important and often puzzling patterns to be interpreted in terms of phylogenetic relationships. In several instances, the result of comparative developmental genetics suggested abandoning well entrenched and long cherished phylogenetic hypotheses, in favour of new ones, or of old alternatives that had been put aside.

## 12.2  MORPHOLOGY TO MOLECULES TO MORPHOLOGY

Comparative developmental genetics offers an opportunity for reducing (if not bridging) the gap between molecular and morphological evidence. Molecular phylogenetics relies mainly on comparisons of nucleotide sequences which are not necessarily identical with protein coding segments, but also include blocks such as ITSs (internal transcribed spacers) and shorter sequences such as SINEs (short interspersed nuclear elements). Correspondingly, nucleotide sequences relevant in development extend well beyond the protein coding units, by including *cis-* and/or *trans-* regulatory regions. Thus, investigating the complex network of control relationships among genes involved in establishing the main features of an animal's organisation involves detailed knowledge of the whole genetic machinery, at the level of nucleotide sequences, as well as from the point of view of control cascades and spatiotemporal patterns of expression. In this way, as we will show through some examples, we are helped in developing comparisons among very divergent and thus hardly comparable body plans, without the need to totally ignore morphology and rely on molecular evidence only. This is critically important at high taxonomic levels, where the heuristic power of comparative morphology has been exploited since Cuvier's pioneering articulation of the animal kingdom into four 'embranchements' (the vertebrates, the articulates, the molluscs and the radiates)[6], in setting up a classification where species are ultimately grouped into phyla, that is, higher rank taxa separated by such deep differences in overall organization, as to cause major problems of comparison at morphological level[7]. On the other hand, molecular phylogenetics has provided tools for assessing phylogenetic relationships among remotely related taxa, thus overcoming the limitation of morphology, and often suggesting unconventional affinities, such as between arthropods and nematodes (in a group named Ecdysozoa[8]) or between hermit crabs and king crabs[9], affinities that are less than obvious at the level of morphology. However, this distance between morphological and molecular evidence may be shortened if we get comparative information about the identity, sequence and expression of genes which are more or less specifically involved in the generation of animal (or plant) form.

Indeed, this approach has been the source of unexpected discoveries, such as the fundamental identity of the genetic control of dorsoventral (DV) patterning in animals that are as different from each other as arthropods and vertebrates. Zoologists are familiar with a fundamental difference between them, as representatives of gastroneuralians (animals with ventral nervous cord) and notoneuralians (animals with dorsal nervous cord) respectively. In his idiosyncratic belief in a unity of plan of all animals, Geoffroy Saint-Hilaire proposed that arthropods might be equated to vertebrates, provided that we regard the dorsal aspect of the former as equivalent to the ventral aspect of the latter, and vice versa. At the time no factual argument could be advanced to support such an unconventional thesis, and the comparison was universally disregarded as a fanciful speculative exercise. However, comparative developmental genetics was eventually to rescue his idea from oblivion, by demonstrating that in *Drosophila* DV patterning is controlled by two genes (*short gastrulation* and *decapentaplegic*) that are homologous to two genes (*chordin* and *bone morphogenetic protein-4*) which perform the same job in vertebrates, but at opposite DV sides[10,11].

In the meantime, a group of genes involved in the AP patterning of the main body axis of all bilaterian metazoans had been discovered[12], thus nourishing hopes that the main traits of the hypothetical ancestor of all bilaterians, the so called Urbilateria, could eventually be inferred from comparative developmental genetic evidence. The list of traits that have been progressively added to this idealised ancestor include AP polarity and DV patterning[13], heart[13], cephalisation[14], brain and brain areas[15], a primitive photoreceptor[16], a skeleton[17], a 'humble appendage or antenna-like outgrowth'[13], hemocoel[18] and even segmentation[19]. Unfortunately, no element in this reconstruction has been evaluated using the tools of phylogenetic systematics. Indeed, doubts have often been raised as to the uniquely derived nature of most of these traits[20,21]. But the main message we want to bring forward here is not about the cavalier methods by which these phylogenetic inferences have been produced, but about the increasing availability of comparative data that can be applied to the evaluation of phylogenetic relationships among higher groups. These data are not limited to pure sequence information about genes and their products, but open vistas into an understanding of the origin and evolution of form[22–24].

### 12.2.1 Phylogenetic Inference from Sequence Data of Developmental Genes

The crudest level at which we can exploit comparative developmental genetics in reconstructing phylogeny is by focusing molecular phylogenetics on sequence data of selected classes of genes which play a key role in establishing the main features of animal body plans during early embryonic development.

Let us consider, for example, those classes of molecules, such as cell adhesion molecules with intracellular signal transduction pathways and gradient-forming morphogens or growth factors[25], that probably played a critical role in the very origin of metazoans from their unicellular ancestors.

These molecules are today inextricably involved in building and patterning the supracellular architecture of the animal body. But homologues of the corresponding genes were probably also present
along the stem lineage of living metazoans; their occurrence is likely in the living representatives
of the metazoans' sister group, even if the function of these genes is different from the function
that evolved in the metazoan branch of the phylogenetic tree. This is indeed a pathway of inquiry
where a detailed knowledge of developmental genetics suggests a class of genes on which to focus,
as a source of information for assessing a charismatic node of the tree of life, namely, the splitting
of metazoans from their sister group. Indeed, King and colleagues[26] have found that choanoflagellates, the most likely candidates to be the closest relatives of metazoans among the living unicells,
express representatives of many cell signalling and adhesion protein families. These include cadherins,
C-type lectins, tyrosine kinases and components of the tyrosine kinase signalling pathway.

    Other phylogenetic studies focusing on developmentally important genes have used the *Hox*
family. The reconstructed duplication event by which a proto*Hox* gene cluster would have given
rise to the *Hox* genes sensu stricto and to the paralogous set of the *ParaHox* genes has been
suggested as relevant to the Cambrian explosion of animal body plans[27]. Another major event in
animal evolution, the origin of the vertebrate lineage, is often thought to have been accompanied
by a quadruplication of the *Hox* gene cluster[28]. Other phylogenetic analyses of *Hox* genes give
support to the hypothesis that insects and crustaceans form a clade to the exclusion of myriapods[29]
and demonstrate the lophotrochozoan affinities of bryozoans[30].

    The role of these genes in controlling basic features of the animal body has probably been
overemphasised. Indeed, criticisms of the mainstream gene-centred view of development are
increasingly frequent[20,31-34], but in addition to these theoretical perspectives there is also experimental evidence pointing in the same direction. For example, the deletion of an entire vertebrate
*Hox* cluster may have little effect on the development of the animal's main body axis[35-37], much
less indeed than the deletion of any individual *Hox* gene in the same model animal. This asks for
a serious rethink of our understanding of the role of these genes in patterning the animal body.

    At any rate, comparing sequences of developmentally important genes does not represent a
major improvement in respect to current practice in molecular phylogenetics. We argue instead that
much more substantial progress is obtained if the basic steps in phylogenetic reconstruction are
approached from an explicit evo-devo perspective.

### 12.2.2 Evo-Devo, Devo-Evo

Researchers involved in the evolutionary or the developmental components of evo-devo may
exchange roles and provide each other with the evidence to be explained, or the framework within
which to look for explanation. One may even claim that we should distinguish between an evo-
devo and a devo-evo biology[38]. Irrespective of the labels we eventually apply to these efforts, such
a two way perspective also applies in the area where evolutionary developmental biology meets
with phylogenetics. One way is easier, but we are not really concerned with it here. This is the use
of independent phylogenetic reconstructions as scenarios against which to evaluate the evolution
of developmental processes. Recent examples include the evolution of segmentation in mecistocephalid centipedes[39], of dentition in basal vertebrate clades[40], of cell lineage in gastropods[41] and
of metamorphosis in nemerteans[42]. The other way is using evidence from comparative developmental biology to get a sounder foundation of our assessments of homology of traits and processes,
and also to discover new useful levels of comparison.

## 12.3  EVO-DEVO INSIGHTS INTO EVOLUTIONARY CHANGE

Independence of traits is, in principle, a prerequisite for their use in coded form in the reconstruction
of phylogeny. Operationally, independence can be read as a sufficiently low level of covariation[43,44].
In biological terms, one may distinguish between developmental independence and functional

independence. The latter is often easier to ascertain, in so far as we are able to single out individual features or structural complexes performing largely distinct functions, thus behaving as minimally overlapping units from the point of view of the selective pressure acting upon the organism. Much easier, but not less important, is the identification of developmentally independent units, due especially to the pervasive pleiotropic effects of the underlying genetic networks. This is certainly one of the areas where evo-devo may offer an especially important contribution.

Evo-devo explicitly addresses the generative mechanisms underlying the evolution of organismal form. An extreme reductionist view of the evolutionary process would argue that at the basis of evolutionary change there is nothing more than a change in the underlying regulatory networks of developmental genes[45]. Alternative views of the evolution of organismal form would maintain that generative mechanisms are not restricted to the genetic circuitry involved in individual development. These mechanisms also arise from the physical properties of biological materials, the self organisational capacities of cells and tissues, and the dynamics of epigenetic interactions among developmental modules[20,46,47]. However, although evo-devo does not coincide with developmental genetics, it must be said that our current understanding of development and the mechanisms of its evolutionary change is much more advanced at the level of the genes.

Due to their pervasive occurrence, the 'nongenetic' components of the developmental processes (physicochemical properties of living matter and epigenetic interactions) are likely to be relevant in the evaluation of character independence. However, even adopting a narrow view of evo-devo, limiting its scope to developmental genetics leaves a lot to say about the assessment of homology, character independence and character polarity.

### 12.3.1 How the Genetic Network Evolves

The traditional view on how the genetic makeup of organisms changes during evolution has centred on the gene's coding sequence. Through a cascade of causal processes, first at the genetic and later at the developmental level, a new allele, or a new allele combination in a new genotype, causes a new ontogenetic trajectory, more precisely, a new reaction norm. However, more recent studies are now challenging this view[48–51]. The evolution of the genetic machinery underlying the diversity of organismal form is mainly due to differences in gene regulation. Activating or repressing the expression of a gene at a new developmental stage or at a new location in the body can produce dramatic variations in the resulting phenotype. Researchers increasingly attribute evolutionary novelties to stretches of DNA with regulatory, rather than coding functions. These sequences, called enhancers, present binding sites for factors that regulate gene transcription. Sometimes the enhancer simply contains multiple copies of the same binding site; at other times, it has sites for different transcription factors. Moreover, recent results suggest that the effect of an enhancer on a gene is determined not just by the combination of transcription factors to which it can bind, but also by the spacing between its different binding sites[48]. Thus the same subset of transcription factors can be used to regulate different genes, by simply changing the spatial configuration with which these proteins bind along the enhancer. Modifications in the order and spacing of binding sites can affect the behaviour of the same gene across different species. Evolutionary change via enhancer rearrangement seems to be simple and effective. For instance, the vertebrate gene *Hoxc8* is involved in defining the number and shape of thoracic vertebrae. There is direct evidence that the enhancer, rather than the coding sequence, plays a pivotal role in generating different axial morphologies amongst species (for example, see Wang et al.[49]). Subtle species-specific differences in the enhancers correlate with the different anterior boundary of *Hoxc8* expression within the embryo. This can contribute to explain the great diversity in the number of thoracic vertebrae among vertebrates[50]. However, there are other ways in which development can be changed through changes at the gene level. Alonso and Wilkins[51] have recently challenged the view that enhancer elements are the only, or the main, sites of gene regulation and therefore the principal players in the evolution of developmental processes. They explored the potential of many different factors involved in the regulation of gene expression, both at the transcriptional and post-transcriptional levels. These 'alternative regulative levels', as they call them, have many of the features, like flexibility, modularity and a combinatorial nature, which make enhancers critical contributors of genetic source materials to the evolution of development. They are sites of 'initial' genetic change, that may be either strengthened or replaced by subsequent 'secondary' genetic changes at other regulatory levels, including the level of the enhancers.

Evolutionary changes at the level of the genome's regulatory elements make it possible that different parts of the same animal are built by exploiting the same gene network or the same developmental module. 'Tinkering', 'multi-functionality', 'redundancy' and 'modularity' are common at the roots of phenotypic variation, but their impact on phenotypic evolution is far from being generally acknowledged. Character independence cannot be inferred from comparative anatomy or descriptive embryology alone and is just one working hypothesis among others in phylogenetic reconstruction. By expanding on the scope of current methods for estimating robustness of phylogenetic trees, it would probably be profitable to develop methods able to cope with an unknown level of character covariation.

## 12.3.2 Tinkering at the Level of Developmental Genes

Reviewing the conceptual framework of evolutionary developmental biology, Arthur[5] lists a number of key concepts that represent the toolbox for evo-devo investigation. Amongst these is a group of patterns and processes related to the 're-use of developmental genes in evolution'. Co-option of developmental genes (or gene networks) for different functional roles, sometimes involving gene duplication, is considered a major source of evolutionary change. Taking this idea even further, co-option can cause the formation of new structures, rather than simple change in old ones.

A more extreme concept is Minelli's axis paramorphism[52]. In metazoans there seems to be a general correspondence between the organisation of the appendages and the organisation of the main body axis of the same animal. Minelli interpreted metazoan appendages (secondary axes) as the product of a duplicate expression of genes already involved in growth and patterning of the main axis, that is, as axial paramorphs of the latter. Following the paramorphism hypothesis, arthropods' potential to produce periodically arranged structures along the main axis was exploited in producing segmented appendages[53].

The hypothesis of axis paramorphism has non-negligible consequences in establishing character polarity for traits related to arthropod appendages. Consider the old question of the mutual relationships between the antenna and the conventional (locomotory) leg of arthropods. The question is whether the antenna is to be regarded as a specialised leg, or vice versa. In other terms, which is the ancestral condition of the arthropod appendage? Different opinions have been expressed recently about the plausibility of the two alternatives (antenna first, or leg first), mainly in the light of developmental genetic evidence. For example, Dong and colleagues[54] favoured the antenna-first hypothesis, whereas Casares and Mann[55] supported the 'leg-first' hypothesis, although in a later paper these two authors have adopted a less clear cut option[56]. Considering paramorphic relationships between the main body axis, already segmented and patterned in AP sequence, and its serial appendages, which to some extent would therefore be already diverse (and segmented) since their very first expression, the whole question of the primacy of the leg versus the antenna would become meaningless[53,57], and the scheme of character transition from one form to the other is no longer applicable.

The great plasticity of the gene network and the non-strictly hierarchical modes of evolutionary change are also attested by the numerous examples of homologous characters that are formed via different developmental pathways[58]. The evolution of simultaneous body segmentation in arthropods may illustrate this point.

Body segmental units originate almost simultaneously in the *Drosophila* embryo. The so called segmentation genes, classified into gap, pair-rule and segment-polarity genes, create a hierarchical cascade of gene activity, leading from the early gap genes to the later expressed pair-rule and segment-polarity genes. These genes encode proteins that are eventually localised in the embryo according to segmental periodicity. In other arthropods, segments originate sequentially in an AP progression from a subterminal region. Simultaneous and sequential segmentation can both occur within the same animal. In many insects with embryo intermediate between short and long germ-band type, the most anterior segments originate synchronously, whereas the remaining segments are sequentially specified from a posterior sub-terminal zone[59]. At least for a significant posterior portion of the main body axis, sequential segmentation is generally considered the primitive condition in arthropods, and mechanisms for the evolutionary change from sequential to simultaneous segmentation have been proposed. These are based on a gradual cellular-to-syncytial transition in the blastoderm where the same segment-forming gene network operates[60], or on a progressive increase (from the anterior) of the number of segmental units falling under the control of gap genes[61].

In this case history of segmentation processes, the downstream developmental processes are conserved, whereas the earliest phase of the segmentation process has changed. This is just one of the many examples that comparative developmental biology can offer in support of a view that homology cannot be a relationship of the all-or-nothing kind. Because evolutionary change is a continuous process, based on the remoulding of pre-existing features, along with the underlying genetic networks that control their development, homology can only be partial[20,62–66]. The view of a character remaining the same (homologue) throughout a number of possible states, defining as many steps in an evolutionary sequence that can be linearly polarised and coded to fill in a phylogenetic data matrix, probably rests on a misrepresentation of how organisms evolve.

### 12.3.3 Evolution of Ontogenies versus Evolution of Characters

The way in which characters originate and change in evolution has fundamental consequences on the patterns of evolutionary change we can reconstruct from character distribution. Organismal form is not the product of one overall hierarchy of developmental modules, either at the level of genetic network or at the level of epigenetic interactions. Some hierarchical relationships occur only locally, both in spatial and temporal senses. This affects the production of variation in evolution. For instance, phylogenetically independent structures in different organisms can exploit the same genetic toolkit, or the same physical properties of living matter, in producing new variants. The remoulding of pre-existing features, genetic networks or developmental trajectories, can operate at any level of organisation. This 'tinkering' can overcome structural and functional boundaries between subsystems within an organism, by exploiting pattern and processes of one subsystem for the use of others[53]. The non-strictly hierarchical nature of evolutionary changes of ontogeny will unavoidably cause homoplasy.

## 12.4 DEALING WITH CHARACTERS FROM AN EVO-DEVO PERSPECTIVE

The new approach to phylogenetic reconstruction based on evolutionary developmental biology begins with the step of articulating the phenotype into a set of manageable and meaningful traits, and proceeds until the step is reached at which a suitable coding for character states can be eventually adopted.

Let us start with the identification of characters to be used in phylogenetic reconstruction. To delimit characters is basically the same as to identify homologies[67,68]. In this respect, evo-devo has broadened the scope of the search for homology, traditionally limited to a comparison of anatomical traits, to encompass also physiological and especially developmental processes. Gilbert and Bolker[69] introduced the term 'homology of process' to describe "the relationship between patterns that are composed of homologous proteins and that are related by common ancestry". However, as one of the present article's authors has remarked elsewhere[20], this still means reducing organ homology to gene homology, something conceptually and methodologically equivalent to reducing species phylogeny to gene phylogeny. As the latter reduction is conceptually unwarranted and must be rejected[70,71], so process homology is not to be reduced to the shared involvement of homologous genes in two developmental sequences, but must be firmly rooted in the shared origin of the developmental pattern itself. This opens several interesting questions, two of which will be dealt with here.

The first question is whether we can formulate hypotheses of homology between developmental stages, rather than between specific developmental events. We think we can, but very cautiously. Nobody will contend that to enter in a data matrix a character 'larva' with states such as 'trochophora', 'caterpillar' and 'tadpole' would be other than plain nonsense, but what about suggesting the grasshopper prelarva as homologous to the larva of beetles or flies? Evidence in favour is admittedly tenuous[72], but the hypothesis cannot be discounted hastily[73,74], and only a thorough exploration of the developmental sequences both upstream of these putatively homologous stages and during the same will hopefully clarify the issue.

Problems with the comparison of developmental stages are sometimes even more subtle. When comparing stages of two closely related insects, with a similar postembryonic developmental schedule but with different number of instars, we may consider whether there is necessarily homology between equally numbered stages of the two species. That is, we may ask whether the fifth and last larval stage of butterfly species A is homologous to the fifth but penultimate larval stage of butterfly species B. In our understanding, there is no universally valid answer to this question, but as a basic rule, we believe that individual instars in a developmentally 'smooth' sequence (that is, one along which moults are only punctuations of the animal's basically continuous growth) cannot be individually treated as homologues. The only meaningful comparison, in the example, would be one between the character state 'larval development through four instars' and 'larval development through five instars', but excluding a direct stage-to-stage comparison between the two species.

The second question is whether we can still rely on the traditional all-or-nothing notion of homology. Our answer is firmly 'no'. All developmental sequences investigated in some detail, and especially those for which a detailed analysis has been performed in terms of genetic control of developmental events, have shown that characters, morphological and developmental alike, are not produced by unique and perfectly well integrated complexes, or networks, of genes. Locally acting dynamics allow the recognition of more or less individualised developmental modules[75–78], but overlaps and cross links are such as to oppose a simply hierarchical dissection of development and, hence, a strictly hierarchical view of homology. A combinatorial approach to homology has been suggested as a viable alternative[63].

### 12.4.1 Segmentation

One of the grand traits of animal organisation on which evolutionary developmental biology offers a renewed perspective is segmentation, a key trait in the taxonomy of some species rich groups, such as the mecistocephalid centipedes[39]. Body segmentation has long been regarded as a character useful in recognising affinities at very high taxonomic levels, for example, as an argument through which zoologists have supported, until recently, Cuvier's[6] pre-evolutionary concept of a taxon Articulata that should include the two major groups of segmented invertebrates, annelids and arthropods. Modern insights into segmentation mechanisms have cast increasing doubt as to the equivalence of segmentation mechanisms in the two groups[79] and the origin of segmentation is now generally regarded as either very deep in animal phylogeny (via a segmented Urbilateria[13,19,80,81]) or, as we prefer to believe, as having evolved in annelids and arthropods convergently[82]. Adopting a broader evo-devo perspective helps with the interpretation of segmented features of animal body architecture in a much more articulated way[83,84]. In this way, we realise that phylogenetically closely related species may differ in their segmentation mechanism to a considerable extent, whereas unexpected similarities in these mechanisms may occur between much more distant lineages. Comparative evidence suggests that the concept of segmentation applies to organs rather than to whole organisms, overall segmentation resulting when independently segmented structures eventually develop to share period and phase of their repetitive patterns. One may even argue that segmentation is a 'generic' property of bilaterians[85], that is, that it depends on basic physicochemical properties of living matter more than on the specific expression patterns of a restricted set of genes. In this perspective, it seems that the traditional distinction between the 'true' segmentation of annelids, arthropods and vertebrates, and the 'pseudosegmentation' of animals such as tapeworms and kinorhynchs should be abandoned. This is obviously important, given the use we may want to make of segmentation in phylogenetic reconstruction.

### 12.4.2 Larval Legs versus 'True' Legs

Atavisms or 'evolutionary Lazarus features'[20] can strongly perturb the process of reconstructing phylogeny, but combining insights from developmental genetics with those from comparative morphology may help avoid pitfalls. One may wonder whether the larval legs of caterpillars are homologous to the other paired appendages of the trunk, a question traditionally studied only in respect to the generalised lack of nongenital abdominal appendages in pterygote insects and the presence of short appendages in the abdomen of more distantly related hexapods. In this case, in addition to providing one more argument in favour of adopting a combinatorial view of homology, developmental genetics has demonstrated that changing from presence to absence of a major anatomical trait may be obtained at relatively low cost. It has been shown, in fact, that the presence of larval legs in the abdomen of lepidopteran larvae is due to the expression of the *distal-less* gene in the abdominal segments, a pattern of expression which is otherwise suppressed in most insect orders, thus determining the absence of limbs on the corresponding segments. A conspicuous morphological difference between lepidopteran larvae and other larval or adult insects is thus basically explained by the de-repression of the expression of this gene[86]. Caterpillar prolegs should be therefore examined more closely for their specific morphology rather than evaluated in plain terms of presence in lepidopterans versus absence in other insect orders.

### 12.4.3 Character States: Discontinuous Variation

One of the largest benefits of the ongoing dialogue between developmental biologists and evolutionary biologists is a better understanding of the causes of discontinuous variation in circumstances where an explanation in terms of natural selection would be at loss. An example is variation in the number of leg-bearing segments in centipedes. This number is always odd in adult centipedes: 15 in Scutigeromorpha, Lithobiomorpha and Craterostigmomorpha, 21 or 23 in Scolopendromorpha and 27 to 191 in Geophilomorpha. Intraspecific variation in segment number is known for one scolopendromorph only (*Scolopendropsis bahiensis* (Brandt 1841))[87] but is widespread among the Geophilomorpha. Only an unbelievably strong selection pressure against centipedes with an even number of leg pairs would explain the complete absence of specimens with 36 or 38 pairs of legs in a population where only specimens with 35, 37 and 39 pairs are recorded. But this is where the evolutionary explanation ends, and developmental biology may enter the scene: only an accurate knowledge of segmentation mechanisms may explain why centipedes with an even number of leg pairs are never produced[88]. The knowledge we have of segmentation in arthropods and in centipedes in particular is limited, but it is only from those studies that we eventually expect to understand why the 35 and the 39 pairs of legs conditions are, indeed, the closest possible to the 37 pairs of legs condition, in both developmental and evolutionary terms, whereas the apparently intermediate conditions, with either 36 or 38 pairs of legs, are 'infinitely' far away. With fairly high numbers of segments, the easiest jump may be even further away than just two segments up or down. For example, in the geophilomorph family Mecistocephalidae, where segment number is nearly always fixed within each species, most evolutionary changes have involved adding four segments in one step[39], such as from 41 to 45 to 49. These evo-devo perspectives on the topology of the ontogenetically accessible morphospace are clearly useful in defining character states.

### 12.4.4 Developmental Processes as Characters for Phylogenetic Analysis

The fact that ascertaining homology among different developmental stages is not a simple thing supports the use of an evo-devo approach so as to allow proper discrimination between true homology and apparent homology. The latter may have a very pervasive effect in phylogenetic reconstruction, particularly when coupled with heterochrony. Heterochrony, the change in developmental timing[89,90], may affect, in different ways, the recognition of homology amongst developmental stages. The order of developmental events may be changed, and this aspect needs be properly addressed, but heterochrony may also stop the development of some taxa at a stage which is not comparable with that reached by other taxa. If this aspect is not taken into account and such characters are included in the phylogenetic reconstructions, the resulting trees are the product of analyses based on noncomparable developmental traits.

This point is well exemplified in phylogenetic reconstructions of salamanders determined on the basis of evidence which included developmental characters affected by paedomorphosis, that is, an arrest of some traits in a larval stage condition[91]. The inclusion of larval characters deeply affects inferences on the higher-level phylogeny of salamanders in three distinct ways. First, such traits, which are homoplastic, are shared by paedomorphic adults of different lineages that thus group together. Second, and a more difficult factor to be removed, the paedomorphic traits destroy the clade-specific synapomorphies that are present in the metamorphosed adults. This phenomenon causes the misplacement of the paedomorphic taxa in the phylogenetic reconstruction with respect to their nonpaedomorphic relatives. Finally, the aquatic life style of the larvae produces parallel adaptive changes that again cause an erroneous placement of paedomorphic taxa. Even worse, the strong bias introduced by heterochrony in the phylogenetic tree based on paedomorphic characters is corroborated by statistical support[91]. This clearly shows the strong value of an evo-devo approach in identifying the traits affected by heterochrony. In this way, we can include them in the data set after suitable corrections, accounting for the heterochronic effects, have been made.

During the last decade the burgeoning evo-devo approach to phylogenetics has moved even further, by trying to produce algorithms that allow the proper handling of the effect of heterochrony. The most widespread approach is based on developmental sequences. "A developmental sequence is a list of different events in the chronological order in which they happen in the ontogeny"[92]. A developmental sequence can be divided into a series of developmental events. "Developmental events may be regarded as series of morphological states which a given embryonic structure passes"[93]. The order of events along the sequence characterises the sequence itself. If we consider two events (A and B) in a sequence, they can only occur in one of the following temporal series: A occurs before B, or simultaneously with it, or after B.

Each of these timing relationships constitutes an event pair that can be given a numerical score[94,95]. For every species having a developmental sequence including N events, there are $1/_2(N^2 - N)$ event pairs. Developmental sequences that have been scored according to the event pairs characterizing them may be assembled in a data matrix. The rows in the matrix are the developmental sequences, while every column of the matrix is occupied by the numerical scores of each event pair previously coded according to procedures such as developed by Smith[94] or Velhagen[95]. This matrix can be used under the criterion of maximum parsimony to reconstruct a phylogenetic tree[96]. Characters may be treated both as unordered or ordered. However, event pairs considered in the developmental sequence are not independent characters and thus violate one of the fundamental requirements of phylogenetic reconstruction, that is the independence of characters analysed.

There are two kinds of nonindependence in the event pairing approach: an ontogenetic dependence due to the fact that some events occurred according to a specific order during ontogeny and a coding dependence due to the fact that scoring of event pairs is not independent. This double nonindependence may lead to highly inconsistent results, or even absurd results, from a logical perspective.

To circumvent this intrinsic flaw in the event pair approach, Schulmeister and Wheeler[92] developed a new method in which every developmental sequence is considered as a single multistate character. A search based optimisation is used to investigate changes within developmental sequences, and step matrices are used to account for changes within each sequence. The new method does not suffer from the nonindependence that characterises the event pair approach and thus appears to be a promising strategy for the use of developmental data in phylogenetic reconstruction when data are affected by heterochrony.

## 12.5 CONCLUSION

In the context of the current dialogue between evolutionary and developmental biology, the value of reliable phylogenetic reconstructions in comparative evaluation of developmental processes has been adequately demonstrated. In contrast, the relevance of evolutionary developmental biology in our effort to reconstruct the tree of life has been very poorly explored until recently. However, as shown in this article, the contribution of an evo-devo approach to the main steps of phylogenetic analysis, such as evaluation of homology, selection of characters and assessment of character polarity, can be critically important.

## ACKNOWLEDGEMENTS

We thank Ronald Jenner for insightful comments on a previous version of this work. The idiosyncratic views we present here are definitely ours. Alessandro Minelli was supported by a grant of the Italian Ministry of Education, University and Research.

## REFERENCES

1. Geoffroy Saint-Hilaire, E., *Philosophie anatomique,* J.B. Ballière, Paris, 2 vols, 1818–1822.
2. von Baer, K.E., *Über Entwicklungsgeschichte der Thiere: Beobachtung und Reflexion,* 1, Bornträger, Königsberg, 1828.
3. de Beer, G.R., *Embryos and Ancestors,* 3rd ed., Clarendon Press, Oxford, 1958.
4. Hall, B.K., *Evolutionary Developmental Biology,* 2nd ed., Chapman and Hall, London, 1998.
5. Arthur, W., The emerging conceptual framework of evolutionary developmental biology, *Nature,* 415, 757, 2002.
6. Cuvier, G., Sur un nouveau rapprochement à établir entre les classes qui composent le règne animal, *Ann. Mus. Natn. Hist. Nat. Paris,* 19, 73, 1812.
7. Minelli, A., *Biological Systematics: The State of the Art,* Chapman and Hall, London, 1993.
8. Aguinaldo, A.M.A. et al., Evidence for a clade of nematodes, arthropods and other moulting animals, *Nature,* 387, 489, 1997.
9. Cunningham, C.W., Blackstone, N.W., and Buss, L.W., Evolution of king crabs from hermit crab ancestors, *Nature,* 355, 539, 1992.
10. Arendt, D. and Nübler-Jung, K., Inversion of dorsoventral axis? *Nature,* 371, 26, 1994.
11. De Robertis, E.M. and Sasai, Y., A common plan for dorsoventral patterning in Bilateria, *Nature,* 380, 37, 1996.
12. Slack, J.M.W., Holland, P.W.H., and Graham, C.F., The zootype and the phylotypic stage, *Nature,* 361, 490, 1993.
13. De Robertis, E.M., The ancestry of segmentation, *Nature,* 387, 25, 1997.
14. Finkelstein, R. and Boncinelli, E., From fly head to mammalian forebrain: the story of *otd* and *Otx, Trends Genet.,* 10, 310, 1994.
15. Arendt, D. and Nübler-Jung, K., Common ground plans in early brain development in mice and flies, *BioEssays,* 18, 255, 1996.

16. Bolker, J. and Raff, R.A., Developmental genetics and traditional homology, *BioEssays*, 18, 489, 1996.
17. Jacobs, D.K. et al., Molluscan *engrailed* expression, serial organization, and shell evolution, *Evol. Dev.*, 2, 340, 2000.
18. Valentine, J.W., Erwin, D.H., and Jablonski, D., Developmental evolution of metazoan bodyplans: the fossil evidence, *Dev. Biol.*, 173, 373, 1996.
19. Kimmel, C.B., Was Urbilateria segmented? *Trends Genet.*, 12, 329, 1996.
20. Minelli, A., *The Development of Animal Form: Ontogeny, Morphology, and Evolution*, Cambridge University Press, Cambridge, 2003.
21. Jenner, R., Towards a phylogeny of the Metazoa: evaluating alternative phylogenetic positions of Platyhelminthes, Nemertea, and Gnathostomulida, with a critical reappraisal of cladistic characters, *Contrib. Zool.*, 73, 3, 2004.
22. Raff, R.A., *The Shape of Life: Genes, Development, and the Evolution of Animal Form*, The University of Chicago Press, Chicago-London, 1996.
23. Arthur, W., *The Origin of Animal Body Plans. A Study in Evolutionary Developmental Biology*, Cambridge University Press, Cambridge, 1997.
24. Valentine, J.W., *On the Origin of Phyla*, The University of Chicago Press, Chicago-London, 2004.
25. Müller, W.E.G., How was metazoan threshold crossed? The hypothetical Urmetazoa, *Comp. Biochem. Physiol. A*, 129, 433, 2001.
26. King, N., Hittinger, C.T., and Carroll, S.B., Evolution of key cell signaling and adhesion protein families predates animal origins, *Science*, 301, 361, 2003.
27. Brooke, N.M., García-Fernàndez, J., and Holland, P.W.H., The *ParaHox* gene cluster is an evolutionary sister of the *Hox* gene cluster, *Nature*, 392, 920, 1998.
28. Bailey, W.J. et al., Phylogenetic reconstruction of vertebrate *Hox* cluster duplication, *Mol. Biol. Evol.*, 14, 843, 1997.
29. Cook, C.E. et al., *Hox* genes and the phylogeny of the arthropods, *Curr. Biol.*, 11, 759, 2001.
30. Passamaneck, Y.J. and Halanych, K.M., Evidence from *Hox* genes that bryozoans are lophotrochozoans, *Evol. Dev.*, 6, 275, 2004.
31. Nijhout, H.F., Metaphors and the role of genes in development, *BioEssays*, 12, 441, 1990.
32. Keller, E.F., *The Century of the Gene*, Harvard University Press, Cambridge, Mass., 2000.
33. Oyama, S., *The Ontogeny of Information*, 2nd ed., Duke University Press, Durham, N.C., 2000.
34. Müller, G.B. and Newman, S.A., *Origination of Organismal Form: Beyond the Gene in Developmental and Evolutionary Biology*, MIT Press, Cambridge, MA, and London, 2003.
35. Suemori, H. and Noguchi, S., Hox C cluster genes are dispensable for overall body plan of mouse embryonic development, *Dev. Biol.*, 220, 333, 2000.
36. Medina-Martinez, O., Bradley, A., and Ramirez-Solis, R., A large targeted deletion of *Hoxb1-Hoxb9* produces a series of single-segment anterior homeotic transformation, *Dev. Biol.*, 222, 71, 2000.
37. Spitz, F. et al., Large scale transgenic and cluster deletion analysis of the *HoxD* complex separate an ancestral regulatory module from evolutionary innovations, *Genes Dev.*, 15, 2209, 2001.
38. Hall, B.K., Evo-devo or devo-evo: does it matter? *Evol. Dev.*, 2, 177, 2000.
39. Bonato, L., Foddai, D., and Minelli, A., A cladistic analysis of mecistocephalid centipedes reveals an evolutionary trend in segment number (Chilopoda: Geophilomorpha: Mecistocephalidae), *Syst. Entomol.*, 28, 539, 2003.
40. Smith, M.M., Vertebrate dentitions at the origin of jaws: when and how pattern evolved, *Evol. Dev.*, 5, 394, 2003.
41. Lindberg, D.R. and Guralnik, R.P., Phyletic patterns of early development in gastropod molluscs, *Evol. Dev.*, 5, 494, 2003.
42. Maslakova, S.A., Martindale, M.Q., and Norenburg, J.L., Vestigial prototroch in a basal nemertean, *Carinoma tremaphoros* (Nemertea; Palaeonemertea), *Evol. Dev.*, 6, 219, 2004.
43. Cheverud, J.M., The genetic architecture of pleiotropic relations and differential epistasis, in *The Character Concept in Evolutionary Biology*, Wagner, G.P., Ed., Academic Press, San Diego, 2001, 411.
44. Klingenberg, C.P. et al., Inferring developmental modularity from morphological integration: analysis of individual variation and asymmetry in bumblebee wings, *Am. Nat.*, 157, 11, 2001.
45. Davidson, E.H., *Genomic Regulatory Systems: Development and Evolution*, Academic Press, San Diego, 2001.
46. Newman, S.A., Generic physical mechanisms of tissue morphogenesis: a common basis for development and evolution, *J. Evol. Biol.*, 7, 467, 1994.

47. Müller, G.B., Six memos for EvoDevo, in *From Embryology to EvoDevo: A History of Embryology in the 20th Century,* Laubichler, M.D., and Maienschein, J., Eds., MIT Press, Cambridge, MA, and London, in press.
48. Pennisi, E., Searching for the genome second code, *Science,* 306, 632, 2004.
49. Wang, W.C.H. et al., Comparative cis-regulatory analyses identify elements of the mouse Hoxc8 early enhancer, *J. Exp. Zool. (Mol. Dev. Evol.),* 302B, 436, 2004.
50. Richardson, M.K. et al., Somite number and vertebrate evolution, *Development,* 125, 151, 1998.
51. Alonso, C.R. and Wilkins, A.S., Developmental evolution: are enhancers the primary source of novelty? *Nat. Rev. Genet.,* 6, 709, 2005.
52. Minelli, A., Limbs and tail as evolutionarily diverging duplicates of the main body axis, *Evol. Dev.,* 2, 157, 2000.
53. Minelli, A. and Fusco, G., Conserved vs. innovative features in animal body organization, *J. Exp. Zool. (Mol. Dev. Evol.),* 304B, 520, 2005.
54. Dong, P.D.S., Chu, J., and Panganiban, G., Proximodistal domain specification and interactions in developing *Drosophila* appendages, *Development,* 128, 2365, 2001.
55. Casares, F. and Mann, R.S., Control of antennal versus leg development in *Drosophila, Nature,* 392, 723, 1998.
56. Casares, F. and Mann, R.S., The ground state of the ventral appendage in *Drosophila, Science,* 293, 1477, 2001.
57. Minelli, A., The origin and evolution of the appendages, *Int. J. Dev. Biol.,* 47, 573, 2003.
58. Schlosser, G. and Wagner, G.P., Introduction: the modularity concept in developmental and evolutionary biology, in *Modularity in Development and Evolution,* Schlosser, G. and Wagner, G.P., Eds., University of Chicago Press, Chicago-London, 2004, 1.
59. Davis, G.K. and Patel, N.H., Short, long and beyond: molecular and embryological approaches to insect segmentation, *Annu. Rev. Entomol.,* 47, 669, 2002.
60. Salazar-Ciudad, I., Solé, R.V., and Newman, S.A., Phenotypic and dynamical transitions in model genetic networks II. Application to the evolution of segmentation mechanisms, *Evol. Dev.,* 3, 95, 2001.
61. Peel, A., The evolution of arthropod segmentation mechanisms, *BioEssays,* 26, 1108, 2004.
62. Shubin, N. and Wake, D., Phylogeny, variation, and morphological integration, *Am. Zool.,* 36, 51, 1996.
63. Minelli A., Molecules, developmental modules and phenotypes: a combinatorial approach to homology, *Mol. Phylogenet. Evol.,* 9, 340, 1998.
64. Abouheif, E., Establishing homology criteria for regulatory gene networks: prospects and challenges, in *Homology,* Bock, G.R. and Cardew, G., Eds., Wiley, Chichester, 1999, 207.
65. Wake, D.B., Homoplasy, homology and the problem of 'sameness' in biology, in *Homology,* Bock, G.R. and Cardew, G., Eds., Wiley, Chichester, 1999, 24.
66. Pigliucci, M., Characters and environments, in *The Character Concept in Evolutionary Biology,* Wagner, G.P., Ed., Academic Press, San Diego, 2001, 363.
67. Nelson, G., Homology and systematics, in *Homology: The Hierarchical Basis of Comparative Biology,* Hall, B.K., Ed., Academic Press, San Diego, 1994, 17.
68. Roth, V.L., Character replication, in *The Character Concept in Evolutionary Biology,* Wagner, G.P., Ed., Academic Press, San Diego, 2001, 81.
69. Gilbert, S.F. and Bolker, J.A., Homologies of process and modular elements of embryonic construction, *J. Exp. Zool. (Mol. Dev. Evol.),* 291B, 1, 2001.
70. Maddison, W.P., Gene trees in species trees, *Syst. Biol.,* 46, 523, 1997.
71. Nichols, R., Gene trees and species trees are not the same, *Trends Ecol. Evol.,* 16, 358, 2001.
72. Heming, B.S., *Insect Development and Evolution,* Comstock Publishing Associates, Ithaca, New York, 2003.
73. Berlese, A., Intorno alle metamorfosi degli insetti, *Redia,* 9, 121, 1913.
74. Truman, J.W., and Riddiford, L.M., The origins of insect metamorphosis, *Nature,* 401, 447, 1999.
75. Wagner, G.P., The biological homology concept. *Annu. Rev. Ecol. Syst.,* 20, 51, 1989.
76. Schlosser, G., The role of modules in development and evolution, in *Modularity in Development and Evolution,* Schlosser, G. and Wagner, G.P., Eds., The University of Chicago Press, Chicago-London, 2004, 519.
77. Nelson, C., Selector genes and the genetic control of developmental genes, in *Modularity in Development and Evolution,* Schlosser, G. and Wagner, G.P., Eds., The University of Chicago Press, Chicago-London, 2004, 17.

78. Cheverud, J.M., Modular pleiotropic effects of quantitative trait loci on morphological traits, in *Modularity in Development and Evolution,* Schlosser, G. and Wagner, G.P., Eds., The University of Chicago Press, Chicago-London, 2004, 132.

79. Minelli, A. and Bortoletto, S., Myriapod metamerism and arthropod segmentation, *Biol. J. Linn. Soc.,* 33, 323, 1988.

80. Holland, L.Z. et al., Sequence and embryonic expression of the amphioxus *engrailed* gene (*AmphiEn*): the metameric pattern of transcription resembles that of its segment-polarity homolog in *Drosophila, Development,* 124, 1723, 1997.

81. Christ. B. et al., Segmentation of the vertebrate body, *Anat. Embryol.,* 197, 1, 1998.

82. Arthur, W., Jowett, T., and Panchen, A., Segments, limbs, homology, and co-option, *Evol. Dev.,* 1, 74, 1999.

83. Budd, G.E., Why are arthropods segmented? *Evol. Dev.,* 3, 332, 2001.

84. Minelli, A. and Fusco, G., Evo-devo perspectives on segmentation: model organisms, and beyond, *Trends Ecol. Evol.,* 19, 423, 2004.

85. Newman, S.A., Is segmentation generic? *BioEssays,* 15, 277, 1993.

86. Panganiban, G., Nagy, L., and Carroll, S.B., The role of the *Distall-less* gene in the development and evolution of insect limbs, *Curr. Biol.,* 4, 671, 1994.

87. Schileyko, A.A., Redescription of *Scolopendropsis bahiensis* (Brandt, 1841), the relations between *Scolopendropsis* and *Rhoda,* and notes on some characters used in scolopendromorph taxonomy (Chilopoda: Scolopendromorpha). *Arthropoda Selecta,* in press.

88. Arthur, W. and Farrow, M., The pattern of variation in centipede segment number as an example of developmental constraint in evolution, *J. Theor. Biol.,* 200, 183, 1999.

89. Gould, S.J., *Ontogeny and Phylogeny,* The Belknap Press of Harvard University Press, Cambridge, MA, 1977.

90. McNamara, K.J., A guide to the nomenclature of heterochrony, *J. Paleontol.,* 60, 4, 1986.

91. Wiens, J.J., Bonnet, R.M., and Chippindale P.T., Ontogeny discombobulates phylogeny: paedomorphosis and higher-level salamander relationships, *Syst. Biol.,* 54, 91, 2005.

92. Schulmeister, S. and Wheeler W., Comparative and phylogenetic analysis of developmental sequences, *Evol. Dev.,* 6, 50, 2004.

93. Bininda-Emonds, O.R.P. et al., From Haeckel to event-pairing: the evolution of developmental sequences, *Theor. Biosci.,* 121, 297, 2002.

94. Smith, K.K., Comparative patterns of craniofacial development in eutherian and metatherian mammals, *Evolution,* 51, 1663, 1997.

95. Velhagen, W.A., Analyzing developmental sequences using sequence units, *Syst. Biol.,* 46, 204, 1997.

96. Jeffery, J.E. et al., Analyzing developmental sequences within a phylogenetic framework, *Syst. Biol.,* 51, 478, 2002.

# Section C

Taxonomy and Systematics of Species Rich Groups (Case Studies)

# 13 Insect Biodiversity and Industrialising the Taxonomic Process: The Plant Bug Case Study (Insecta: Heteroptera: Miridae)

*G. Cassis*
Australian Museum, Research and Collections Branch, Sydney, Australia

*M. A. Wall*
Department of Entomology, San Diego Natural History Museum,
San Diego, California, USA

*R. T. Schuh*
American Museum of Natural History, Division of Invertebrate Zoology,
New York City, USA

## CONTENTS

## ABSTRACT

Insects are the most diverse higher taxon of organisms, comprising more than half of all described species. The rate and scale of species extinction and ecosystem degradation, the so called biodiversity crisis, demands an urgent response by the taxonomic community to comprehensively document global organismal diversity. For 'megadiverse families' within insects, the establishment of predictive classifications that are global in scope and the description of 'all species' are hampered by the scale of the task. To answer this challenge, we support previous calls for industrialising the taxonomic process, involving astronomy-like international collaboration, infrastructural investment, capacity building and taking full advantage of information technology developments. We strongly argue that this unitary approach can be implemented without compromising the hypothesis-driven nature of taxonomic science. The plant bug family Miridae is presented as a case study of this approach.

## 13.1  INTRODUCTION

If we could visualise a tree of life, insects would form the canopy, overshadowing the rest of life (Figure 13.1; insects are the major component of Hexapoda). Nearly a million species of insects make up the 1.7 million species of organisms so far described. Despite their omnipresence, insects are not a recent explosive radiation, nor mere variations on a theme. Insects have a minimum history of 400 million years, and most modern insect orders have been in existence for around 250 million years[1]. Insects are the most dominant and diverse group of terrestrial metazoans by almost all possible measures. Aside from submerged marine habitats, there are few ecological niches that insects have not exploited. In terms of abundance and biomass, insects dominate most terrestrial ecosystems. For instance, arthropods (primarily insects) reach extraordinary biomasses (23.6 kilograms per hectare) and abundances (23.9 million individuals per hectare) in Borneo[2]. Termites alone can reach abundances of up to 10,000 individuals per metre squared[3] and biomasses of 100 g per metre squared[4]. Insects are crucial to terrestrial ecosystem processes such as nutrient cycling[5], seed dispersal[6] and pollination[7].

By virtue of scale alone, no other group epitomises the challenges that species rich taxa present to taxonomy and systematics quite like the insects do. How many of the 350,000 described beetles does one include when reconstructing the phylogeny of Holometabola? How do we reduce duplication of effort so that we do not repeat historic levels of up to 80%[8] synonymy within the Insecta? How do we describe the four to nine million undescribed insects (Table 13.2) within a time frame that meets the demands of scientifically informing the biodiversity crisis?

It is with the biodiversity crisis in mind that taxonomists and the taxonomic method are increasingly faced with questions about relevance. In this chapter we outline the issues that face entomologists in documenting this remarkable diversity of insects. We present the case study of plant bugs (Insecta: Heteroptera: Miridae) as a model group for preserving the taxonomic method but incorporating advances in technology and global cooperation as a means to expediting the documentation process.

## 13.2  ESTIMATES AND DRIVERS OF INSECT DIVERSITY

### 13.2.1  Insect Diversity and Classification

Insects are the most species rich class in Arthropoda, a phylum of considerable diversity even without Insecta included. The vast number of insects and other arthropods suggests that the combination of an exoskeleton, a segmented body plan and jointed appendages have been a recipe for zoological success. The phylogenetic position of the insects within Arthropoda is

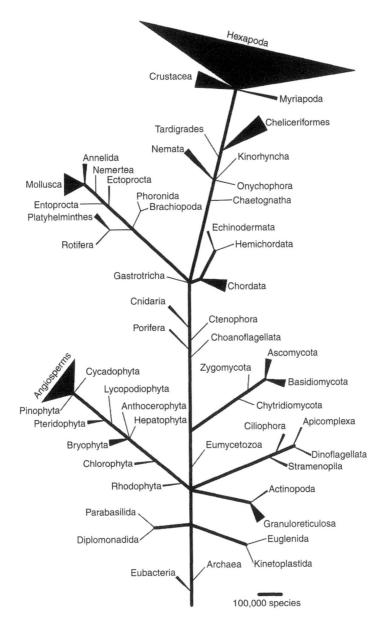

**FIGURE 13.1** Tree of life with terminal branches expanded to represent number of described species in each taxon. Hexapods include insects, springtails, diplurans and proturans. (Tree structure is based on Pennisi[64] and taxon species richness estimates are from Brusca and Brusca[65], Margulis and Schwartz[66] and DSMZ[67].)

contentious, but has mostly come down to arguments about where the root of the phylogenetic tree lies. Most contemporary analyses that include DNA sequence data suggest that insects have arisen from within a paraphyletic Crustacea[9–11], although some authors have suggested that Hexapoda and Crustacea are mutually paraphyletic[12,13]. Within Insecta, interordinal relationships (Figure 13.2) are in some ways poorly resolved, although a great deal of progress has been made in the last quarter of a century. While some clades, such as Holometabola (Figure 13.2), have been well supported since before the time of Hennig, other problematic taxa such as Plecoptera have caused considerable instability in the deep level branching of insect phylogenetic reconstructions.

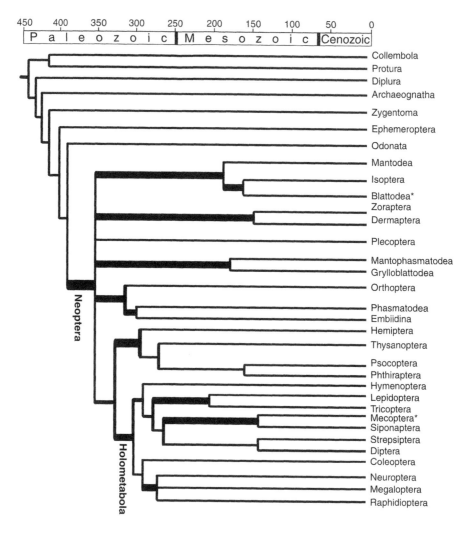

**FIGURE 13.2** Interordinal phylogenetic tree of Hexapoda. Geological time scale is indicated on the left side. Wide branches indicate robust support in contemporary analyses.

The insect phylogenetic tree presented here (Figure 13.2) is a summary tree of recent work[14–16] in the field.

Approximately 925,000 described species[1] are represented in just 32 orders of insects, a tractable higher classification when compared to the approximately 100 orders of vertebrates. Yet, species diversity in the insects is dramatically uneven in distribution. The majority of species are found in just five orders (Table 13.1). Four of these orders (Coleoptera, Diptera, Hymenoptera and Lepidoptera ) belong to Holometabola, a clade characterised by development via complete metamorphosis (Figure 13.2). Together with Hemiptera, these orders represent close to 90% (c. 825,000 species) of the described insect diversity. This uneven distribution of described species diversity extends to every taxonomic level. In fact, just 20 families (Table 13.1) of insects contain a little over 45% of all described insect diversity. Most of these hyperdiverse families are primarily herbivorous, such as Chyrsomelidae, Miridae and Noctuidae, with some notable predaceous (for example, Staphylinidae) and parasitic (for example, Ichneumonidae and Tachinidae) exceptions.

**TABLE 13.1**
**Described Species Diversity within the Hexapods**

| Order | Family | Species | Order | Species |
|-------|--------|---------|-------|---------|
| Coleoptera | | 350,000 | Orthoptera | 20,000 |
| | Curculionidae[68] | 50,000 | Trichoptera | 11,000 |
| | Staphylinidae[69] | 47,000 | Collembola | 9,000 |
| | Cerambycidae[68] | 35,000 | Neuroptera | 6,500 |
| | Chrysomelidae[68] | 35,000 | Odonata | 5,500 |
| | Carabidae[68] | 30,000 | Thysanoptera | 5,000 |
| | Scarabaeidae[68] | 25,000 | Phthiraptera | 4,900 |
| | Tenebrionidae[68] | 18,000 | Psocoptera | 4,400 |
| | Buprestidae[68] | 15,000 | Blattodea | 4,000 |
| | | | Ephemeroptera | 3,100 |
| Lepidoptera | | 150,000 | Phasmatodea | 3,000 |
| | Noctuidae[70] | 25,000 | Isoptera | 2,900 |
| | Geometridae[70] | 21,000 | Siphonaptera | 2,500 |
| | Crambidae[70] | 11,630 | Dermaptera | 2,000 |
| | Arctiidae[70] | 11,000 | Plecoptera | 2,000 |
| | | | Mantodea | 1,800 |
| Diptera[71] | | 120,000 | Diplura | 1,000 |
| | Tipulidae[71] | 10,203 | Protura | 600 |
| | Tachinidae[71] | 9,451 | Mecoptera | 600 |
| | Chironomidae[71] | 7,739 | Strepsiptera | 550 |
| | | | Archaeognatha | 500 |
| Hymenoptera | | 125,000 | Embiidina | 500 |
| | Ichneumonidae[72] | 15,000 | Zygentoma | 400 |
| | Braconidae[72] | 15,000 | Zoraptera | 32 |
| | Formicidae[73] | 11,839 | Grylloblattodea | 26 |
| | | | Mantophasmatodea | 14 |
| Hemiptera | | 90,000 | | |
| | Cicadellidae[74] | 20,000 | | |
| | Miridae[50] | 10,040 | | |

*Note:* Described species diversity within the hexapods. All orders and the 20 largest families are shown.

*Source:* Species richness estimates from Grimaldi and Engel[1] unless otherwise indicated.

## 13.2.2 Drivers of Diversity

The remarkable diversity found in Insecta has been attributed to several extrinsic and intrinsic drivers. Intrinsically, the development of certain 'key innovations' has been associated with increased rates of diversification within specific lineages of insects. Insects were the first animal lineage to evolve powered flight, and wings are considered to be one of their greatest morphological innovations. Specifically, the development of the wing flexion is associated with increased rates of diversification in Neoptera[17]. At a finer taxonomic scale, innovations such as pollen-collecting tentacles in yucca moths[18], and elongated snouts in weevils for preparation of oviposition sites[19] have been associated with species richness in those groups.

The coevolution of insects and plants is undoubtedly a key driver of diversity for many groups of insects. At the broadest scale, the explosion of insect diversity in the fossil record corresponds with evolution and diversification of seed plants[1]. More specifically, several studies[20,21] have shown that origins of angiosperm feeding within the beetles are associated with increased rates

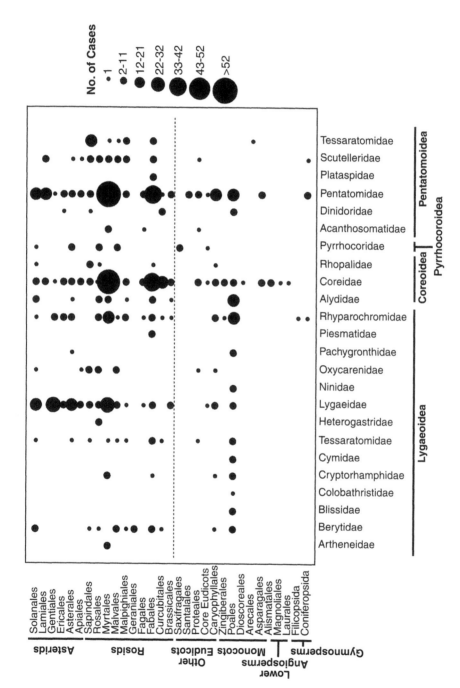

**FIGURE 13.3** Host plant affinities of families of land bugs belonging to the infraorder Pentatomomorpha (Heteroptera) with ordinal taxa of land plants. Dashed line separates asterids and rosids from remainder of land plants.

of speciation. Another example of insect/plant coradiation can be found within Australian Heteroptera. The landbug infraorder Pentatomomorpha is primarily phytophagous and associated with a broad range of vascular plants, but primarily flowering plants. The documentation of their hosts[22] reveals that the majority of Australian pentatomomorphans are associated with rosid and asterid angiosperms (Figure 13.3). Most extant pentatomomorphan families appear in the fossil record in the early Cenozoic[23], corresponding with increases in the dominance and diversification of rosid and asterid angiosperms[24,25]. Although there is a paucity of insect-plant cospeciation case studies (see Rønsted, *Chapter 9*), and stratigraphic correlations as proposed in the previous example are very coarse, the abundance of relationships between insects and plants is fundamental in explaining the adaptive radiation of insect herbivores.

### 13.2.3 ESTIMATES OF INSECT SPECIES RICHNESS

Understanding of the relationship of insects and plants has been central to developing estimates of global insect species richness. Contemporary debate over global insect species diversity began with Erwin's[26] estimation of about 30 million species of terrestrial arthropods globally. Erwin arrived at this figure by fogging a single species of tropical forest tree with insecticide; determining the number of species in the 'beetle rain'; estimating the number of host-specific beetles; multiplying this by the number of tropical tree species; and extrapolating to a total number of arthropods based on the known proportions of other arthropod groups relative to beetles. Many of Erwin's assumptions have been criticised in the literature[27,28], particularly his estimates of plant host specificity for insect herbivores. In fact, much of the variation we see in estimates of global insect species diversity (Table 13.2) can be attributed to variation in estimates of host specificity in insects[27,29–32].

Nonetheless, most contemporary authors agree on estimates of between 5 and 10 million species of insects (Table 13.2). But with the exception of an early conservative estimate by Hodkinson and Casson[33], the general trend has been that of decreasing estimates for insect species richness (Table 13.2). This trend, combined with high levels of synonymy in existing names in some groups[34,35], suggests that global insect species richness may be at the low end of 5–10 million species. However the pendulum may now be beginning to swing back towards higher estimates. Ødegaard[32], in a recent study incorporating phylogenetic relatedness, suggested a shift from the low end to the high end of the 5- to 10-million range[27,31]. Even higher estimates can be anticipated if DNA taxonomy or barcoding[36] and phylogenetic species concept[37] approaches take a more central role in species delineation (see Seberg and Petersen,

**TABLE 13.2**
**Estimates of Insect Species Diversity in the Scientific Literature and the Year in Which They Were Published**

| Estimated Number of Species (in millions) | Reference | Year |
|---|---|---|
| 30 | Erwin[26] | 1982 |
| 7–80 | Stork[28] | 1988 |
| 5 | Gaston[75] | 1991 |
| 1.8–2.6 | Hodkinson and Casson[33] | 1991 |
| 12.5 | Hammond[76] | 1992 |
| 5 | Ødegaard et al.[27] | 2000 |
| 2.0–3.4 | Dolphin and Quicke[77] | 2001 |
| 4–6 | Novotny et al.[29] | 2002 |
| 10 | Ødegaard et al.[32] | 2005 |

*Chapter 3*). For example, Agapow et al.[38] found a 48% increase in species names after phylogenetic revision and application of the phylogenetic species concept. In considering these adjustments, there is the potential for 7.4–14.8 million phylogenetically defined species of insects in the world.

Leaving all of these hypothetical assumptions, abstractions and extrapolations behind, there is a clear message in all of these estimates: insect taxonomists have a considerable job to meet the challenge of developing an encyclopedia of life[39]. In a survey of the Zoological Record from 2000 through 2004, we found an average of 8,500 new insect species described per year. At this rate, it will take 480 to 1,070 years to describe the world insect fauna (based on estimates of 5–10 million insect species). Clearly this rate of species description is not adequate in meeting the contemporary needs of society, including ameliorating the alarming decline in biodiversity. This rate must be multiplied by a factor of 10–24 in order to document scientifically the world's undescribed insect fauna in the next 100 years, and 100 times that if the fauna had to be described in the next 25 years, as some people have suggested.

## 13.3  DEALING WITH DIVERSITY: FROM THE COTTAGE TO THE FACTORY

The description of nearly one million insects in 250 years is not a meagre effort, but the issue of completing the task in a given time is compelling because of the universally recognised biodiversity crisis. The world is quickly approaching a sixth major extinction event[40], with species extinction vastly exceeding prehistorical rates[41]. In such an environment, the onus on completing biotic documentation is globally accepted, as evidenced by the development of international instruments and frameworks (such as the Convention on Biological Diversity and the Global Biodiversity Information Facility and Global Taxonomic Initiative) and funding schemes such as various US National Science Foundation programmes (for example, Partnerships for Enhancing Expertise in Taxonomy, Assembling the Tree of Life, Planetary Biodiversity Inventory), as well as the repeated call to arms[39,42,43] in the scientific literature.

The task, however, is not just a question of increasing resources and infrastructure, including arresting the alarming decline in training the next generation of taxonomists (see Schram, *Chapter 2*). Scientists such as Godfray[42] and Wheeler[43] have independently called for a 'new taxonomy,' where the process is global in scope, with emphasis on a new astronomy-like culture of cooperative research, while taking best advantage of information technology. Although there is not a consensus on a new methodology, amongst many taxonomic entomologists there is a tacit agreement that conventional, hypothesis-driven and morphologically based descriptive taxonomy remains at the core of the task. However, the question remains, if we retain a traditional taxonomic core, can international cooperation and computer technology alone increase the rate of species description by two orders of magnitude within a generation?

There are a number of other interlocking factors, such as classification and collection shortfalls (see Utteridge and de Kok, *Chapter 18*), that seriously impede the process of documenting all the species in nature. For taxa, such as the 'Big 20' families of insects, suprageneric classifications are often in contention and lack stability. For example, in the leaf beetles, family Chrysomelidae, there is polarisation in the suprageneric classifications being utilised, with some workers accepting the phylogenetic classification and recognition of 11 subfamilies of Reid[44], while others follow the previous, and more traditional classification[45], recognising 18 subfamilies. Likewise in the plant bug family Miridae, Schuh's[46] phylogenetic classification recognises 13 tribes, whereas the main alternative of Carvalho[47] lists 26 tribes, and some tribes in both classifications are placed in different subfamilies. This classificatory instability can result in uncertain species placement, increased rates of synonymy and the erection of many unnecessary monotypic taxa.

In addition, much has been made of the value of existing collections[48], but it is also clear that many cryptic groups of insects require specialised collecting. In the past 50 years there has been an ever increasing trend of taxonomists working on insect families rather than groups of families or even orders. This has resulted in a concomitant change from general to specialised collecting. These new survey efforts have led to the discovery of large numbers of species that were not represented in existing collections. For example, in a recent revision[49] of Australian barkbugs (Aradidae: Mezirinae), 45 of the 93 species represented in Australia were described as new, based on material collected primarily by the revision's author.

In summary, if the intent is to describe all insects in nature, and not just those in existing collections, then the taxonomic impediment is not merely a shortfall in species description. Codependent classification and collection impediments require parallel attention. In the following sections we provide a case study that documents the methodological transition from the single investigator to an industrial model of taxonomy that strives to overcome taxonomic classification and collection impediments within the plant bug family Miridae.

## 13.4 PLANT BUG DIVERSITY, BIOLOGY AND CLASSIFICATION

The plant bugs, or family Miridae (Figure 13.4), are the most speciose family in the hemipteran suborder Heteroptera, with 1,507 genera and 10,040 species described[50]. This has long been considered to be an underestimate of the number of plant bugs worldwide. In fact, the number of species currently described is, at most, half of that to be found in nature[51]. This makes Miridae one of the most species rich families of organisms known (Table 13.1).

Although mirids are known colloquially as plant bugs, they exhibit broader ecological diversity than their common name would suggest. Numerous taxa are largely ground dwelling in their habits (for example, Cylapinae: *Vannius* complex[52]; Phylinae: Hallodapini). Likewise, though most species are phytophagous, a significant number of taxa are predaceous (for example, Isometopinae, Deraeocorinae: Termatophylini[53]), and information is gathering that numerous species are zoophytophagous (for example, Bryocorinae: Dicyphini[54]). The biology of the basal taxa is not well known, but it is apparent that members of Cylapinae, such as the type genus *Cylapus,* feed on fungal mycelia[55]. For the phytophagous plant bug species, there is accumulated evidence that the majority of species show a high degree of host specificity, with most restricted to a single host plant[56–58].

Divided into seven subfamilies, 75% of the diversity of plant bugs occurs in three subfamilies; Mirinae, Orthotylinae and Phylinae. Mirinae are the most diverse subfamily of plant bugs, with over 40% of all mirid species. However, Orthotylinae and Phylinae (Figure 13.5) together contain an equivalent number (35%). One of the more outstanding features of the Orthotylinae and Phylinae is the multiple evolutionary development of ant mimetic taxa (Figure 13.4A), with

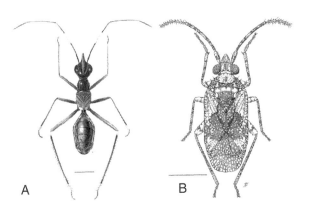

A                    B

**FIGURE 13.4** Two undescribed species of Miridae from Australia: *Mymecorides* sp. an ant mimic (A) and *Peritropis* sp. (B). Scale bars = 1 mm.

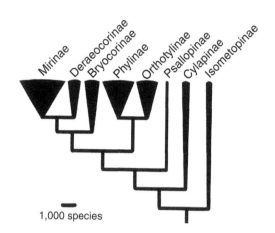

**FIGURE 13.5** Phylogeny of Miridae with terminal branches expanded to represent number of described species in each taxon.

myrmecomorphy being found in hundreds of species and many genera. The majority of Orthotylinae and Phylinae is highly host specific, occurring primarily on meristematic growth of developing flowers and/or shoots.

## 13.5 PLANT BUGS AS A COTTAGE INDUSTRY

The taxonomic history of the group has modest beginnings, with Linnaeus[59] describing 17 species in 1758. A species description accumulation curve for Miridae subfamilies Orthotylinae and Phylinae (Figure 13.6) indicates that the description of world fauna was largely gradual until the last quarter of the nineteenth century. Around this time, numerous European scientists began describing new species from the Southern hemisphere, the Indian subcontinent and Central America. The twentieth century saw the increased saturation of species descriptions for the Nearctic and Palaearctic regions, as well as the continued enhancement of the plant bug faunas of Latin America and Africa. Plant bug alpha taxonomy was transformed between 1957 and 1960 when Carvalho

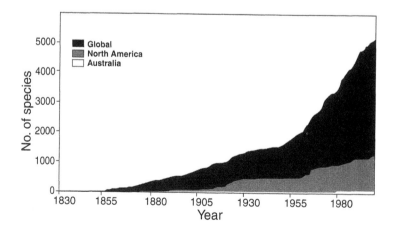

**FIGURE 13.6** Global and regional chart of species description accumulation for the Orthyotylinae and Phylinae from 1830 to present.

published the multipart *Catalogue of the Miridae of the World*. An updated version of the catalogue was published in 1995 by Schuh[58], who now maintains an up-to-date online version[50]. Since Carvalho's catalogue, species have been described at a rate of approximately 145 species per year. This represents a doubling of the rate of the previous 50 years (73 species per year) to the publication of the catalogue.

Plant bug taxonomy has historically been a cottage industry in which single investigators have worked on regional collections, producing a modest list of species names over a lifetime. In total, some 340 authors (excluding junior authors) have published 13,048 species group names in Miridae. Although synthetic 'global' taxonomists have emerged throughout history, the vast majority of plant bug taxonomists (73%) have described fewer than 15 species. Until recently, most plant bug taxonomists worked alone. Only 13% of plant bug names are described in multi-author papers, suggesting that the image of taxonomists as lone investigators working in isolation is an apt description of past behaviour. However, when these data are partitioned by decade of description, a different image emerges. Since the 1970s, the proportion of plant bug species described in multi-authored papers has steadily increased by an average of 5% a decade. In fact, in the last ten years, 40% of plant bug species names are the product of a collaborative effort.

What are the benefits of collaborative efforts? Collaboration almost invariably increases taxonomic and/or geographic breadth. In particular the collaboration of global authorities with regional experts reduces redescription of geographically widespread species. For example, Carvalho, the most prolific global plant bug worker in history, occasionally collaborated with regional experts to produce works on geographically restricted faunas. Broadening taxonomic and geographic breadth through collaboration has the potential to increase the stability, universality, and predictive value of classifications. On the other hand, reclusive approaches may produce deleterious results, as for example the near simultaneous but independent work of Knight[60] and Kelton[61] on the genus *Reuteroscopus*. Within Miridae, levels of synonymy are approximately 23%, suggesting that there is room for improvement, with collaboration offering an obvious possibility.

Although many taxonomists have contributed to the plant bug taxonomic literature, just 22 taxonomists have described 75% of plant bug species. Do these 'über-taxonomists' represent the ideal for which we should strive? Obviously, the introduction of species names should not be the only measure by which we judge the output of taxonomists. For instance, the value of Stichel's 316 species names of Miridae is markedly decreased by the subsequent treatment of 285 (90%) of those names as junior synonyms. In contrast, the American entomologist Lattin is not the primary author for any plant bug name; however, just five of his students have produced approximately 750 species group names, with very low rates of synonymy. Nonetheless if we are to attain the rates of taxonomic output necessary to chronicle the diversity of life on Earth, then creating the infrastructure and resources for efficient networks of collaborating taxonomists has the greatest potential for advancing the cause.

## 13.6 TAXONOMIC, COLLECTIONS AND CLASSIFICATION IMPEDIMENTS

As with many insect taxa, the taxonomic impediment for Miridae exists primarily in the Southern hemisphere, particularly in Australia and Southern Africa. A representation of species richness by country (Figure 13.7) indicates that the plant bug faunas of continental United States, followed by parts of the Palaearctic, Latin America and Sub-Saharan Africa, are apparently the most diverse areas for Miridae in the world. Despite the undoubted high species diversity of plant bugs in these regions, this map is more a representation of sampling bias and the in-country presence of mirid specialists in the twentieth century, rather than a true representation of global species diversity patterns. For example, in Australia the plant bug fauna is represented by about 200 described species, which would signal a depauperate fauna. However, between 1995 and 2001, we have collected at over 400 sites across Australia, resulting in the accumulation of about 100,000 new specimens. These have

**FIGURE 13.7** Map of plant bug species richness. The patterns suggest a high degree of sampling bias and correlation with distribution of plant bug taxonomists.

been roughly sorted into 2,000 species, which equates to an order of magnitude increase on published knowledge.

Based on these figures alone, the Australian plant bug fauna would be categorised as one of the most species rich in the world. However, the sampling of the Australian flora is far from adequate. In recent surveys of the Australian Miridae, we have sampled just over 1,200 species of flowering plants and found that 75% of Australian plant bugs are known from only one or two hosts. Although we do not keep records for host plant species sampled without plant bugs, our sampling efforts are to this stage only a fraction of the 18,000 known species of plants comprising the Australian flora. In addition, most localities have only been visited once, and temporal turnover patterns for plant bugs at these localities is largely unknown. In the few cases where there has been repeat sampling, a highly significant temporal turnover in plant bug species has been found[62].

Other factors also contribute greatly to the low rate of species description of insect faunas in the Southern hemisphere, and for plant bugs the lack of adequate generic classifications is a fundamental issue. A historical overview of the description of the Australian Miridae indicates that Northern hemisphere generic concepts were often applied to what we are finding to be a highly endemic Australian plant bug fauna. For instance, *Melanotrichus australianus* Carvalho (Orthotylinae) is the only representative of this genus in the Southern hemisphere. Cursory examination of the species indicates that it is clearly misplaced, and in fact belongs to an undescribed genus in Phylinae. These determinations can often only be made in hindsight; however, it emphasises the importance of quickly building classificatorial frameworks for poorly described faunas.

## 13.7 PLANT BUGS IN THE TWENTY-FIRST CENTURY: INDUSTRIAL CYBER-TAXONOMY

### 13.7.1 PLANT BUG PLANETARY BIODIVERSITY INVENTORY

In considering the history and current status of plant bug taxonomy, eroding the taxonomic impediment at an accelerated rate requires an enhancement of collaborative arrangements and applying methods from the information technology revolution. This approach is not unique to the plant bugs and is in line with the strategic rethinking of taxonomy as proposed by others[40,41]. To realise this new taxonomic vision does, however, require significant financial investment, as provided by programmes such as those funded by the US National Science Foundation. One such programme, the Planetary Biodiversity Inventory (PBI), established both a funding programme and articulated goals for undertaking and accelerating taxonomic research. The programme was established in 2003 with the goal of documenting on a global scale the diversity of species rich monophyletic taxa. There were recommendations under the programme guidelines that projects would focus on species description, phylogenetic classifications, global cooperation and information technology.

The Plant Bug Inventory project funded under the PBI programme focuses on plant bug sister subfamilies (Figure 13.5), Orthotylinae and Phylinae. These two subfamilies were chosen as a model group for several reasons. There are significant taxonomic, collection and classification impediments to overcome in this group. Yet there is sufficient existing taxonomic expertise in plant bugs that allows for both generational and international capacity building. Moreover, the documentation of key biological attributes, such as host plant associations, ant mimicry and distribution, allows the outcomes of the plant bug taxonomic research to inform evolutionary and biodiversity research more broadly. In particular, it is envisaged that comprehensive species documentation and the establishment of phylogenetic classifications will allow for studies in coevolution, biodiversity surrogacy and conservation planning, and evolution of ant mimicry. The strategic planning for the first two years of this plant bug project has involved the establishment and development of human, collection and information resources and infrastructure. The key elements that have been implemented and a description of the lessons are as follows.

### 13.7.2  Human Resources

Large-scale taxonomic efforts can only be accomplished through coordinated effort. Global cooperation guarantees that all participants are working towards unified classification and data standards.
In addition, strategic planning allows for a structured division of labour so as to avoid duplication
of effort. The Plant Bug Inventory team comprises an international assemblage of established
workers, postdoctoral fellows, postgraduate and undergraduate students. These individuals are
located in institutions situated in five countries on three continents. As part of this arrangement,
dual research hubs were established at the American Museum of Natural History and the Australian
Museum, where the development of the information technology, student training and specimen
preparation are focused. This degree of centralisation has enabled project management and focused
student training. However, it is also crucial that the 'satellite institutions' involved in the project
have equivalent access to research tools. This was accomplished through the development of
Internet-based research tools as described below.

### 13.7.3  Specimen Resources and Field Work

The most important existing specimen sources for information on plant bug diversity reside in
approximately 20 institutional collections, mostly in Europe and North America. The collections
contain approximately 500,000 specimens of Orthotylinae and Phylinae. Although much information
derived from these collections exists in the published literature, that information is not easily recovered
digitally from the literature for further use and evaluation. To enable this, the development of a
specimen-level database was necessary to incorporate the published information with new survey data.
    The need for additional collecting in Miridae, as with most groups of insects, is still great in
many parts of the world. Collection priorities were established through the evaluation of existing
institutional collections, and coarse-scale mapping of the associated data indicated broad survey
gaps. The PBI project collecting has focused particularly on Australia and South Africa, because
of the paucity of described species in the face of known high plant diversity[63]. These efforts have
produced more than 150,000 specimens in addition to those already available in collections.

### 13.7.4  Producing Descriptions

The Plant Bug PBI team has adopted the approach of producing well documented descriptions, as
opposed to totally uniform descriptions across all investigators. This involved the identification of
minimal attributes to be recorded, without enslaving investigators to an overly uniform style of
description. Because of past knowledge of plant bug systematics, male specimens were used as the
primary gender for species delineation. For each species the following attributes have been documented as the minimum data set: illustrated male genitalia; illustrated body form in dorsal view;
morphometric measurements in a common format; and scanning and light micrographic images of
any additional diagnostic morphological features. This 'minimum set of attributes' can also be
tagged and databased to generate succinct web-based 'species pages' that complement the formally
published descriptions. In addition to the intrinsic attributes of the species, a minimum set of
extrinsic attributes is recorded, which include: host plant species and families; host plant specimen
herbarium voucher numbers; collection event data; and point location information and associated
hierarchical distributional descriptors.

### 13.7.5  Technical Resources

One of our major arguments in addressing the taxonomic impediment both generally and specifically
for plant bugs involves the utilisation of web-based information technology. From a research infrastructural point of view, the successful incorporation of an integrated set of information technology tools
is regarded as the 'silver bullet' to the global cooperative framework proposed by us, amongst others.

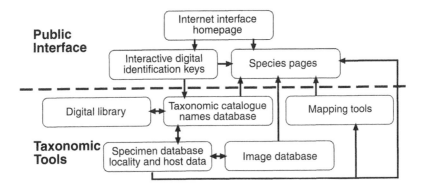

**FIGURE 13.8** Information technology infrastructure for the plant bug PBI.

In Figure 13.8 we outline the information technology framework for the Plant Bug PBI project, which is divided into the cyber-based taxonomic tools, and the overarching Internet interface that is designed to provide universal and immediate access to the generated taxonomic outputs. The key cyber-taxonomic tools that are implemented or in development are described below:

**Web-based systematic catalogue.** Within the confines of available funding and technological understanding, the Plant Bug PBI team has chosen to place as many research tools and as much research information as possible on the Internet. At the core of this approach is a systematic catalogue of Miridae. This source, in the form of a relational database, provides an up-to-date bibliographic history for all taxa in the group under study. It provides a powerful tool for organising and retrieving information on nomenclature, classification, host associations and geographical distributions. The relational database allows for potentially continuous updating and the rapid delivery of identical results to users anywhere in the world, thus maintaining a contemporary species list for Miridae. Beyond its capacity to serve catalogue data, the systematic catalogue serves as a platform for the retrieval of pages from the digital library and other key information from the specimen and image databases.

**Digital library.** Taking advantage of the relational structure of the systematic catalogue, a digital library of relevant literature, comprising some 30,000 pages has been uploaded to the web in searchable PDF format (http://research.amnh.org/pbi/catalog). These pages relate to the taxonomy, morphology and natural history of Orthotylinae and Phylinae. The most obvious limitation of this approach is that permission for copyrighted material published during the last 70 years could not always be secured, and in such cases this literature is not incorporated into the digital library. Publications that are already available on the web, especially those published very recently, can be included through the use of linking Uniform Resource Locators (URLs). The rewards produced by this digital archive go well beyond its relatively modest production costs. It provides access to a near comprehensive body of primary literature, including access to the older literature, which often has restricted availability, particularly to scientists in developing countries.

**Web-based specimen database.** Although the structural attributes of specimen data have been widely agreed upon for some time (for example, Darwin Core schema), the approaches to acquiring and retrieving those data are less well settled. In an effort to accommodate the international partners on the Plant Bug PBI team, the project implemented a web-based approach to the acquisition of specimen data. This approach takes advantage of high speed Internet connections and has the desirable property of allowing for centralised geo-referencing, real-time data entry, and the security of using a centralised enterprise level computer server with regular backups and institutional support.

**Matrix code unique specimen identification.** Whilst unique specimen identification has long been used in vertebrate collections, the 'barcoding' (not to be confused with DNA barcoding) of insect specimens has become a relatively common practice only in the last few years. Unique specimen identification allows for the tracking of information otherwise not possible, and particularly for the rapid retrieval of database records. Yet the codes may require handling of the specimens in order to be read or might inordinately increase the amount of space required to house collections. The Plant Bug PBI team has adopted the use of 'matrix code' labels, which provide the benefits of unique specimen identification. These labels are relatively small and only increase the total amount of space occupied by the collection by one third. Their small size does not, however, increase specimen handling, as the specimens are machine readable without removing them from the collection.

**Real-time mapping/host data from labels.** The integration of the specimen database with the systematic catalogue allows for the real-time mapping of species distributions and the assessment of host specificity from actual specimen data. Geo-referenced specimen data can also be easily exported to GIS software for generation of distribution maps for taxonomic manuscripts. Also, voucher material of plants is collected in the field, determined by botanists, digitally scanned and deposited in various herbaria. The host plant data are then linked to specimens through herbaria accession numbers and the unique identifiers associated with plant bug specimens.

**High-resolution digital imaging.** The description and documentation of taxa can be greatly enhanced through the use of effective imaging and illustration techniques. The Plant Bug PBI team has adopted the use of digital imaging systems that allow for the rapid capture of high-resolution images. These images are supplemented with scanning electron micrographs of specialised morphology. All of these images are databased and linked with specimens, resulting in an image morphological databank for Miridae.

**Species pages and integration of information on the web.** Using the digitised information sources described above, taxon information is combined into 'species pages' on the web. These displays incorporate real-time nomenclatural, descriptive, host plant, distributional and bibliographic information, as well as morphological imagery. This allows for a comprehensive perspective on the attributes of individual plant bug species. End users may arrive at these pages through Internet search engines, the plant bug online catalogue, or multi-entry web-based identification keys that are being developed by the Plant Bug PBI team.

## 13.8 CONCLUSIONS

To complete the tree of life we must assemble all of the pieces. The possibility of an endpoint in this task may appear remote, particularly within the lifespan of existing taxonomists. To achieve this goal, the comprehensive documentation of species is a necessary objective, worthy of strategic planning and investment. The insects are a dauntingly diverse taxon, whose complete description and cataloguing in a short time span will take a Herculean effort. In reaching for this outcome, it is important to overcome misconceptions that taxonomy is nothing more than a cataloguing process. Wheeler[41] has made the necessary defence of taxonomy, that it is a hypothesis-driven science. The outputs of taxonomy (such as character homology, taxa and classifications) are the foundation upon which most of biological science rests, and cannot be tossed out for expediency. The speed with which individual investigators can recognise and diagnose new taxa without doubt will occupy some minimum period of time. This is simply part of the analytic process and the fact that species as we understand them are concepts, not self-identifying entities in nature. So the question remains, how do we maintain the cornerstones of traditional taxonomy and ramp up the effort? In this chapter we have argued that the tools of industrial taxonomy must derive from the proper mix of human power, collaboration and technology.

Almost everyone agrees that the World Wide Web and digital technology have the potential to accelerate taxonomy in the twenty-first century. However, the acceleration of taxonomy is not simply increasing the rate of species descriptions, but also greatly enhancing the availability of data to end users. In line with the original PBI objectives, the Plant Bug team has developed and implemented technology that simultaneously assists plant bug taxonomists and enhances broader accessibility of taxonomic data (Figure 13.8). Through the use of relational databases and custom web applications, users can generate 'material examined' lists for taxonomic manuscripts, track collection data for rare taxa, query morphological and molecular data for phylogenetic analyses, examine image libraries for studies in comparative morphology, or generate the data necessary to build a species richness map for a biological preserve. Furthermore, the integration of these tools with a team-based approach allows for the division of taxonomic effort, while at the same time being able to focus on the larger problem of producing an up-to-date classification for monophyletic groups on a worldwide basis.

Combining traditional taxonomic approaches with global collaboration, centralised web-accessible plant bug data and targeted biological survey work, the Plant Bug PBI has a model for high taxonomic output in Miridae. The success of this project will be judged by those within the taxonomic community, but also by other stakeholders, stretching from the biologically curious to environmental decision makers. Whilst no single investigator can possibly master any megadiverse group, applying suitable web-based technology can effectively couple the skills of multiple investigators working towards a common goal.

## ACKNOWLEDGEMENTS

Sincere appreciation is extended to Hannah Finlay for illustrations in Figure 13.4. Additional thanks are given to Lance Wilkie, Michael Elliot and Gareth Carter for assistance with Figures 13.3, 13.6 and 13.7. This paper was supported in part by the NSF Planetary Biodiversity Inventory grant DEB-0316495.

## REFERENCES

1. Grimaldi, D. and Engel, M.S., *Evolution of the Insects,* Cambridge University Press, New York, 2005.
2. Dial, R.J. et al., Arthropod abundance, canopy structure, and microclimate in a Bornean lowland tropical rainforest, *Biotropica,* 38, 643, 2006.
3. Watt, A.D. et al., Impact of forest loss and regeneration on insect abundance and diversity, in *Forests and Insects,* Watt, A.D., Stork, N.E., and Hunter, M.D., Eds., Chapman and Hall, London, 1997, 273.
4. Eggleton, P. et al., The diversity, abundance, and biomass of termites under differing levels of disturbance in the Mbalmayo Forest Reserve, southern Cameroon, *Philos. Trans. R. Soc. Lond. B,* 351, 51, 1996.
5. Bignell, D.E. et al., Termites as mediators of carbon fluxes in tropical forest: budgets for carbon dioxide and methane emissions, in *Forests and Insects,* Watt, A.D., Stork, N.E., and Hunter, M.D., Eds., Chapman and Hall, London, 1997, 109.
6. Berg, R.Y., Myrmecochorous plants in Australia and their dispersal by ants, *Aust. J. Bot.,* 23, 475, 1975.
7. Buchmann, S.L. and Nabhan, G.P., *The Forgotten Pollinators,* Island Press, Washington, DC, 1996.
8. Gaston, K.J. and Mound, L.A., Taxonomy, hypothesis testing and the biodiversity crisis, *Proc. R. Soc. Biol. Sci. B,* 251, 139, 1993.
9. Regier, J.C., Shultz, J.W., and Kambic, R.E., Pancrustacean phylogeny: hexapods are terrestrial crustaceans and maxillopods are not monophyletic, *Proc. R. Soc. Biol. Sci. Ser.,* 272, 395, 2005.
10. Babbitt, C.C. and Patel, N.H., Relationships within the Pancrustacea: examining the influence of additional Malacostracan 18S and 28S rDNA, in *Crustacea and Arthropod Relationships,* Koenemann, S. and Jenner, R.A., Eds., CRC Press, Boca Raton, 2005, 275.
11. Giribet, G. et al., The position of crustaceans within Arthropoda: evidence from nine molecular loci and morphology, in *Crustacea and Arthropod Relationships,* Koenemann, S. and Jenner, R.A., Eds., CRC Press, Boca Raton, 2005, 307.
12. Cook, C.E., Yue, Q.Y. and Akam, M., Mitochondrial genomes suggest that hexapods and crustaceans are mutually paraphyletic, *Proc. R. Soc. Biol. Sci. B,* 272, 1295, 2005.
13. Carapelli, A. et al., Relationships between hexapods and crustaceans based on four mitochondrial genes., in *Crustacea and Arthropod Relationships,* Koenemann, S. and Jenner, R.A., Eds., CRC Press, Boca Raton, 2005, 295.
14. Willmann, R., Phylogenetic relationships and evolution of insects, in *Assembling the Tree of Life,* Cracraft, J. and Donoghue, M.J., Eds., Oxford University Press, New York, 2004, 330.

15. Whiting, M.F., Phylogeny of the holometabolous insects: the most successful group of terrestrial organisms, in *Assembling the Tree of Life*, Cracraft, J. and Donoghue, M.J., Eds., Oxford University Press, New York, 2004, 345.

16. Terry, M.D. and Whiting, M.F., Mantophasmatodea and phylogeny of the lower neopterous insects, *Cladistics*, 21, 240, 2005.

17. Mayhew, P.J., Shifts in hexapod diversification and what Haldane could have said, *Proc. R. Soc. Lond. B*, 269, 969, 2002.

18. Pellmyr, O. and Krenn, H.W., Origin of a complex key innovation in an obligate insect-plant mutualism, *Proc. Natl. Acad. Sci. USA*, 99, 5498, 2002.

19. Anderson, R.S., Weevils and plants: phylogenetic versus ecological mediation of evolution of host plant association in Curculioninae (Coleoptera: Curculionidae), *Mem. Entomol. Soc. Can.*, 165, 197, 1993.

20. Farrell, B.D., 'Inordinate fondness' explained: why are there so many beetles? *Science*, 281, 555, 1998.

21. Marvaldi, A.E. et al., Molecular and morphological phylogenetics of weevils (Coleoptera, Curculionoidea): do niche shifts accompany diversification? *Syst Biol*, 51, 761, 2002.

22. Cassis, G. and Gross, G.F., *Hemiptera: Heteroptera (Pentatomomorpha)*, CSIRO Publishing, Melbourne, 2002.

23. Schcherbakov, D.E. and Popov, Y.A., Superorder Cimicidae Laicharting, 1781 Order Hemiptera Linné, 1758. The bugs, cicadas, plantlice, scale insects, etc., in *History of Insects*, Rasnitsyn, A.P. and Quicke, D.L.J., Eds., Kluwer Academic Publishers, Dordrecht, 2002, 143.

24. Niklas, K.J., Tiffney, B.H., and Knoll, A.H., Patterns in vascular plant diversification, *Nature*, 303, 614, 1983.

25. Davies, T.J. et al., Darwin's abominable mystery: insights from a supertree of the angiosperms, *Proc. Natl. Acad. Sci. USA*, 101, 1904, 2004.

26. Erwin, T.L., Tropical forests: their richness in Coleoptera and other arthropod species, *Coleopterists' Bulletin*, 36, 74, 1982.

27. Ødegaard, F., How many species of arthropods? Erwin's estimate revised, *Biol. J. Linn. Soc.*, 71, 583, 2000.

28. Stork, N.E., Insect diversity: facts, fiction and speculation, *Biol. J. Linn. Soc.*, 35, 321, 1988.

29. Novotny, V. et al., Low host specificity of herbivorous insects in a tropical forest, *Nature*, 416, 841, 2002.

30. Novotny, V. and Basset, Y., Review: host specificity of insect herbivores in tropical forests, *Proc. R. Soc. Biol. Sci. B*, 272, 1083, 2005.

31. Ødegaard, F. et al., The magnitude of local host specificity for phytophagous insects and its implications for estimates of global species richness, *Conserv. Biol.*, 14, 1182, 2000.

32. Ødegaard, F., Diserud, O.H., and Ostbye, K., The importance of plant relatedness for host utilization among phytophagous insects, *Ecol. Lett.*, 8, 612, 2005.

33. Hodkinson, I.D. and Casson, D., A lesser predilection for bugs — Hemiptera (Insecta) diversity in tropical rainforests, *Biol. J. Linn. Soc.*, 43, 101, 1991.

34. Alroy, J., How many named species are valid? *Proc. Natl. Acad. Sci. USA*, 99, 3706, 2002.

35. Solow, A.R., Mound, L.A., and Gaston, K.J., Estimating the rate of synonymy, *Syst. Biol.*, 44, 93, 1995.

36. Tautz, D. et al., A plea for DNA taxonomy, *Trends Ecol. Evol.*, 18, 70, 2003.

37. Nixon, K.C. and Wheeler, Q.D., An amplification of the phylogenetic species concept, *Cladistics*, 6, 211, 1990.

38. Agapow, P.M. et al., The impact of species concept on biodiversity studies, *Q. Rev. Biol.*, 79, 161, 2004.

39. Wilson, E.O., The encyclopedia of life, *Trends Ecol. Evol.*, 18, 77, 2003.

40. Thomas, J.A. et al., Comparative losses of British butterflies, birds, and plants and the global extinction crisis, *Science*, 303, 1879, 2004.

41. May, R.M., Lawton, J.H., and Stork, N.E., Assessing extinction rates, in *Extinction Rates*, Lawton, J.H. and May, R.M., Eds., Oxford University Press, New York, 1995, 1.

42. Godfray, H.C.J., Challenges for taxonomy — the discipline will have to reinvent itself if it is to survive and flourish, *Nature*, 417, 17, 2002.

43. Wheeler, Q.D., Taxonomic triage and the poverty of phylogeny, *Philos. Trans. R. Soc. Lond. B.*, 359, 571, 2004.

44. Reid, C.A.M., Spilopyrinae Chapuis: a new subfamily in the Chrysomelidae and its systematic placement, *Invertebr. Taxon*, 14, 837, 2000.

45. Gressit, J.L. and Kimoto, S., Chyrsomelidae (Coleopt.) of China and Korea, Part 1, *Pacific Insects Monographs,* 1A, 1, 1961.

46. Schuh, R.T., Pretarsal structure in the Miridae (Hemiptera) with a cladistic analysis of relationships within the family, *Am. Mus. Novit.,* 2585, 1, 1976.

47. Carvalho, J.C.M., On the major classification of the Miridae (Hemiptera): with keys to subfamilies and tribes and a catalogue of the world genera, *An. Acad. Bras. Cienc.,* 24, 31, 1952.

48. Suarez, A.V. and Tsutsui, N.D., The value of museum collections for research and society, *Bioscience,* 54, 66, 2004.

49. Monteith, G.B., Revision of the Australian flat bugs of the subfamily Mezirinae (Insecta: Hemiptera: Aradidae), *Mem. Queensl. Mus.,* 41, 1, 1997.

50. Schuh, R.T., Plant Bug Inventory Systematic Catalog, http://research.amnh.org/pbi/catalog/, 2005.

51. Henry, T.J. and Wheeler, A.G., Family Miridae Hahn, 1833 (=Capsidae Burmeister, 1835), in *Catalog of the Heteroptera, or the True Bugs, of Canada and the Continental United States,* Henry, T.J. and Froeschner, R.C., Eds., Brill, New York, 1988, pp. 251.

52. Cassis, G., Schwartz, M.D., and Moulds, T., Systematics and new taxa of the *Vannius* complex (Hemiptera: Miridae: Cylapinae) from the Australian region, *Mem. Queensl. Mus.,* 49, 123, 2003.

53. Cassis, G., A reclassification and phylogeny of the Termatophylini (Heteroptera: Miridae: Deraeocorinae), with a taxonomic revision of the Australian species, and a review of the tribal classification of the Deraeocorinae, *Proc. Entomol. Soc. Wash.,* 97, 258, 1995.

54. Sanchez, J.A., Gillespie, D.R., and McGregor, R.R., Plant preference in relation to life history traits in the zoophytophagous predator *Dicyphus hesperus, Entomol. Exp. Appl.,* 112, 7, 2004.

55. Wheeler, Q.D. and Wheeler, A.G., Mycophagous Miridae? Associations of Cylapinae (Heteroptera) with pyrenomycete fungi (Euascomycetes: Xylariaceae), *J. N. Y. Entomol. Soc.,* 102, 114, 1994.

56. Wheeler, A.G., *Biology of the Plant Bugs (Hemiptera: Miridae),* Cornell University Press, Ithaca, NY, 2001.

57. Schuh, R.T. and Slater, J.A., *True Bugs of the World (Hemiptera: Heteroptera): Classification and Natural History,* Cornell University Press, Ithaca, NY, 1995.

58. Schuh, R.T., *Plant Bugs of the World (Insecta: Heteroptera: Miridae),* New York Entomological Society, New York, 1995.

59. Linnaeus, C., *Systema Naturae per Regna tria Naturae, secundum Classes, Ordines, Genera, Species, cum Characteribus, Differentis, Synonymis, Locis. Editio decima, reformata* Laurentii Salvii, Holmiae, 1758.

60. Knight, H.H., A new key to species of *Reuteroscopus* Kirk. with descriptions of new species (Hemiptera, Miridae), *Iowa State J. Sci.,* 40, 101, 1965.

61. Kelton, L.A., Revision of the genus *Reuteroscopus* Kirkaldy 1905 with descriptions of eleven new species (Hemiptera: Miridae), *Can. Entomol.,* 96, 1421, 1964.

62. Major, R.E. et al., The effect of habitat configuration on arboreal insects in fragmented woodlands of south-eastern Australia, *Biol. Conserv.,* 113, 35, 2003.

63. Myers, N. et al., Biodiversity for conservation priorities, *Nature,* 403, 2000.

64. Pennisi, E., Modernizing the tree of life, *Science,* 300, 1692, 2003.

65. Brusca, R.C. and Brusca, G.J., *Invertebrates,* 2nd ed., Sinauer Associates, Sunderland, MA, 2002.

66. Margulis, L. and Schwartz, K.V., *Five Kingdoms: An Illustrated Guide to the Phyla of Life on Earth,* 3rd ed., W.H. Freemand and Company, New York, 1998.

67. DSMZ, *Bacterial Nomenclature Up-to-Date,* Deutsche Sammlung von Mikroorganismen und Zellkulturen GmbH, Braunschweig, Germany, 2005.

68. Minelli, A., *Biological Systematics: The State of the Art,* Chapman and Hall, London, 1993.

69. Thayer, M.K. and Newton, A.F., What is a Staphylinid? http://www.fieldmuseum.org/peet_staph/whatisstaph.html, 2003.

70. Wagner, D.L., Moths, in *Encyclopedia of Biodiversity,* Levin, S., Ed., Academic Press, San Diego, 2001, 249.

71. Thompson, F.C., Biosystematic Database of World Diptera, http://www.sel.barc.usda.gov.diptera/biosystem.htm 2000.

72. Goulet, H. and Huber, J.T., *Hymenoptera: An Identification Guide to Families,* Research Branch Agriculture Canada, Ottawa, 1993.

73. Agosti, D. and Johnson, N.F., Antbase, http://www.antbase.org, version, 2005.

74. Dietrich, C.H., Phylogeny of the leafhopper subfamily Evacanthinae with a review of Neotropical species and notes on related groups (Hemiptera: Membracoidea: Cicadellidae), *Syst. Entomol.*, 29, 455, 2004.
75. Gaston, K.J., The magnitude of global insect species richness, *Conserv. Biol.*, 5, 283, 1991.
76. Hammond, P.M., Species inventory, in *Global Biodiversity. Status of the Earth's Living Resources*, Groombridge, B., Ed., Chapman and Hall, London, 1992, 17.
77. Dolphin, K. and Quicke, D.L.J., Estimating the global species richness of an incompletely described taxon: an example using parasitoid wasps (Hymenoptera: Braconidae), *Biol. J. Linn. Soc.*, 73, 279, 2001.

# 14 Cichlid Fish Diversity and Speciation

*J. R. Stauffer, Jr. and K. E. Black*
School of Forest Resources, Penn State University,
University Park, Pennsylvania, USA

*M. Geerts*
Swalmen, The Netherlands

*A. F. Konings*
Cichlid Press, El Paso, Texas, USA

*K. R. McKaye*
Appalachian Laboratory, University of Maryland System,
Frostburg, Maryland, USA

## CONTENTS

## ABSTRACT

Cichlids are one of the most species rich families of vertebrates, with conservative estimates citing more than 2,000 extant species. Although native to tropical areas of the world, with the exception of Australia, some 70–80% of cichlids are found in Africa, with the greatest diversity found in the Great Lakes (lakes Victoria, Tanzania and Malawi). Their highly integrated pharyngeal jaw apparatus permits cichlids to transport and process food, thus enabling the oral jaws to develop specialisations for acquiring a variety of food items. This distinct feature has allowed cichlids to achieve great trophic diversity, which in turn has lead to great species diversity. The high species diversity of this vertebrate family is not accompanied by an appropriately high genetic diversity. The combination of rapid radiation of the group and relatively low genetic diversity has confounded attempts to diagnose species and discern phylogenetic relationships. Behavioural traits appear to be important characters for diagnosing many cichlid species.

## 14.1  INTRODUCTION

Cichlids (Cichlidae) are a species rich group of fish from the lowland tropics[1] and are indigenous primarily to the fresh waters of Africa, South America and Central America, with one species extending its range to the Rio Grande River in southern North America. In addition, cichlids are found in Madagascar, the Levant and India and have also been introduced into nearly all tropical and subtropical regions of the world, either through escapes from aquaculture or ornamental fish operations or intentionally to provide sport fishing opportunities or to control exotic plants[2]. They have established breeding populations in warm waters of industrial effluents in temperate areas[3] and have been introduced into some marine environments[4]. Numerous investigators[5–14] have focused on cichlid fishes for their ecological, evolutionary and behavioural research.

Without doubt, the cichlids' explosive speciation, unique feeding specialisations, diverse mating systems and great importance as a protein source in tropical countries have been factors stimulating research interest in this group[15–18]. In fact, Greenwood[19] referred to the cichlid species flocks as "evolutionary microcosms repeating on a small and appreciable scale the patterns and mechanisms of vertebrate evolution". Many of these research efforts, however, have been slowed and results often confused because of the uncertain systematic status of some of the cichlids being studied[20,21]. The conservative bauplan of cichlids[1] and relatively low genetic divergence is coupled with a great morphological diversity that makes it difficult to diagnose species using morphological criteria alone (Figure 14.1). Systematic confusion exists within Cichlidae and also within and between its higher taxonomic ranks such as suborders. Such relationships are currently being debated. The reasons why cichlids have managed to speciate so successfully, often within a restricted geographic range such as the Great Lakes of Africa, have also been under investigation[14,15,22].

## 14.2  CICHLID PHYLOGENY

### 14.2.1  Higher Level Taxonomic Relationships of Cichlids

Morphological studies have provided insight into the phylogenetic position of Cichlidae. The resultant relationships, however, differ considerably from that of molecular investigations. Kaufman and Liem[23] (see also Stiassny and Jensen[24]) grouped cichlid fishes (Acanthopterygii: Perciformes) with Embiotocidae, Labridae and Pomacentridae in the suborder Labroidei on the basis of the following three pharyngeal characters: fusion of the fifth ceratobranchial bones into one unit; contact of the upper pharyngeal jaws and the basicranium; and lack of a dorsal subdivision of the sphincter oesophagi muscle. As it is now recognised, this suborder includes some 1,800 species and represents some 5–10% of all extant fishes[24]. Müller[25] was the first to group the above families together, and morphological data support the view that Labroidei has phyletic integrity[24] based on the following synapomorphies: bladelike keel on the lower pharyngeal bone and the change in insertion of at least part of the transversus ventralis onto the keel; division of the transversus dorsalis anterior muscle[26]; bony facets of the third pharyngobranchials of the upper pharyngeal jaws are exposed; ventrally projecting rounded form of the neurocranial apophysis; and no subdivision of the sphincter oesophagi muscle. The cichlids appear to be the sister group of all the other labroids[1].

The monophyly of Labroidei is, however, confused, because many homoplasies exist in the morphological data sets[24,27]. Moreover, studies of nuclear DNA suggest that Labroidei are not monophyletic, and that embiotocids and pomacentrids are more closely related to more basal perciforms[28]. This view is corroborated by Sparks and Smith[29] who believe that the sister group to the cichlids "might comprise a large assemblage of diverse perciform lineages, including but

**FIGURE 14.1** *(A colour version of this figure follows page 240)* Cichlid fishes. Cichlids have a conservative bauplan, and specialised attributes, such as hypertrophied lips are the result of parallel evolution, thus making species and higher level diagnoses difficult. (a) *Amphilophus* sp. 'fatlip' in Lake Xiloa, Nicaragua; (b) *Placidochromis milomo* at Nkhomo Reef, Lake Malawi, Malawi; (c) *Lobochilotes labiatus* at Nkondwe Island, Lake Tanganyika, Tanzania. (Photos reproduced with permission from A.F. Konings.)

presumably not limited to the other 'labroid' lineages, sparids, anabantids-nandids, haemulids, percids, moronids, and kyphosids". Westneat and Alfaro[27] reported maximum *RAG2* DNA sequence divergence between wrasses and outgroups ranging as high as 23% between parrotfishes and some of the cichlids they examined. Nevertheless, they supported the inclusion of the parrotfish as a subgroup of Labridae; thus, it is doubtful if these families belong to the same suborder.

Boulenger[30] first speculated that the cichlids form a natural group within the perciform Acantho-pterygians. Stiassny[1] recognised Cichlidae as a monophyletic group based on five apomorphic characters:

- Loss of a structural association between parts $A_2$ and $A_{10}$ of the adductor mandibulae muscle and the attachment of a large ventral section of $A_2$ onto the posterior border of the ascending process of the anguloarticular
- An extensive cartilaginous cap on the front margin of the second epibranchial bones
- An expanded head of the fourth epibranchial bones
- Presence of characteristically shaped and distributed microbranchiospines on the gill arches
- The subdivision of the traversus dorsalis anterior muscle into three distinct parts as described by Liem and Greenwood[26]

The monophyly of Cichlidae is further supported by the morphology of otoliths and configuration of the digestive tracts. Gaemers[31], based on the structural configuration of the sagitta, also hypothesised that the cichlids are monophyletic. The sagitta is usually the largest otolith in most teleosts, including cichlids. The sagitta of cichlids is strong, thick, with a more or less oval, short, elliptical to rounded pentagonal shape[31]. If the pseudocolliculum in the sagitta of cichlids is a synapomorphic character, it supports other evidence of monophyly of the family[26,32]. Finally, three structural attributes of a cichlid's digestive tract support cichlid monophyly: the stomach's extendible blind pouch; the left hand exit to the anterior intestine; and the position of the first intestinal loop on the left side[33].

### 14.2.2  INTRAFAMILIAL RELATIONSHIPS OF CICHLIDS

Many subfamilial names have been used to indicate groupings within Cichlidae, but do not necessarily represent hierarchical patterns[34]. Etroplinae (that is *Etroplus* and *Paretroplus*) has been proposed to be the sister group of all other cichlids[29]. Following this separation, Ptychochrominae were considered to be the sister group to the remaining cichlids. What then is left has been divided further into two purportedly monophyletic groups, the African Pseudocrenilabrinae and the New World Cichlinae. It should be noted that Schliewen and Stiassny[35] used the term Haplotilapiines to refer to non-*Heterochromis* cichlids from Africa and the Levant.

Within the family, a Congolese genus, *Heterochromis,* was proposed to be the most primitive of the African cichlids[36]. Stiassny[37] diagnosed *Heterochromis* to be the sister group of all the other cichlids, based on the presence of a single ligament attaching the lower pharyngeal bone to the cartilaginous fourth basibranchial. The monophyly of Neotropical cichlids was questioned by Cichocki[38], who postulated that *Cichla*, a Neotropical genus, was grouped with the cichlids of Africa. This was debated by Stiassny[1], who favoured a *Cichla*-crenicichline clade within the Neotropics. Stiassny[1] postulated a close etropline relationship with *Heterochromis,* while Kullander[34] regarded *Heterochromis* as the sister group of the majority of Neotropical cichlids. Despite this relationship, Kullander[34] emphasised that the dichotomy of Old and New World cichlids was well supported by the following character states:

- Short anterior arm of epibranchial, which is a reduction that occurs in several lineages, including the Etroplinae and *Astatotilapia,* and which is reversed in groups such as Cichlasomatinae
- Interdigitating suture connecting the vomerine shaft and the parasphenoid bar (requires independent development in *Ptychochromis* and reversal in *Biotoecus, Dicrossus* and *Nannacara*)
- Presence of an anterior palatoethmoid ligament, which occurs in all Neotropical cichlids and *Heterochromis,* but in no other Old World cichlids

*Retroculus* is regarded as the earliest diverging lineage of Neotropical cichlids and the sister group of *Cichla-Crenicichla,* which is placed in a new subfamily, Cichlinae[34]. Kullander[34] recognises the following Neotropical subfamilies: Astronotinae (*Astronotus, Chaetobranchus*), Geophaginae (for example, *Geophagus, Apistogramma*) and Cichlasomatinae (Cichlasomines, Heroines *Acaronia*).

Mitochondrial DNA supports the monophyly of Cichlidae, but differs in its interpretation of intrafamilial relationships[39]. In particular, *Heterochromis* is sister to the remaining African cichlids, and *Retroculus* is the most basal taxon of the Neotropical cichlids. Farias et al.[40] believed that *Astronotus, Cichla* and *Retroculus* formed three independent basal lineages, even though one of their trees favoured *Astronotus* as the sister group to *Cichla.* Therefore, Astronotinae, sensu Kullander, is not accepted by molecular biologists, whereas Chaetobranchus and Chaetobranchopsis are grouped together as Chaetobranchines. Farias et al.[40] remove *Crenicichla* and *Teleocichla* from Cichlinae (sensu Kullander) and transfer that group to Geophaginae. In the molecular phylogeny, *Acaronia* is no longer regarded as the sister group of the *Heroini*/Cichlasomatini group[40].

## 14.3  CICHLID DISTRIBUTION

Apart from Australia and Antarctica, cichlids occur on all the major fragments of the former supercontinent Gondwana. The extant distribution of cichlids and phylogenetic relationships suggest that the family was well established before the separation of Gondwanaland. Thus, the family should be at least 160 million years old[41,42]. This view, however, is not supported by studies based on the fossil record, and the time of origin of this taxon is still hotly debated. For example, many authors believe that the cichlids are not old enough to have been present on that supercontinent[43]. India and Madagascar drifted away during the Late Jurassic (160 million years before present, mybp) and not during the Middle Cretaceous, as Van Couvering[44] apparently believed. Molecular evidence presented by Vences et al.[45] suggests that the extant distribution pattern of the cichlids came into being well after Gondwana fell apart. In that respect, it should be mentioned that Murray[46] believed that a Gondwanan origin is not needed to explain the present distribution of cichlids.

The oldest known fossil cichlids, the species of the genus *Mahengochromis,* have an age of 46.3 mybp, which means that they date from the Middle Eocene[46]. Fossil cichlids from the Oligocene of Africa have been described by Van Couvering[44], and because of the highly specialised dental features of these fishes, Stiassny[1] concluded that the origin of the family predates the earliest fossils considerably and that the cichlids probably arose in the Early Cretaceous, some 135 mybp.

This view is incompatible with the fossil record. In a recent paper, Arratia et al.[47] have emphasised, once more, that thus far no perciform fishes have been found that are older than the latest Cretaceous (85–65 mybp). That age differs dramatically from that postulated by molecular studies (Salmoniformes and Gadiformes versus Perciformes divergence of some 285 mybp). In this respect, it should be remembered that the cichlids are regarded as advanced perciforms, which means that they should be much younger than the primitive forms that have been found in the Late Cretaceous sediments. Cichocki[38] postulated that the cichlids with the most plesiomorphic characters are found in Madagascar and India, which was endorsed by Kullander[34], discussed by Stiassny[1], and supported by mtDNA analysis by Farias et al.[39].

## 14.4  CICHLID DIVERSITY AND SPECIATION

Between 70 and 80% of all cichlid fishes are native to African freshwaters, of which the majority are part of the major lacustrine radiations; for example, Lake Victoria (c. 500 species), Lake Tanganyika (c. 200 species) Lake Malawi (c. 850 species)[15,48]. The remainder occur in the freshwaters of the New World, with the exception of a single genus, *Etroplus,* which is found in the coastal waters of Southern India and Sri Lanka; five genera, which are found on Madagascar; the genus *Tristramella,* which is endemic to the Levant; and one genus, *Iranocichla,* which is endemic to Southern Iran.

By far the greatest radiation of cichlids is found in the Great Lakes of Africa, with Lake Malawi alone having as many as 850 species[49]. The phylogenetic diversity ranges from the single invasion of Lake Malawi, which resulted in the endemism of all but a few species, to multiple invasions in Lake Tanganyika, which resulted in the presence of 12 different tribes[50]. The rich fauna of these lakes is primarily attributable to the explosive adaptive radiation and speciation[51–53] of the haplochromines sensu lato (see Schliewen and Stiassny[35]). The driving mechanism for these speciation events is unknown. The two most widely proposed methods are allopatric speciation[5,15,54–57] and intrinsic isolating mechanisms[14,22,58–64]. Furthermore, biologists generally agree that female mate choice can act as a strong driving force in runaway speciation where the average female preference for a specific male trait differs between two allopatric populations[65–69]. Thus, behavioural traits are important tools for the diagnosis of these African cichlids, primarily because behavioural traits played a very important role in and facilitated the rapid radiation of these fishes, which may not always be accompanied by discernable morphological changes[70].

Certainly, there are fewer species of cichlids in South and Central America than in the Old World. Greenwood[19] noted that the focus on African cichlids has distracted attention from the South American Cichlidae, a fact which he regarded as 'sad but understandable'. Nevertheless, the neotropical cichlid fauna is varied and diverse[1], comprising some 50 genera and 450 species[34], with new species still being discovered[71]. In the last decade, on average more than 20 species of cichlids have been formally described each year. The existence of literally hundreds of species awaiting description is likely to continue this trend.

## 14.5  CICHLID ADAPTIVE RADIATION

The cichlids of the Great Lakes of Africa have undergone one of the most rapid radiations of any known vertebrate group[47,72–74]. This rapid speciation rate, however, is not correlated with a high genetic diversity[74], as discussed earlier. Conversely, the neotropical cichlids, especially the geophagines, have a significantly higher rate of genetic divergence than their African counterparts[39,75], but this does not seem to be expressed in the more diversified group. This makes the taxonomy and systematics of the group challenging. The high genetic diversity of the Neotropical cichlids, when compared to the ones from Africa, is somewhat surprising considering the greater species diversity of Old World cichlids. A number of the following adaptations may be associated with cichlid speciation.

### 14.5.1  Feeding Adaptations

High trophic diversity of the Old World cichlids results in the consumption of virtually every food type available in the environment[76,77]. The successful radiation of cichlids in the rift valley (for example, in Lake Malawi and Lake Tanganyika) is purported to be a result of cichlids' differential ability to acquire food[78]. In cichlids, mouth structure, dentition, gill raker number[79] and jaw structure vary tremendously, and this variation in structure seems to be tied to a variety of feeding techniques[15,80]. Documented feeding strategies of Lake Malawi fishes, for example, illustrate numerous feeding specialisations including death feigning[81], paedophagy[82,83], lepidiophagy[84], cleaning[85] and scraping and raking of an algal, diatomaceous and detrital biolayer (aufwuchs) from the rock surfaces[5]. Phenotypes, such as reverse counter shading, are associated with such bizarre strategies as hunting upside down[86].

Primarily through the work of Liem and his coworkers[26,76,77,87] it became evident that the trophic diversification of cichlids does not require major structural adaptations. Liem[72] attributes the great colonising success of cichlids to their possession of a highly integrated pharyngeal jaw apparatus. Thus, the pharynx has been a particular focus of modification in Labroidei, including the cichlids[1]. Liem[72] further states that this specialised innovation allows cichlids to transport and prepare food, enabling the premaxillary and mandibular jaws to develop specialisations for collecting diverse food items. It is this distinct feature of cichlids that has permitted them to dominate colonisation of many habitats and adopt feeding opportunities available in lacustrine environments[72,76].

Phenotypic plasticity is defined as the possible environmental modification of the phenotype[88]. The degree of phenotypic plasticity that cichlids exhibit is congruent with the ability of cichlids to take advantage of many habitats and feeding opportunities. Despite the morphological plasticity observed in other fishes[89–91], the morphology of African cichlids initially was thought to be rigid[92]; however, much morphological variability has been observed[93–98]. In particular, the observed phenotypic plasticity in some instances has involved the pharyngeal jaw apparatus. For example, the phenotypic diversity of the pharyngeal jaws of the New World *Herichthys minckleyi*[99,100] has been documented extensively. Furthermore, differences in bone structure of the lower pharyngeal jaw of *Astatoreochromis alluaudi* resulted from different diets[101–103]. In addition to these observations, Meyer[94] and Wimberger[97,98] have experimentally demonstrated the effects of diet on plasticity of head morphology in New World cichlids, but Meyer[94] hypothesised that the plasticity of mouth brooding Old World cichlids may not be as pronounced due to constraints on jaw morphology for mouth brooding. Stauffer and Van Snik Gray[104] effected significant differences in head morphology of Lake Malawi rock-dwelling cichlids by experimentally manipulating diets. The magnitude of plasticity in these mouth brooding Lake Malawi cichlids, however, was not as pronounced as that observed for the New World substrate brooder *Herichthys cyanoguttatum*. They did postulate, however, that phenotypic plasticity might have contributed to the extensive trophic radiation and subsequent explosive speciation observed in Old World haplochromine cichlids. Lewontin[105] postulated that colonising species, possessing a high degree of phenotypic plasticity, may have a selective advantage because of their ability to exploit additional resources in differing environments.

## 14.5.2 BREEDING ADAPTATIONS

Cichlids have also a number of reproductive strategies associated with their speciation. Barlow[65,66] and Keenleyside[16,17] give an excellent review of reproductive strategies and parental care in cichlid fishes. The breeding tactics of cichlids can be separated into two major groups: substrate brooders and mouth brooders[65,106–109]. Substrate brooding cichlids are generally monogamous and exhibit biparental care of the eggs and brood, although some species are polygynous[110]. Substrate brooding appears to be the plesiomorphic character state. The primitive cichlids of Asia and Madagascar (that is Etroplinae and Ptychochrominae), most of the New World species, Old World *Tilapia* spp., and selected Old World pseudocrenilabrines are substrate brooders[109,111–113].

Most mouth brooders are polygamous, although monogamy is found in a few species[107,114]. Some New World cichlid mouth brooders begin as substrate brooders, but then gather either the eggs (ovophilous mouth brooders) or the hatched fry (larvophilous mouth brooders) into their mouths, including some *Aequidens*, *Geophagus*, *Gymnogeophagus* and *Satanoperca*[115,116]. Some *Chromidotilapia* from Africa employ a similar delayed mouth brooding strategy[117,118], while male *Chromidotilapia guentheri* brood the eggs and larvae. *Sarotherodon galilaeus* exhibits an intermediate form of biparental care, where both the male and female gather the eggs into their mouths after fertilisation and then separate[65].

The vast majority of African mouth brooders are polygamous. The rock dwelling cichlids of lakes Malawi and Tanganyika defend territories and attract females to spawn in either rock crevices, small caves, on the rock surface, or in the water column above rocks. Males of these fishes presumably rely on their brilliant and diverse colour patterns to attract females[70]. In addition, many of the Lake Malawi sand dwelling cichlids construct a wide variety of bower forms in leks with a hundred to many thousand individuals[119–123]. These bowers can be broadly divided into 10 types[121] and range in size from giant craters three metres in diameter[124] to small depressions in the sand. The females move over the arena and lay eggs with several males. The entire egg laying process takes 25–60 minutes from the time a female lays her first egg[125] until she leaves the arena with eggs in her mouth. Genetic studies of paternity for several Lake Malawi species show that females mate with as many as six males per brood[126].

The process of female choice is complex. Female *Otopharynx* c.f. *argyrosoma* selectively chose males that occupy bowers in the centre of the arena[121], while female *Mehenga conophoros* and *Lethrinops* c.f. *parvidens* chose males that build the largest bowers[122,125,127]. If the rapid radiation of the Lake Malawi cichlid flock was accelerated by sexual selection, the observed differences in behaviour might be the best way to distinguish between sibling species that differ little in morphology[71].

## 14.6  FUTURE DIRECTIONS

Barlow[66] dedicated an entire chapter in his excellent book, *The Cichlid Fishes: Nature's Grand Experiment in Evolution,* to the problems associated with the Earth's growing population. Whilst he discussed the risks of burgeoning human populations and the associated pattern of diminishing resources on the Earth's biota, he focused on cichlids in particular. Certainly, one of the detrimental aspects of human populations is the decrease in genetic diversity connected to the increase in the rate of extinction of natural populations. Probably the most drastic incident of mass extinction in our lifetimes occurred with the destruction of the cichlid species flock in Lake Victoria that was associated with the introduction of the Nile Perch[128,129]. The relatively small and colourful cichlids endemic to the lake were suddenly confronted with a huge predator, which, a few decades after its introduction, had severely decimated the cichlids and established itself as the new food source for the local populace. In addition to the loss of cichlids due to introductions, the introduction of cichlids has had detrimental impacts on native fish faunas throughout the world[2,3,130].

The high morphological diversity of the cichlids coupled with their conservative bauplan and the relative low genetic divergence of the Old World cichlids makes it difficult to diagnose species using morphological criteria alone. The extremely recent diversification of several cichlid lineages presents serious challenges for the use of genetic techniques for alpha and beta taxonomy. Initially, the attempts to use allozymes to resolve phylogenetic relationships were limited[131]. Kocher et al.[132] have discovered a large number of polymorphisms in the tilapia genome, and Arnegard et al.[133] and Markert et al.[134] have used microsatellites to study population structure of several Lake Malawi rock dwelling cichlids. The use of amplified fragment length polymorphisms by Albertson et al.[135] was promising for the determination of supraspecific relationships and species diagnoses. As stated previously, behavioural traits are important characters for diagnosing many of these cichlids and determining phylogenetic relationships[70]. Thus, Stauffer and McKaye[71] recommended that a combination of morphological, genetic and behavioural data be used to diagnose these species and determine phylogenetic relationships.

## REFERENCES

1.  Stiassny, M.L.J., Phylogenetic intrarelationships of the family Cichlidae: an overview, in *Cichlid Fishes Behaviour, Ecology and Evolution.* Keenleyside, M.H.A., Ed., Chapman and Hall, New York, 1991, chap. 1.
2.  Courtenay Jr., W.R. and Stauffer Jr., J.R., *Biology, Distribution and Management of Exotic Fishes,* Johns Hopkins University Press, Baltimore, 1, 1984.
3.  Stauffer Jr., J.R., Boltz, S.E., and Boltz, J.M., Thermal tolerance of the blue tilapia, *Oreochromis aureus,* in the Susquehanna River, *N. Am. J. Fisheries Management,* 8(3), 329, 1988.
4.  Lobel, P.S., Invasion by the Mozambique tilapia (*Sarotherodon mossambicus;* Pisces: Cichlidae) of a Pacific atoll marine ecosystem, *Micronesia,* 16, 349, 1980.
5.  Fryer, G., The trophic interrelationships and ecology of some littoral communities of Lake Nyasa with special reference to the fishes, and a discussion of the evolution of a group of rock-frequenting Cichlidae, *Proc. Zool. Soc. Lond.,* 132, 153, 1959.

6.  Jackson, P.B.N. et al., *Report on the Survey of Northern Lake Nyasa 1954–1955,* Government Printing Office, Zomba, 1, 1963.
7.  Holzberg, S., A field and laboratory study of the behaviour and ecology of *Pseudotropheus zebra* (Boulenger) an endemic cichlid of Lake Malawi (Pisces: Cichlidae), *Z. Zool. Syst. Evol.,* 16, 171, 1978.
8.  Marsh, A.C., Ribbink, A.J., and Marsh, B.A., Sibling species complexes in sympatric populations of *Petrotilapia* Trewavas (Cichlidae, Lake Malawi), *Zool. J. Linn. Soc.,* 71, 253, 1981.
9.  Kocher, T.D. et al., Similar morphologies of cichlid fish in lakes Tanganyika and Malawi are due to convergence, *Mol. Phylogenet. Evol.,* 2, 158, 1993.
10. Sato, T. and Gashagaza, M.M., Shell-brooding cichlid fishes of Lake Tanganyika: their habitats and mating system, in *Fish Communities in Lake Tanganyika,* Kawanabe, H., Hori, M., and Nagoshi, M., Eds., Kyoto University Press, Kyoto, 1997, 219.
11. Kawanabe, H.M., Hori, M., and Nagoshi, M., *Fish Communities in Lake Tanganyika,* Kyoto University Press, Kyoto, 1997.
12. Land, R., Seehausen, O., and van Alphen, J.J.M., Mechanisms of rapid sympatric speciation by sex reversal and sexual selection in cichlid fish, *Genetica,* 112, 435, 2001.
13. Genner, M.J. et al., How does the taxonomic status of allopatric populations influence species richness within African cichlid fish assemblages? *J. Biogeography,* 31, 93, 2004.
14. Genner, M.J. and Turner, G.F., The mbuna cichlids of Lake Malawi: a model for rapid speciation and adaptive radiation, *Fish and Fisheries,* 6, 522, 2005.
15. Fryer, G. and Iles, T.D., *The cichlid fishes of the Great Lakes of Africa,* Oliver and Boyd, London, 1972.
16. Keenleyside, M.H.A., Ed., *Cichlid Fishes: Behaviour, Ecology and Evolution,* Chapman and Hall, London, 1991.
17. Keenleyside, M.H.A., Parental care, in *Cichlid Fishes: Behaviour, Ecology and Evolution,* Keenleyside, M.H.A., Ed., Chapman and Hall, New York, 1991, 191.
18. Kornfield, I., African cichlid fishes: model systems for evolutionary biology, *Ann. Rev. Ecol. Syst.,* 31, 163, 2000.
19. Greenwood, P.H., The cichlid fishes of Lake Victoria, East Africa: the biology and evolution of a species flock, *Bull. Br. Mus. Nat. Hist. (Zool.),* 6, 1, 1974.
20. Stauffer Jr., J.R. and McKaye, K.R., The naming of cichlids, *J. Aquaricult. Aquatic Sci.,* 9, 1, 2001.
21. Turner, G.F. et al., Identification and biology of *Diplotaxodon, Rhamphochromis,* and *Pallidochromis,* in *The Cichlid Diversity of Lake Malawi/Nyasa/Niassa: Identification, Distribution, and Taxonomy,* Snoeks, J., Ed., Cichlid Press, El Paso, 2004, 198.
22. McKaye, K.R. and Stauffer Jr., J.R., Description of a gold cichlid (Teleostei: Cichlidae) from Lake Malawi, Africa, *Copeia,* 1986, 870, 1986.
23. Kaufman, L. and Liem, K.F., Fishes of the suborder Labroidei (Pisces: Perciformes): phylogeny, ecology and evolutionary significance, *Breviora,* 472, 1, 1982.
24. Stiassny, M.L.J. and Jensen, J.S., Labroid intrarelationships revisited: morphological complexity, key innovations, and the study of comparative diversity, *Bull. Mus. Comp. Zool.,* 151, 269, 1987.
25. Müller, J., Beiträge zur Kenntniss der natürlichen Familien der Fische, *Arch. Naturgesch.,* 9, 381, 1844. (published 1846)
26. Liem, K.F. and Greenwood, P.H., A functional approach to the phylogeny of the pharyngognath teleosts, *Amer. Zool.,* 21, 83, 1981.
27. Westneat, M.W. and Alfaro, M.E., Phylogenetic relationships and evolutionary history of the reef fish family Labridae, *Mol. Phylogen. Evol.,* 36, 370, 2005.
28. Streelman, T.J. and Karl, S.A., Reconstructing labroid evolution with single-copy nuclear DNA, *Proc. R. Soc. Lond. B,* 264, 1011, 1997.
29. Sparks, J. and Smith, W., Phylogeny and biogeography of cichlid fishes (Teleostei: Perciformes: Cichlidae), *Cladistics,* 20, 501, 2004.
30. Boulenger, G.A., A revision of the African and Syrian fishes of the family Cichlidae, Part 1, *Proc. Zool. Soc. Lond.,* 132, 1898.
31. Gaemers, P.A.M., Taxonomic position of the Cichlidae (Pisces, Perciformes) as demonstrated by the morphology of their otoliths, *Neth. J. Zool.,* 34, 566, 1984.
32. Stiassny, M.L.J., The phyletic status of the family Cichlidae (Pisces, Perciformes): a comparative anatomical investigation, *Neth. J. Zool.,* 31, 275, 1981.
33. Zilhler, F., Gross morphology and configuration of digestive tracts of Cichlidae (Teleostei, Perciformes): phylogenetic and functional significance, *Neth. J. Zool.,* 32, 544, 1982.

34. Kullander, S.O., A phylogeny and classification of the South American Cichlidae (Teleostei: Perciformes), in *Phylogeny and Classification of Neotropical Fishes,* Malabarba, L.R. et al., Eds., EDIPUCRS, Porto Alegre, 1998, 461.

35. Schliewen, U. and Stiassny, M.L.H., *Etia nguti,* a new genus and species of cichlid fish from the River Mamfue, Upper Cross River Basin in Cameroon, West Central Africa, *Ichthyol. Explor., Freshwaters,* 14, 61, 2003.

36. Oliver, M.K., *Systematics of African cichlid fishes: determination of the most primitive taxon, and studies of the Haplochromines of Lake Malawi (Teleostei:Cichlidae),* Ph.D. thesis, Yale University, New Haven, CT, 1984.

37. Stiassny, M.L.J., Cichlid familial intrarelationships and the placement of the neotropical genus *Cichla* (Perciformes, Labroidei), *J. Nat. Hist.,* 21, 1311, 1987.

38. Cichocki, F.P., Cladistic history of the cichlid fishes and reproductive strategies of the American genera *Acarichthys, Biotodoma,* and *Geophagus,* Ph.D. thesis, University of Michigan, Ann Arbor, 1976.

39. Farias, I.P. et al., Mitochondrial DNA phylogeny of the family Cichlidae: monophyly and fast molecular evolution of the neotropical assemblage, *J. Mol. Evol.,* 48, 703, 1999.

40. Farias, I.P., Orti, G., and Meyer, A., Total evidence molecules, morphology, and the phylogenetics of cichlid fishes, *J. Exp. Zool.,* 288, 76, 2000.

41. Chakrabarty, P., Cichlid biogeography: comment and review, *Fish and Fisheries,* 5, 97, 2004.

42. Sparks, J. and Smith, W., Freshwater fishes, dispersal ability, and non-evidence: 'Gondwana life rafts' to the rescue, *Syst. Biol.,* 54, 158, 2005.

43. Lundberg, J., African-South American freshwater clades and continental drift; problems with a paradigm, in *Biological Relationships between Africa and South America,* Goldblatt, P., Ed., Yale University Press, New Haven, 1993, 156.

44. Van Couvering, J.A.H., Fossil cichlid fish of Africa, *Spec. Pap. Palaeont. Lond.,* 29, 1982.

45. Vences, M. et al., Reconciling fossils and molecules: Cenozoic divergence of cichlid fishes and the biogeography of Madagascar, *J. Biog.,* 28, 1091, 2001.

46. Murray, A., The oldest fossil cichlids (Teleostei: Perciformes): indication of a 45 million year old species flock, *Proc. R. Soc. Lond. B,* 268, 679, 2001.

47. Arratia, G.A. et al., Late Cretaceous-Paleocene percomorphs (Teleostei) from India — early radiation of Perciformes, in *Recent Advances in the Origin and Early Radiation of Vertebrates,* Arratia, G., Wilson, M.V.H., and Cloutier, R., Eds., Pfeil Verlag, Munich, 2004, 635.

48. Greenwood, P.H., Speciation, in *Cichlid Fishes: Behaviour, Ecology and Evolution,* Keenleyside, M.H.A., Ed., Chapman and Hall, New York, 1991, 86.

49. Konings, A., *Malawi Cichlids in Their Natural Habitat,* 3rd edition, Cichlid Press, El Paso, TX, 2001.

50. Poll, M., Deuxième serie de Cichlidae nouveaux recueillis par la mission hydrobiologique belge au lac Tanganyika (1946–1947), *Bull. Inst. Sci. Nat. Belg.,* 25, 1, 1949.

51. Regan, C.T., The cichlid fishes of Lake Nyassa, *Zool. Soc. Lond.,* 21, 675, 1922.

52. Trewavas, E., A synopsis of the cichlid fishes of Lake Nyassa, *Ann. Mag. Nat. Hist.,* 16, 65, 1935.

53. Greenwood, P.H., Towards a phyletic classification of the 'genus' *Haplochromis* (Pisces: Cichlidae) and related taxa, Part I, *Bull. Br. Mus. Nat. Hist. (Zool.),* 35, 265, 1979.

54. Greenwood, P.H., African cichlids and evolutionary theories, in *Evolution of Fish Species Flocks,* Echelle, A.A. and Kornfield, I., Eds., University of Maine at Orono Press, Orono, 1984, 141.

55. Greenwood, P.H., *The haplochromine fishes of the east African lakes,* Kraus International Publications, Munchen, and British Museum Natural History, London, 1981.

56. Marlier, G., Observations sur la biologie littorale du lac Tanganyika, *Rev. Zool. Bot. Afr.,* 59, 16, 1959.

57. Matthes, H., Note sur la reproduction des poissons au lac Tanganyika, *C.S.A. 3rd Symp. Hydrobiol. Major Lakes,* 107, 1962.

58. Bush, G.L., Modes of animal speciation, *Ann. Rev. Ecol. Syst.,* 6, 339, 1975.

59. Kosswig, C., Selective mating as a factor for speciation in cichlid fish of east African lakes, *Nature,* 159, 604, 1947.

60. Kosswig, C., Ways of speciation in fishes, *Copeia,* 1963, 238, 1963.

61. McKaye, K.R., Explosive speciation: the cichlids of Lake Malawi, *Discovery,* 13, 24, 1978.

62. McKaye, K.R., Seasonality in habitat selection by the gold color morph of *Cichlasoma citrinellum* and its relevance to sympatric speciation in the family Cichlidae, *Environ. Biol. Fish.,* 5, 75, 1980.

63. Takahashi, K. et al., A novel family of short interspersed repetitive elements (SINEs) from cichlids: the patterns of insertion of SINEs at orthologous loci support the proposed monophyly of four major groups of cichlid fishes in Lake Tanganyika, *Mol. Biol. Evol.*, 15, 391, 1998.

64. Takahashi, K. et al., Phylogenetic relationships and ancient incomplete lineage sorting among cichlid fishes in Lake Tanganyika as revealed by analysis of the insertion of retroposons, *Mol. Biol. Evol.*, 18, 2057, 2001.

65. Barlow, G.W., Mating systems among cichlid fishes, in *Cichlid Fishes: Behaviour, Ecology, and Evolution*, Keenleyside, M.H.A., Ed., Chapman and Hall, New York, 1991, 173.

66. Barlow, G.W., *The Cichlid Fishes: Nature's Grand Experiment in Evolution*, Perseus Publishing, Cambridge, 2000.

67. Clutton-Brock, T.H., *The Evolution of Parental Care*, Princeton University Press, Princeton, 1991.

68. Johnsgard, P.A., *Arena Birds: Sexual Selection and Behavior*, Smithsonian Institution Press, Washington, 1994.

69. Hogland, J. and Alatalo, R.V., *Leks*, Princeton University Press, Princeton, 1995.

70. Stauffer Jr., J.R., McKaye, K.R., and Konings, A.F., Behaviour: an important diagnostic tool for Lake Malawi cichlids, *Fish and Fisheries*, 3, 213, 2002.

71. Stauffer Jr., J.R. and McKaye, K.R., Descriptions of three new species of cichlid fishes (Teleostei: Cichlidae) from Lake Xiloá, Nicaragua, *Cuadernos Invest., UCA*, 12, 1, 2002.

72. Liem, K.F., Evolutionary strategies and morphological innovations: cichlid pharyngeal jaws, *Syst. Zool.*, 22, 425, 1974.

73. Echelle, A.A. and Kornfield, I., Eds., *Evolution of Fish Species Flocks*, University of Maine Press, Orono, 1984.

74. Meyer, A. et al., Monophyletic origin of Lake Victoria cichlid fishes suggested by mitochrondrial DNA sequences, *Nature*, 347, 550, 1990.

75. Zardoya, R. et al., Evolutionary conservation of microsatellite flanking regions and their use in resolving the phylogeny of cichlid fishes (Pisces: Perciformes), *Proc. R. Soc. Lond. B*, 263, 1589, 1996.

76. Liem, K.F. and Osse, J.W.M., Biological versatility, evolution, and food resource exploitation in African cichlid fishes, *Am. Zool.*, 15, 427, 1975.

77. Liem, K.F., Adaptive significance of intra- and interspecific differences in the feeding repertoires of cichlid fishes, *Am. Zool.*, 20, 295, 1980.

78. Axelrod, H.R. and Burgess, W., *African Cichlids of Lakes Malawi and Tanganyika*, 8th ed., T.F.H. Publications, Neptune City, 1979.

79. Stauffer Jr., J.R. et al., Evolutionarily significant units among cichlid fishes: the role of behavioral studies, *Am. Fisheries Soc. Symp.*, 17, 227, 1995.

80. Kocher, T.D., Adaptive evolution and explosive speciation: the cichlid fish model, *Nature Rev. Genetics*, 5, 288, 2004.

81. Fryer, G., Biological notes on some cichlid fishes of Lake Nyasa, *Rev. Zool. Bot. Afr.*, 54, 1, 1956.

82. McKaye, K.R. and Kocher, T.D., Head ramming behaviour by three paedophagous cichlids in Lake Malawi, Africa, *Anim. Behav.*, 31, 206, 1983.

83. Stauffer Jr., J.R. and McKaye, K.R., Description of a paedophagous deep-water cichlid (Teleostei: Cichlidae) from Lake Malawi, Africa, *Proc. Biol. Soc. Wash.*, 99, 29, 1986.

84. Ribbink, A.J., The feeding behavior of a cleaner, scale and skin and fin eater of Lake Malawi (*Docimodus evelynae*, Pisces: Cichlidae), *Neth. J. Zool.*, 34, 182, 1984.

85. Stauffer Jr., J.R., Description of a facultative cleanerfish (Teleostei: Cichlidae) from Lake Malawi, Africa, *Copeia*, 1991, 141, 1991.

86. Stauffer Jr., J.R., Posner, I., and Seltzer, R., Hunting strategies of a Lake Malawi cichlid with reverse countershading, *Copeia*, 1999, 1108, 1999.

87. Liem, K.F., Evolutionary strategies and morphological innovations: cichlid pharyngeal jaws, *Syst. Zool.*, 22, 425, 1973.

88. Bradshaw, A.D., Evolutionary significance of phenotypic plasticity in plants, *Adv. Genet.*, 13, 115, 1965.

89. Barlow, G.W., Causes and significance of morphological variation in fishes, *Syst. Zool.*, 10, 105, 1961.

90. Behnke, R.J., Systematics of salmonid fishes of recently glaciated lakes, *J. Fisheries Res. Board Canada*, 29, 639, 1972.

91. Chernoff, B., Character variation among populations and the analysis of biogeography, *Am. Zool.*, 22, 425, 1982.

92. van Oijen, M.J.P., Ecological differentiation among the piscivorous haplochromine cichlids of Lake Victoria (East Africa), *Neth. J. Zool.*, 32, 336, 1982.

93. Hoogerhoud, R.J.C., A taxonomic reconsideration of the Haplochromine genera *Gaurochromis* Greenwood, 1980 and *Labrochromis* Regan, 1920 (Pisces, Cichlidae), *Neth. J. Zool.*, 34, 539, 1984.

94. Meyer, A., Phenotypic plasticity and heterochromy in *Cichlasoma managuense* (Pisces: Cichlidae) and their implications for speciation in cichlid fishes, *Evolution*, 41, 1357, 1987.

95. Meyer, A., Ecological and evolutionary consequences of the trophic polymorphism in *Cichlasoma citrinellum* (Pisces: Cichlidae), *Biol. J. Linn. Soc.*, 39, 279, 1990.

96. Witte, F., Barel, C.D.N., and Hoogerhoud, R.J.C., Phenotypic plasticity of anatomical structures and its ecomorphological significance, *Neth. J. Zool.*, 40, 278, 1990.

97. Wimberger, P.H., Plasticity of jaw and skull morphology in the neotropical cichlids *Geophagus brasiliensis* and *G. steindachneri*, *Evolution*, 45, 1545, 1991.

98. Wimberger, P.H., Plasticity of body shape: the effects of diet, development, family and age in two species of *Geophagus* (Pisces: Cichlidae), *Biol. J. Linn. Soc.*, 45, 197, 1992.

99. Kornfield, I. and J.N. Taylor, J.N., A new species of polymorphic fish, *Cichlasoma minckleyi* from Cuatro Cienegas, Mexico, (Teleostei: Cichlidae), *Proc. Biol. Soc. Wash.*, 96, 253, 1983.

100. Trapani, J., Morphological variability in the Cuarto Cienegas cichlid, *Cichlasoma minckleyi*, *J. Fish Biol.*, 62, 276, 2003.

101. Huysseune, A., Sire, J.-Y., and Meunier, F.J., Comparative study of lower pharyngeal jaw structure in two phenotypes of *Astatereochromis alluaudi* (Teleostei: Cichlidae), *J. Morph.*, 221, 25, 1994.

102. Smits, J.D., Witte, F., and Povel, D., Differences between inter- and intraspecific architectonic adaptations to pharyngeal mollusk crushing in cichlid fishes, *Biol. J. Linn. Soc.*, 59, 367, 1996.

103. Smits, J.D., Witte, F., and van Veen, F.G., Functional changes in the anatomy of the pharyngeal jaw apparatus of *Astatoreochromis alluaudi* (Pisces, Cichlidae), and their effects on adjacent structures, *Biol. J. Linn. Soc.*, 59, 389, 1996.

104. Stauffer Jr., J.R. and van Snik Gray, E., Phenotypic plasticity: its role in trophic radiation and explosive speciation in cichlids (Teleostei: Cichlidae), *Anim. Biol.*, 54, 137, 2004.

105. Lewontin, R.C., Selection for colonizing ability, in *The Genetics of Colonizing Species*, Baker, H.G. and Stebbins, G.L., Eds., Academic Press, New York, 1965, 77.

106. Baerends, G.P. and Baerends-van Roon, J.M., An introduction to the study of the ethology of cichlid fishes, *Behaviour*, 1, 1, 1950.

107. Kuwamura, T., Parental care and mating systems of cichlid fishes in Lake Tanganyika: a preliminary field survey, *J. Ethol.*, 4, 146, 1986.

108. Lowe-McConnell, R.H., The breeding behaviour of *Tilapia* species (Pisces; Cichlidae) in natural waters: observations on *T. karomo* Poll, and *T. variabilis* Boulenger, *Behaviour*, 9, 140, 1956.

109. Trewavas, E., *Tilapiine Fishes of the genera Saratherodon, Oreochromis, and Danakilia*, British Museum (Natural History), London, 1983.

110. van den Berghe, E. and McKaye, K.R., Reproductive success of maternal and biparental care in a Nicaraguan cichlid fish, *Parachromis dovii*, *J. Aquaricult. Aquatic Sci.*, 9, 49, 2001.

111. Barlow, G.W., A test of appeasement and arousal hypotheses of courtship behaviour in a cichlid fish, *Etroplus maculates*, *Z. Tierpsychol.*, 27, 779, 1970.

112. Ward, J.A. and Wyman, R.L., Ethology and ecology of cichlid fishes of the genus *Etroplus* in Sri Lanka: preliminary findings, *Environ. Biol. Fishes*, 2, 137, 1977.

113. Konings, A., *Tanganyika cichlids*, Verduijn Cichlids, Zevenhuizen, 1988.

114. Apfelback, R., Vergleichende quantitative Untersuchungen des Fortpflanzungs- und Brutpflegeverhalten von mono- und dimorpher Tilapien (Pisces, Cichlidae), *Z. Tierpsychol.*, 26, 692, 1969.

115. Loiselle, P.V., *The Cichlid Aquarium*, Tetra-Press, Melle, 1985.

116. Timms, A.M. and Keenleyside, M.H.A., The reproductive behaviour of *Aequidens paraguayensis*, *Z. Tierpsych.*, 39, 8, 1975.

117. Linke, H. and Staeck, W., *African Cichlids I: Cichlids of West Africa*, Tetra Press, Melle, 1981.

118. Myrberg Jr., A.A., A descriptive analysis of the behaviour of the African cichlid fish, *Pelmatochromis guentheri* (Sauvage), *Anim. Behav.*, 13, 312, 1965.

119. McKaye, K.R., Ecology and breeding behavior of a cichlid fish, *Cyrtocara eucinostomus*, on a large lek in Lake Malawi, Africa, *Environ. Biol. Fish.*, 8, 81, 1983.

120. McKaye, K.R., Behavioural aspects of cichlid reproductive strategies: patterns of territoriality and brood defense in Central American substratum spawners versus African mouth brooders, in *Fish Reproduction: Strategies and Tactics,* Wooton, R.J. and Potts, G.W., Eds., Academic Press, New York, 1984, 245.

121. McKaye, K.R., Sexual selection and the evolution of the cichlid fishes of Lake Malawi, Africa, in *Cichlid Fishes: Behavior, Ecology and Evolution,* Keenleyside, M.H.A., Ed., Chapman and Hall, London, 1991, 241.

122. McKaye, K.R., Louda, S.M., and Stauffer Jr., J.R., Bower size and male reproductive success in a cichlid fish lek, *Am. Naturalist,* 135, 597, 1990.

123. Stauffer Jr., J.R., LoVullo, T.J., and McKaye, K.R., Three new sand-dwelling cichlids from Lake Malawi, Africa, with a discussion of the status of the genus *Copadichromis* (Teleostei: Cichlidae), *Copeia,* 1993, 1017, 1993.

124. McKaye, K.R. and Stauffer Jr., J.R., Seasonality, depth, and habitat distribution of breeding males, *Oreochromis* spp., 'Chambo', in Lake Malawi National Park, *J. Fish Biol.,* 33, 825, 1988.

125. Kellogg, K.A., Stauffer Jr., J.R., and McKaye, K.R., Characteristics that influence male reproductive success on a cichlid lek, *Behav. Ecol. Sociobiol.,* 47, 164, 2000.

126. Kellogg, K.A. et al., Microsatellite variation demonstrates multiple paternity in lekking cichlid fishes from Lake Malawi, Africa, *Proc. R. Soc. Lond. B,* 260, 79, 1995.

127. Stauffer Jr., J.R., Kellogg, K.A., and McKaye, K.R., Experimental evidence of female choice in Lake Malawi cichlids, *Copeia,* 2005, 656, 2005.

128. Witte, F. et al., The destruction of an endemic species flock: quantitative data on the decline of the haplochromine cichlids of Lake Victoria, *Environ. Biol. Fishes,* 34, 1, 1992.

129. Goldschmidt, T. and Witte, F., Explosive speciation and adaptive radiation of haplochromine cichlids from Lake Victoria: an illustration of the scientific value of a lost species flock, *Mitt. Internat. Verein. Limnol.,* 23, 101, 1992.

130. McKaye, K.R. et al., African tilapia in Lake Nicaragua: ecosystem in transition, *BioScience,* 45, 406, 1995.

131. Kornfield, I., McKaye, K.R., and Kocher, T.D., Evidence for the immigration hypothesis in the endemic cichlid fauna of Lake Tanganyika, *Isozyme Bull.,* 15, 76, 1985.

132. Kocher, T.D. et al., A genetic linkage map of a cichlid fish, the tilapia (*Oreochromis niloticus*), *Genetics,* 148, 1225, 1998.

133. Arnegard, M.E. et al., Population structure and colour variation of the cichlid fish *Labeotropheus fuelleborni* Ahl along a recently formed archipelago of rocky habitat patches in southern Lake Malawi, *Proc. R. Soc. Lond. B,* 266, 1, 1999.

134. Markert, J.A. et al., Biogeography and population genetics of the Lake Malawi cichlid *Melanochromis auratus*: habitat transience, philopatry, and speciation, *Mol. Ecol.,* 8, 1013, 1999.

135. Albertson, R.C. et al. Phylogeny of a rapidly evolving clade: the cichlid fishes of Lake Malawi, East Africa, *Proc. Natl. Acad. Sci.,* 96, 5107, 1999.

# 15 Fungal Diversity

## A. M. C. Tang, B. D. Shenoy and K. D. Hyde
Centre for Research in Fungal Diversity, Department of Ecology and Biodiversity, The University of Hong Kong, P. R. China

## CONTENTS

## ABSTRACT

Fungi are ubiquitous, beneficial, harmful and mutualistic. They perform some of the most important basic roles in life and have some of the greatest potential for biotechnology, yet as few as 7% of the total estimated fungal species on Earth are described. There are thought to be 1.5 million fungal species, but there are huge problems in obtaining estimates of fungal diversity. These include: species recognition, as there are usually few useful characters to distinguish species; separate taxonomic binomials for asexual and sexual states of the same species; lack of specialist mycologists;

and the unfortunate downward trend for mycological biodiversity funding. Estimates of fungal
diversity are discussed for selected plant groups, insects and species rich genera with more than
1,000 species. We conclude that it is important to identify habitats and substrates where a greater
fungal diversity may occur in order to offer maximum protection to fungal resources. The large
variation in estimates of fungal diversity means that considerable data are required before we can
produce a reliable estimate of the number of species of fungi.

## 15.1 INTRODUCTION TO THE FUNGI

### 15.1.1 GENERAL CHARACTERISTICS

Fungi are extremely important organisms, as they have beneficial, harmful, mutualistic and basic
roles in life. They cause diseases of animals, including humans, and particularly of plants and can
have deleterious effects on crop yields. They also provide food in the form of mushrooms and are
used in numerous biotechnological applications, including wine and bread production and as
flavourings. Their role in biodegradation is basic to life through nutrient cycling, but this ability
itself causes problems, such as in wood decay and food moulds.

The motivation for a better understanding of fungal diversity results from the need for knowledge on their ecological functioning, evolutionary relationships, physiological and biochemical
properties, and biotechnological and pharmaceutical potential. The decline in fungal diversity
following habitat destruction in tropical forests has prompted mycologists to stress the need to
accelerate exploration and characterisation of fungal resources in order to use them sustainably and
protect them from destruction[1,2]. Fungal resources have been extensively screened for novel compounds, and bioexploitation has been undertaken by numerous biotechnological and pharmaceutical
companies[3]. Six of the top 20 best-selling drugs are of fungal origin, and it has been estimated that
the overall value of the fungal bioprospecting market may range between US$100 and 200 million[4-6].
In order to offer maximum protection to fungal resources and optimise the potential for biologically
active novel compound discovery, it is important to identify habitats and substrates where a greater
fungal diversity may occur[3].

Fungi are eukaryotic organisms that lack chlorophyll and are saprobic on dead organic matter.
They are generally microscopic, and their cell walls are composed primarily of chitin and glucans[7].
These organisms encompass a huge range of forms from microscopic single celled yeasts to large
macrofungi such as truffles and puffballs. Moulds are composed of long filaments of cells, termed
hyphae, joined together. When moulds grow, the hyphae intertwine to form the mycelium.

Fungi are primarily responsible for the recycling of mineral nutrients in forest ecosystems[8,9]. During
the decomposition process, nutrients that are immobilised in the detritus are mineralised and released
into the soil in a form suitable for plant uptake[10]. The role of fungi is crucial in the decomposition
process, since they can degrade the lignocellulose matrix in forest litter, whilst other organisms
cannot[11,12]. Fungi also form mutualistic relationships with other organisms, such as lichens, mycorrhizae
and endophytes. Lichens are formed from the symbiosis of algae or cyanobacteria with certain fungi,
mostly ascomycetes. It is estimated that about one fifth of all known extant fungal species form these
obligate symbiotic relationships, and major Ascomycota lineages today were once derived from lichen-
forming ancestors[13]. Fungi exist as beneficial symbionts with plants, occurring within their roots systems
and aerial parts. Mycorrhizal symbiosis within the root systems are ubiquitous, ancient and essential
for plants to survive in natural ecosystems, and molecular evidence suggests that successful colonisation
of land by plants probably was facilitated by mycorrhizal symbiosis[14]. Endophytic symbiosis within
plants is also ubiquitous, and endophytes have been isolated from all plants that have been examined,
and every individual plant is probably host to at least two to four endophytes[6,15,16].

Fungi are well known for causing diseases of plants, animals and other fungi. Notable phyto-
pathological examples include chestnut blight caused by *Cryphonectria parasitica,* Dutch elm disease
by *Ceratocystis ulmi,* ergot of sorghum by *Claviceps africana* and the devastating late blight of potato

by *Phytophthora infestans,* which was responsible for the epidemics that contributed to the Irish famine in 1845[17]. Contrary to fungal diseases in plants, fungal diseases in animals are more specific, and probably every species of animal has some specific fungal parasites; *Beauveria* and *Metarhizium* are examples of well studied insect parasites. Fungi have been tested and formulated for application in insect pest management systems as important biocontrol agents[18]. For example, *Cordyceps* is a pathogenic fungus which produces fruiting bodies from caterpillars after killing the host. It is well known for its ability to produce numerous bioactive metabolites, including cyclosporins and efrapeptins that have been used in medicine for the immunosuppressive capabilities[19,20].

Fungi form one of the six kingdoms of life (Animalia, Bacteria, Chromista, Fungi, Plantae and Protozoa; but see Hodkinson and Parnell, *Chapter 1* for a more recent phylogenetic interpretation of the major groups of life)[21]. Surprisingly, fungi are a group more closely related to animals than plants according to ribosomal DNA and protein coding gene sequences, but this theory is still controversial[22–24]. Fungi are subdivided into four phyla, namely Ascomycota, Basidiomycota, Chytridiomycota and Zygomycota[25]. Molecular data suggest that some phyla that were once considered as fungi, such as the plasmodial and cellular slime moulds (Myxomycota and Dictyosteliomycota) and the water moulds (Oomycota), should now be excluded from the kingdom[23]. The following is a summary of the characteristic features of the four phyla within the Fungi.

### 15.1.2 Ascomycota

The phylum Ascomycota, or sac fungi (Greek (hereafter Gr.) *ascus,* sac; *mycetos,* fungi) is a group in which the sexual process involves the production of eight (or multiples of eight) haploid ascospores through the meiosis of a diploid nucleus in an ascus (Figure 15.1a–d)[26]. It is the largest phylum of fungi, with approximately 45,000 described species, and it represents 65% of the known species of fungi[27]. It includes many notable members such as *Claviceps purpurea,* the natural hallucinogen producer which grows on the grains of grasses, *Penicillium notatum* and *P. chrysosogenum* used in the production of antibiotic penicillin, *Saccharomyces cerevisiae* responsible for fermentation in the production of alcohol, and *Neurospora,* the model organism for genetic studies, as well as morels (*Morchella esculenta*) and truffles, such as *Tuber melanosporum,* used in Western cuisines. As well as reproducing sexually, Ascomycetes also sporulate asexually, with the formation of conidia (spores) on conidiophores (Hyphomycetes) or inside a conidiomata (Coelomycetes). The sexual stage of an ascomycete is termed the teleomorph, and the asexual stage is the anamorph. Three subphyla are designated in Ascomycota according to our recent classification[28]. They are the subphylum Pezizomycotina (Euascomycetes), Saccharomycotina (Hemiascomycetes) and Taphrinomycotina (Archiascomycetes).

Pezizomycotina (Euascomycetes) are a group comprising more than 90% of Ascomycota, and 98% are lichenised. Members of Pezizomycotina are designated into two groups: ascohymenial and ascolocular. Ascohymenial relates to an ascocarp that forms after nuclear pairing. The ascohymenial type ascomata may be closed (cleistothecium) (Figure 15.1a), provided with an opening (perithecium) (Figure 15.1b), or open as a cup (apothecium) (Figure 15.1c). Ascolocular relates to a mode of ascocarp growth in which a perithecium (flask-shaped fruiting body) develops within a cushioning hollow of cells (stroma) in a depression of the hymenium (locule). Notable examples of ascohymenial ascomycetes include *Aspergillus* and *Penicillium* (class Eurotiomycetes), *Ascobolus* and *Morchella* (Pezizomycetes), and *Claviceps, Cordyceps* and *Neurospora* (Sordariomycetes). Examples of ascolocular ascomycetes include *Pleospora, Pyrenophora* and *Venturia* (Dothideomycetes).

Saccharomycotina (Hemiascomycetes) are a small subphylum but of tremendous importance. They are characterised by the absence of ascoma so that the asci are naked. They include the 'true yeast' *Saccharomyces cerevisiae* (Figure 15.1d), which is important in the processing of bread and alcoholic beverages. *Saccharomyces* was also the first eukaryote to have its genome completely sequenced[29,30].

Taphrinomycotina (Archiascomycetes) are a diverse group including saprobic and parasitic forms that have been grouped primarily on the basis of rDNA sequence analysis[31,32]. In some

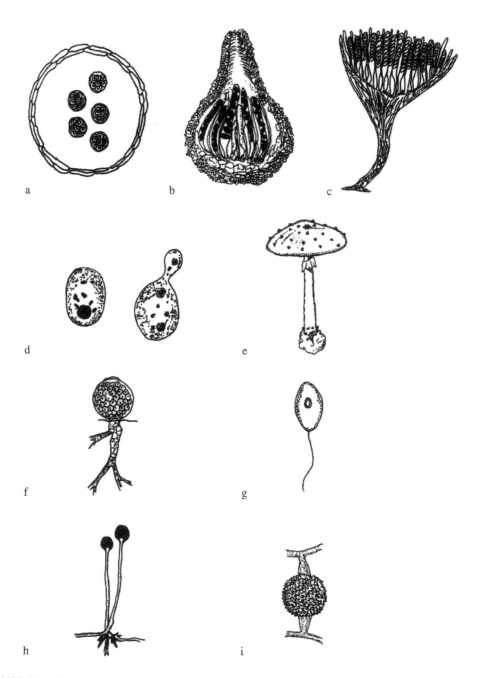

**FIGURE 15.1** Fungi representing different fungal classes. (a) Cleistothecium fruiting body (ascomycete); (b) Perithecium fruiting body (ascomycete); (c) Apothecium fruiting body (ascomycete); (d) *Saccharomyces cerevisiae* (ascomycete); (e) *Amanita* species (basidiomycete); (f) Zoosporangium stage of *Chytriomyces* species (chytridiomycete); (g) Zoospore stage of *Chytriomyces* species (chytridiomycete); (h) Sporangium (asexual) stage of *Rhizopus stolonifer* (zygomycete); (i) Zygosporangium (sexual) stage of *Rhizopus stolonifer* (zygomycete). (Drawings by Alvin M.C. Tang.)

species, such as *Schizosaccharomyces pombe,* the fission yeast was surprisingly separated from *Saccharomyces cerevisiae* (budding yeast, subphylum Saccharomycotina) based on molecular data[33]. *Pneumocystis carinii,* an extracellular biotroph of alveoli in infected lungs of mammals, was once thought to be a protozoan, but is now classified to this subphylum based on DNA sequences[34].

### 15.1.3 BASIDIOMYCOTA

The phylum Basidiomycota (Gr. *basidion,* small base or pestal; *mykes,* fungi) is a group in which the sexual process involves the production of haploid basidiospores borne on a basidium in which a diploid nucleus undergoes meiosis (Figure 15.1e; Figure 15.2)[26]. It contains approximately 30,000 described species, which accounts for about 35% of the known species of fungi[27]. Molecular analyses have defined three lineages from Basidiomycota, two without fruiting bodies, Urediniomycetes and Ustilaginomycetes, and one with fruiting bodies, Hymenomycetes[35].

Urediniomycetes contain approximately 7,400 (34%) of the described species of Basidiomycota[27,36]. It includes the plant pathogenic fungi, the rusts (Uredinales) and the yeasts (Sporidiales), which are saprotrophs and pathogens of plants, animals and fungi. Ustilaginomycetes contain approximately

**FIGURE 15.2** *(A colour version of this figure follows page 240)* Basidiomycete fungi. (a) *Dacryopinax spathularia,* (b) *Pseudocoprinus disseminatus.* (Photos reproduced with permission from Edward Grand, Chiang Mai, Thailand.)

1,300 (6%) of the described species of Basidiomycota[27,37]. It includes the smut fungi (Ustilaginales), which form black and dusty masses of teliospores in diseased plants. Smuts are notorious as they cause millions of dollars of damage to important food crops and ornamentals. Hymenomycetes consists of about 13,500 (60%) of the described species of Basidiomycota[27,35]. There are two basal evolutionary branches (sister groups to the rest of Hymenomycetes), one leading to Tremellales (jelly fungi) and the other to Dacrymycetales, Auriculariales (tree ear fungi), Agaricales (mushrooms) and Aphyllophorales (shelf fungi). Agaricales contains many names that have been known since humans started to collect mushrooms. *Amanita* (Figure 15.1e) is a genus commonly associated with mushroom poisoning, whilst *Agaricus bitorquis* (button mushroom), *Flammulina velutipes* (Enokitake), *Lentinula edodes* (shiitake), and *Pleurotus ostreatus* (oyster mushroom) are widely known as food.

### 15.1.4 CHYTRIDIOMYCOTA

The Chytridiomycota is the only fungal phylum that produces motile zoospores and requires water for dispersal (Figure 15.1f–g). They have been classified in the Protista and Protoctista[38,39], but based on SSU rDNA sequence they were recently included in the kingdom Fungi[40]. Chytridiomycota are probably a very ancient group, with extant forms possibly having changed little since the early periods of eukaryotic evolution[41]. They are commonly found in lakes, streams, ponds, roadside ditches and coastal marine environments, as well as in soil. As members of terrestrial and aquatic microbial communities, chytrids play an important ecological role in decomposition of chitin, cellulose, keratin and hemicellulose[42]. Notable plant pathogenic species include *Synchytrium endobioticum* (potato black wart), *Physoderma maydis* (corn brown spot) and *Urophlytis alfalfae* (alfalfa crown wart). As a representative of lower fungi, *Allomyces macrogynus* has become fashionable in molecular biology for the comparative study of the primitive genetic features with other higher fungi[43–45].

### 15.1.5 ZYGOMYCOTA

The phylum Zygomycota (Gr. *zygos,* yoke of marriage; *mykes,* fungi) is principally characterised by the presence of nonseptate (coenocytic) mycelium and the production of dark, thick-walled, ornamented sexual spores, called zygospores (Figure 15.1h–i). Members of the phylum are generally morphologically and ecologically diverse, with some species not possessing zygospores[46]. Zygomycota has been subdivided into two classes: Trichomycetes and Zygomycetes[47,48]. Trichomycetes are symbionts in the gut of arthropods, while Zygomycetes are saprobic, haustorial or nonhaustorial parasites of animals, plants or fungi[48]. The position of Trichomycetes within the kingdom Fungi remains controversial, since members of Amoebidiales in this class have recently grouped with protists by molecular data, and phylogenetic relationships of Eccrinales and Asellariales are still unresolved[49,50]. Members of Mucorales in the class Zygomycetes, on the other hand, are the most well known group. *Rhizopus* and *Mucor* are the most notable fungi, because they cause fruit rots and bread moulds. Clinically, members of this order, such as *Cokeromyces, Cunninghamella, Rhizomucor, Rhizopus* and *Saksenaea,* are potential human or animal pathogens, especially in immunosuppressed patients during organ transplants and patients with immunological disorders[48,51]. Members of Glomales in Zygomycetes are the most important order ecologically, as they form mycorrhizae with the majority of plants worldwide.

## 15.2 PROBLEMS IN ESTIMATING FUNGAL DIVERSITY

### 15.2.1 FUNGAL SPECIES CONCEPTS AND NOMENCLATURE

The concepts of species recognition in fungi have often been problematic. Attempts to derive a universally applicable concept have been difficult, with several concepts currently in use[52–54]. Species definitions have been based on phenotypic similarity, ecological parameters, reproductive isolation or cohesion and evolutionary principles. In practice, fungal species concepts are always the combinations of morphological, biological and phylogenetic species concepts[55–58]. There are often very few useful characters to separate fungal species, and therefore visual observation may not always be definitive. Cultures are usually required to ascertain species status using morphological and biological species concepts, and yet only 16% of the approximately 100,000 known fungal species are in culture collections worldwide[59]. This makes progress of species determination difficult.

Further complications are caused by the presence of asexual forms (anamorphs) of Ascomycota and Basidiomycota. A single fungal genome always has separate taxonomic binomials for the teleomorph (sexual stage), multiple anamorphs (asexual stage), the chlamydospores, the sclerotia and even the vegetative mycelium[60]. Some Ascomycota may have two or more anamorphs, whilst others seem to be strictly asexual, as sexual reproduction has not clearly been observed[60]. Article 59 of the International Code of Botanical Nomenclature was specifically written for fungi to allow dual or multiple names for a single fungal genome. Whether a single name or multiple names should be used has been controversial; however, it remains an unavoidable complication for many fungal groups, due to the rarity with which multiple morphs are encountered and technical difficulties in linking stages of life cycles[60,61].

## 15.2.2 USE OF BIODIVERSITY MEASURES

Biodiversity measures are quantitative expressions of community structure. The species richness (number of species of a given taxon) and evenness (how similar species are in their abundances) are important measures of diversity[62]. There are many indices available for estimating these measures, and each index emphasises different components of diversity so that no unified diversity index is available[62,63]. For example, Menhinick's index and Margalef's diversity index[64] are measures of species richness. The use of species richness indices to interpret biodiversity may cause variations in results due to factors such as sampling size, scope and duration[65]. For example, the most frequently used Shannon index has problems of confounding species richness and evenness; changes of either richness or evenness can result in changes in the index[62]. The Brillouin index is a more appropriate choice when the randomness of the sample cannot be guaranteed, or if the community is completely censused[66–69]. However, the Brillouin index has been less popular than the Shannon index, since it is more time consuming to produce and is dependent on sample size[62].

## 15.2.3 KNOWLEDGE OF FUNGAL DIVERSITY

The significance of microorganisms as a component of biological diversity had not been fully acknowledged until the recent growth in bioprospecting research (the search for novel compounds or organisms that might have potential economic uses)[70–75]. The major problem in obtaining reliable estimates of fungal diversity is the lack of specialist mycologists, especially in most developing countries where there are few careers available and no training is offered. The other problem is the lack of funds. A recent trend is the diversion of resources from basic inventories and monographic works to molecular systematic studies[76]. The huge sums supplied by the National Science Foundation (NSF) to support ATOL (Assembling the Tree of Life) (US$8 million in 2002; US$12 million in 2003) or the Deep Hypha Project for fungi for molecular work have inevitably de-emphasised the theoretical basis for taxonomy and undermined the intellectual contents of species identification[76]. Taxonomy, an already weakened component of science due to decades of negligence, is now suffering from the loss of positions and funding[77]. The rate of description of new fungal species has been declining since the 1960s, from about 1,400–1,500 species per year to 1,097 species per year in 1990s[78,79]. New species being catalogued in Index of Fungi (http://www.cabi.org) are documented at the rate of about 800 each year. So far, we know as little as 7% of the estimated 1.5 million world fungal diversity. If we calculate according to recent cataloguing rates, we will probably need about 800 more years before we can know 50% of all the fungal species. The unfortunate fact is that our knowledge is already 70–100 years behind that of vascular plants[59,73].

## 15.3  GLOBAL FUNGAL DIVERSITY ESTIMATE: DESCRIBED AND UNDESCRIBED

Fungal names have been catalogued in a variety of printed records, such as: Index of Fungi (1940–present, published twice yearly by CAB International), Sylloge Fungorum (1882–1931, 1972 by P.A. Saccardo) and Petrak's List (1920–1939 by F. Petrak). Searching for early records of fungal names is a daunting task and is effectively restricted to the several institutions where a range of publications are available. Since the mid-1980s, efforts have been made in the International Myco-logical Institute (IMI; now incorporated into CABI Bioscience) to convert all the printed records into an open and accessible computerised database[80], and it is now available both in the fungal database Index Fungorum (http://www.indexfungorum.org/Names/Names.asp) and the world species database Species 2000 (http://www.sp2000.org/). In June 2005, there were 382,808 records in Index Fungorum, which probably include less than 10% of duplicates that could not be easily or unambiguously identified by automatic means, and also perhaps >50% synonyms. Hawksworth[59] and Kirk[80] tried to estimate the actual number of known fungal species by using a synonym percentage of 65%, derived from 15 fungal monographs[81], and concluded that about 105,000 species of fungi may already be known. This figure was a little lower than the estimation of 120,000 from the same data set when only directly ascribed synonyms and no excluded taxa were considered[79]. Nevertheless, these figures are higher than the 72,065 in the eighth edition of *Ainsworth and Bibsy's Dictionary of the Fungi* and 80,060 of the ninth edition in which about 5–10% of the duplicated names have not been adjusted[27,81,82].

The question of how many fungal species there are in the world has challenged mycologists for nearly half a century. Information has been analysed through different attempts, such as plant-to-fungus ratios, insect-to-fungus ratios, inferences from intensive sampling and extrapolation from numbers of new species being found in a group[73,84,85]. The 1.5 million species estimate of Hawksworth was the first in-depth analysis of the magnitude of fungal diversity[73]. It was based on extrapolations from three independent data sets: ratios of the numbers of fungi in all habitats to plants in the British Isles, number of species on native plants, and the number of species in an alpine community study. Nonetheless, the figure was considered a conservative estimate due to the modest 270,000 figure used for the world number of vascular plants, the fact that no separate allowances were made for fungi on insect species and on unstudied plants, and the fact that no special account was taken of potentially hyperdiverse tropical and polar regions[73,79].

Since Hawksworth's estimation[73], keen discussion and arguments have produced figures ranging from 0.5 to 9.9 million (Table 15.1). Amongst the 15 estimates under review, only one author accepted Hawksworth's figure as moderately accurate[86], three proposed estimates lower than 1.5 million[2,87,88], and eleven proposed higher estimates[70,74,84,88–95]. Those that proposed higher estimates tend to draw particular attention to species richness in tropical forests[74,94,96]. Empirical evidence appears to substantiate the view that tropical fungi are hyperdiverse and that newly visited sites and substrates should yield a high percentage of species new to science[74,92,95,97]. For example, during surveys of palm fungi, Hyde et al.[98] discovered as many as 75% of the fungal species were new to science, and Fröhlich and Hyde yielded a very high plant to fungus ratio of 1:33 for palm fungi in the tropics[84]. Particular groups, such as endophytes, insect fungi and macromycetes are considered to be hyperdiverse and may comprise 1–1.5 million species each[91–93,95]. For example, Cifuentes Blanco et al.[93] found that there were 1,300 species of macromycetes associated with 450 plants in Mexico; such a macromycete to plant ratio of about 3.5:1 yields a total of 1 million macromycete species worldwide.

May[88], who adopted a 0.5 million figure for fungi, however, stressed the problems of scaling up from local to global totals and considered that his figure was more likely for fungal species, as they tend to have a wider geographic distribution than plant species. Rossman produced an overall figure of about 1 million by using information from the US National Fungus Collection Database, literature, discussions with other mycologists and personal experience of Rossman[2].

**TABLE 15.1**
**Published Fungal Diversity Estimates Since 1990**

| Author | Year | Estimated Species (Millions) |
|---|---|---|
| Pascoe[89] | 1990 | 2.7 |
| May[74] | 1991 | 1.5+ |
| Hawksworth[73] | 1991 | 1.5 |
| Hammond[87] | 1992 | 1 |
| Smith and Waller[90] | 1992 | 1 (on tropical plants alone) |
| Hywel-Jones[91] | 1993 | 1.5 (insect fungi alone) |
| Dreyfuss and Chapela[92] | 1994 | 1.3 (endophytes only) |
| Rossman[2] | 1994 | 1 |
| Hammond[83] | 1995 | 1.5 |
| Aptroot[100] | 1997 | 0.04–0.07 (ascomycetes alone) |
| Cannon[70] | 1997 | 9.9 |
| Cifuentes Blanco et al.[93] | 1997 | 1 (macromycetes alone) |
| Shivas and Hyde[94] | 1997 | 0.27 (plant pathogens alone) |
| Fröhlich and Hyde[84] | 1999 | 1.5+ |
| Arnold et al.[95] | 2000 | 1.5+ |
| May[88] | 2000 | 0.5 |

*Source:* Modified from Hawksworth[79].

With such large variations in estimates of fungal diversity, it is important that work is carried out in selected research topics to provide data for poorly understood diversity questions. To achieve that, we need more detailed information on particular sites, fungus to plant and fungus to insect ratios and sustained increased attention on the fungi associated with particular plants or groups of insects, especially in the tropics[79]. Several researchers have been carrying out such research, and their data provide more insights into fungal diversity, and some of these data are discussed below. Selected groups and hosts have also been proposed for rapid biodiversity assessments, such as macromycetes, Xylariaceae, lichen-forming fungi, endophytes, palms, bamboos, *Pandanus* species, freshwater fungi and pathogens[99].

## 15.4  EXAMPLES OF FUNGAL DIVERSITY FROM SELECTED HOSTS

### 15.4.1  FUNGI ON POACEAE, CYPERACEAE AND JUNCACEAE

We select grasses, as they are the world's most important agricultural plants and because fungal diversity on grasses has rarely been reviewed[101]. Poaceae (Gramineae) includes cereals, sugar cane, forage grasses for farm animals, ornamental grasses and bamboos and comprises about 10,000 species in 650 genera[101,102] (see Hilu, *Chapter 11*; Hodkinson et al., *Chapter 17*). Grasses (especially cereal grasses) provide favourable substrates for fungal colonisation, as evident from fungal records on various grasses available from the fungal databases of the Systematic Botany and Mycology Laboratory (SBML), Agricultural Research Service, United States Department of Agriculture (http://nt.ars-grin.gov/fungaldatabases/index.cfm)[103]. As many as 14 well studied grass genera support more than 800 fungal taxa, and these are listed in Table 15.2.

There are at least 30,000 records of fungi in the SBML database. These records cover terrestrial habitats[104–109], freshwater habitats[107], estuarine regions[110–112] and marine regions[113,114]. However, they are not exhaustive. Previous studies were biased towards economically important plants, and estimates used small sample sizes and a limited number of sampling sites. Increased sampling,

**TABLE 15.2**
**Fungal Records on Selected Grass Genera**

| Genus | No. of Fungal Records | Genus | No. of Fungal Records |
|---|---|---|---|
| *Zea* | 4,788 | *Oryza* | 1,719 |
| *Triticum* | 3,731 | *Bambusa* | 1,333 |
| *Panicum* | 3,328 | *Pennisetum* | 1,303 |
| *Sorghum* | 2,985 | *Digitaria* | 1,232 |
| *Saccharum* | 1,894 | *Cynodon* | 1,170 |
| *Paspalum* | 1,894 | *Phragmites* | 898 |
| *Setaria* | 1,771 | *Sporobolus* | 928 |

*Source:* Data from Farr et al.[103].

longer study periods, new habitats and unexplored sites tend to yield new data. For example, the number of different saprobic fungi on one well studied cosmopolitan reed, *Phragmites australis,* based on seven studies, can be more than 300[107,109,112,115,116]. Intensive survey of smut fungi (microscopic Basidiomycota) also yielded surprising results. More than 350 species of smut fungi were isolated from nine grasses (including *Bothriochloa, Capillipedium, Chrysopogon, Cynodon, Dichanthium, Hyparrhenia, Muhlenbergia, Saccharum* and *Sorghum*) in New Zealand[117–135]. If we accept 10% of the fungi associated with the *P. australis* above to be host specific, a high ratio of 30:1 results. This fungi to host ratio will increase when endophytes, mycorrhizal fungi, pathogens, rusts and smut fungi are included in the estimation, as these groups are more host specific[136].

Our knowledge of bamboo fungi is still at the cataloguing stage, and new species are often described after field sampling[137–143]. Eriksson and Yue[144] provided an annotated checklist of bambusicolous fungi, and Hyde et al.[143] provided a review of bambusicolous fungi recorded worldwide. There have been some taxonomic or ecological studies on bamboo fungi, but these are limited to France[145], Hong Kong[137–139,143,146] and Japan[147–155]. In June 2005, there were in total 3,222 records of fungi associated with 11 of the most common bamboo genera (*Arundinaria, Bambusa, Chusquea, Dendrocalamus, Gigantochloa, Guadua, Phyllostachys, Pleioblastus, Pseudosasa, Schizostachyum* and *Sinobambusa*) in the SBML database[103]. After correction (allowing about 30–40% for duplicated names and multiple records of single species), there are at least 1,933 fungal species known for bamboo. This figure is much higher than the 1,100 species reviewed by Hyde and coworkers in 2002[143]. It is obvious that this figure will continue to increase as more field studies are conducted.

Fungal diversity on sedges has not been well studied in comparison with fungi reported on grass hosts. The obvious reason may be that sedges have less economic value[156,157]. There are 9,585 records of fungi associated with Cyperaceae in the SBML database[103]. Most of the records were contributed from studies with the genera *Carex* and *Cyperus* (Table 15.3), while the records on

**TABLE 15.3**
**Fungal Records on Selected Sedge Genera**

| Genus | No. of Fungal Records | Genus | No. of Fungal Records |
|---|---|---|---|
| *Carex* | 5,836 | *Fimbristylis* | 304 |
| *Cyperus* | 1,075 | *Eriophorum* | 110 |
| *Scirpus* | 634 | *Kyllinga* | 91 |
| *Rhynchospora* | 563 | *Uncinia* | 86 |
| *Eleocharis* | 318 | *Schoenoplectus* | 83 |

*Source:* Data from Farr et al.[103].

other genera are below 800. This disparity between Poaceae and Cyperaceae may also be attributed to the more diverse morphology and anatomy of the former family than the latter[156].

Juncaceae (rushes) are the sister group of Cyperaceae and are a family of eight genera and about 400 species. They are distributed mainly in temperate climates or the montane regions of the tropics. *Juncus* is one of the dominant genera in estuarine marshes of the American East coast. This genus has received fairly intensive studies in relation to fungi. One endemic species, *Juncus roemerianus* (needle rush)[158,159], was reported to harbour 117 fungal species[160]. If we adopt 10% of host specific fungi, the fungi to host ratio will be 11:1, which is much higher than an estimated average of 5.7 to 8.5[73]. The 117 species (66 Ascomycota, one Basidiomycota and 50 anamorphic taxa) include 48 novel species, 14 novel genera and one novel family[160].

## 15.4.2 LEAF LITTER FUNGI

Degradation of leaves in an aquatic ecosystem involves a diverse assemblage of bacteria, fungi and invertebrates[161,162]. Among aquatic fungi, aquatic hyphomycetes, especially Ingoldian fungi, play an important role in the decomposition of submerged leaf in streams[163]. In 1942, Ingold described 16 species of Ingoldian fungi; in 1981 Webster and Descals described 150 species[163,164]. However, about 300 species of Ingoldian fungi have now been described, mostly from temperate regions[165]. Chan et al.[166] reviewed the Ingoldian fungi from Hong Kong and listed 51 species in 37 genera.

There has been some research on fungi occurring on leaves in tropical forests in different parts of the world. Bills et al.[167] and Polishook et al.[168] have isolated large numbers of rare species and a few common species from leaf litter in Costa Rica and Puerto Rico. Parungao et al.[169] studied the fungi degrading leaf litter of 13 tree species in Australia and identified two to three unique fungi from leaves of each species, with overlap in only 40% of species. Paulus et al.[170] studied diversity of microfungi on decaying leaves of *Ficus pleurocarpa,* and 104 species were identified. Promputtha et al.[171] who have been studying fungal communities on leaves of *Magnolia liliifera,* also recently reported some new species from the host[172,173]. Photita et al.[174] examined the large decaying leaves of *Musa* sp. (banana) in Hong Kong and found 27 taxa at two sites.

## 15.4.3 FUNGI ON INVERTEBRATES

Few studies have addressed fungal numbers on invertebrates. In fact, since the important paper of Weir and Hammond[175] there have been relatively little data on biodiversity of invertebrate fungi. If invertebrate fungi were host specific and occurred in most insects this would have extreme implications for fungal numbers. Weir and Hammond suggest that between 5 and 7% of beetle species may act as hosts for Laboulbeniales (ascomycetous obligate ectoparasites of Arthropoda) and speculated that at least 20,000 and possibly 50,000 species of Laboulbeniales await description[175]. Trichomycetes (symbiotic gut fungi) numbers were also shown to be dependent on host diversity, and host specificity was shown to be a crucial factor in trichomycete diversity[176]. Much work is still needed to address fungal numbers in this area.

## 15.5 SPECIES RICH GENERA OF FUNGI

One of the largest problems in estimating fungal diversity is the genera containing more than 1,000 species. Are these genera really so diverse, or considering the paucity of characters that can be used to separate species, have their species numbers been exaggerated? Plant pathogenic genera such as *Colletotrichum, Pestalotiopsis* and *Mycosphaerella* comprise numerous species, but generally, these species have been described based on the host on which they were found, with little reference to their own characteristics. Whether these genera are megadiverse requires investigation, and molecular techniques are now available to help address this question.

### 15.5.1 *Colletotrichum*

The species rich genus *Colletotrichum* causes various plant diseases often known as anthracnose and is worldwide in distribution[177]. *Colletotrichum* species cause major damage to crops in tropical, subtropical and temperate regions. Cereal, vegetables, legumes, ornamentals and fruit trees may be seriously affected by this pathogen[178]. *Colletotrichum* species are also commonly isolated as endophytes, and latent and quiescent infections by these species on several hosts have been reported[16]. Their ability to cause latent infection, that is, infection without visible symptoms, makes them one of the most successful pathogens causing postharvest disease in a wide range of crop species[177].

*Colletotrichum* is the anamorphic stage of several species of *Glomerella* and has a taxonomic history of about 200 years[179]. There are 17 acknowledged generic synonyms for *Colletotrichum*, and two further names are tentatively included, and there are about 900 species names assigned to this genus[177,180]. The identification and characterisation of *Colletotrichum* species are mainly based on morphological and cultural criteria or a combination of both. It has become apparent that the classification system presently used has limited scope, since some species names assigned to collections and isolates lack the precision required by users. The numbers of morphological characters derived from growth in culture are limited, and growth conditions have rarely been standardised. Moreover, the inherent phenotypic plasticity of individual isolates creates confusion in identification. There are group species or species complexes such as *C. dematium, C. gloeosporioides* and *C. lindemuthianum,* which are known to be represented by at least nine distinct subtaxa[177].

At least nine different Colletotrichum species (*C. capsici, C. coccodes, C. crassipes, C. dematium, C. destructivum, C. gloeosporioides, C. lindemuthianum, C. trifolii* and *C. truncatum*) have been reported on economically important legumes in tropical and temperate regions[181]. All of these species are reported to infect at least two hosts, and *C. capsici, C. gloeosporioides* and *C. lindemuthianum* are reported to have the widest host ranges amongst these nine. *C. gloeosporioides* is a particularly large complex comprising taxa that cause diseases of a wide range of crops. The taxa have been isolated as pathogens, endophytes and saprobes, and it is not clear whether these different lifestyles are associated with specific lineages or have evolved many times. It is therefore particularly important that we gain an understanding of the diversity of organisms within this complex.

Under these circumstances the species name has limited practical significance to the plant pathologist involved in disease management and quarantine and the breeder involved in resistance breeding. The development of different systems for identification of species over time has largely been the result of subtle changes in species concept involving different aspects of morphology combined with ideas about host range and host–pathogen relationships for particular taxa. Despite these amendments the current species concept used in *Colletotrichum* systematics is still very broad, unreliable and unpredictable, being based on the combination of classical criteria such as conidial shape and size, presence, absence and morphology of setae, presence of sclerotia and appressoria and symptom expression on host. Moreover, the current classification system for *Colletotrichum* in general is unsatisfactory because the constituent species are inadequately defined[61]. With further research we may expect to uncover significant levels of synonymy but also discover new species in complexes such as *C. gloeosporioides.*

### 15.5.2 *Pestalotiopsis*

*Pestalotiopsis* species commonly cause diseases on a variety of plants and are commonly isolated as endophytes or occur as saprobes[182]. The genus contains about 205 named species with many named after their hosts in much the same way as *Colletotrichum.* The understanding of species relationships within this weakly parasitic genus is complicated by the lack of morphological characters to differentiate species, and in many cases host association has provided a convenient means to separate species. Jeewon et al.[182] used DNA data from a number of *Pestalotiopsis* isolates to test whether isolates from the same host are phylogenetically related. They also investigated the validity of naming species based on host association. Their results indicated that there was a close phylogenetic relationship between isolates possessing similar morphological characteristics, but isolates from the same host were not necessarily closely related. They advised that, when describing new *Pestalotiopsis* species, morphological characteristics should be taken into account rather than host association. They considered that the high numbers of *Pestalotiopsis* species named in the literature was an overestimate given that naming species based on host is not valid.

### 15.5.3 *MYCOSPHAERELLA*

Species of *Mycosphaerella* (and their anamorphs) are commonly associated with leaf spots or stem cankers[183]. As in the previous two genera, many species have been described based on host association. However, unlike those genera, as well as being host specific, most of these taxa are also highly tissue specific, to the degree that some cercosporoids will sporulate on either the upper or lower leaf surface. Since 1993, Crous and coworkers have described nearly 40 new species of *Mycosphaerella* and associated anamorphs from *Eucalyptus*[183] which appear to be highly specific to this host. An exception to the rule is the *Mycosphaerella tassiana* complex, as well as other species with *Cladosporium* anamorphs.

## 15.6 AN ERA OF GENOMICS AND MOLECULAR BIOLOGY

### 15.6.1 APPLICATIONS OF MOLECULAR BIOLOGY IN MYCOLOGICAL SYSTEMATICS

Classification of fungi in the past 250 years has been largely based on morphological characters. Our current knowledge of phylogenetics and classification mainly stems from morphological studies[184]. However, the dependence of morphological characters to infer the evolutionary history has a number of limitations, the major one of which is the difficulty of recognising homology[185]. Morphology is generally more susceptible to directional selection pressures and often exhibits high levels of homoplasy[186]. In the case of Ascomycota, problems are particularly bad due to the relatively small number of morphological characters that appear to be phylogenetically informative and with the complication of the anamorphic stage[185,187]. As a result, conflicting classification schemes have been proposed for many fungal classes[28,46,188–191].

With the availability of molecular techniques, studies of mycology have entered a new era in the last two decades. These techniques can be applied in many disciplines, and some of these are summarised in Table 15.4. In studies of fungal diversity, DNA-based techniques can provide a comprehensive measure of diversity and composition of fungal communities, as they review both the culturable and often predominant nonculturable members of a community[192,193]. Results from DNA sequencing of genes of different species can be compared and analysed with the sequences from the other studies downloaded from the world DNA sequence database such as GenBank (http://www.ncbi.nlm.nih.gov) or the Assembling Fungal Tree Of Life Project (AFTOL) (http://aftol.org/data.php). Large-scale comparisons are now possible with other distant organisms such as plants and animals to elucidate patterns of organism diversifications and relationships[22,23,194,195]. For instance, the exclusion of the morphologically similar phyla Myxomycota and Dictyosteliomycota from kingdom Fungi was made after molecular data were available. On the other hand, application of specific statistical testing and mathematical models in DNA analyses on different genetic levels can elucidate differences in rates and modes of evolution and estimate the time for diversification in ecological characters and will result in a deeper understanding of biological diversity[196,197].

**TABLE 15.4**
**Synopsis of Potential Applications of Molecular Biology**

| Areas of Study | Application of Molecular Techniques |
|---|---|
| Fungal systematics and evolutionary studies | Provide semi-quantitative measures of relationships<br>Elucidate evolutionary pathways<br>Define particular systematic groupings |
| Fungal biotechnology | Predict protein structures and determine gene function<br>Identify genes with particular properties<br>Modify the genes in target fungi |
| Medical mycology | Develop diagnostic tests for human fungal infections<br>Understand the pathogenicity of fungal diseases |
| Agriculture | Characterise and identify plant pathogens<br>Develop and genetically modify biocontrol fungi |
| Food | Trace and detect spoilage fungi |
| Environmental studies | Measure the fungal diversity<br>Analyse the composition of fungal communities |

*Source:* Modified from Bridge[196].

## 15.6.2 MOLECULAR STUDY ON FUNGAL DIVERSITY IN SOIL

Microorganisms in soil are critical to the maintenance of soil function and quality because of their involvement in soil structuring, decomposition of organic matters, recycling of nutrients and promoting plant growth[198,199]. Traditionally, cultivation and isolation were used to analyse the soil microbial communities. Unfortunately, the number and range of fungal species present in the soil are not accurately represented by those methods, as some of the fungi are either nonculturable or suppressed by other fungi in the cultivation based system[199,200]. It is increasingly common to use molecular approaches to study fungal communities in soil. These molecular techniques are generally based on PCR or RT-PCR of specific or generic targets in the soil DNA or RNA. The use of specific PCR primers further enables specific groups of fungi, such as Ascomycota and Basidiomycota, to be amplified in the presence of other groups of organisms, such as algae or bacteria[201]. PCR products amplified using specific fungal primers yield a mixture of DNA fragments representing a number of PCR-accessible species present in the soil[199]. These mixed PCR products can be used for preparing the clone libraries and a range of fingerprinting techniques such as denaturing or temperature gradient gel electrophoresis (DGGE/TGGC)[202–204], amplified rDNA restriction analysis (ARDRA), terminal restriction fragment length polymorphism (T-RFLP)[205], and ribosomal intergenic spacer length polymorphism (RISA)[206].

## 15.6.3 UNRAVELLING SPECIES RICH GENERA USING MOLECULAR TECHNIQUES

Various DNA-based systems have been used to study phylogeny, systematics, genetic diversity and population structure of species rich fungi. The molecular markers include restriction fragment length polymorphisms (RFLP), random amplified polymorphic DNA (RAPD), amplified fragment length polymorphisms (AFLP), rDNA internal transcribed spacer (ITS-1 to ITS-2) and small subunit ribosomal RNA (18S rDNA sequences). They have been utilised successfully in differentiating populations of different *Colletotrichum* species and to reliably assign correct names to morphologically different but genetically similar species. For instance, *C. orbiculare* from cucumber, *C. trifolii* from alfalfa, *C. malvarum* from prickly sida and *C. lindermuthianum* from bean were found to be closely related[207–209]. Sheriff et al.[210] further proposed that *C. lindermuthianum*, *C. malvarum*, *C. orbiculare* and *C. trifolii* should be considered as a single species based

**COLOUR FIGURE 14.1** Cichlid fishes. Cichlids have a conservative bauplan, and specialised attributes, such as hypertrophied lips are the result of parallel evolution, thus making species and higher level diagnoses difficult. (a) *Amphilophus* sp. 'fatlip' in Lake Xiloa, Nicaragua; (b)   *Placidochromis milomo* at Nkhomo Reef, Lake Malawi, Malawi; (c) *Lobochilotes labiatus* at Nkondwe Island, Lake Tanganyika, Tanzania. (Photos reproduced with permission from A.F. Konings.)

a

b

**COLOUR FIGURE 15.2** Basidiomycete fungi. (a) *Dacryopinax spathularia,* (b) *Pseudocoprinus disseminatus.* (Photos reproduced with permission from Edward Grand, Chiang Mai, Thailand.)

**COLOUR FIGURE 16.4** Flowers and fruit of species of the *Syzygium* group. (A) Flowers of *Syzygium malaccense* (L.) Merr. and L.M. Perry; (B–D) flowers, inflorescence and fruit, respectively, of *Acmena* cf. *divaricata* Merr. and L.M. Perry; (E) fruit of *Piliocalyx bullatus* Brongn. and Gris; (F) buds and flowers of *Syzygium longifolium* (Brongn. and Gris) J.W. Dawson; (G–H) fruit and flowers, respectively, of *Syzygium aqueum* (Burm. f.) Alston; (I) flowers of *Syzygium jambos* (L.) Alston; (J) buds (note calyptras) of *Syzygium kuebiniense* J.W. Dawson; (K) fruit of *Syzygium rubrimolle* B. Hyland; (L–M) flowers and fruit, respectively, of *Syzygium glenum* Craven. (Reproduced with permission from G. Sankowsky (A–D, G, H, K–M), L. Craven. (F, I) and E. Biffin (J).)

**COLOUR FIGURE 16.5** Flowers, fruit and foliage of species of the *Syzygium* group. (A–B) buds and flowers, and fruit, respectively, of *Acmenosperma pringlei* B. Hyland; (C) *Syzygium wilsonii* subsp. *cryptophlebium* (F. Muell.) B. Hyland; (D) fruit of *Syzygium elegans* (Brongn. and Gris) J.W. Dawson; (E–G) habit, young leaves, and buds and flowers, respectively, of *Syzygium acre* (Pancher ex Guillaumin) J.W. Dawson; (H) fruit of *Syzygium cormiflorum* (F. Muell.) B. Hyland; (I) flowers of *Syzygium boonjee* B. Hyland; (J) flower of *Syzygium* sp.; (K) flowers of *Syzygium balansae* (Guillaumin) J.W. Dawson; (L) fruit of *Syzygium maraca* Craven and Biffin; (M) young fruit of *Syzygium* sp. (Reproduced with permission from A. Ford (A–B), G. Sankowsky (C, H, I, L), E. Biffin (D), L. Craven. (E–G, K, M) and J. Dowe (J).)

on rDNA sequence analysis. In a study of *Colletotrichum* from almond, avocado and strawberry, Freeman et al.[211] found that, although morphological criteria indicated that the Israeli isolates of almond are unique, the population was grouped within the *C. acutatum* species according to molecular analyses. It is obvious that further studies of other species rich genera at the molecular level are necessary before we can obtain a conclusion of the effects of these genera on fungal numbers.

### 15.6.4  FUNGAL ENDOPHYTES: UNDERESTIMATED?

The term 'endophyte' refers to the symptomless, mutualistic fungi which grow within aerial parts of plants[27]. Almost all vascular plant species examined have been found to harbour endophytic fungi, and each species of vascular plant harbours at least two to four endophyte species unique to that plant species[16,95,206]. Some taxa, especially coelomycetous anamorphic taxa, such as *Colletotrichum, Coniothyrium, Cylindrosporium, Hendersonia, Phoma, Phomopsis, Phyllosticta, Septoria* and *Stagonospora,* have been found to occur on nearly all vascular plants examined[16].

Traditional isolation techniques involve plating healthy, surface-sterilised plant tissues on agar media and observing the outgrowth of fungi, or incubation of washed plant tissues under humid conditions and subsequent collection and plating of discharged spores, as summarised in Kirk et al.[27]. In most of the studies on the diversity of fungal endophytic communities in the tropics[211–220], as well as in temperate regions[212,218,221–223], a large number of fungi did not sporulate in culture. The nonsporulating fungi, commonly categorised as *mycelia sterilia,* are grouped into 'morphological species' on the basis of similarity in colony surface textures, hyphal pigments, exudates, margin shapes and growth rate[16]. This concept of 'morphospecies' has provided a practical means to estimate endophytic fungal diversity by incorporating the nonsporulating fungal isolates. Lacap et al.[224] summarised the proportions of *mycelia sterilia* found in some of the earlier studies of fungal communities. For a given host, *mycelia sterilia* can comprise an average of 20% of the population of fungal endophytes and may be as high as 54%[222,224]. Lacap et al.[224] also conducted a study to verify the concept of morphospecies based on DNA sequence data on *mycelia sterilia* isolated from *Polygonum multiforum* and to identify the valid taxonomic groups. With the aid of phylogenetic analyses, those morphospecies can also be grouped into their respective orders and families, or even assigned to genera or species. Guo et al.[225] performed a phylogenetic analysis of 5.8S DNA gene sequences on the morphospecies isolated from palm *Livistona chinensis* and found that the morphospecies were mostly ascomycetes, belonging to the Loculoascomycetes and Sordariomycetes.

The culture-based methods above, however, have their limitations in that they do not reveal nonculturable endophytic fungi. In a litter decomposition study, Nikolcheva et al.[226] investigated the early stages of colonisation of five leaf species and birch wood by aquatic hyphomycetous taxa using both the traditional methods and DGGE. They extracted the total DNA from senescent leaves, and DNA fragments were amplified using PCR with specific fungal primers. The mixture of PCR products of 18S rDNA was then separated by DGGE, based on differences in the ease of denaturation arising from sequence variability, and the separated bands sequenced and identified.

In a modern endophytic fungal community study, a multipronged approach will often be used. This includes the use of traditional microscope-based methods after plating of sterilised plant tissues on agar medium[215], DNA sequence analysis to identify fungi on nonsporulating culture plates after designation to morphospecies, and DGGE to reveal the nonculturable endophytic fungi[225,226]. This combined approach provides a comprehensive measure of endophytic fungal diversity and reliably reveals the diversity of endophytic fungal communities.

## 15.7  CONCLUDING REMARKS

In order to gain a better understanding of fungal diversity, we should continue to concentrate on studying the fungal communities in selected habitats and substrates, especially those that appear to support high diversity and also explore understudied or unstudied habitats and substrates.

For example, palms have been shown to be hyperdiverse substrates for fungi[84,98,214,227–229], and recent studies in palm swamps in Thailand have yielded numerous new taxa[230,231]. Palms and other unstudied substrates in other areas should be investigated to establish if new species discovery will continue unabated. To advance our knowledge, we must prioritise funding for inventory and monographic studies simultaneously with funding for molecular biology. This will provide an invaluable legacy of data for conservation evaluation and biotechnological and pharmaceutical utilisation.

## REFERENCES

1. Reid et al., Ed., *Biodiversity Prospecting: Using Genetic Resources for Sustainable Development,* World Resources Institute, Washington, DC, 1993.
2. Rossman, A.Y., A strategy for an all-taxa inventory of fungal diversity, in *Biodiversity and Terrestrial Ecosystems,* Monograph Series No. 14, Peng, C.I. and Chen, C.H., Eds., Institute of Botany, Academia Sinica, Taipei, 1994, 169.
3. Hyde, K.D., Increasing the likelihood of novel compound discovery from filamentous fungi, in *Bio-Exploitation of Filamentous Fungi,* Fungal Diversity Research Series, Pointing, S.B. and Hyde, K.D., Eds., Fungal Diversity Press, Hong Kong, 6, 77, 2001.
4. Chapela, I.H., Bioprospecting: myths, realities and potential impact on sustainable development, in *Mycology in Sustainable Development: Expanding Concepts, Vanishing Borders,* Palm, M.E. and Chapela, I.H., Eds., Parkway Publisher, Boone, NC, 1997, 238.
5. Concepcion, G.P., Lazaro, J.E., and Hyde, K.D., Screening for bioactive novel compounds, in *Bio-Exploitation of Filamentous Fungi,* Fungal Diversity Research Series, Pointing, S.B. and Hyde, K.D., Eds., 6, 93, 2001.
6. Strobel, G.A., Endophytic fungi: new sources for old and new pharmaceuticals, *Pharm. News,* 3, 7, 1996.
7. Madigan, M.T., Martinko, J.M., and Parker, J., *Brock Biology of Microorganisms,* 10th ed., Prentice Hall and Pearson Education, Upper Saddle River, NJ, 2003.
8. Deacon, J.W., *Fungal Biology,* Blackwell, UK, 2005.
9. Swift, M.J., Heal, O.W., and Anderson, J.M., *Decomposition in Terrestrial Ecosystems,* Blackwell, Oxford, UK, 1979.
10. Cotrufo, M.F., Miller, M., and Zeller, B., Litter decomposition, in *Carbon and Nitrogen Cycling,* Schulze, E.D., Ed., Springer-Verlag, Heidelberg, 2000.
11. Risna, R.A. and Suhirman, Lignolytic enzyme production by *Polyporaceae* from Lombok, Indonesia, *Fung. Divers.,* 9, 123, 2002.
12. Urairuj, C., Khanongnuch, C., and Lumyong, S., Ligninolytic enzymes from tropical endophytic Xylariaceae, *Fung. Divers.,* 13, 209, 2003.
13. Lutzoni, F., Pagel, M., and Reeb, V., Major fungal lineages are derived from lichen symbiotic ancestors, *Nature,* 411, 937, 2001.
14. Heckman, D.S. et al., Molecular evidence for the early colonization of land by fungi and plants, *Science,* 293, 1129, 2001.
15. Rodrigues, K.F., Fungal endophytes of palms, in *Endophytic Fungi of Grasses and Woody Plants,* Redlin, S.C. and Carris, L.M., Eds., APS Press, St. Paul, MN, 1996.
16. Bills G.F., Isolation and analysis of endophytic fungal communities from woody plants, in *Endophytic Fungi in Grasses and Woody Plants,* Redlin, S.C. and Carris, L.M., Eds., APS Press, St. Paul, MN, 1996, 31.
17. Agrios, G.N., *Plant Pathology,* 4th ed., Academic Press, San Diego, CA, 1997.
18. Faria, M. and Wright, S.P., Biological control of *Bemisia tabaci* with fungi, *Crop Protection,* 20, 767, 2001.
19. Hodge, H.T., Krasnoff, S.B., and Humber, R.A., *Tolypocladium inflatum* is the anamorph of *Cordyceps subsessilis, Mycologia,* 88, 715, 1996.
20. Bandani, A.R. et al., Production of efrapeptins by *Tolypocladium* species and evaluation of their insecticidal and antimicrobial properties, *Mycol. Res.,* 104, 537, 2000.
21. Cavalier-Smith, T., A revised six-kingdom system of life, *Biol. Rev.,* 73, 203, 1998.
22. Baldauf, S.L. and Palmer, J.D., Animals and fungi are each other's closest relatives: congruent evidence from multiple proteins, *Proc. Natl. Acad. Sci. USA,* 90, 11558, 1993.
23. Wainright, P.O. et al., Monophyletic origins of the Metazoa: 'an evolutionary link with fungi', *Science,* 260, 340, 1993.

24. Wang, D.Y.C., Kumar, S., and Hedges, S.B., Divergence time estimates for the early history of animal phyla and the origin of plants, animals and fungi, *Proc. R. Soc. Lond. B,* 266, 163, 1999.
25. Bruns, T.D. et al., Evolutionary relationships within the fungi: analyses of nuclear small subunit RNA sequences, *Mol. Phylogenet. Evol.,* 1, 231, 1992.
26. Carlile, M.J. and Watkinson, S.C., *The Fungi,* Academic Press, London, 1994.
27. Kirk, P.M. et al., *Ainsworth and Bisby's Dictionary of the Fungi,* 9th ed., CAB International, Oxon, UK, 2001.
28. Eriksson, O.E. et al., Eds., Outline of Ascomycota 2003, *Myconet,* http://www.umu.se/myconet/M9. html, 2003.
29. Kim, J.M. et al., Transposable elements and genome organization: a comprehensive survey of retrotransposons revealed by the complete *Saccharomyces cerevisiae* genome sequence, *Genome Res.,* 8, 464, 1998.
30. Sánchez, R. and Sali, A., Large-scale protein structure modeling of the *Saccharomyces cerevisiae* genome, *Proc. Natl. Acad. Sci. USA,* 95, 13597, 1998.
31. Nishida, H. and Sugiyama, J., Phylogenetic relationships among *Taphrina, Saitoella,* and other higher fungi, *Mol. Biol. Evol.,* 10, 431, 1993.
32. Nishida, H. and Sugiyama, J., Archiascomycetes: detection of a major new linage within the Ascomycota, *Mycoscience,* 35, 361, 1994.
33. Taylor, J.W. et al., Fungal model organisms: phylogenetics of *Saccharomyces, Aspergillus,* and *Neurospora, Syst. Biol.,* 42, 440, 1993.
34. Taylor, J.W., Swann, E., and Berbee, M.L., Molecular evolution of ascomycete fungi: phylogeny and conflict, in *First International Workshop on Ascomycete Systematics,* Hawksworth, D.L., Ed., NATO Advanced Science Institutes Series, Plenum Press, New York, 1994, 201.
35. Swann, E.C. and Taylor, J.W., Higher taxa of basidiomycetes: an 18S rRNA gene perspective, *Mycologia,* 85, 923, 1993.
36. Swann, E.C., Frieder, E.M., and McLaughlin, D.J., Urediniomycetes, in *The Mycota VII: Systematics and Evolution Part B,* McLaughlin, D.J., McLaughlin, E.G., and Lemke, P.A., Eds., Springer-Verlag, Berlin, 2001, 37.
37. Bauer, R. et al., Ustilaginomycetes, in *The Mycota VII Systematics and Evolution Part B,* McLaughlin, D.J., McLaughlin, E.G., and Lemke, P.A., Eds., Springer-Verlag, Berlin, 2001, 57.
38. Whittaker, R.H., New concepts of kingdoms of organisms, *Science,* 163, 150, 1969.
39. Margulis, L. et al., *Handbook of Protoctista,* Jones and Bartlett, Boston, 1990.
40. Bowman, B.H. et al., Molecular evolution of the fungi: relationship of the Basidiomycetes, Ascomycetes, and Chytridiomycetes, *Mol. Biol. Evol.,* 9, 285, 1992.
41. Barr, D.J.S., Chytridiomycota, in *The Mycota VII Systematics and Evolution Part B,* McLaughlin, D.J., McLaughlin, E.G., and Lemke, P.A., Eds., Springer-Verlag, Berlin, 2001, 93.
42. Barr, D.J.S., Phylum Chytridiomycota, in *Handbook of Protoctista,* Margulis L. et al., Ed., Jones and Bartlett, Sudbury, MA, 1990, 454.
43. Paquin, B. and Lang, F., The mitochondrial DNA of *Allomyces macrogynus*: the complete genomic sequence from an ancestral fungus, *J. Mol. Biol.,* 255, 688, 1996.
44. Ribichich, K.F. et al., Gene discovery and expression profile analysis through sequencing of expressed sequence tags from different developmental stages of the chytridiomycete *Blastocladiella emersonii, Eukaryot. Cell,* 4, 455, 2005.
45. Rocha, C.R.C. and Gomes, S.L., Characterization and submitochondrial localization of the alpha subunit of the mitochondrial processing peptidase from the aquatic fungus *Blastocladiella emersonii, J. Bacteriol.,* 181, 4257, 1999.
46. Alexopoulos, C., Mims, C., and Blackwell, M., *Introductory Mycology,* Wiley and Sons, New York, 1996.
47. Benny, G.L., Zygomycota: Trichomycetes, in *The Mycota VII Systematics and Evolution Part B,* McLaughlin, D.J., McLaughlin, E.G., and Lemke, P.A., Eds., Springer-Verlag, Berlin, 2001, 147.
48. Benny, G.L., Humber, R.A., and Morton, J.B., Zygomycota: Zygomycetes, in *The Mycota VII Systematics and Evolution Part B,* McLaughlin, D.J., McLaughlin, E.G., and Lemke, P.A., Eds., Springer-Verlag, Berlin, 2001, 113.
49. Benny, G.L. and O'Donnell, K.O., *Amoebidium parasiticum* is a protozoan, not a Trichomycete, *Mycologia,* 92, 1133, 2000.
50. Ustinova, I., Krienitz, L., and Huss, V.A.R., *Hyaloraphidium curvatum* is not a green alga, but a lower fungus: *Amoebidium parasiticum* is not a fungus, but a member of the DRIPs, *Protist,* 151, 253, 2000.

51. Guarro, J., Gene, J., and Stchigel, A.M., Developments in fungal taxonomy, *Clin. Microbiol. Rev.*, 12, 454, 1999.

52. Harrington, T.C. and Rizzo, D.M., Defining species in the fungi, in *Structure and Dynamics of Fungal Populations,* Worrall, J.J., Ed., Kluwer Press, Dordrecht, Netherlands, 1999, 43.

53. Mayden, R.L., A hierarchy of species concepts: the dénouement in the saga of the species problem, in *Species: The Units of Biodiversity,* Claridge, M.F., Dawah, H.A., and Wilson, M.R., Eds., Chapman and Hall Ltd., London, UK, 1997, 381.

54. Taylor, J.W. et al., Phylogenetic species recognition and species concepts in fungi, *Fun. Genet. Biol.*, 31, 21, 2000.

55. Blackwell, M., Phylogenetic systematics of ascomycetes, in *The Fungal Holomorph,* Reynolds, D.R. and Taylor, J.W., Eds., CAB International, Wallingford, UK, 1993, 93.

56. Hibbett, D.S., et al., Phylogenetic diversity in shiitake inferred from nuclear ribosomal DNA sequences, *Mycologia,* 87, 618, 1995.

57. Vilgalys, R., Speciation and species concepts in the *Collybia dryophila* complex, *Mycologia,* 83, 758, 1991.

58. Vilgalys, R. and Sun, B.L., Ancient and recent patterns of geographic speciation in the oyster mushroom *Pleurotus ostreatus* revealed by phylogenetic analysis of ribosomal DNA, *Proc. Nat. Acad. Sci. USA,* 91, 4599, 1994.

59. Hawksworth, D.L., Fungal diversity and its implications for genetic resource collections, *Stud. Mycol.,* 50, 9, 2004.

60. Seifert, K.A. and Samuels, G.J., How should we look at anamorphs? *Stud. Mycol.,* 45, 5, 2000.

61. Cannon, P.F. and Kirk, P.M., The philosophy and practicalities of amalgamating anamorph and teleomorph concepts, *Stud. Mycol.,* 45, 19, 2000.

62. Magurran, A.E., *Measuring Biological Diversity,* Blackwell Science, Oxford, 2004.

63. Clarke, K.R. and Warwick, R.M., A further biodiversity index applicable to species lists: variation in taxonomic distinctness, *Mar. Ecol. Prog. Ser.,* 216, 265, 2001.

64. Clifford, H.T. and Stephenson, W., *An Introduction to Numerical Classification,* Academic Press, London, 1975.

65. Gaston, K.J., Species richness: measure and measurement, in *Biodiversity: A Biology of Numbers and Difference,* Gaston, K.J., Ed., Oxford University Press, Oxford, UK, 1996, 77.

66. Ito, A. and Imai, S., Ciliates from the cecum of capybara (*Hydrocheorus hydrochaeris*) in Bolivia 2: the family Cycloposthiidae., *Eur. J. Protist.,* 2000, 36, 169.

67. Pielou, E.C., *An Introduction to Mathematical Ecology,* Wiley, New York, 1969.

68. Pielou, E.C., *Ecological Diversity,* Wiley InterScience, New York, 1975.

69. Southwood, R. and Henderson, P.A., *Ecological Methods,* Blackwell Science, Oxford, UK, 2000.

70. Cannon, P.F., Diversity of *Phyllachoraceae* with special reference to the tropics, in *Biodiversity of Tropical Microfungi,* Hyde, K.D., Ed., Hong Kong University Press, Hong Kong, 1997, 255.

71. Cannon, P.F., Strategies for rapid assessment of fungal diversity, *Biodiver. Conserv.,* 6, 669, 1997.

72. Colwell, R.R. et al., The microbial species concept and biodiversity, in *Microbial Diversity and Ecosystem Function,* Allsopp, D., Colwell, R.R., and Hawksworth, D.L., Eds., Cambridge University Press, Cambridge, UK, 1995, 3.

73. Hawksworth, D.L., The fungal dimension of biodiversity: magnitude, significance, and conservation, *Mycol. Res.,* 95, 641, 1991.

74. May, R.M., A fondness for fungi, *Nature,* 352, 475, 1991.

75. Hyde, K.D., Where are the missing fungi? Does Hong Kong have any answers? *Mycol. Res.,* 105, 1514, 2001.

76. Korf, R.P., Reinventing taxonomy: a curmudgeon's view of 250 years of fungal taxonomy, the crisis in biodiversity, and the pitfalls of the phylogenetic age, *Mycotaxon,* 93, 407 2005.

77. Wheeler, Q.D., Taxonomic triage and the poverty of phylogeny, *Phil. Trans. R. Soc. Lond. B,* 359, 571, 2004.

78. Ainsworth, G.C., The number of fungi, in *The Fungi: An Advanced Treatise,* Vol. 3, Ainsworth, G.C. and Sussman, A.S., Eds., Academic Press, New York, 1968, 505.

79. Hawksworth, D.L., The magnitude of fungal diversity: the 1.5 million species revisited, *Mycol. Res.,* 105, 1422, 2001.

80. Kirk, P.M., World catalogue of 340K fungal names on-line, *Mycol. Res.,* 104, 516, 2000.

81. Hawksworth, D.L., The need for a more effective biological nomenclature for the 21st century, *Bot. J. Linn. Soc.,* 109, 543, 1992.

82. Hawksworth, D.L. et al., *Ainsworth and Bisby's Dictionary of the Fungi,* 8th ed., CAB International, Wallingford, 1995.
83. Hammond, P.M., Described and estimated species numbers: an objective assessment of current knowledge, in *Microbial Diversity and Ecosystem Function,* Allsopp. D., Colwell, R.R., and Hawksworth, D.L., Eds., Cambridge University Press, Cambridge, UK, 1995, 29.
84. Fröhlich, J. and Hyde, K.D., Biodiversity of palm fungi in the tropics: are global fungal diversity estimates realistic? *Biodiver. Conserv.,* 8, 977, 1999.
85. Sipman, H.J.M. and Aptroot, A., Where are the missing lichens? *Mycol. Res.,* 105, 1433, 2001.
86. Hammond, P.M., The current magnitude of biodiversity, in *Global Biodiversity Assessment,* Heywood, V.H., Ed., Cambridge University Press, Cambridge, UK, 1995, 113.
87. Hammond, P.M., Species inventory, in *Global Biodiversity: Status of the Earth's Living Resources,* Groombridge, B., Ed., Cambridge University Press, Cambridge, UK, 1992, 113.
88. May, R.M., The dimensions of life on earth, in *Nature and Human Society: The Quest for a Sustainable World,* Raven, P.H. and Williams, T., Eds., National Academy Press, Washington, DC, 2000, 30.
89. Pascoe, I.G., History of systematic mycology in Australia, in *History of Systematic Mycology in Australia,* Short, P.S., Ed., Australian Systematic Botany Society, South Yarra, 1990, 259.
90. Smith, D. and Waller, J.M., Culture collections of microorganisms: their importance in tropical plant pathology, *Fitopatol. Brasil.,* 17, 1, 1992.
91. Hywel-Jones, N.L., A systematic survey of insect fungi from natural, tropical forest in Thailand, in *Aspects of Tropical Mycology,* Issac, S. et al., Eds., Cambridge University Press, Cambridge, UK, 1993, 300.
92. Dreyfuss, M.M. and Chapela, I.H., Potential of fungi in the discovery of novel, low-molecular weight pharmaceuticals, in *The Discovery of Natural Products with Therapeutic Potential,* Gullo, V.P., Ed., Butterworth-Heinemann, London, UK, 1994, 49.
93. Cifuentes Blanco, J. et al., Diversity of macromycetes in pine-oak forests in the neovolcanic axis, Mexico, in *Mycology in Sustainable Development: expanding concepts, vanishing borders,* Palm, M.E. and Chapela, I.H., Eds., Parkway Publishers, Boone, NC, 1997, 111.
94. Shivas, R.G. and Hyde, K.D., Biodiversity of plant pathogenic fungi in the tropics, in *Biodiversity of Tropical Microfungi,* Hyde, K.D., Ed., Hong Kong University Press, Hong Kong, 47, 1997.
95. Arnold, A.E. et al., Are tropical fungal endophytes hyperdiverse? *Ecol. Lett.,* 3, 267, 2000.
96. Hawksworth, D.L., Rossman, A.Y., Where are all the undescribed fungi? *Phytopathol.,* 87, 888, 1997.
97. Lodge, D.J., Ed., A survey of patterns of diversity in non-lichenized fungi, *Mitteilungen der Eidgenössischen Forschungsanstlalt für Wald, Schnee und Landschaft,* 70, 157, 1995.
98. Hyde, K.D., Fröhlich, J., and Taylor, J., Diversity of ascomycetes on palms in the tropics, in *Biodiversity of Tropical Microfungi,* Hyde, K.D., Ed., Hong Kong University Press, Hong Kong, 1997, 141.
99. Hyde, K.D. et al., Estimating the extent of fungal diversity in the tropics, in *Nature and Human Society: The Quest for a Sustainable World,* Raven, P.H. and Williams, T., Eds., National Academy Press, Washington, DC, 2000, 156.
100. Aptroot, A., Species diversity in tropical rainforest ascomycetes: lichenized *versus* non-lichenized; foliicolus verus corticolous, *Abstracta Botanica,* 21, 37, 1997.
101. Chapman, G.P. and Peat, W.E., *An Introduction to Grasses (Including Bamboos and Cereals),* Redwood Press Ltd., UK, 1992.
102. Younger, V.B. and McKell, C.M., *The Biology and Utilization of Grasses,* Academic Press, New York and London, 1972, 426.
103. Farr, D.F. et al., Fungal Databases, Systematic Botany and Mycology Laboratory, ARS, USDA, http://nt.ars-grin.gov/fungaldatabases/index.cfm, 2005.
104. Apinis, A.E., Chester, C.G.C., and Taligoola, H.K., Colonization of *Phragmites communis* leaves by fungi, *Nova Hedwigia,* 23, 113, 1972.
105. Barr, M.E., Huhndorf, S.M., and Rogerson, C.T., The Pyrenomycetes described by J.B. Ellis, *Memoirs of the New York Botanical Garden,* 79, 1, 1996.
106. Piepenbring, M., Ecology, and seasonal variation, and altitudinal distribution of Costa Rican smut fungi, Basidiomycetes: Ustilaginales and Tilletiales, *Rev. Biol. Trop.,* 44, 115, 1996.
107. Poon, M.O.K. and Hyde, K.D., Evidence for the vertical distribution of saprophytic fungi on senescent *Phragmites australis* culms at Mai Po Marshes, Hong Kong, *Bot. Mar.,* 41, 285, 1998.
108. Sivanesan, A., Graminicolous species of *Bipolaris, Curvularia, Drechslera, Exserohilum* and their teleomorphs, *Mycol. Pap.,* 158, 1, 1987.

109. Wong, M.K.M. and Hyde, K.D. Diversity of fungi on six species of *Gramineae* and one species of *Cyperaceae* in Hong Kong, *Mycol. Res.,* 105, 1485, 2001.
110. Poon, M.O.K. and Hyde, K.D., Biodiversity of Intertidal estuarine fungi on *Phragmites* at Mai Po Marshes, Hong Kong, *Bot. Mar.,* 41, 141, 1998.
111. Sabada, R.B. et al., Observations on vertical distribution of fungi associated with standing senescent *Acanthus ilicifolius* stems at Mai Po Mangrove, Hong Kong, *Hydrobiol.,* 295, 119, 1995.
112. Shearer, C.A., The freshwater ascomycetes, *Nova Hedwigia,* 56, 1, 1993.
113. Lee, S.Y., Net aerial productivity, litter production and decomposition of the *Phragmites australis* in a nature reserve in Hong Kong: management implications, *Marine Ecology Progress Series,* 66, 161, 1990.
114. Newell, S.Y., Decomposition of shoots of a salt-marsh grass, in *Advances in Microbial Ecology,* 13, Jones, J.G., Ed., Plenum Press, New York, 1993, 1.
115. Farr, D.F. et al., *Fungi on Plants and Plant Products in the United States,* APS Press, St. Paul, MN, 1989, 1.
116. van Ryckegem, G. and Verbeken, A., Fungal diversity and community structure on common reed (Phragmites australis) along a salinity gradient in the Scheldt-estuary, Nova Hedwigia, 80, 173, 2005.
117. Shivas, R.G. and Vánky, K., The smut fungi on *Cynodon,* including *Sporosorium normanensis* sp. nov. from Australia, *Fung. Divers.,* 8, 149, 2001.
118. Vánky, K., Ten new species of Ustilaginales, *Mycotaxon,* 18, 319, 1983.
119. Vánky, K., Taxonomic studies on Ustilaginales XII, *Mycotaxon,* 54, 215, 1995.
120. Vánky, K., Taxonomic studies on Ustilaginales XIII, *Mycotaxon,* 56, 197, 1995.
121. Vánky, K., Taxonomic studies on Ustilaginales XV, *Mycotaxon,* 62, 127, 1997.
122. Vánky, K., Taxonomic studies on Ustilaginales XVI, *Mycotaxon,* 65, 133, 1997.
123. Vánky, K., Taxonomic studies on Ustilaginales XX, *Mycotaxon,* 74, 161, 2000.
124. Vánky, K., The smut fungi on *Sacchraum* and related grasses, *Aust. Pl. Pathol.,* 29, 155, 2000.
125. Vánky, K., Taxonomic studies on Ustilaginales XXI, *Mycotaxon,* 78, 265, 2001.
126. Vánky, K., Taxonomic studies on Ustilaginales XXII, *Mycotaxon,* 81, 367, 2002.
127. Vánky, K., The smut fungi (Ustilaginomycetes) of *Hyparrhenia* (Poaceae), *Fung. Divers.,* 12, 179, 2003.
128. Vánky, K., Taxonomic studies on Ustilaginales XXIII, *Mycotaxon,* 85, 1, 2003.
129. Vánky, K., The smut fungi (Ustilaginomycetes) of *Sporobolus* (Poaceae), *Fung. Divers.,* 14, 205, 2003.
130. Vánky, K., The smut fungi (Ustilaginomycetes) of *Bothriochloa, Capillipedium* and *Dichanthium* (*Poaceae*), *Fung. Divers.,* 15, 219, 2004.
131. Vánky, K., Taxonomic studies on Ustilaginales 24, *Mycotaxon,* 89, 55, 2004.
132. Vánky, K., The smut fungi (Ustilaginomycetes) of *Boutelouinae* (Poaceae), *Fung. Divers.,* 16, 167, 2004.
133. Vánky, K., The smut fungi (Ustilaginomycetes) of *Muhlenbergia* (Poaceae), *Fung. Divers.,* 16, 199, 2004.
134. Vánky, K., The smut fungi (Ustilaginomycetes) of *Chrysopogon* (Poaceae), *Fung. Divers.,* 18, 177, 2005.
135. Vánky, K. and Shivas, R.G., Smut fungi (Ustilaginomycetes) of *Sorghum* (Gramineae) with special regard to Australia, *Mycotaxon,* 80, 339, 2001.
136. Zhou, D.Q. and Hyde, K.D., Host-specificity, host-exclusivity and host-recurrence in saprobic fungi, *Mycol. Res.,* 105, 1449, 2001.
137. Hyde, K.D. et al., Saprobic fungi on bamboo culms, *Fung. Divers.,* 7, 35, 2001.
138. Zhou, D.Q., Biodiversity of saprobic microfungi associated with bamboo in Hong Kong and Kunming, China, Ph.D. thesis, The University of Hong Kong, 2000.
139. Zhou, D.Q. and Hyde, K.D., Fungal succession on bamboo in Hong Kong, *Fung. Divers.,* 10, 213, 2002.
140. Zhou, D., Cai, L., and Hyde, K.D., *Astrosphaeriella* and *Roussoella* species on bamboo from Hong Kong and Yunnan, China, including a new species of *Roussoella, Cryptogam. Mycol.,* 24, 191, 2003.
141. Tanaka, K., and Harada, Y., Bambusicolous fungi in Japan (1): four Phaeosphaeria species, *Mycosystema,* 45, 377, 2004.
142. Shenoy, B.D., Jeewon, R., and Hyde, K.D., *Oxydothis bambusicola,* a new ascomycete with a huge subapical ascal ring found on bamboo in Hong Kong, *Nova Hedwigia,* 80, 511, 2005.
143. Hyde, K.D. et al., Vertical distribution of saprobic fungi on bamboo culms, *Fung. Divers.,* 11, 109, 2002.
144. Eriksson, O.E. and Yue, J.Z., Bambusicolous pyrenomycetes, an annotated check-list, *Myconet,* 1, 25, 1998.
145. Petrini, O., Candoussau, F., and Petrini, L.E., Bambusicolous fungi collected in southern western France 1982-1989, *Mycologica Helvetica,* 3, 263, 1989.

146. Goh, T.K. and Hyde, K.D., Fungi on submerged wood and bamboo in the Plover Cove Reservoir, Hong Kong, *Fung. Divers.*, 3, 57, 1999.

147. Hino, I., *Icones Fungorum Bambusicolorum Japonicolorum,* The Fuji Bamboo Garden, Japan, 1961.

148. Rehm, H., Ascomycetes philippinenses collecti a clar. D.B. Baker, *Philippine Journal of Science,* 8, 181, 1913.

149. Rehm, H., Ascomycetes philippinenses IV, *Leaflets of Philippine Botany,* 6, 1935, 1913.

150. Rehm, H., Ascomycetes philippinenses V, *Leaflets of Philippine Botany,* 6, 2191, 1914.

151. Rehm, H., Ascomycetes philippinenses VI, *Leaflets of Philippine Botany,* 6, 2257, 1914.

152. Rehm, H., Ascomycetes philippinenses VIII, *Leaflets of Philippine Botany,* 8, 2935, 1916.

153. Sydow, H. and Sydow, P., Enumeration of Philippine fungi with notes and descriptions of new species, I: Micromycetes, *Philippine Journal of Science, Section C, Botany,* 8, 265, 1914.

154. Sydow, H. and Sydow, P., Enumeration of Philippine fungi with notes and descriptions of new species, II, *Philippine Journal of Science, Section C, Botany,* 8, 475, 1914.

155. Cai, L. et al., Freshwater fungi from bamboo and wood submerged in the Liput River in the Philippines, *Fung. Divers.*, 13, 1, 2003.

156. Cannon, P.F. and Hawksworth, D.L., The diversity of fungi associated with vascular plants, *Adv.Plant Pathol.,* 11, 277, 1995.

157. Mabberley, D.J., *The Plant Book: A Portable Dictionary of the Higher Plants,* Cambridge University Press, Cambridge, 1987.

158. Eleuterius, L.N., The distribution of *Juncus roemerianus* in the salt marshes of North America, *Chesapeake Science,* 17, 289, 1976.

159. Snogerup, S., A revision of *Juncus* subgen. *Juncus* (Juncaceae), *Willdenowia,* 23, 23, 1993.

160. Kohlmeyer, J. and Volkmann-Kohlmeyer, B., The biodiversity of fungi on *Juncus roemerianus, Mycol. Res.,* 105, 1411, 2001.

161. Gessner, M.O., Aquatische Hyphomyceten, in *Methoden der Biologischen Wasseruntersuchung — Biologische Gewässeruntersuchung,* von Tümpling, W. and Friedrich, G., Eds., Gustav Fischer Verlag, Jena, 1999, 185.

162. Gessner, M.O., Bärlocher, F., and Chauvet, E., Qualitative and quantitative analyses of aquatic hyphomycetes in streams, in *Freshwater Mycology,* Fungal Diversity Research Series, Tsui, C.K.M. and Hyde, K.D., Eds., 2003, 10, 127.

163. Ingold, C.T., Aquatic hyphomycetes of decaying alder leaves, *Trans. Br. Mycol. Soc.,* 25, 339, 1942.

164. Webster, J. and Descals, E., Morphology, distribution and ecology of conidial fungi in freshwater habitat, in *Biology of Conidial Fungi,* Vol. 1, Cole, G.T. and Kendrick, B., Eds., Academic Press, New York, 1981, 295.

165. Goh, T.K., Tropical freshwater hyphomycetes, in *Biodiversity of Tropical Microfungi,* Hyde, K.D., Ed., Hong Kong University Press, 1997, 189.

166. Chan, S.Y., Goh, T.K., and Hyde, K.D., Ingoldian fungi in Hong Kong, *Fung. Divers.*, 5, 89, 2000.

167. Bills, G.F. and Polishook, J.D., Abundance and diversity of microfungi in leaf litter of a lowland rain forest in Costa Rica, *Mycologia,* 86, 187, 1994.

168. Polishook, J.D., Bills, G.F., and Lodge, D.J., Microfungi from decaying leaves of two rainforest trees in Puerto Rico, *J. Ind. Microbiol.,* 17, 284, 1996.

169. Parungao, M.M., Fryar, S.C., and Hyde, K.D., Diversity of fungi on rainforest litter in North Queensland, Australia, *Biodiver. Conserv.*, 11, 1185, 2002.

170. Paulus, B., Gadek, P., and Hyde, K.D., Estimation of microfungal diversity in tropical rain forest leaf litter using particle filtration: the effects of leaf storage and surface treatment, *Mycol. Res.,* 107, 748, 2003.

171. Promputtha, I. et al., Fungal succession on senescent leaves of *Manglietia garrettii* in Doi Suthep-Pui National Park, northern Thailand, in *Fungal Succession, Fungal Diversity,* Hyde, K.D. and Jones, E.B.G., Eds., Fungal Diversity Press, Hong Kong, 10, 89, 2002.

172. Promputtha, I. et al., *Dokmaia monthadangii* gen. et sp. nov. a synnematous anamorphic fungus on *Manglietia garettii., Sydowia,* 55, 99, 2003.

173. Promputtha, I. et al., Fungi on *Manglietia garettii: Cheiromyces manglietiae* sp. nov. from dead branches. *Nova Hedwigia,* 80, 527, 2005.

174. Photita, W. et al., Fungi on *Musa acuminata* in Hong Kong, *Fung. Divers.*, 6, 99, 2001.

175. Weir, A. and Hammond, P.M., A preliminary assessment of species-richness patterns of tropical, beetle-associated Labuolbeniales (Ascomycetes), in *Biodiversity of Tropical Microfungi,* Hyde, K.D., Ed., Hong Kong University Press, Hong Kong, 1997, 121.

176. Cafaro, M.J., Species richness patterns in symbiotic gut fungi (Trichomycetes), *Fung. Divers.,* 9, 47, 2002.

177. Sutton, B.C., The genus *Glomerella* and its anamorph *Colletotrichum,* in *Colletotrichum: Biology, Pathology and Control,* Bailey, J.A. and Jeger, M.J., Eds., CAB International, Wallingford, 1992, Chap. 1.

178. Freeman, S. et al., Molecular analyses of *Colletotrichum* species from almond and other fruits, *Phytopathol.,* 90, 608, 2000.

179. Corda, A.C.I., Die Pilze Deutschlands, in *Deutschlands Flora in Abbildungen nach der Natur mit Beschreibungen,* Sturm, J., Ed., Nürnberg, Sturm, 1837, 3, 1.

180. Sutton, B.C., *The Coelomycetes: Fungi Imperfecti with Pycnidia, Acervula and Stromata,* Commonwealth Mycological Institute, Kew, Surrey, England, 1980.

181. Lenne, J.M., *Colletotrichum* diseases of legumes, in *Colletotrichum: Biology, Pathology and Control,* Bailey, J.A. and Jeger, M.J., Eds., CAB International, Wallingford, 1992, 134.

182. Jeewon, R., Liew, E.C.Y., and Hyde, K.D., Phylogenetic evaluation of species nomenclature of *Pestalotiopsis* in relation to host association, *Fung. Divers.,* 17, 39, 2004.

183. Crous, P.W., Mycosphaerella spp. and their anamorphs associated with leaf spot diseases of *Eucalyptus,* *Mycol. Mem.,* 21, 1, 1998.

184. Platnick, N.I., Philosophy and the transformation of cladistics, *Syst. Zool.,* 28, 537, 1979.

185. Lutzoni, F. and Vilgalys, R., Integration of morphological and molecular data sets in estimating fungal phylogenies, *Can. J. Bot.,* 73, S649, 1995.

186. Baker, R.H. and Gatesy, J., Is morphology still relevant? in *Molecular Systematics and Evolution: Theory and Practice,* DeSalle, R., Giribet, G. and Wheeler, W., Eds., Birkhauser Verlag, Basel, 2002.

187. Liu, Y., Whelen, S., and Hall, B.D., Phylogenetic relationships among ascomycetes: evidence from an RNA polymerase II subunit, *Mol. Biol. Evol.,* 16, 1799, 1999.

188. Barr, M.E., The ascomycetes connection, *Mycologia,* 75, 1, 1983.

189. Barr, M.E., *Prodomus to Class Loculoascomycetes,* Newell, Amherst, Mass, 1987.

190. Barr, M.E., Prodomus to nonlichenized, pyrenomycetous members of Class Hymenoascomycetes, *Mycotaxon,* 39, 43, 1990.

191. Eriksson, O. and Hawksworth, D.L., Outline of the Ascomycetes-1993, *Systema Ascomycetum,* 12, 1, 1993.

192. van Elsas, J.D. et al., Analysis of the dynamics of fungal communities in soil via fungal-specific PCR of soil DNA followed by denaturing gradient gel electrophoresis, *J. Microbiol. Meth.,* 43, 133, 2000.

193. May, L.A., Smiley, B., and Schmidt, M.G., Comparative denaturing gradient gel electrophoresis analysis of fungal communities associated with whole plant corn silage, *Can. J. Microbiol.,* 47, 829, 2001.

194. Berbee, M.L. and Taylor, J.W., Fungal molecular evolution: gene, trees and geologic time, in *The Mycota VII: Systematics and Evolution Part B,* McLaughlin, D.J., McLaughlin, E.G., and Lemke, P.A., Eds., Springer-Verlag, New York, 2001, 229.

195. Lang, B.E. et al., The closest unicellular relative to animals, *Curr. Biol.,* 12, 1773, 2002.

196. Bridge, P., The history and application of molecular biology, *Mycologist,* 16, 90, 2002.

197. Tautz, D. et al., A plea for DNA Taxonomy, *Trends Ecol. Evol.,* 18, 70, 2003.

198. Doran, J.W., Sarrantonio, M., and Liebig, M.A., Soil health and sustainability, *Adv. Agron.,* 56, 2, 1996.

199. Garbeva, P., van Veen, J.A., and van Elsas, J.D., Microbial diversity in soil: selection of microbial populations by plant and soil type and implications for disease suppressiveness, *Annu. Rev. Phytopathol.,* 42, 243, 2004.

200. Bridge, P. and Spooner, B., Soil fungi: diversity and detection, *Pl. Soil,* 232, 147, 2001.

201. Crespo, A., Bridge, P.D., and Hawksworth, D.L., Amplification of fungal rDNA-ITS regions from non-fertile specimens of the lichen-forming genus *Parmelia, Lichenol.,* 29, 275, 1997.

202. Heuer, H. et al., Analysis of actinomycete communities by specific amplification of gene encoding 16S rDNA and gel-electrophoretic separation in denaturing gradient, *Appl. Environ. Microbol.,* 63, 3233, 1997.

203. Muyzer, G., de Waal, E.C., and Uitterlinden, A.G., Profiling of complex microbial populations by denaturing gradient gel electrophoresis analysis of polymerase chain reaction-amplified genes coding for 16S rRNA, *Appl. Environ. Microbiol.,* 59, 695, 1993.

204. Muyzer, G. and Smalla, K., Application of denaturing gradient gel electrophoresis (DGGE) and temperature gradient gel electrophoresis (TGGE) in microbial ecology, *Antonie Van Leeuwenhoek,* 73, 127, 1998.

205. Liu, W.T. et al., Characterization of microbial diversity by terminal restriction fragment length polymorphisms of genes encoding 16S rRNA. *Appl. Environ. Microbiol.,* 63, 4516, 1997.

206. Ranjard, L. et al., Characterization of bacterial and fungal soil communities by automated ribosomal intergenic spacer analysis fingerprints: biological and methodological variability, *Appl. Environ. Microbiol.,* 67, 4479, 2001.

207. Bailey, J.A. et al., Molecular taxonomy of *Colletotrichum* species causing anthracnose of Malvaceae, *Phytopathol.,* 86, 1076, 1996.

208. O'Neill, N.R., Application of amplified restriction fragment polymorphism for genetic characterization of *Colletotrichum* pathogens of alfalfa, *Phytopathol.,* 87, 745, 1997.

209. Pain, N.A. et al., Monoclonal antibodies which show restricted binding of four *Colletotrichum* species: *C. lindermuthianum, C. malvarum, C. orbiculare,* and *C. trifolii, Physiol. Mol. Pl. Pathol.,* 40, 111, 1992.

210. Sheriff, C. et al., Ribosomal DNA sequence analysis reveals new species groupings in the genus *Colletotrichum, Exp. Mycol.,* 18, 121, 1994.

211. Freeman, S., et al., Molecular analyses of *Colletotrichum* species from almond and other fruits, *Phytopathol.,* 90, 608, 2000.

212. Brown, K.B., Hyde, K.D., and Guest, D.I., Preliminary studies on endophytic fungal communities of *Musa acuminata* species complex in Hong Kong and Australia, *Fung. Divers.,* 1, 27, 1998.

213. Fisher, P.J. et al., A study of fungal endophytes in leaves, stem and roots of *Gynoxis oliefolia* Muchler (*Compositae*) from Ecuador, *Nova Hedwigia,* 60, 589, 1995.

214. Fröhlich, J. et al., Endophytic fungi associated with palms, *Mycol. Res.,* 104, 1202, 2000.

215. Kumar, D.S.S., and Hyde, K.D., Biodiversity and tissue-recurrence of endophytic fungi in *Tripterygium wilfordii, Fung. Divers.,* 17, 69, 2004.

216. Lodge, D.J., Fisher, P.J., and Sutton, B.C., Endophytic fungi of *Manilkara bidentata* leaves in Puerto Rico, *Mycologia,* 85, 733, 1996.

217. Rodrigues, K.F., The foliar endophytes of the Amazonian palm *Euterpe oleracea, Mycologia,* 86, 376, 1994.

218. Taylor, J.E., Hyde, K.D., and Jones, E.B.G., Endophytic fungi associated with the temperate palm *Trachycarpus fortunei* both within and outside of its natural geographic range, *New Phytol.,* 142, 335, 1999.

219. Toofanee, S.B. and Dulymamode, R., Fungal endophytes associated with *Cordemoya integrifolia, Fung. Divers.,* 11, 169, 2002.

220. Umali, T., Quimio, T., and Hyde, K.D., Endophytic fungi in leaves of *Bambusa tultoides, Fung. Sci.,* 14, 11, 1999.

221. Dreyfuss, M.M. and Petrini, O., Further investigations on the occurrence and distribution of endophytic fungi in the tropic plants, *Botanica Helvetica,* 94, 33, 1984.

222. Fisher, P.J. et al., Fungal endophytes from the leaves and twigs of *Quercus ilex* L. from England, Majorca and Switzerland, *New Phytol.,* 127, 133, 1994.

223. Pereira, J.O., Azevedo, J.L., and Petrini, O., Endophytic fungi of *Stylosanthes*: a first report, *Mycologia,* 85, 362, 1993.

224. Lacap, D.C., Hyde, K.D., and Liew, E.C.Y., An evaluation of the fungal 'morphotype' concept based on ribosomal DNA sequences, *Fung. Divers.,* 12, 53, 2003.

225. Guo, L.D., Hyde, K.D., and Liew, E.C.Y., Identification of endophytic fungi from *Livistona chinensis* based on morphology and rDNA sequences, *New Phytol.,* 147, 617, 2000.

226. Nikolcheva, L. et al., Determining diversity of freshwater fungi on decaying leaves: comparison of traditional and molecular approaches, *Appl. Env. Microbiol.,* 69, 2548, 2003.

227. Hyde, K.D., Taylor, J.E., and Fröhlich, J., *Genera of Palm Ascomycetes,* Fungal Diversity Research Series 3, Fungal Diversity Press, Hong Kong, 2000.

228. Fröhlich, J. and Hyde, K.D., *Palm Microfungi,* Fungal Diversity Research Series 3, Fungal Diversity Press, Hong Kong, 2000.

229. Taylor, J.E. and Hyde, K.D., *Microfungi of Tropical and Temperate Palms,* Fungal Diversity Research Series 12, Fungal Diversity Press, Hong Kong, 2003.

230. Pinnoi, A. et al., *Submersisphaeria palmae* sp. nov. with a key to species and notes on *Helicoubisia, Sydowia,* 56, 72, 2004.

231. Pinruan, U. et al., Three new species of *Craspedodidymum* from palm in Thailand, *Mycoscience,* 45, 177, 2004.

# 16 Matters of Scale: Dealing with One of the Largest Genera of Angiosperms

*J. A. N. Parnell*
Department of Botany, School of Natural Sciences,
Trinity College Dublin, Ireland

*L. A. Craven*
Australian National Herbarium, Centre for Plant Biodiversity Research,
CSIRO Plant Industry, Canberra, Australia

*E. Biffin*
Division of Botany and Zoology, Australian National University,
Canberra, Australia

## CONTENTS

## ABSTRACT

*Syzygium*, with about 1,200 species, is one of the largest generic groupings of Myrtaceae. Conventionally, it is considered to be taxonomically difficult due to its previous confusion with another large genus of the family (*Eugenia*), the seeming lack of 'good' diagnostic characters, and the uncertainty as to the delimitation of genera within the *Syzygium* complex per se. Current divergent taxonomic approaches are discussed, and the taxonomic history of *Syzygium* is summarised.

Present research includes floristic and reproductive biological studies, and active studies into morphological, anatomical and molecular aspects are in progress. The structural, ecological and biological diversity of the group, together with its economic and biodiversity significance, point to *Syzygium* being a challenging but rewarding subject for future research.

## 16.1  INTRODUCTION

This chapter aims to bring together past and present taxonomic and systematic research on the very large and taxonomically perplexing angiosperm genus *Syzygium* Gaertner (Myrtaceae) and to outline and stimulate further work; therefore, it is both retrospective and prospective. We show that *Syzygium* poses many problems, including its delimitation as a genus, documentation of its species and understanding of many aspects of its biology. Nevertheless, we suggest that ongoing floristic and phylogenetic studies have the potential to significantly improve our current understanding of the genus.

### 16.1.1  INTRODUCTION TO THE MYRTACEAE AND *SYZYGIUM*

The Myrtaceae are a mostly Southern hemisphere family of moderate size, containing between 130 and 155 genera and 3,675 and 5,000 species[1–7]. Despite its modest size, the family poses a disproportionately large number of complex taxonomic problems evident at many levels in the taxonomic hierarchy, for example, concepts at and above species level within *Eucalyptus* L'Hér.[8] and within *Syzygium*[9–12], or the separation of *Syzygium* from *Eugenia* L.[13]. McVaugh[14] described the species of American Myrtaceae as "distressingly alike in aspect and in most individual characters, making identification and classification of both genera and species a correspondingly difficult and tedious matter". It is evident that species of South East Asian *Syzygium* fall into this category and are not generally clear and easy to distinguish, but Craven[9] suggests that Australian-Papuasian and New Caledonian species are.

Within the Myrtaceae two unequally sized subfamilies were recognised by Niedenzu[15]: the Leptospermoideae and Myrtoideae. The Myrtoideae contain only 60 genera[16] but approximately two thirds (2,375–3,400) of the species known in the family[6]. These species occur in the New and Old World tropics, are mainly shrubs or trees of wet forests, usually have an inferior ovary and almost always possess opposite broad leaves and a fleshy indehiscent fruit. Niedenzu's[15] subfamilial divisions are convenient and still used[16,17]. However, several recent studies indicate that Niedenzu's classification[15] is unsatisfactory[6,13,18,19]. Johnson and Briggs developed an informal system of alliances and suballiances through cladistic analysis of a fairly large and comprehensive data set[18,19]. They concluded that Niedenzu's division[15] of the family on the basis of fruit characteristics was not phylogenetically supportable, as their alliances and suballiances crossed traditional subfamilial boundaries.

Wilson et al.[20], based on *matK* sequence data but with a strong morphological and anatomical backgound, have proposed a classification containing two subfamilies and 17 tribes. They showed that *Syzygium* together with *Acmena* DC., *Acmenosperma* Kausel, *Anetholea* Peter G. Wilson and *Waterhousea* B. Hyland form a tribe (the Syzygieae Peter G. Wilson), with *Eugenia* L. in a separate tribe (the Myrteae DC.) along with a large number of other genera including *Myrtus* L., *Psidium* L., *Rhodamnia* Jack and *Rhodomyrtus* (DC.) Reich. This result reinforces the work of Johnson and Briggs[19] and Johnson et al.[2]. These latter workers discussed the two alliances of particular concern in this chapter, the *Acmena* and Myrtoideae *s.s.* alliances[19], or more specifically, the *Acmena* and *Eugenia* alliances[2]. Traditionally these alliances were placed near one another in the fleshy fruited Myrtoideae, but they are far apart phylogenetically. Indeed, they could scarcely be much further apart in Johnson and Briggs's cladogram[19], wherein the *Eugenia* alliance is in a clade which incorporates the *Leptospermum* alliance, whilst the *Acmena* alliance forms an altogether separate clade with, amongst others, the *Eucalyptus* alliance. However, in a later compromise classification both alliances are placed together in the subfamily

Myrtoideae[2], now defined by Wilson et al.[20] to encompass more or less the entire Myrtaceae as previously recognised.

*Syzygium* and *Eugenia* are two of the most taxonomically confused genera in the Myrtaceae, and there are many other genera that have, at one time or another, been cleaved off from them or been reunited with them. Schmid[13] pointed out that there were about 35 generic names which have been or could be reduced to *Syzygium s.l.* and at least another 30 assignable to *Eugenia s.l.* Since Schmid's publication the number of segregates has increased with, for example, the description of *Waterhousea* B. Hyland and *Monimiastrum* A.J. Scott. In addition, several species have been placed within *Syzygium* on the basis that accurate subdivision or description of segregate genera is currently not possible (such as Craven[9]). Together, these genera form a 'vast array of more or less closely allied species' (Ashton[21]). This 'array', dominated by *Eugenia* and *Syzygium,* is very large. The standard printed work, *Index Kewensis*, has over 3,000 species listed under *Eugenia* and over 1,000 under *Syzygium.* Undoubtedly, this does not reflect the true balance in numbers of species between these genera when they are considered in the strict sense, as even now many authors prefer, because of historical precedent and because of the enormous number of consequent nomenclatural changes, to ignore the differences between them.

Schmid[13] provides a review of the status of *Syzygium s.l.* and makes clear why *Eugenia* and *Syzygium* were confused. Schmid's work summarises many of the relevant references and arguments, and is therefore not repeated here in detail. Essentially, Schmid showed that *Eugenia* and *Syzygium* were not closely related, differing most evidently in respect of the substitution of the transeptal vascular supply to the ovule of *Eugenia* with an axile one in *Syzygium.* Kochummen[3], Kostermans[22] and others have criticised this work on the basis that very few species were studied; however, as contrary data have not been forthcoming, we accept Schmid's conclusions.

### 16.1.2 *SYZYGIUM:* DERIVATION AND MATTERS OF SIZE

*Syzygium* has long proved taxonomically difficult ever since its initial establishment with four species. Even the derivation of the name is unclear; the *Oxford English Dictionary*[23] indicates that the word Syzygium is most likely derived from the late Latin *syzygia* and hence from Greek (taken from *syn.* = together and *zygon* = yoke). In the case of the genus *Syzygium* this may indicate a reference to the paired arrangement of the leaves; however, this is not certain.

Some idea of the size of *Syzygium* relative to other members of the family in terms of numbers of species can be obtained from Figure 16.1 and Figure 16.2. Figure 16.1 shows that the Myrtaceae contains a large number of genera with very few species and very few genera with many species. This is typical of the 'hollow curve structure' discussed by Willis[24], Minelli[25] and many others (see Hilu, *Chapter 11*) and which may, for obscure reasons, fit a fractal pattern[26]. Figure 16.2 shows that by far the largest genus in the Myrtaceae is *Syzygium.* As pointed out by Minelli[25] and Frodin[27], *Syzygium* is clearly one of the very large genera of vascular plants. Frodin, utilising an estimate of 1,041 *Syzygium* species, places *Syzygium* sixteenth in his list of the 57 largest genera of flowering plants. However, estimating the number of species of *Syzygium* is difficult. The most up-to-date source, the International Plant Names Index (IPNI) (http://www.ipni.org/index.html), lists 1,507 specific epithets under *Syzygium.* To this current list of names must be added an estimate for those as yet undiscovered and, obviously, unnamed species. Work in Thailand[28] has shown that previously undescribed species form at least 6–7% of the total species number for that country, and work in Malaya[3,29] suggests that c. 33% of *Syzygium* species in that Flora may be unnamed. In addition, many validly described species await transfer to *Syzygium.* For example, based on Chantaranothai and Parnell[28,30], up to 90% of specific epithets in *Syzygium* are derived from transfers from other genera. However, as a counter to these trends, which act to increase the number of available epithets, a large proportion (maybe as much as 90%) of the currently available species epithets under *Syzygium* are likely to be synonyms of other species within *Syzygium.* On the basis of these arguments, it appears that the reduction in the number of valid epithets due to synonymy may be

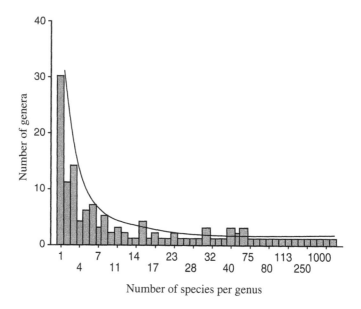

**FIGURE 16.1** Histogram and hand fitted trend line showing a plot of the number of genera against the number of species within each genus for the Myrtaceae.

balanced by the combination of the number of new species awaiting description and transfers into *Syzygium* of valid species wrongly placed in other genera. Conservatively, therefore, we estimate that the total number of species of *Syzygium* is likely to be more than 1,000 but less than 1,500. On this basis, *Syzygium* may be positioned higher up Frodin's[27] table than currently, possibly even within the top 10 largest plant genera in the world.

In summary of this section, *Syzygium* is an extremely large genus wherein many species await description.

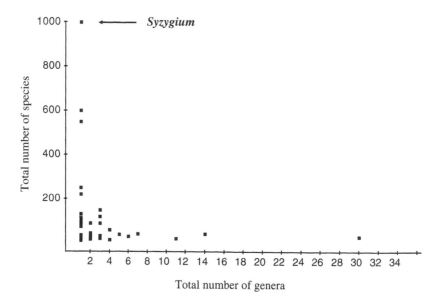

**FIGURE 16.2** Frequency diagram of the total number of genera containing a certain number of species for the Myrtaceae.

### 16.1.3 *Syzygium*: Ecology, Distribution and Other General Information

*Syzygium* is an Old World genus with most species found in southern and South East Asia, Australia, Southern China, Malesia and New Caledonia. Fewer species occur in Africa, Malagasy, the Mascarenes and the remainder of the South Western Pacific islands, and there is one indigenous species each in Hawaii and New Zealand. Mostly, the species are found in rainforests, but they occur in nearly all vegetation types, from littoral communities at sea level through swamp, dry deciduous, bamboo, peat swamp, lowland, evergreen and montane forests, up to subalpine shrubberies. They also occur in open grassy savannah, gallery forest and heathland. Few species appear highly specialised in terms of their ecological requirements, but exceptions do occur with the rheophytic species and those species that occur obligately on ultramafic substrates. Gamage et al.[31] suggested that forest disturbance and hydrology were important environmental factors influencing the distribution of *Syzygium* in Sri Lanka and that leaf physiology and structure were related to each other and to shade tolerance and water use. Parnell[12], using numerical methods, showed that Thai *Syzygium* could be divided into two morphologically definable groups: one widespread and lowland with larger, broader, longer petioled leaves with a wider midrib, secondary veins that are relatively close to each other, intramarginal veins relatively far from the margin, larger calyces, petals (and these have more gland dots), ovules and longer styles; the other, essentially only found above 1,000 m, containing many endemics and with the opposite suite of morphological characteristics. Whether this pattern is ecologically determined or not requires investigation.

Their habit ranges from canopy emergent trees to canopy trees, understorey trees, treelets, shrubs, rheophytes, and rarely even prostrate or semiscandent shrubs. Species mostly occur as scattered individuals but can be locally common. In rare cases, a species can be clonal, producing a patch of stems, but this may be restricted to savannah environments and be associated with frequent natural fires. Some species are widespread, such as *Syzygium grande* (Wight) Walpers which is relatively common throughout much of mainland South East Asia, and depending on how *Syzygium* is defined, *Cleistocalyx nervosum* (Roxb.) I.M. Turner (= *Cleistocalyx nervosus* (DC.) Kosterm. and *Syzygium nervosum* DC.) that occurs from southern and South East Asia-Southern China to Northern Australia, but there are many point endemics, for example *Syzygium kerrii* P. Chantaranothai and J. Parn.

Relatively little is known of the importance of *Syzygium* to other organisms, but it is highly likely the genus as a whole has high ecosystem significance. The often massed nectariferous flowers and typically fleshy fruit (Figure 16.4 and Figure 16.5) are food sources for a wide range of animals, from small insects through to large birds and primates. In one Thai study, 26% of hornbill nests located in 302 sample plots were in *Syzygium* (as *Eugenia*) trees, despite *Syzygium* representing only 3% of all large trees in these plots[32].

A number of species are cultivated. One of the few that is economically important in world trade is *S. aromaticum* (L.) Merr. and L.M. Perry. It is the source of clove and clove oil and a vital component of the large Indonesian cigarette industry. In 2003, 59,328 tons of cloves were exported, these being worth US$115,473,000[33]. Clove oil is one of the 20 most important essential oils in the world; one whose production value in 1993 was US$7,000,000[34]. By contrast, although *S. polyanthum* (Wight) Walpers, the source of Indonesian bay leaf, is hugely important as a condiment in mainland South East Asia, it is scarcely traded on a world scale[35]. Several other species are of local commercial importance, being only sparsely traded. *Syzygium malaccense* (L.) Merr. and L.M. Perry (Malay apple) and *S. samarangense* (Blume) Merr. and L.M. Perry (Java apple), although very commonly cultivated as fruit trees in South East Asia, are almost unknown in world trade[36]. The potential for *S. aqueum* (Burm.f.) Alston (Jambu Air, water-, rose apple, etc.) to become a major crop is, however, considerable, as Taiwan market figures indicate[37], and the commercial prospects for several other little known species, for example, *S. maraca* Craven and Biffin[11] or *S. rubrimolle* B. Hyland[38], also appear good. Equally, medicinal prospects for species in the genus appear high. The bark of *S. jambos* (L.) Alston provides effective antimicrobial activity[39], as have the essential oils of a number of other species[40]. For a genus with so many tree species, *Syzygium* is not regarded as a major timber resource. Lemmens[4] indicates that only small amounts of the locally important timber Kelat (a South East Asian trade name that covers timber produced by a number of species of *Syzygium*) are exported. Eddowes[41] classed water gum (the Papua New Guinea trade name for *Syzygium* timber) as a major exportable hardwood, although it does not comprise a large proportion of the timber exported[4].

To summarise this section, *Syzygium* is a widespread Old World genus of considerable ecological and economic importance and therefore, pragmatically, a predictive taxonomic classification will be of wide utility.

### 16.1.4 *SYZYGIUM:* TAXONOMY AND DIFFERENTIAL MORPHOLOGICAL CHARACTERISTICS

Taxonomically important features of *Syzygium* at species level include leaf size and shape, venation, inflorescence position, flower shape and size and fruit colour and size (Figure 16.3 to 16.5).

Various characteristics have been used to distinguish species of *Syzygium*. For example, Ashton's key to *Syzygium* in Sri Lanka[21] makes extensive use of characters of the leaf, in particular its size and shape, the number of veins and the number of intramarginal veins, but makes almost no mention of floral characteristics. Kochummen[3] also makes almost no use whatsoever of floral characters, preferring instead features of the leaf (notably its shape, size and the number of veins). By contrast, Chantaranothai and Parnell working in Thailand[28], Kostermans working in Sri Lanka[22], Hartley and Perry working in Papuasia[42], Craven and Matarczyk working in Australia[43] and Chen and Craven working in China[44], make extensive use of floral characters. These characters include the shape of the hypanthial cup and pseudostipe, size of calyx lobes, number of gland dots on the corolla, number and size of the stamens and style, ovule and placentation features, as well as the number of intramarginal veins and the number of secondary veins and inflorescence position. Parnell's analysis of Thai *Syzygium*[12] demonstrated that many morphological features show overlap between species or groups of species and are therefore 'traits' (that is, features which may assume the value observed in either of the groups being investigated). However, there are some non-overlapping features (at least in terms of mean values) which do not, or cannot by definition, overlap and are not polymorphic and which can be termed 'characters'. Most authors working on *Syzygium* have failed to rigorously separate traits and characters. In mitigation, we note that Stevens[45] suggests that such separation may, for many features, be more subjective than absolute.

In summary of this section, both vegetative and floral features have been stressed by different authors as being taxonomically important in *Syzygium*. Such features are either traits or characters; rigorous investigation and definition of these features is awaited.

## 16.2 TAXONOMIC HISTORY

As might be expected with such a species rich and geographically widespread group, many authors have contributed to our taxonomic, morphological, anatomical and floristic understanding of *Syzygium s.l.*. Its taxonomic history has been discussed by Schmid[13], Parnell[12] and Craven[9], and only the more significant publications are detailed in this section.

### 16.2.1 NINETEENTH CENTURY

Some of the species later included in *Syzygium* were known to pre-Linnaean authors, and Linnaeus treated the species then known to him under *Caryophyllus* L., *Eugenia* and *Myrtus*[46]. Plant classifications in the eighteenth century were not always intended by their authors to reflect natural relationships, and it was not until the early nineteenth century that more natural systems were commonly adopted. This taxonomically higher-level work was stimulated by the need to accommodate the many unusual plants being discovered in Africa, the Americas, South and East Asia and

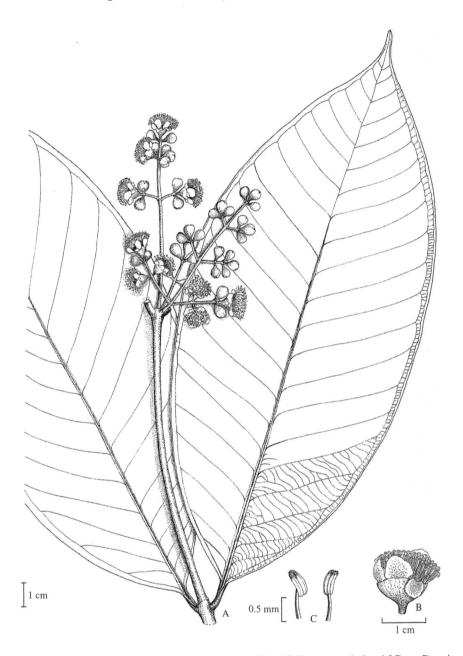

**FIGURE 16.3** Illustration of *Syzygium pergamentaceum* (King) P.Chantaranothai and J.Parn. Drawing shows (A) opposite leaves each with two intramarginal veins and inflorescence, (B) hypanthial cup, petals and gland dots on the petals and (C) stamens. (Reproduced from Chantaranothai and Parnell[28] with permission.)

Australasia. For Myrtaceae, the work of De Candolle[47] stands out as the most comprehensive classification of the early nineteenth century. De Candolle[47] placed the fleshy fruited, one to few large seeded species in one of five genera: *Acmena, Caryophyllus, Eugenia, Jambosa* Adans. and *Syzygium*. The New World species were accommodated by De Candolle[47] in *Eugenia* and the Old World species classified in one of the remaining four genera.

Wight[48] proposed the merging of the five genera recognised by De Candolle[47] into one on the basis that the floral features demonstrated continuous variation, the internal structure of the flowers

**FIGURE 16.4** *(A colour version of this figure follows page 240)* Flowers and fruit of species of the *Syzygium* group. (A) Flowers of *Syzygium malaccense* (L.) Merr. and L.M. Perry; (B–D) flowers, inflorescence and fruit, respectively, of *Acmena* cf. *divaricata* Merr. and L.M. Perry; (E) fruit of *Piliocalyx bullatus* Brongn. and Gris; (F) buds and flowers of *Syzygium longifolium* (Brongn. and Gris) J.W. Dawson; (G–H) fruit and flowers, respectively, of *Syzygium aqueum* (Burm. f.) Alston; (I) flowers of *Syzygium jambos* (L.) Alston; (J) buds (note calyptras) of *Syzygium kuebiniense* J.W. Dawson; (K) fruit of *Syzygium rubrimolle* B. Hyland; (L–M) flowers and fruit, respectively, of *Syzygium glenum* Craven. (Reproduced with permission from G. Sankowsky (A–D, G, H, K–M), L. Craven. (F, I) and E. Biffin (J).)

**FIGURE 16.5** *(A colour version of this figure follows page 240)* Flowers, fruit and foliage of species of the *Syzygium* group. (A–B) buds and flowers, and fruit, respectively, of *Acmenosperma pringlei* B. Hyland; (C) *Syzygium wilsonii* subsp. *cryptophlebium* (F. Muell.) B. Hyland; (D) fruit of *Syzygium elegans* (Brongn. and Gris) J.W. Dawson; (E–G) habit, young leaves, and buds and flowers, respectively, of *Syzygium acre* (Pancher ex Guillaumin) J.W. Dawson; (H) fruit of *Syzygium cormiflorum* (F. Muell.) B. Hyland; (I) flowers of *Syzygium boonjee* B. Hyland; (J) flower of *Syzygium* sp.; (K) flowers of *Syzygium balansae* (Guillaumin) J.W. Dawson; (L) fruit of *Syzygium maraca* Craven and Biffin; (M) young fruit of *Syzygium* sp. (Reproduced with permission from A. Ford (A–B), G. Sankowsky (C, H, I, L), E. Biffin (D), L. Craven. (E–G, K, M) and J. Dowe (J).)

and structure of the fruit were very uniform, and the habit of the plants themselves was generally uniform. Wight did acknowledge the practical difficulty of not having some taxonomic substructure for such a species rich genus as *Eugenia* had now become. Therefore, he recognised each of the five genera he had merged into *Eugenia* as subgenera using the same epithets. Wight's solution for classifying all the large seeded species also had the effect of stabilising nomenclature, and from the comment made by Bentham[49], this appeared to be one consideration for adoption of the same taxonomy by Bentham and Hooker[50] in their influential Genera Plantarum. Meanwhile researchers, mainly Dutch botanists working in the Malesian region, were continuing to describe new species in *Jambosa* and/or *Syzygium* (for example, Blume[51] and Miquel[52]). As more novel morphological variation was encountered, new genera were also described from Malesia and the South West Pacific (for example, *Acicalyptus* A. Gray, *Aphanomyrtus* Miq., *Clavimyrtus* Blume, *Cleistocalyx* Blume, *Cupheanthus* Seem., *Pareugenia* Turrill and *Piliocalyx* Brongn. and Gris.). Where the various segregate genera were known to Bentham and Hooker[50], they were all reduced to *Eugenia* in Genera Plantarum and therein assigned to one of the three sections they recognised, sect. *Jambosa*, sect. *Syzygium* or sect. *Eugenia*. Although Bentham and Hooker's circumscription of *Eugenia* was followed by many taxonomists for over 100 years, it was not universally accepted.

Late in the century, Niedenzu's account of Myrtaceae was published in *Die Natürlichen Pflanzenfamilien*[15]. This work is similar to that of De Candolle[46] in that *Eugenia* is retained for the New World species (with a very few Old World species that clearly were part of this grouping) and the very great majority of the Old World species were assigned to four other genera. The Old World genera he recognised were *Acicalyptus*, *Jambosa* (including *Cleistocalyx* among others), *Piliocalyx* and *Syzygium* (including *Acmena* among others). The narrower generic concepts for the *Eugenia* group were welcomed by those taxonomists of the following century who believed Bentham and Hooker's broad circumscription[50] to be unsatisfactory.

In summary of this section, Bentham and Hooker's 'all in one' concept of genus[50] provided nomenclatural stability but was, by the end of the nineteenth century, often deemed unsatisfactory.

## 16.2.2 TWENTIETH CENTURY

As noted above, the word Syzygium is believed to refer to a joining or yoking together, but the derivation is unclear. If nothing else the genus certainly has two yoked histories, as twentieth-century authors can be grouped into two schools of thought as to its classification. One school used the all-inclusive Bentham and Hooker concept of a single genus, *Eugenia*, and the other school accepted *Eugenia* for the New World centred species and differing numbers of genera for the Old World syzygioid species. Typically, the latter taxonomic school recognised *Syzygium*, *Acmena*, *Cleistocalyx* and often *Jambosa*, with some additionally recognising *Acicalyptus*, *Acmenosperma*, *Cupheanthus* and *Piliocalyx* for some South West Pacific species. The twentieth-century champion of the *Eugenia* school in the Indo-Pacific was Henderson, who published a comprehensive account of the species occurring in the Malay Peninsula[53]. In this work, Henderson arranged the Malayan species in four sections: *Acmena*, *Cleistocalyx*, *Fissicalyx* and *Syzygium*, the latter of which he split into five groups. His circumscription of sect. *Syzygium* included the concepts of *Caryophyllus*, *Jambosa* and *Syzygium* that had been adopted by previous workers at genus or subgenus level.

Of the authors advocating the recognition of separate genera for the syzygioid species, the most significant have been Merrill and his coworker Perry. These two authors published accounts of *Acmena* and *Cleistocalyx*[54,55], floristic treatments of *Syzygium* and its allies for Indo-China, China and Borneo[56–58], and important contributions on the Papuasian[59] and Philippine[60] species. Interestingly, Merrill had accepted more genera than just *Acmena*, *Cleistocalyx* and *Syzygium* early in his career but moved to a more conservative position when he was most active floristically. Relatively late in his career he returned to the narrower generic position and recognised *Acmena*, *Aphanomyrtus*, *Caryophyllus*, *Cleistocalyx*, *Pareugenia*, *Syzygium*, *Tetraeugenia* Merrill and tentatively *Jambosa*[60].

Investigations of *Syzygium* and *Eugenia* by other researchers in the second half of the century have contributed significantly to the debate. Ingle and Dadswell[61] studied the wood anatomy of Myrtaceae in the South West Pacific and concluded that the *Eugenia s.l.* species sampled fell into two distinct groups. A few species agreed anatomically with the New World species of *Eugenia s.s.* but the majority were distinct from these and comprised species of *Acmena, Cleistocalyx* and *Syzygium.* Pike[62] found that pollen morphology supported the conclusions of Ingle and Dadswell[61]. Pike further noted that the pollen of the *Eugenia s.s.* species examined resembled that of the subtribes Myrtinae and Myrciinae, a finding of significance in the light of the recent work of Wilson et al.[20], in which *Eugenia* and *Syzygium* are placed in different tribes, that is, Myrteae and Syzygieae, respectively.

Floral anatomical investigations by Schmid[13] provided strong evidence that *Eugenia s.s.* and the *Syzygium* group were not as closely related as believed by many earlier workers. Schmid[13] considered that neither *Eugenia* nor *Syzygium* were directly ancestral to the other and that their divergence occurred long ago. This view was supported by the phylogenetic analysis of morphological and anatomical data by Johnson and Briggs[19]. This study indicated that *Eugenia* formed a clade with other Myrtoideae genera (for example, *Austromyrtus* (Nied.) Burret, *Myrcia* DC. ex Guill., *Myrtus* and *Psidium* L.), whereas *Syzygium* was in a clade with *Acmena* and other Old World species remote from the Myrtoideae *s.s.* clade. Leaf anatomy has been studied in Malay Peninsula species of *Eugenia* sects. *Acmena, Cleistocalyx, Fissicalyx* and *Syzygium* by Khatijah et al.[63]. The results supported the recognition of sect. *Acmena* but not of sects. *Cleistocalyx* and *Fissicalyx*, which were found to be similar to sect. *Syzygium.* Haron and Moore[64] in a study of leaf micromorphology of Old and New World *Eugenia s.l.* species, that is, species referable to *Syzygium* and *Eugenia s.s.*, found that there were differences in foliar features between the two groups.

In summary of this section, twentieth-century authors have either adopted Bentham and Hooker's 'all in one' concept of genus[50] or accepted *Eugenia* for the New World centred species and varying numbers of genera for the Old World syzygioid species.

## 16.3  CURRENT RESEARCH

Research into the distinction between *Eugenia s.s.* and the *Syzygium* group has not been the focus of current studies; that the two groups are amply distinct appears to be an accepted fact by all current workers as shown above.

### 16.3.1  MORPHOLOGICAL, DEVELOPMENTAL AND CHEMICAL STUDIES

Current supraspecific work on *Syzygium* and its allies has centred on developing an understanding of the relationships between and within the genera. Phenetic and phylogenetic analysis of morphological data derived from Thai species of *Acmena, Cleistocalyx* and *Syzygium* provided support for the taxonomic reinstatement of *Jambosa* at some level but did not give strong resolution to other possible taxonomic groups[12]. Pollen studies of Thai *Syzygium* did not support any known taxonomic groupings of the species, although the pollen of the jambosoid species tended to be larger, suggesting that there might be differences in the breeding biology of the studied species[65].

Floral development has been studied in one species of *Acmena* and one species of *Syzygium* by Belsham and Orlovich[66]. Development of the hypanthium in *Acmena* was similar to that in some dry-fruited Myrtaceae but the androecial development was similar to that of the fleshy-fruited *Luma* A. Gray, whereas the reverse was the case in the *Syzygium* studied. Unfortunately, the number of species examined was small and further work is warranted. The distribution of polyhydroxyalkaloids (PHAs) in 217 species of Myrtaceae was studied by Porter et al.[67], but the taxonomic significance of PHAs in the *Syzygium* group appears inconclusive, and further work may be required.

Khatijah et al.[63], on the basis of 25 Malesian species surveyed, showed that Henderson's[53] groups 2 and 3 within *Syzygium* have paracytic stomata, whilst those in his group 4 are anisocytic; further investigation of this promising line of research has yet to be undertaken. The intrusive material present in the seeds of *Acmena, Acmenosperma* and *Piliocalyx* is an intriguing phenomenon. Hartley and Craven[68], in studies on *Acmena,* reported that the intrusive tissue was of placental origin. Work in progress by Biffin indicates that the tissues in *Acmena* may be derived from the chalaza and be homologous with tissues that surround the seed in several species of *Syzygium s.s.* The embryology of the *Syzygium* group is another area of research that may be of systematic significance. Both unitegmic and bitegmic ovules have been observed[69,70], and this promising work is being continued.

In summary of this section, new morphological and anatomical analysis has brought forward promising characters that may be of considerable taxonomic importance. However, in the majority of cases, further analysis is needed before their significance can be adequately assessed.

### 16.3.2 MOLECULAR SEQUENCE STUDIES

Harrington and Gadek[71] utilised sequence data from the internal transcribed spacers (ITS) and external transcribed spacer (ETS) of nuclear rDNA from 65 Australian and the one Lord Howe Island species of the *Syzygium* group. Their sample included species representative of *Acmena, Acmenosperma, Anetholea, Cleistocalyx, Syzygium* and *Waterhousea,* together with six unnamed species at that time not placed to genus[71]. The results of their analyses were not congruent with current taxonomic circumscriptions, indicating that the separation of the Old World genera into two groups, the *Acmena* suballiance and the *Syzygium* suballiance, proposed by Briggs and Johnson[18] was unjustified. There was also no support for the conventional concepts of *Acmena, Acmenosperma, Cleistocalyx, Jambosa, Syzygium* and *Waterhousea,* and *Anetholea* was also nested in *Syzygium*[71]. Biffin et al.[72] analysed cpDNA sequences from the *matK* and *ndhF* genes and the *rpl16* intron from 87 species of the *Syzygium* group. The sampling was comprehensive, covering the taxic and morphological diversity sampled by Harrington and Gadek[71], with the addition of species from the South West Pacific that represented the generic concepts of *Acicalyptus* A. Gray, *Cupheanthus* Seem. and *Piliocalyx* Brongn. and Gris, a few species from Africa and Malesia and some widely cultivated species of uncertain geographic origin. As with the study by Harrington and Gadek[70], there was no support for conventional concepts but there were some major clades that are being further investigated. It is possible that these clades may be 'cryptic', that is, not readily diagnosed morphologically, and future directions for research into this aspect are discussed by Biffin et al.[73].

Figure 16.6 and Figure 16.7 are summaries of present knowledge as to the phylogeny of *Syzygium s.l.* inferred from analysis of molecular sequence data. The *Syzygium s.l.* species that were analysed to obtain these two trees are representative samples drawn from the genera accepted by various workers in recent times, that is, *Acmena, Acmenosperma, Anetholea, Cleistocalyx, Piliocalyx, Syzygium* and *Waterhousea*. Figure 16.6 shows a strict consensus tree derived from combined *ndhF* and *matK* data. Four well supported clades are evident with two species, *Anetholea anisata* Peter G. Wilson and *Syzygium wesa* B. Hyland, each comprising a monospecific lineage that is not well supported as members of any of these clades. *Jambosa* and *Syzygium,* as it is traditionally circumscribed, species are in Group I; this clade therefore includes the very great majority of species of the complex and has a correspondingly large geographic range. Group II, comprising *Acmena, Piliocalyx* and *Waterhousea* species, along with *Syzygium gustavioides* (F.M. Bailey) B. Hyland, *S. glenum* Craven and *S. monimioides* Craven, is characterised by the majority of its species possessing intrusive material within the cotyledons. *Syzygium gustavioides* and *S. monimioides,* however, differ in that they have seeds typical of *Syzygium s.s.* Groups III and IV and may equate to Henderson's Groups 4 and 5, respectively, of his *Eugenia* sect. *Syzygium*[53]. *Acmenosperma claviflorum* (Roxb.) Kausel, a member of Group IV, has intrusive material in the cotyledons as in many Group II species, but it is not yet known if the tissues are homologous. Work is in progress to identify practical macromorphological features that can be used to support the recovered clades and in classification.

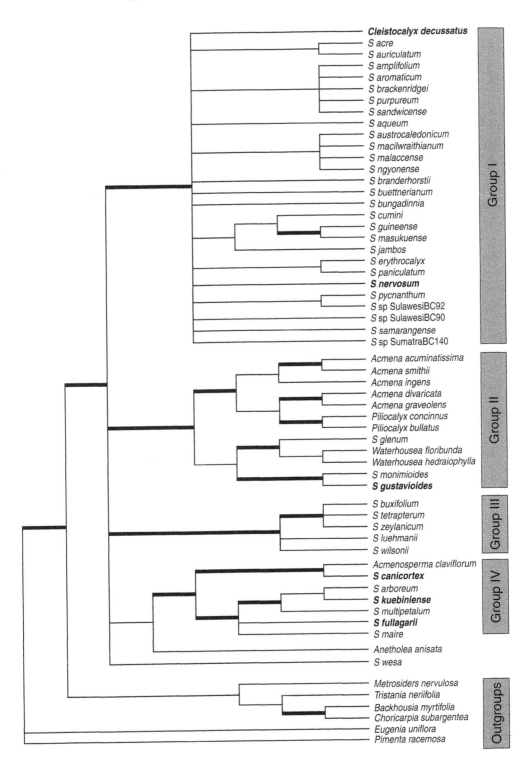

**FIGURE 16.6** Strict consensus tree of 10,000 trees derived from a combined *ndhF, matK* and *rpl16* dataset from a representative sampling of *Syzygium s.l.* Bold branches have BS ≥ 90%; length 877; CI = 0.831; RI = 0.829. Names in bold indicate species referrable to *Cleistocalyx* sensu Merrill and Perry[54].

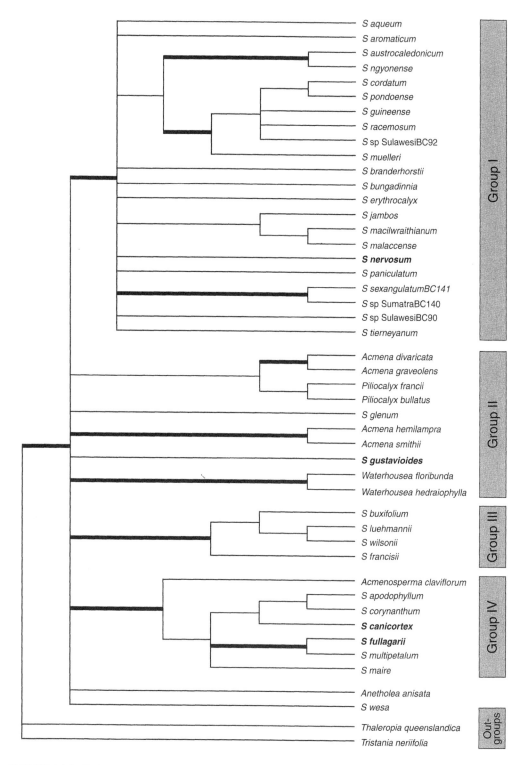

**FIGURE 16.7** 50% bootstrap consensus tree for the ITS data set from a representative sampling of *Syzygium s.l.* Data analysed under parsimony with transversions receiving four times the weight of transition substitutions. Bold branches have BS ≥ 90%. Names in bold indicate species referable to *Cleistocalyx* sensu Merrill and Perry[54].

The ITS data set shows a strong bias towards transition substitutions (CT, AG), and the tree (Figure 16.7) was derived with transversion substitutions receiving four times the weight of transitions. The ITS data provides moderate to strong support for clades consistent with Groups I, III and IV in the chloroplast data, although Group II is not resolved as monophyletic, and the relationships of *Anetholea* and *S. wesa* are also unresolved. It is important to note, however, that areas of disagreement between the ITS and chloroplast data are only weakly, or are not statistically supported in the ITS data, consistent with the hypothesis that these data are uninformative regarding some relationships within *Syzygium s.l.,* rather than suggesting an alternative, conflicting resolution. For instance, we note that, at moderate to high levels of sequence divergence, transition substitutions are saturated (that is, there is a high probability of unobserved substitutions occurring at some nucleotide positions), and as such, the historical signal may be obscured by 'noise'. On the other hand, congruence between our data sets increases our confidence in the recognition of Groups I, III and IV. Additional lines of evidence, including further sequences of nuclear DNA and from morphology, will be required to confidently resolve relationships of the 'acmenoid' taxa (Group II), *S. wesa* and *Anetholea*.

In summary of this section, molecular data have suggested that current, largely morphologically derived, generic characterisations are flawed.

### 16.3.3 FLORISTIC STUDIES

Floristic activity is presently high, and several projects have been completed recently or are underway (Table 16.1). This type of research is both undervalued and underfunded. In large part, the current advances made in molecular studies by Harrington and Gadek[71] and Biffin et al.[72,73] are an undeniably vital component in exploring the evolution and relationships of this large genus. Their foundations were built upon the knowledge of the morphological diversity occurring in Australia and Melanesia that is encapsulated in the revisionary and floristic studies of Hyland[38], Smith[74], Dawson[75] and Craven and Matarczyk[42]. Despite such activity, coverage is far from even and still utterly inadequate in a number of countries. For example, Utteridge[76], in relation to a checklist of woody plants of Sulawesi, states: "This highlights how some groups, for example *Syzygium* (Myrtaceae), are badly in need of specialist systematic work. For example, only four species of *Syzygium* are recorded in the checklist, but approximately 350 un-named collections are listed".

### TABLE 16.1
### Current and Recently Published Floristic Research in *Syzygium s.l.*

| Region | Project | Status | Author(s) |
|---|---|---|---|
| Australia | Revision | Completed | Hyland[38] |
| | Flora of Australia | Completed (in press) | Craven and Matarczyk[43] |
| East Africa | Flora of Tropical East Africa | Completed | Verdcourt[84] |
| Fiji | Flora of Fiji | Completed | Smith[74] |
| Mascarenes | Flore des Mascareignes | Completed | Scott[83] |
| Sri Lanka | Flora of Ceylon | Completed | Ashton[21] |
| | Revision | Completed | Kostermans[22] |
| Thailand | Flora of Thailand | Completed | Chantaranothai and Parnell[28] |
| New Caledonia | Flore de la Nouvelle-Calédonie | Part completed, balance in preparation | Dawson[75] |
| Borneo | Tree Flora of Sabah and Sarawak | In preparation | Ashton[86] |
| China | Flora of China | In preparation | Chen and Craven[44] |
| Papuasia | Revision | In preparation | Craven[87] |
| Indo-China | Flore du Cambodge du Laos et du Viêtnam | Initiated | Parnell and Chantaranothai[85] |

There are significant impediments to floristic and systematic research on *Syzygium*. The group is badly undercollected in many parts of its geographic range. Parnell[77] showed that the distribution of even the most common species of Thai *Syzygium* showed significant false gaps, which could be filled in by subsequent collecting, a process that has not yet been even closely approached in Thailand (and therefore most of South East Asia). Furthermore, Parnell et al.[78] showed that Thailand was severely undercollected, with a low collecting density and low rate of collecting activity. Such undercollecting is typical of most countries where *Syzygium* is native. In addition to this lack of floristic survey, *Syzygium* species are infrequent flowerers, and nonflowering material is generally abhorred by tropical collectors, as usually it cannot be named to species. Therefore, even when areas are thoroughly sampled over a one- to two-year period, species are passed over. Another limitation is due to the inadequate representation in herbaria of the reproductive stages necessary for complete descriptions and key preparation, for rarely does a species carry both flowers and fruit at the same time. Other limitations equally applicable to all plant groups but especially critical for groups of exceptional size are that the world's herbaria are understaffed and that the largest are located, through historical accident, in Europe[79–82]. The major collections are therefore removed from the centres of diversity of *Syzygium,* and this does not aid field study. Current projects that will result in the easier exchange of data through imaging will mitigate this problem.

Despite the research activity described in Table 16.1, there are a number of countries or areas where no adequate floristic account exists or is realistically projected, including the Philippines, Sulawesi and Sumatra. Areas where much further detailed work is needed include Peninsular Malaysia, Kalimantan and the Andaman and Nicobar Islands, and these therefore could form the focus for involvement of new workers on the genus. The presence of novel taxonomic data retrievable only from the above areas cannot be ruled out. Such data, if it exists, could have a dramatic impact on the structure of any phylogenetic hypothesis.

In summary of this section, current floristic activity is high, but nevertheless there are significant gaps, which may be of evolutionary significance.

### 16.3.4 BREEDING BIOLOGY

The breeding biology of *Syzygium* is underinvestigated, and this is a considerable impediment to understanding the delimitation and evolution of its species. Nic Lughadha and Proença[16] review the literature for the Myrtoideae and show that dichogamy is likely in *Syzygium* and that bird and mammal pollination occurs, with nectar being the primary reward. Boulter et al.[88] show that *S. sayeri* (F.M. Muell.) B. Hyland is visited for its nectar both during the day and at night by a wide range of pollinators, including bats, birds, bees, wasps, moths, thrips and the occasional ant or spider. Hopper[89] observed that the majority of interflower movements of pollinators on *S. tierneyanum* (F. Muell.) Hartley and L.M. Perry were within the same plant. Lack and Kevan[90] found that *S. syzygioides* (Miq.) Merr. and L.M. Perry was self-incompatible; consequently the spatial arrangement of flowering trees and pollinator behaviour become issues in maintaining reproductive success. Boulter et al.[88] indicate that *S. sayeri* is xenogamous, with low selfcompatibility and moderate levels of outcrossing. Chantaranothai and Parnell[91] record that visitor numbers to *S. jambos, S. megacarpum* (Craib) Rathakr. and N.C. Nair and *Syzygium samarangense* (Blume) Merr. and L.M. Perry 'See-nak', are often very low, although Free[92] indicates that there are no reports that pollination is inadequate in *S. aromaticum* (as *Eugenia caryophyllus* Bullock and Harrison). Taken as a whole, these data do not conclusively indicate whether inbreeding or out-breeding is the norm in *Syzygium*. It may be that part of the complexity of structure and species richness of the genus is a result of inbreeding. Chantaranothai and Parnell[91] showed that three breeding systems may operate in *Syzygium*. All species tested, *S. samarangense* 'See-nak,' *S. jambos, S. formosum* (Wall.) Masam. and *S. megacarpum,* showed self-compatibility; *S. samarangense* also showed enhanced seed set through self pollination, and it and *S. jambos* exhibited nonobligate apomixis with both fruits and seeds having the potential to develop without fertilisation.

Free[92] indicated that *S. aromaticum* may be a nonobligate apomict. The data in Boulter et al.[88] indicate that *S. sayeri* may be able to act as an agamosperm, but that agamospermy is very much less successful in terms of successfully pollinated buds than outcrossing. A further complication is the inconsistent exhibition of adventitous polyembryony in certain species, for example *Syzygium cumini* (L.) Skeels, wherein it is combined with reportedly varying levels of polyploidy[16].

There are no data on the frequency of occurrence of apomixis, nor varying ploidy levels, nor the frequency of adventitious embryony in *Syzygium*. Nevertheless, it is clear that the description of species based on few collections, when combined with a very low collecting rate and density over much of the range of *Syzygium,* and the potential for apomixis and ploidy variation might result in the false delimitation of many taxa as new species which are, at best, microspecies. As far as we are aware, the suggestion that a significant number of microspecies may exist in *Syzygium* is novel.

The anthers in many *Syzygium* species have conspicuous, although small, glands associated with the connective which appear secretory. In addition, the petals of *Syzygium* are also often glandular. Discussion of the function, if any, of these glands in *Syzygium* is almost nonexistent, and details of the chemical composition of the glands' secretions is unknown. However, the secretions are clearly variable in quantity and — probably — composition and function; and gland density has been used as a taxonomic characteristic (Chantaranothai and Parnell[28]). Their further study may offer novel taxonomic data and insights on breeding biology as has been suggested for another Myrtaceous genus, *Verticordia* DC. (Ladd et al.[93,94]).

In summary of this section, the breeding biology of *Syzygium* is underinvestigated. Various systems ranging from inbreeding to outbreeding occur, and the lack of information on their frequency of occurence and distribution is a considerable impediment to understanding the delimitation and evolution of species in the genus.

## 16.4 FUTURE PROSPECTS

We have shown that *Syzygium* has never been revised as a unit at species level outside of the nineteenth century and have also shown that current estimates of phylogeny suggest that *Syzygium* comprises several well supported clades. Whether or not these clades are indicative that *Syzygium* is nonmonophyletic is dependent upon one's delimitation of that genus per se. In addition, we have shown that there is considerable floristic activity dealing with *Syzygium,* by definition on a regional basis, and we are confident that the genus is simply too large to facilitate a worldwide species-level monograph. We acknowledge that current activity will leave significant unexplored geographical gaps. Especially in a university, but also in a research institution, monographic activity on the scale needed is likely to be unsustainable, as high performance indicators are difficult to maintain when undertaking a monograph. We do not wish to rehearse, yet again, arguments contrasting monographs with floras but are aware that current regional work in *Syzygium* has the potential to produce inchoate accounts that do not mesh with each other, and that to some extent this has already occurred. We believe that this is undesirable, but admit that it is nevertheless almost inevitable, as the only practical way forward at species level is a regional one. To mitigate the detrimental potential of local working, we believe that a World Wide Web-based site for workers on *Syzygium* might stimulate interest and simplify exchange of material and ideas. Such a site would clearly interdigitate with the activities of the Flora Malesiana project (the majority of species of *Syzygium* occur in the Malesian region) and the initiative coordinated by the Royal Botanic Gardens Kew, U.K., which aims to produce a world checklist of Myrtaceae.

There are many species new to science awaiting description within *Syzygium s.l.*. At present, there are two main approaches taken to the placement of such species. Craven[9] favours placement of all such new species, which are of course always described on the basis of morphology, in *Syzygium s.l.*. The core argument advanced here is that to split off anomalous variation will necessarily lead to the splitting of *Syzygium* into a plethora of genera, and that these will be both ambiguous and impossible to recognise without both flower and fruit. In addition, Craven implies that novel data, especially molecular data, are likely to suggest splitting of *Syzygium* in unforeseen ways which will then allow the erection of robust, phylogenetically defined genera, and that it is unwise to set up new genera in the interim. Based upon a synthesis of the presently available molecular and morphological evidence, however, Craven and Biffin consider that the species under discussion constitute a single, natural group, and that all should be classified in *Syzygium* with an infrageneric classification that reflects the evolutionary relationships of the constituent clades. By contrast, Parnell believes that the inclusion of the majority of new species within an expanded concept of *Syzygium* might make it more polyphyletic and overstretch the genus boundaries. He argues that knowledge of phylogeny will always be imperfect and favours the erection of separate genera to accommodate such new species based on a sufficiency of current evidence. He believes that any degree of predictability which could be derived from the current classification will be diluted by cramming all the currently split off genera (for example, *Acmena, Acmenospermum, Cleistocalyx Piliocalyx* and *Waterhousea*) along with new, probably generically distinct taxa, into an ever-expanding *Syzygium*. Neither Craven's nor Parnell's methodology eliminates the necessity for future species transfer between genera — rather, both admit that it will be necessary — however, they have not agreed which procedure will be minimally disruptive, producing the smallest number of intergeneric transfers. It does not appear that Article 34 of the International Code of Botanical Nomenclature[95] can be stretched to resolve this problem, as neither Craven or Parnell suggest that the species described are invalid, nor do they suggest that the new species might not belong to *Syzygium*.

This debate raises the question as to whether strict monophyly should be the overwhelming consideration for classification of such a species rich group, and if it is, how (that is, on what basis) it is to be established. Clearly, it is unlikely that sufficient numbers of strict monophyletic lineages can be established in the short to medium term in such a species rich, widespread and poorly known genus as *Syzygium,* where molecular data are limited and phylogenetically promising morphological data are still being discovered. However, if major clades can be identified, then the task of classifying the genus will be facilitated because researchers will be able to narrow down the number of species included in their studies. Despite attempts to utilise morphological data cladistically by Parnell[12] and Craven[96], the lack of resolution suggests that, as in many other genera, morphology by itself will be inadequate for the task. It may be, as Olmsted and Scotland[97] argue, that molecular data offer 'more and better data' to reconstruct phylogeny. Results of analysis of the chloroplast *ndhF, matK* and *rpl16* data are summarised in Figure 16.6 and are generally congruent with the ITS data (Figure 16.7). The reliability of ITS data for phylogenetic reconstruction has been questioned by Álvarez and Wendel[98], and Biffin[99] is presently investigating the utility of the nuclear encoded large subunit of RNA polymerase (*rpb2*) as a source of data for a second nuclear region. Whether general congruence is a sufficient measure indicating accurate reconstruction of phylogeny requires further debate.

Strict adherence to the concept of monophyly may also be operationally infeasible. Our work suggests that the variation patterns of *Syzygium s.s.* species in South East Asia and Australasia are different, that there are many species awaiting description and naming, and that a uniform species concept may, in part due to different breeding systems, be inapplicable. We believe that such basic descriptive is best undertaken in a phylogenetic framework. However it is clear that in *Syzygium,* although the overall framework is being developed as a result of the studies by Harrington and Gadek[71] and Biffin et al.[72,73], the detail will take longer and requires much work. It is important that this work be undertaken for the extremely large *Jambosa-Syzygium* clade (Group I), not only because of its extreme size (c. 1,000 species are involved) but also because this group contains the economically important fruit and spice species. Although morphology may lack the strength of molecular sequence data for phylogenetics in *Syzygium,* it still has much to offer the *Syzygium* systematist. Apart from its obvious significance for identification, morphology will be important in characterising the clades recovered from analysis of sequence data.

If monophyly is not given prime place, then this raises the issue of what drives classification and nomenclature. We do not believe that historical precedent and convenience should be the sole pilots of classification. So, for example, we welcome the transfer of inappropriately placed species from *Eugenia* to *Syzygium,* as no rational systematist now argues that they are closely related. Unsatisfactory as it might seem, we believe that what will drive classification and nomenclature in *Syzygium,* and the segregation of related genera etc., will be similar arguments, essentially based on an 'unassailable mass' of evidence. We differ, however, in our consideration of what is sufficient mass. Guidance may well be provided by molecular data. Where phylogenetic trees of large genera constructed on the basis of a few exemplar species, or only a single gene, challenge currently accepted patterns, we certainly suggest that those patterns and their underlying causes are re-examined and the testing expanded. Taxonomic change should not be undertaken rashly. We have shown that DNA sequence data do offer new insights into *Syzygium,* especially at the higher levels; however, it is unclear at present how robust those insights are. Advances in phylogenetic reconstruction methods may be needed for large datasets which might derive from large groups such as *Syzygium.* One promising development is continuous jackknife function analysis[100], and this appears unutilised for large datasets which might derive from large groups such as *Syzygium.* Its application may be an important tool allowing assessment of the stability of large group phylogenies and impartial assessment of the achievement of 'unassailable mass'.

Another difficulty is the question of whether locally distinctive species groups (for example, the Fijian species assigned to *Cleistocalyx,* the Papuasian species of the *Syzygium furfuraceum* Merr. and L.M. Perry group and the trimerous New Caledonian species) should be given recognition at some level. If this is done, it is likely to result in a paraphyletic classification with a very large number of comparably ranked taxa that had to be established merely to 'balance' the classification. In part, we are here concerned with a conflict between the operability and utility of classifications and their predictability and monophyly. In general, we accept the thrust of the letter coordinated by Nordal and Stedje[101] which advocates the acceptance of paraphyletic taxa (at least for Floras) and on this basis, there is no reason not to allow for the recognition of locally distinctive species groups.

Clearly, *Syzygium* is unusual in size and an obvious question is 'why is it so big?'. For example, we may want to know if there are any key innovations that can correlate with, or explain, diversification patterns (see Davies and Barraclough, *Chapter 10*; Hodkinson et al., *Chapter 17*). In some other large genera, there appears to be an uniting apomorphy of great importance in driving speciation; in *Solanum* it may be buzz pollination, in *Euphorbia* it may be the cyathium, in *Ficus* it may be the fig (i.e., the syconium), in the Compositae it may be either a specialised incompatibility mechanism linked to specialised pollination mechanisms or the development of chemical poisons. In *Syzygium* it may be invidious to single out only one key innovation; perhaps, it is better to consider *Syzygium's* combination of features as innovative.

Research areas that we believe will be personally rewarding to study, and which are important to pursue from the biodiversity perspective include the following:

- Resolving interrelationships of the 80–90% of the genus that comprises Group I (as defined on the basis of molecular analysis), that is the *Jambosa-Syzygium s.s.* clade. This will be a major task, given the sampling issues posed by the geographical distribution of the group, let alone the identification of suitable DNA sequence regions for analysis.
- Completing floristic surveys of the major regions not yet investigated adequately, especially Myanmar, Peninsula Malaysia, Kalimantan, Sulawesi and the Philippines.
- Developing an understanding of the biogeography of the major clades, especially of their prehistorical biogeography.
- Investigating the breeding systems to establish to what extent, if any, there are implications for taxonomy from factors such as apomixis, hybridisation and introgression.
- Studying evolutionary phenomena, such as r and K adaptive strategies.

- Examining novel, or understudied but promising, morphological characters, including stomatal type and especially those associated with characteristics of the placenta and ovule, including vascular anatomy, the development and types of intrusive tissue and the number of integuments.
- Examining the chemical composition of the secretions of the anther gland connective and petals.
- Determining the factors, including 'key innovation(s)', that drive diversification of *Syzygium*.

In conclusion, we regard the size of *Syzygium* as a positive, even though we acknowledge there are caveats on logistical grounds, as it offers opportunities for the initiation of major and stimulating research projects well into the twenty-first century and beyond. The enormous structural diversity embodied in the plants themselves, their habit, foliage, their often highly attractive flowers and fruit, their manifestly diverse ecology and wide geography, their biotic and abiotic interactions with other animals including man, all ensure that exciting and meaningful research is limited only by money and imagination.

## ACKNOWLEDGEMENTS

We wish to thank various agencies and individuals whose data and support have contributed to this chapter. John Parnell thanks the EU for support under the Marie Curie Scheme for various post-doctoral fellows and under the Human Capital and Mobility Scheme, the Trinity Trust and Trinity College Dublin (TCD) for sponsorship of various postgraduate students, especially Professor Pranom Chantaranothai, and all of the herbaria, especially TCD and all others listed in their publications, without whose collections and support this chapter would have been unconstructible. Lyn Craven and Ed Biffin acknowledge support from the Pacific Biological Foundation, CSIRO and ANU, and the many individuals and institutions who generously have provided material, information and field and other assistance. Ed Biffin holds an ABRS Postgraduate Scholarship from the Australian Biological Resources Study and a Scholarship from the Australian National University.

## REFERENCES

1. George, A.S., Myrtaceae, Family description, in *Fl. Australia 19*, George, A.S., Ed., Australian Government Publishing Service, Canberra, 1988, 1.
2. Johnson, L.A.S. et al., Myrtaceae, in *Flowering Plants in Australia*, Morley, B.D. and Toelken, H.R., Eds., Rigby, Willoughby, 1988, 175.
3. Kochummen, K.M., Eugenia, in *Tree Flora of Malaya* 3, Ng, F.S.P., Ed., Longman, London, 1995, 172.
4. Lemmens, R.H.M.J., *Syzygium*, in PROSEA *(Plant Resources of South East Asia) 5 Timber Trees: Minor Commercial Timbers,* Eds. Lemmens, R.H.M.J., Soerianegara I., and Wong, W.C., Backhuys, Leiden, 1995, 441.
5. Mabberley, D.J., *The Plant Book,* 2nd ed., Cambridge University Press, Cambridge, 1997.
6. Schmid, R., Comparative anatomy and morphology of *Psiloxylon* and *Heteropyxis,* and the subfamilial and tribal classification of Myrtaceae, *Taxon,* 29, 559, 1980.
7. Craven, L.A., Myrtaceae of New Guinea, in *Ecology of Papua,* Conservation International, in press.
8. Chippendale, G.M., *Eucalyptus,* in *Fl. Australia 19,* George, A.S., Ed., Australian Government Publishing Service, Canberra, 1988, 1.
9. Craven, L.A., Unravelling knots or plaiting rope: what are the major taxonomic strands in *Syzygium* sens. Lat. (Myrtaceae) and what should be done with them? in *Taxonomy: The Cornerstone of Biodiversity Proc. Fourth Fl. Males. Symp.,* Saw, L.G., Chua, L.S.L., and Khoo, K.C., Eds., Forest Research Institute, Malaysia, Kuala Lumpur, 2001, 75.
10. Craven, L.A., Four new species of *Syzygium* (Myrtaceae) from Australia, *Blumea,* 48, 479, 2003.
11. Craven, L.A. and Biffin, E., *Anetholea anisata* transferred to, and two new Australian taxa of, *Syzygium* (Myrtaceae), *Blumea,* 50, 157, 2005.

12. Parnell, J., Numerical analysis of Thai members of the *Eugenia-Syzygium* group (Myrtaceae), *Blumea,* 44, 351, 1999.

13. Schmid, R., A resolution of the *Eugenia-Syzygium* controversy Myrtaceae, *Amer. J. Bot.,* 59, 423, 1972.

14. McVaugh, R., The genera of American Myrtaceae: an interim report, *Taxon,* 17, 354, 1968.

15. Niedenzu, F., Myrtaceae, in *Die Natürlichen Pflanzenfamilien 3,* Abteilung 7, Engler, A. and Prantl, K., Eds., Engelmann, Leipzig, 1893, 57.

16. Nic Lughadha, E. and Proença, C., A survey of the reproductive biology of the Myrtoideae Myrtaceae, *Ann. Missouri Bot. Gard.,* 83, 480, 1996.

17. Hora, F.B., Myrtaceae, in *Flowering Plants of the World,* Heywood, V.H., Ed., Oxford University Press, Oxford, 1978, 161.

18. Briggs, B.G. and Johnson, L.A.S., Evolution of the Myrtaceae: evidence from inflorescence structure, *Proc. Linn. Soc. New South Wales,* 102, 157, 1979.

19. Johnson, L.A.S. and Briggs, B.G., Myrtales and Myrtaceae: a phylogenetic analysis, *Ann. Missouri Bot. Gard.,* 71, 700, 1984/5.

20. Wilson, P.G. et al., Relationships within Myrtaceae sensu lato based on a *mat*K phylogeny, *Pl. Syst. Evol.,* 251, 3, 2005.

21. Ashton, P.S., Myrtaceae, in *A Rev. Handbook Fl. Ceylon* 2, Dassanayake, M.D., Ed., Balkema, Rotterdam, 1981, 403.

22. Kostermans, A.J.G.H., *Eugenia, Syzygium* and *Cleistocalyx* (Myrtaceae) in Ceylon, *Quart. J. Taiwan Mus.,* 34, 117, 1981.

23. Simpson, S.C. and Weiner, E.S.C., *The Oxford English Dictionary,* Book Club Associates for Oxford University Press, London, 1989.

24. Willis, J.C., *Age and Area,* Cambridge University Press, Cambridge, 1922.

25. Minelli, A., *Biological Systematics: The State of the Art,* Chapman and Hall, London, 1993.

26. Minelli, A., Fusco, G., and Sartori, S., Self-similarity in biological classifications, *Biosystems,* 26, 89, 1991.

27. Frodin, D.G., History and concepts of big plant genera, *Taxon,* 53, 753, 2004.

28. Chantaranothai, P. and Parnell, J., *Syzygium,* in *Fl. Thailand 7,* Santisuk, T. et al., Eds., Forest Herbarium, Royal Forest Department, Bangkok, 2002, 811.

29. Turner, I.M., Myrtaceae, in *A Catalogue of the Vascular Plants of Malaya: Gardens' Bull. Singapore,* 47, 370, 1995.

30. Chantaranothai, P. and Parnell, J., A revision of *Acmena, Cleistocalyx, Eugenia s.s.* and *Syzygium* (Myrtaceae) in Thailand, *Thai Forest Bull.,* 21, 1, 1994.

31. Gamage, H.K., Ashton, M.S. and Signhakumara, B.M.P., Leaf structure of *Syzygium* spp. (Myrtaceae) in relation to site affinity within a tropical rain forest, *Bot. J. Linn. Soc.,* 141, 365, 2003.

32. Poonswad, P., Nest site characteristics of four sympatric species of hornbills in Khao Yai National Park, Thailand, *Ibis,* 137, 183, 1995.

33. FAOSTAT, *Food and Agricultural Organisation Statistical Data,* http://faostat.fao.org/faostat, 2005.

34. Oyen, L.P.A. and Xuan, Dung, N., Introduction, in *PROSEA (Plant Resources of South East Asia) 19: Essential Oils,* Oyen, L.P.A. and Xuan Dung, N., Eds., Backhuys, Leiden, 1999, 15.

35. Sardjono, S., *Syzygium polyanthum* (Wight) Walpers, in *PROSEA (Plant Resources of South East Asia) 13: Spices,* de Guzman, C.C. and Siemonsma, J.S., Eds., Backhuys, Leiden, 1999, 218.

36. Panggabean, G., *Syzygium,* in *PROSEA (Plant Resources of South East Asia) 2: Edible Fruits and Nuts,* Oyen, L.P.A. and Xuan Dung, N., Eds., Backhuys, Leiden, 1992, 292.

37. Vinning, G. and Moody, T., Wax apple, in *A Market Compendium of Tropical Fruit,* RIRDC, Barton, ACT, 1997, 267.

38. Hyland, B.P.M., A revision of *Syzygium* and allied genera (Myrtaceae) in Australia, *Austral. J. Bot. Suppl. Ser.,* 9, 1, 1983.

39. Djadjo Djipaa, C., Delmée, M., and Quetin-Leclercq, J., Antimicrobial activity of bark extracts of *Syzygium jambos* (L.) Alston (Myrtaceae), *J. Ethnopharmacol.,* 71, 307, 2000.

40. Shafi, P.M., et al., Antibacterial activity of *Syzygium cumini* and *Syzygium travancoricum* leaf essential oils, *Fitoerapia,* 73, 414, 2002.

41. Eddowes, P.J., Water gum, in *Commercial Timbers Papua New Guinea: Their Properties and Uses,* Office of Forests, Port Moresby, 1977, 20.

42. Hartley, T.G. and Perry, L.M., A provisional enumeration of species of *Syzygium* Myrtaceae from Papuasia, *J. Arnold Arb.,* 54, 160, 1973.

43. Craven, L.A. and Matarczyk, C.A., *Acmena, Acmenosperma, Eugenia, Syzygium, Waterhousea*, in *Fl. Australia,* Wilson, A., Ed., in press.

44. Chen, J. and Craven, L.A., Myrtaceae, in *Fl. China,* Zhengyi, W. Raven, P.H. and Deyuan, H., Eds., in press.

45. Stevens, P.F., On characters and characters states: do overlapping and non-overlapping variation, morphology and molecules all yield data of the same value, in *Homology and Systematics,* Scotland, R, and Pennington, T., Eds., Taylor and Francis, London, 2000, 80.

46. Linnaeus, C., *Species Plantarum,* Impensis Laurentii Salvii, Stockholm, 1753.

47. De Candolle, A.P., Myrtaceae, in *Prodr. Syst. Nat. Reg. Veg. 3,* Treuttel and Würz, Paris, 1828, 207.

48. Wight, R., Myrtaceae, in *Illustrations of Indian Botany 2,* American Mission Press, Madras, 1841, 6.

49. Bentham, G., Notes on Myrtaceae, *J. Linn. Soc., Bot.* 10, 101, 1869.

50. Bentham, G. and Hooker, J.D., Myrtaceae, in *Genera Plantarum 1,* Reeve and Co., London, 1865, 690.

51. Blume, C.L., Myrtaceae, in *Mus. Bot. Lugduno-Batavum,* Brill, Leiden, 1850, 66.

52. Miquel, F.A.W., Myrteae, in *Fl. Ned. Indië 1,* Post, Amsterdam, Post, Utrecht, Fleischer, Leipzig, 1855, 407.

53. Henderson, M.R., The genus *Eugenia* (Myrtaceae) in Malaya, *Gardens' Bull. Singapore,* 12, 1, 1949.

54. Merrill, E.D. and Perry, L.M., Reinstatement and revision of *Cleistocalyx* Blume (including *Acicalyptus* A. Gray), a valid genus of the Myrtaceae, *J. Arnold Arb.,* 18, 322, 1937.

55. Merrill, E.D. and Perry, L.M., A synopsis of *Acmena* DC., a valid genus of the Myrtaceae, *J. Arnold Arb.,* 19, 1, 1938.

56. Merrill, E.D. and Perry, L.M., On the Indo-Chinese species of *Syzygium* Gaertner, *J. Arnold Arb.,* 19, 99, 1938.

57. Merrill, E.D. and Perry, L.M., The Myrtaceae of China, *J. Arnold Arb.,* 19, 191, 1938.

58. Merrill, E.D. and Perry, L.M., The myrtaceous genus *Syzygium* Gaertner in Borneo, *Mem. Amer. Acad. Arts Sci.,* 18, 135, 1939.

59. Merrill, E.D. and Perry, L.M., Plantae Papuanae Archboldianae, IX, *J. Arnold Arb.,* 23, 233, 1942.

60. Merrill, E.D., Readjustments in the nomenclature of Philippine *Eugenia* species, *Phil. J. Sci.,* 79, 351, 1951.

61. Ingle, H.D. and Dadswell, H.E., The anatomy of the timbers of the South-west Pacific area, *Austral. J. Bot.,* 1, 353, 1953.

62. Pike, K.M., Pollen morphology of Myrtaceae from the South-west Pacific area, *Austral. J. Bot.,* 4, 3, 1956.

63. Khatijah, H.H., Cutler, D.F., and Moore, D.M., Leaf anatomical studies of *Eugenia* L. (Myrtaceae) species from the Malay Peninsula, *Bot. J. Linn. Soc.,* 110, 137, 1992.

64. Haron, N.W. and Moore, D.M., The taxonomic significance of leaf micromorphology in the genus *Eugenia* L. (Myrtaceae), *Bot. J. Linn. Soc.,* 120, 265, 1996.

65. Parnell, J., Pollen of *Syzygium* (Myrtaceae) from S.E. Asia, especially Thailand, *Blumea,* 48, 303, 2003.

66. Belsham, S.R. and Orlovich, D.A., Development of the hypanthium in *Acmena smithii* and *Syzygium australe* (*Acmena* alliance, Myrtaceae), *Austral. Syst. Bot.,* 16, 621, 2003.

67. Porter, E.A., Nic Lughadha, E., and Simmonds, M.S.J., Taxonomic significance of polyhydroxyalkaloids in the Myrtaceae, *Kew Bull.,* 55, 615. 2000.

68. Hartley, T.G. and Craven, L.A., A revision of the Papuasian species of *Acmena* (Myrtaceae), *J. Arnold Arb.,* 58, 325, 1977.

69. Biffin, E., unpublished data, 2005.

70. Tobe, H., unpublished data, 2005.

71. Harrington, M.G. and Gadek, P.A., Molecular systematics of the *Acmena* alliance (Myrtaceae): phylogenetic analyses and evolutionary implications with reference to Australian taxa, *Austral. Syst. Bot.,* 17, 63, 2004.

72. Biffin, E., Craven, L.A., Crisp, M.D., and Gadek, P.A., Molecular systematics of *Syzygium* and allied genera (Myrtaceae): evidence from the chloroplast genome, *Taxon,* 55, 79, 2006.

73. Biffin, E. et al., Evolutionary relationships within *Syzygium s.l.* (Myrtaceae): molecular phylogeny and new insights on morphology, *Proc. Sixth Fl. Males. Symp.,* in press.

74. Smith, A.C., Myrtaceae, in *Fl. Vitiensis Nova 3,* Pacific Tropical Botanical Garden, Lawai, Hawaii, 1985, 289.

75. Dawson, J.W., Myrtaceae, Myrtoideae I: *Syzygium, Fl. Nouvelle-Calédonie,* 23, 1, 1999.

76. Utteridge, T., Review of checklist of woody plants of Sulawesi, Indonesia, Blumea Supplement 14, *Kew Bull.,* 59, 174, 2004.

77. Parnell, J., The conservation of biodiversity: aspects of Ireland's role in the study of tropical plant diversity with particular reference to the study of the flora of Thailand and *Syzygium,* in *Biodiversity: The Irish Dimension,* Rushton, B.S., Ed., Royal Irish Academy, Dublin, 2000, 205.

78. Parnell, J.A.N. et al., Plant collecting spread and densities; their potential impact on biogeographical studies in Thailand, *J. Biogeogr.,* 30, 1, 2003.

79. Parnell, J., European plant systematics and the European Flora, in *Systematics Agenda 2000: The Challenge for Europe,* Blackmore, S. and Cutler, D., Eds., Samara Publishing for the Linnean Society of London, London, 1996, 31.

80. Roos, M.C., Charting tropical plant diversity: Europe's contribution and potential, in *Systematics Agenda 2000: The Challenge for Europe,* Blackmore, S., and Cutler, D., Eds., Samara Publishing for the Linnean Society of London, London, 1996, 54.

81. Schram, F.R. and Los, W., Training systematists for the 21st century, in *Systematics Agenda 2000: The Challenge for Europe,* Blackmore, S., and Cutler, D., Eds., Samara Publishing for the Linnean Society of London, London, 1996, 89.

82. Walmsley, Baroness et al., *What on Earth?* House of Lords Select Committee on Science and Technology Third Report, London, http://www.publications.parliament.uk/pa/ld200102/ldselect/ldsctech/118/11802.htm, 2002.

83. Scott, A.J., *Syzygium,* in *Fl. Mascareignes 92,* Bosser, J. et al., Eds., MSIRI, Mauritius; ORSTOM, Paris; RBG, Kew, 1990, 28.

84. Verdcourt, B., *Syzygium,* in *Fl. Trop. East Africa,* Beentje, H.J., Ed., Balkema, Rotterdam, Brookfield, 2001, 67.

85. Parnell, J. and Chantaranothai, P., *Myrtaceae,* in *Fl. Laos, Cambodg. Viet.,* Vidal, J. and Hull, S., Eds., Muséum National d'Histoire Naturelle et Association Botanique Tropicale, Paris, in prep.

86. Ashton, P.S., *Myrtaceae,* in *Tree Fl. Sabah Sarawak,* Forest Research Institute, *Malaysia,* in prep.

87. Craven, L.A., *Syzygium* (Myrtaceae) in Papuasia, in prep.

88. Boulter, S.L. et al., Any which way will do: the pollination biology of a northern Australian rainforest canopy tree (*Syzygium sayeri;* Myrtaceae), *Bot. J. Linn. Soc.,* 149, 69, 2005.

89. Hopper, S.D., Pollination of the rainforest tree *Syzygium tierneyanum* (Myrtaceae) at Kuranda, Northern Queensland, *Aust. J. Bot.,* 28, 223, 1980.

90. Lack, A.J. and Kevan, P.G., On the reproductive biology of a canopy tree, *Syzygium syzygioides* (Myrtaceae), in a rain forest in Sulawesi, Indonesia, *Biotropica,* 16, 31, 1984.

91. Chantaranothai, P. and Parnell, J., The breeding biology of some Thai *Syzygium species, Trop. Ecol.* 35, 199, 1994.

92. Free, J.B., Myrtaceae, in *Insect Pollination of Crops,* 2nd ed., Academic Press, London, 1993, 383.

93. Ladd, P.G., Parnell, J.A.N., and Thomson, G., Anther diversity and function in *Verticordia* DC. (Myrtaceae), *Pl. Syst. Evol.,* 219, 79, 1999.

94. Ladd, P.G., Parnell, J., and Thompson, G., The morphology of pollen and anthers in an unusual myrtaceaous genus (*Verticordia*), in *Pollen and Spores: Morphology and Biology,* Harley M.M., Morton C.M. and Blackmore S., Eds., Royal Botanic Gardens Kew, London, 2000, 325.

95. Greuter, W. et al., *Int. Code Bot. Nomenclature (Saint Louis Code),* Koeltz Scientific Books, Königstein, 2000.

96. Craven, L.A., unpublished data, 2002.

97. Olmsted, R.G. and Scotland, R.W., Molecular and morphological datasets, *Taxon,* 54, 7, 2005.

98. Álvarez, I. and Wendel, J.F., Ribosomal ITS sequences and plant phylogenetic inference, *Mol. Phyl. Evol.,* 29, 417, 2003.

99. Biffin, E., unpublished data, 2005.

100. Miller, J.A., Assessing progress in systematics with continuous jackknife function analysis, *Syst. Biol.,* 52, 55, 2003.

101. Nordal, I. and Stedje, B., Paraphyletic taxa should be accepted, *Taxon,* 54, 5, 2005.

# 17 Supersizing: Progress in Documenting and Understanding Grass Species Richness

*T. R. Hodkinson*
Department of Botany, School of Natural Sciences, Trinity College Dublin, Ireland

*V. Savolainen*
Molecular Systematics Section, Jodrell Laboratory, Royal Botanic Gardens, Kew, Richmond, Surrey, England

*S. W. L. Jacobs*
National Herbarium, Royal Botanic Gardens Sydney, NSW, Australia

*Y. Bouchenak-Khelladi and M. S. Kinney*
Department of Botany, School of Natural Sciences, Trinity College Dublin, Ireland

*N. Salamin*
Department of Ecology and Evolution, University of Lausanne, Switzerland

## CONTENTS

## ABSTRACT

This paper reviews the progress in documenting and understanding species richness for one of the most diverse and economically important groups of plants (the grasses; Poaceae). It discusses the value of modern taxonomic resources and large phylogenetic trees for macro-evolutionary studies. More specifically, it discusses the use of phylogenetic trees for detecting and dating major lineages, investigating biogeographical origins, identifying patterns of diversification and investigating factors leading to species richness. Theoretical and practical issues regarding the production of large phylogenetic trees and supertrees of the grass family (c. 650 genera and 10,000 species) are also discussed. It asks how far we are from complete tribal, generic and species phylogenetic trees of the grasses.

## 17.1  INTRODUCTION

If you can confidently say you know the grass family (Poaceae) you are misled. You are almost certainly talking about a geographical area or one of its taxonomic groups (genera or tribes) because, in terms of species richness, the family is vast. Even a specialist with lifelong devotion will only just have begun to understand the diversity that exists within this family. It is the fifth most species rich angiosperm family, ranking only behind Asteraceae (daisies), Fabaceae (beans), Orchidaceae (orchids) and Rubiaceae (coffee family)[1]. Despite its size (651 genera and 10,000 species sensu Clayton and Renvoize[1]; 635 genera and 9,000 species sensu Mabberley[2]), advances in grass taxonomy and systematics have occurred faster than in most groups of plant because of their socioeconomic and ecological importance. They cover, chiefly as grasslands or bamboo forests, more than one third of the world's land surface[3] and provide staple cereal, sugar crops and reeds (such as *Arundo, Avena, Hordeum, Oryza, Phragmites, Saccharum, Secale, Sorghum, Triticum* and *Zea*). They also include many noncommercial and commercially bred forage and lawn species (such as the temperate species in *Alopecurus, Cynosurus, Dactylis, Festuca, Lolium, Phleum* and *Poa,* or the tropical species in *Cynodon, Digitaria, Panicum, Paspalum, Pennisetum, Stenotaphrum, Urochloa* and *Zoysia*).

Despite recent advances in grass systematics few large phylogenetic trees of the family have been produced. Grass phylogenetics is, in many ways, still in its infancy and lags behind classical taxonomy in its coverage of species and genera. Phylogenetic studies such as those by the Grass Phylogeny Working Group (GPWG)[4] are helping to shape taxonomic treatments and better define genera and species[5,6] but are often based on limited sampling. Large phylogenetic trees are required also for accurate inferences of macro-evolutionary processes[7-9]. It is desirable to sample most of the diversity of taxa within a study group to reduce the risk of incorrect phylogenetic tree reconstruction[10-12] and to include most of the relevant information to make optimal use of the evolutionary trees obtained[7,13-15].

This paper focuses on the problems and prospects of documentation and furthering systematic understanding of species rich groups using the grasses as a case study. The first part of the chapter outlines progress that has been made in classification, in monographic/floristic studies, and in the dissemination of taxonomic information via electronic resources and informatics. The second part reviews the current state of grass phylogenetics for the study of patterns and processes in grass evolution. The final part examines future prospects in documenting and understanding grass species richness and discusses some of the theoretical and practical issues regarding the production of large phylogenetic trees.

## 17.2  TAXONOMY AND CLASSIFICATION OF THE GRASSES

### 17.2.1  CLASSIFICATION

Many grass classifications have been produced, and the most recent ones have been influenced heavily by phylogenetic studies. Past classifications were based largely on gross morphology and anatomy, such as Clayton and Renvoize[1], Renvoize and Clayton[16] or Watson and Dallwitz[17]. However, these are being revised by studies based on additional molecular evidence. For an historical account of grass classification see Clark et al.[18] or GPWG[4]. Perhaps the most significant recent subfamily classification of the grasses was made by the GPWG[4] and was based on combined analyses of anatomical, molecular (sequences and structural characters) and morphological data. Figure 17.1 summarises the number of genera and species in each subfamily of the GPWG[4]. The classification has twelve subfamilies, which is double the number included by Clayton and Renvoize[1] (Figure 17.2). The additional subfamilies were recognised (or created in the case of Danthonioideae) to accommodate the non-monophyly of Arundinoideae and Bambusoideae, and Pooideae were expanded. Arundinoideae were divided into three subfamilies (Aristidoideae, Arundinoideae and Danthonioideae) and Bambusoideae split into five subfamilies (Anomochlooideae, Bambusoideae, Ehrhartoideae, Pharoideae and Puelioideae). Despite this robust classification, much work remains on grass classification at lower taxonomic ranks and the dissemination of this information via the World Wide Web. The GPWG[4] included tribes in their classification but recognised that considerable phylogenetic work was required to confidently define and reach consensus for many of these.

### 17.2.2  TAXONOMIC LITERATURE AND ELECTRONIC RESOURCES

A taxonomist may spend many years working on a group of organisms to deposit the results in an obscure, poorly accessible journal, or worse still, an unpublished Ph.D. thesis or report. This is clearly far from an ideal situation for the advancement of the research field and is not good value for the sponsoring organisation. This situation has prompted many systematists to push for an improvement in readily available electronic resources for taxonomists[19,20]. The general availability of journal articles for all scientific disciplines has improved with electronic publishing, but many

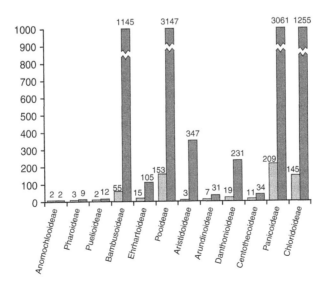

**FIGURE 17.1** Distribution of species and genera in subfamilies of grasses. Light grey bars represent number of genera and dark grey bars represent number of species. (Source: GPWG[4].)

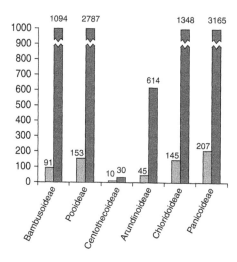

**FIGURE 17.2** Distribution of species and genera in subfamilies of grasses. Light grey bars represent number of genera and dark grey bars represent number of species. (Source: Clayton and Renvoize[1].)

taxonomic journals are not yet available in this format. The main stumbling block is that, as yet, new names are not recognised as legitimate if only published electronically. Digitally available books are still a rarity. However, the trend over the next few decades will undoubtedly be towards increased electronic publishing. A study commissioned by the British Library predicts that by 2020, approximately 80% of UK book output will be available in electronic form, and approximately 40% (including research monographs) will only be available in this form[21]. There is also the potential to digitise material that was not 'born digital' such as highly valuable existing taxonomic literature. For example, the British Library has scanned various such works, including Shakespeare and nineteenth-century newspapers[21].

Taxonomists have also undertaken a number of major initiatives to provide web-based taxonomic resources for all organisms, but this task is not small. Indeed, Wheeler et al.[22] used the term 'terascale taxonomy' to describe the mountainous task of compiling information on trillions of observations, for 10 or more millions of species in museum collections. These include observations on classification, nomenclature, phylogeny, morphology, physiology, ontogeny, ecology, behaviour, geography and genome. Global initiatives include: the Catalogue of Life (a consortium involving Species 2000; http://www.sp2000.org); the Integrated Taxonomic Information System (ITIS; http://www.itis.usda.gov); the Global Taxonomy Initiative (GTI; http://www.biodiv.org); and the Global Biodiversity Information Facility (GBIF; http://www.gbif.org). The Catalogue of Life consortium aims to catalogue all known organisms and construct a web-based freely accessible synonymic index of species and associated data[23]. The GTI of the Convention on Biological Diversity is involved in building taxonomic capacity and making taxonomic information available. The GBIF is developing a network that links together dispersed but electronically available taxonomic information[24]. The meshing together of informatics and taxonomy has therefore begun, and much progress is being made in making taxonomic information available over the web[23,25].

Electronic resources available to grass systematists include general plant bibliographic or nomenclatural databases such as the Kew Bibliographic Databases (including the Kew Record of Taxonomic Literature, the Plant Micromorphological Bibliographic Database and the Economic Botany Bibliographic Database; http://www.kew.org/kbd/searchpage.do), Index Kewensis (available on CD-ROM), the International Plant Names Index (IPNI; http://www.ipni.org) and W3 Tropicos (http://mobot.mobot.org/W3T/search/vast.html). Useful sources of grass taxonomic literature can also be obtained electronically from J.F. Veldkamp, National Herbarium of the Netherlands, Leiden (veldkamp@nhn.leidenuniv.nl), and a list of links to many web-based electronic resources for grass systematists can be found at http://mobot.mobot.org/W3T/search/nwgc.html. Web-based databases are also available for the family. Notable among these are the near complete descriptive treatments of all grass genera, species and their synonymy such as the World Grass Species database and the World Grass Species Synonymy database (http://www.rbgkew.org.uk/data/grasses), the Grass Genera of the World database (http://delta-intkey.com/grass[17]) and the Catalogue of New World Grasses (CNWG; http://mobot.mobot.org/W3T/search/nwgc.html).

There is also a move towards digitising herbarium specimens in the form of scanned images. A photograph can never replace a specimen, but virtual herbaria have many uses and can be particularly useful for type specimens. For example, the US National Herbarium's Botanical Type Specimen Register (http://ravenel.si.edu/botany/types), W3 Tropicos and the CNWG have specimen information and digital images of many grass species including type specimens. Digital Floras are also being produced. Examples include AusGrass[26], an interactive key with maps, species descriptions and line illustrations of Australian grasses, and the Grasses for North America project (http://herbarium.usu.edu/webmanual) that has descriptions, keys, maps and illustrations.

One challenge to the global grass taxonomy community will be to produce a list of accepted names (and synonyms) and to standardise taxonomic treatments. For example, W3 Tropicos and the World Grass Species database are not fully congruent. There is clearly great potential to further develop web-based resources, and these will help alleviate some of the problems associated with access to herbarium collections and type specimens in particular. There is also a need for monographic work at lower taxonomic rank such as genus. Monographs are steadily produced for the grasses, but progress is slow, especially for large genera. The task of producing monographs is far from equal for all grass genera (and angiosperms in general) because species are not distributed randomly among genera. The frequency distribution of genera approximates a logarithmic curve (the hollow curve) and is typical of angiosperm families (Hilu, *Chapter 11*[1,27]). The distribution is skewed toward monotypic genera and those with few species[28–30]. Furthermore there is a disproportionate number of species in certain genera. A high percentage (c. 30%) of all Poaceae species are in a few genera such as *Agrostis, Bambusa, Digitaria, Eragrostis, Festuca, Poa, Panicum, Paspalum* and *Stipa*[17], although these genera are slowly diminishing as new taxonomic treatments become available and species are moved elsewhere. Furthermore, c. 3% of the genera contain 50% of the grass species (Hilu, *Chapter 11*). Therefore, the concept of average generic size is almost meaningless[1]. The biological and evolutionary reasons for these patterns are discussed in Frodin[30] and Hilu (*Chapter 11*). These large genera clearly need to be revised if diversity within the family is to be fully understood. The monophyly of many genera has also been questioned such as *Cortaderia*[31], *Eragrostis*[32], *Miscanthus*[5,6], *Panicum*[33–35], *Pennisetum*[36], *Setaria*[36,37] and *Sorghum*[38,39] (but see Dillon et al.[40] for contrary evidence). Apomicts, such as many species in the large genus *Poa*, also have their own taxonomic challenges (see Frodin[30]), as do polyploidy complexes, which often include apomixes[41,42].

The rate of progress in production of taxonomic monographs is slow. For example, Steussy[43] estimates that a typical Ph.D. revision project of three to four years might include 10 to 40 species, depending on the amount of macromolecular work. He also suggested that in professional life a reasonable pace is only two species per year. Some believe that DNA taxonomy (or an alternative, DNA barcoding) offers a full or partial solution to taxon recognition (discovery) and identification. It has been suggested that DNA sequences be used to identify species; the sequenced DNA is placed in a web-based databases such as DDBJ/EMBL/GenBank and linked to a verified herbarium specimen from which it was taken. This may have some potential for grass identification and taxon recognition, but many reservations apply. These have been discussed at length (Seberg and Petersen, *Chapter 3*; Wheeler[20,47]; Tautz et al.[44,46]; Lipscomb et al.[45]; Chase[48]; Kristiansen[49]) and include concerns about sequence quality, insufficient sampling within and among species, pseudogenes, herbarium specimen quality and availability, type specimen use and common occurrence of hybridisation and introgression and associated DNA exchange (capture) between closely related species.

However, it is premature to dismiss DNA taxonomy, as it undoubtedly has high potential. DNA barcoding is seen by many as a better alternative (see Seberg and Petersen, *Chapter 3*) and uses DNA sequences to aid identification but is not all prevailing when it comes to identification. DNA sequences also have huge potential for phylogenetics, classification and for providing a phylogenetic framework for developing meaningful monographic studies.

## 17.3  PHYLOGENETICS OF THE GRASSES

Phylogenetic studies have been central in the study of patterns and processes in evolution[13,15]. Major advances have been made in molecular biology (especially automated DNA sequencing), the theory of phylogenetic reconstruction and computer technology (to allow computationally difficult phylogenetic reconstructions to be made). Molecular phylogenetic studies of angiosperms have also made good progress[50,51] and strongly support[52–55] the order Poales, a grouping of 18 families of which the dominant two are Cyperaceae (c. 4,000 species[56]) and Poaceae (c. 10,000 species). Within Poales, a graminoid clade[4] can be defined that contains the closest relatives to the grasses. The graminoid clade includes Anarthriaceae, Centrolepidaceae, Ecdeiocoleaceae, Flagellariaceae, Joinvilleaceae, Poaceae and Restionaceae. Cyperaceae group with Juncaceae and Thurniaceae, the cyperoid clade, that is sister to the graminoid clade[57]. The grasses are not as closely related to Cyperaceae as previously thought[58].

### 17.3.1  MAJOR CLADES OF THE GRASSES

The monophyly of Poaceae is strongly supported by a number of synapomorphies[4] including the caryopsis (with the outer integument developmentally fused to the inner wall of the ovary), a highly differentiated embryo, a laterally positioned embryo and the presence of intraexinous channels in the pollen wall[59,60]. A wealth of DNA characters also strongly support their monophyly (see below).

In contrast, the designation of the sister group to Poaceae has been controversial. It is likely to be Ecdeiocoleaceae[57,61,62], Joinvilleaceae[18,60,63–65] or a group containing them both[66]. The GPWG[4] recognised Joinvilleaceae as sister to the grasses, but a comprehensive combined analysis of *rbcL* and *atpB* genes across Poales by Bremer[57] identified a strongly supported sister group relationship between Poaceae and Ecdeiocoleaceae. Joinvilleaceae were strongly supported as sister to this pair. *Joinvillea* (Joinvilleaceae) and *Ecdeiocolea* (Ecdeiocoleaceae) share a 6-kilobase plastid DNA inversion with the grasses and not with other closely allied families[63,67]. Furthermore, they both lack the *trnK* inversion that occurs in the grasses[67]. Michelangeli et al.[67] resolved Ecdeiocoleaceae (but with low bootstrap support) as sister to the grasses in a combined analysis of multiple datasets including plastid and mitochondrial DNA sequences, plastid DNA inversions and morphological data. The molecular data alone produced an unresolved trichotomy of Ecdeiocoleaceae, Joinvilleaceae and Poaceae.

The first phylogenetic studies of the grass family, based on DNA sequences, were by Hamby and Zimmer[68] using nuclear ribosomal DNA and Doebley et al.[69] using the plastid *rbcL* gene. Both resolved a group known as the PACC clade containing Panicoideae, Arundinoideae, Centothecoideae and Chloridoideae. Davis and Soreng[70] subsequently used plastid DNA restriction site variation on a sample of taxa that included all subfamilies of Clayton and Renvoize[1] to generate phylogenetic trees and also found support for the PACC clade. Many single gene analyses from all genomes, but mainly plastid and nuclear, have subsequently been produced. These include studies using nuclear DNA (*gbssI*[71]; ITS[72]; *PHYB*[73,74]), plastid DNA (*ndhF*[18]; *matK*[65,75]; *rbcL*[76,77]; *rpl16*[78]; *rpoC2*[79]; *rps4*[80]; *trnL*[33,66]) and mitochondrial DNA (*atpA*[67]).

Of these, the study by Clark et al.[18] using the plastid gene *ndhF* was the first to include a thorough sampling of grass diversity and many previously poorly sampled Bambusoideae and Ehrhartoideae taxa. Their results supported two major lineages within the grasses, the BEP clade (Bambusoideae, Ehrhartoideae and an expanded Pooideae) and the PACC clade. *Anomochloa* and *Streptochaeta* (Anomochlooideae) were sister to the rest of the grasses (the earliest diverging lineage relative to the rest of the grasses, hereafter 'earliest diverging lineage'). These are distributed in the neotropics and are broad leaved forest genera. The next earliest diverging lineage was *Pharus* (Pharoideae).

In contrast to single gene analyses there have been relatively few reports of combined analyses or multi-gene studies of the entire family. The most significant combined data analysis included 62 grasses sampling approximately 8% of the genera[4]. Data sets of DNA sequences (nuclear *PHYB*, ITS2 and *gbssI*; plastid *ndhF, rbcL, rpoC2*), plastid restriction site data and morphological data were analysed alone and in combination. Relatively well resolved and well supported trees were obtained and allowed for major reclassification of the family. Other studies include the combined *ndhF, rbcL* and *PHYB* dataset of Clark et al.[8] and the morphological, chromosomal, biochemical and plastid DNA character set of Soreng and Davis[64,82].

## 17.3.2 Supersizing: Large Phylogenetic Trees of Poaceae

Recently, supertree approaches have allowed combination of phylogenetic trees produced from different genes and other data sources[83–85]. These have allowed much larger trees to be produced. A near complete tribal-level tree of the grasses, modified from Hodkinson et al.[86] is shown in Figure 17.3. This is an enlargement, in terms of the number of genera, of the supertree of Salamin et al.[9] It contains 426 genera (approximately two thirds of all grass genera) and 39 out of the 42 tribes recognised by the GPWG[4]. It is representative of most published phylogenetic studies of the grass family, including GPWG[4]. In general, only poorly supported nodes from previous phylogenetic studies are incongruent with nodes found in the supertrees (soft incongruence). In accordance with the GPWG[4], Poaceae are monophyletic and Joinvilleaceae are their sister group (the supertree lacks Ecdeiocoleaceae). The 12 subfamilies, as defined by the GPWG[4] (Figure 17.1), were all resolved in the supertree (Figure 17.3). A group including Anomochlooideae was sister to the rest of the grasses. Pharoideae and Puelioideae[81], in that order and consistent with GPWG[4] and Clark et al.[18], were the next to diverge. Within the group making up the rest of the grasses, the PACCAD clade (Panicoideae, Arundinoideae, Centothecoideae, Chloridoideae, Arundinoideae and Danthonioideae) of the GPWG was resolved. The relationships of subfamilies within the PACCAD clade are generally not well supported by molecular evidence and require further study. Relationships of subfamilies and tribes within the PACCAD clade are discussed in GPWG[4] and Hodkinson et al.[86].

The GPWG[4] tentatively recognises the BEP clade sister to PACCAD, but a clear pattern has yet to emerge. The supertree (Figure 17.3) shows Pooideae grouping, in a polytomy, with the PACCAD clade. We suggest that it is more prudent to recognise a PACCAD, B-E, P clade instead of a PACCAD-BEP clade because single gene analyses conflict in the placement of Bambusoideae, Ehrhartoideae and Pooideae relative to each other and the PACCAD clade. Studies on *ndhF*[4,18] and *PHYB*[74] resolve the BEP clade with moderate to high support, respectively. However, studies using chloroplast restriction sites[70], ITS sequences[72], *rbcL* sequences[76,77] and morphology[4] place the pooids as sister to the PACCAD clade. We have demonstrated random and systematic error in some of these single gene analyses[87]. Our simulations showed that in most genes, this error could be removed by the addition of more characters, indicating that only random error is involved. However, in some genes inconsistency was detected; that is, increasing the number of characters compounded the error. Encouragingly, our simulation studies have also shown that if taxon sampling is improved by judiciously breaking long branches, then even this error can be reduced (see below for more theoretical issues regarding the reconstruction of large phylogenies the size of the grass family).

## 17.3.3 Genome-Wide Phylogenomic Studies

Instead of sequencing large numbers of taxa to infer phylogenetic relationships, an alternative approach is to compare a high amount of genome information from a restricted number of individuals (from the relevant taxa under investigation). In the grasses we are fortunate that a large

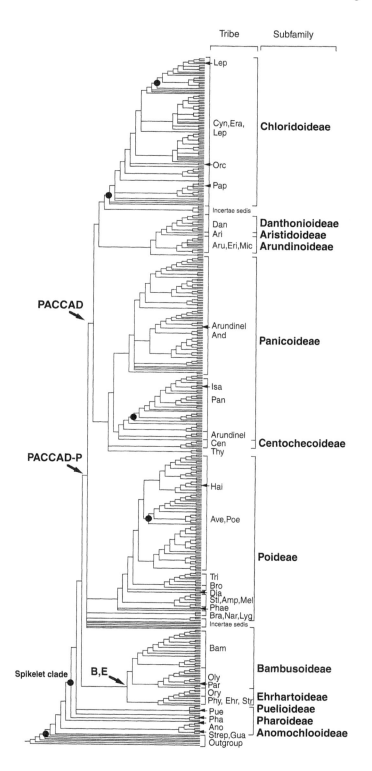

FIGURE 17.3

amount of genomic information exists for a number of model species belonging to the largest subfamilies. For example, the fully sequenced plastid genomes of maize (*Z. mays* ssp. *mays;* Panicoideae), rice (*O. sativa;* Ehrhartoideae) and wheat (*T. aestivum;* Pooideae) are available[88,89] and have been compared by Matsuoka et al.[90] with tobacco (*Nicotiana*) as an outgroup. The three cereals contained 106 genes, and eight of these were invariable between species. Analyses using neighbour joining (NJ) (parameters not given) and 84 to 98 genes, depending on the inclusion or exclusion of genes with significant rate heterogeneity, always resolved wheat as sister to rice (consistent with a BEP clade hypothesis). However, bootstrap values varied in the range of 52–87%, and the branch connecting maize to the wheat-rice grouping was very short in comparison to the terminal branches. Ogihara et al.[91] have examined structural alterations (caused by replication slippage and intra-molecular recombination facilitated by microsatellite regions) in the plastid genomes of the wheat, maize and rice. They found that the structure of the grass plastid chromosome was highly similar, but that some hot spots for structural mutation were present. By examining the deletion patterns of open reading frames in the inverted repeat (IR) regions and the junctions between the IR and the small single copy region (SSC) they concluded that wheat is much more similar to rice than to maize.

It seems, therefore, that results of whole genome analysis, including structural rearrangements[91] and sequence nucleotide variation[90], of three cereal species would be more consistent with the BEP hypothesis than the PACCAD-P hypothesis. However, three taxon comparisons like those of Matsuoka et al.[90] and Ogihara et al.[91] tell us little about broad phylogenetic relationships, and this may not be the final word in the BEP, PACCAD-P (or alternatives) debate. More genomes will need to be added to represent all major clades. To test the monophyly of the BEP clade, it is necessary to include a Bambusoideae s.s. taxon (for example, a woody or herbaceous bamboo). It would also be fruitful to establish what patterns are found in the early diverging lineages of grasses to help determine when the alterations described above took place. Further advances in phylogenomic methods are expected in the near future[92], and the grasses are well positioned as a model group for studies in this direction.

## 17.3.4 ORIGINS AND PREHISTORICAL BIOGEOGRAPHY

Fossil evidence indicates that the angiosperms originated approximately 125 million years before present (mybp). The divergence of Nymphaeales and Archaefructaceae is likely to have occurred first[93,94], and the divergence of the monocots occurred shortly after that[50,95]. The earliest Poales fossils date back to the Cretaceous. For example, several types of Poales pollen matching Poaceae or Restionaceae morphology have been discovered from the late Cretaceous in Maastrichtian deposits of Africa and South America (>65 mybp[96]). The material could not, however, differentiate these two families, as it was insufficiently preserved to identify grass diagnostic channels in their outer walls.

Until recently, unequivocal recordings of grass fossils date to 55 mybp (Paleocene/Eocene boundary[97]). The macrofossils discovered included entire plants with inflorescences, spikelets,

**FIGURE 17.3** Supertree of grass genera indicating positions of significant shifts in diversification rate. A semistrict consensus of 1,000 trees is shown. Black shaded circles represent nodes where significant or marginally significant shifts in diversification rate have occurred (*P*-values < 0.06). Subfamilies and tribes follow GPWG[1] and the tribes are labelled as follows: Amp, Ampelodesmeae; And, Andropogoneae; Ano, Anomochloeae; Ari, Aristideae; Arundinel, Arundinelleae; Aru, Arundineae; Ave, Aveneae; Bam, Bambuseae; Bra, Brachyelytreae; Bro, Bromeae; Cen, Centotheceae; Cyn, Cynodonteae; Dan, Danthonieae; Dia, Diarrheneae; Era, Eragrostideae; Ehr, Ehrharteae; Eri, Eriachneae; Gua, Guaduelleae; Hai, Hainardeae; Isa, Isachneae, Lep; Leptureae; Lyg, Lygeeae; Mel, Meliceae; Mic, Micraireae; Nar, Nardeae; Oly, Olyreae; Orc, Orcuttieae; Ory, Oryzeae; Pan, Paniceae; Pap, Pappophoreae; Par, Parianeae; Phae, Phaenospermatideae; Pha, Phareae; Phy, Phyllorachideae; Poe, Poeae; Pue, Puelieae; Sti, Stipeae; Str, Streptogynaeae; Strept, Streptochaeteae; Thy, Thysanolaeneae; Tri, Triticeae.

anthers and pollen. However, further identification is problematic with this material, and even confident assignment to subfamily level has not been possible. They have raceme-like panicles and two florets per spikelet. Morphological characters could place them in Aveneae (Pooideae) or Arundineae (Arundinoideae). Recently, Prasad et al.[98] reported that they had detected phytoliths from at least five taxa from extant grass subclades on the Indian continent dating from c. 70–65 mybp (Late Cretaceous). They postulate a Gondwanan origin for the grasses (before 125 mybp). The earliest clearly identifiable macrofossil is of *Pharus* preserved in amber and trapped in mammalian hair. It dates to 45–30 mybp (Late Eocene/Early Oligocene[99]).

It is possible to conclude therefore that, on the basis of fossil evidence, grasses can be dated back to 55 mybp with some confidence and to 70 mybp if grass-like pollen and phytoliths are considered. In contrast to these dates, paleobotanical (grass macrofossils and grass pollen, known generically as *Monoporites annulatus*) and indirect paleofaunal evidence (from herbivore morphology) indicate that widespread grass-dominated ecosytems did not evolve until the early to middle Miocene (20–15 mybp[100]), by which time all major lineages of grass were likely to have evolved (that is, all currently recognised subfamilies were present). It seems therefore, that the major divergences detected within Poaceae PACCAD, B, E, P (but excluding Pharoideae, Anomochlooideae and Puelioideae, which were probably already present) occurred considerably earlier than the 25–15 mybp suggested by the GPWG[4] but still a long time following the origin of the family (a conservative 70–50 mybp). All grasses known to have $C_4$ metabolism belong to the PACCAD clade. The first fossil $C_4$ grass has been dated at 12.5 mybp, and isotopic evidence for $C_4$ dates from approximately 15 mybp (see Kellogg[101] for a discussion of the parallel evolution of this important trait within the grasses). This suggests that the PACCAD clade is at least 15 million years old; however, it is likely to be considerably older, and molecular clock estimations of dates support this. For example, Bremer[57] indicates that the major diversifications including the PACC clade date to about 55–35 mybp, and the divergence of maize and *Pennisetum* has been dated at 25 mybp[102].

Molecular phylogenetic dating is complicated by a range of issues[103–105] including the paucity of the fossil record, the uncertainty in calibrating reference nodes with fossils, the uncertainty of the proposed phylogeny and deviations of its branches from the molecular clock. Numerous molecular phylogenetic dating methods exist, and recent advances have removed the assumption of a molecular clock[103,106–109]. However, relatively few attempts have been made to date major clades in Poales or Poaceae using molecular phylogenetic trees. Molecular phylogenetic dating of the monocots has estimated the origin of Poales to the mid-Cretaceous (115 ± 11 mybp[110]). Bremer[110] using the *rbcL* gene and NPRS dates the origin of Poaceae at between 80 and 70 mybp, depending on whether the root or crown node is considered for calibration. Gaut[111] used a phylogenetic tree based on *rbcL* and *ndhF* gene sequences to date major radiations in the grasses and calibrated the tree on a single point that assumed rice and maize diverged 50 mybp. This showed the grasses to originate at 77 mybp and Ehrhartoideae (rice) and Pooideae (oats, barley and wheat) to diverge at 46 mybp. The divergence of Panicoideae from Ehrhartooideae and Pooideae occurred at 50 mybp. The studies[110,111] were based on small sample numbers (both included only nine grass species). It is clear that there is a need for large, accurately dated trees of the grass family for macroevolutionary studies.

Inferences can sometimes be made directly about the prehistorical biogeography of species by examining the geographical distribution of fossils and the present day distribution of species. However, the GPWG[4] believes that few conclusions regarding the biogeographical origin of grasses can be made using such an approach. This is because the earliest diverging lineage of the grasses (Anomochlooideae) is in Central and South America, the next earliest diverging lineage (Pharoideae) is pantropical, and the next (Puelioideae) is restricted to tropical Africa. The paucity of the fossil record cannot help interpret this complex multicontinental distribution pattern of early diverging lineages. The sister groups to the grasses also vary considerably in biogeography. For example, Joinvilleaceae is in Borneo, New Caledonia and throughout the Pacific, but Ecdeiocoleaceae is restricted to Australia.

However, if the time scale is pushed back with recent fossil discoveries[98,112] and the interpretation of the Australian tectonic plate first proposed by Audley-Charles[113] is considered, then perhaps these distributions can provide us with information. The sister group is concentrated on or near the Australian plate, the earliest diverging grass lineage (Anomochlooideae) is in Central and South America, the next earliest diverging lineage (Pharoideae) is pantropical, the next (Puelioideae) is restricted to tropical Africa, and the earliest phytolith fossils from central India at c. 70 mybp were diverse in their subfamily composition. This distribution certainly seems to suggest a Gondwanan origin for the whole clade.

An alternative method of prehistorical biogeography is to use phylogenetic trees that encompass adequate taxon sampling (including representatives of clades with broadly differing geographical origins). In these studies attempts are made to optimise the geography of taxa on trees (that is, on nodes throughout the tree). Various types of parsimony reconstruction and related methods such as vicariance dispersal analysis[114] are commonly used for this purpose. Bremer[57] used vicariance dispersal analysis (using DIVA[115]) on phylogenetic trees of Poales to explore prehistorical biogeographic patterns. His reconstructions suggest that Poaceae originated in South America, but the graminoid clade and Poales originated in Australia. Ecdeiocoleaceae were sister to the grasses in his reconstructions and these two families shared a common ancestor about 76 mybp. The connection between Australia and South America broke up about 35 mybp, so land migration was still possible between these continents. Such a hypothesis would also help explain why reconstructions show a South American origin for Poaceae, but an Australian origin for Ecdeiocoleaceae and the graminoid clade. It should be noted, however, that some of his reconstructions yielded alternative optimisations for the origin of Poaceae, but the origin of the graminoid clade was always Australia. Despite this, the origin of the grasses looks most likely to be Gondwanan and probably South American; a hypothesis that needs further exploration using large phylogenetic trees and other methods of biogeographic reconstruction.

### 17.3.5 PATTERNS AND PROCESSES OF DIVERSIFICATION

The tree presented in Figure 17.3 does not appear to be fully symmetrical (sister clades are not of equal size), but it is not possible to judge by eye whether this imbalance is due to variation in diversification rate or whether it requires deterministic explanations. The significance of imbalance in phylogenetic trees can be assessed in many ways, including topological or temporal methods[8,116,117] or a combination of the two. Temporal methods require branch lengths and topology; topological methods require only the topological branching pattern[116]. If a tree is asymmetrical, we may like to identify where on the trees the imbalance lies, that is, detect in which clades significant shifts in diversification have occurred. We may then like to ask what correlates with these shifts to help identify factors and processes that may have led to these diversification patterns. Hodkinson et al.[86] used the topological $M$ statistics of Moore et al.[117] and Chan and Moore[118] to assess diversification rate variation in grasses. They showed that the supertree shown in Figure 17.3 was imbalanced ($M$ statistic $P$-values $< 0.002$). This was significant even if the supertree was corrected to have genera numbers in proportion to tribe size (to reduce sampling bias on the statistics). They also showed that significant shifts (or marginally significant shifts) in diversification rate variation had occurred a number of times (Figure 17.3; bold circles). The nodes where shifts in diversification rate variation have occurred are spread relatively evenly across the tree. The deepest shift in the grasses corresponded to the group recognised as the spikelet clade. This is a shift above the earliest diverging lineages of the grasses (Anomochlooideae) including Pharoideae, Puelioideae and all other grasses. Higher diversification rates are therefore associated with genera that have typical grass spikelets and structures homologous to glumes, lemmas, paleas and lodicules. The statistic also suggests that a shift in diversification occurred some time after the origin of the family, but certainly by approximately 40–35 mybp, because the earliest macrofossil of this group is of *Pharus* dating to that time[99]. Other later shifts in diversification rate corresponded to species rich subfamilies or parts of subfamilies, including the chloridoids, a clade within the chloridoids, a clade in the panicoids (Paniceae) and a clade in the pooids (an Aveneae-Poeae subgroup). Shifts in diversification rate have therefore occurred some time after the origin of these tribes.

Factors influencing diversification rate variation include prolific cladogenesis, adaptive radiations, mass extinctions, key innovations, global change and coevolution. The relative importance of each of these factors as causal agents to the diversification patterns we describe above is unknown. Further study is required, diversification rate studies will need to incorporate branch length information and dates will need to be found for the diversification rate shifts. This will also facilitate the examination of correlations between diversification rate variation and past environmental and global change factors such as temperature, aridity and $CO_2$ levels. The factors correlating to these shifts are currently being investigated (Bouchenak-Khelladi et al., in prep.). The study by Hodkinson et al.[86] used genera and not species because of practical data/tree availability limitation. However, we envisage future studies trying to investigate species-level variation in this and other ways. For example, we would like to know why relatively few grass genera, such as *Agrostis, Bambusa, Digitaria, Eragrostis, Festuca, Poa, Panicum, Paspalum* and *Stipa,* account for a considerable percentage of all Poaceae species (possibly over 30%[12]; see also Hilu, *Chapter 11*).

Identifying certain traits, key innovations or other factors that might influence the rate of evolution and production of new species is a challenge to evolutionary biologists[119,120]. The observed differences in species richness between certain clades can be correlated with the presence of particular factors. However, to identify correlates of species richness, the hierarchical nature of evolutionary history has to be taken into account in order to avoid erroneous inferences[13]. Comprehensive phylogenetic trees that contain, when possible, estimates of divergence dates are therefore required to not only assess the patterns of diversification but also to assess the processes leading to that diversification[121]. Kellogg[101] identified $C_4$ metabolism as a possible key innovation influencing species richness in the grasses by simply comparing lineages with known species numbers. However, a more powerful approach to assess the potential causes (correlations) of diversification and species richness is to use sister clade comparison tests[7,122]. In these, comparisons are made between sister taxa, one possessing a trait (or factor) and the other not. For example, Salamin and Davies[123] mapped traits from Watson and Dallwitz[17] onto a grass supertree and identified all sister clades with contrasting traits (for example, bisexual versus monoecious breeding system). They then made a comparison between the number of species in each sister clade against the null hypothesis of equal speciation rates (following the methods of Slowinski and Guyer[122] and Goudet[124]). Their results indicated that herbaceous habit and an annual life cycle have a significant correlation with species richness. The results were also consistent with the hypothesis that annuals might be better able to fit new niches and become more species rich[125] and that generation time is a factor influencing species richness in the grasses. Recombination and genetic change will be facilitated by the short generation time of annuals. Woodiness is also linked to generation time in grasses, with many of the woody bamboos having long generation times in comparison to other grasses[1]. A link between speciation rates and nucleotide substitution rates has been established in other taxonomic groups[126–128], but evidence is inconclusive in the grasses[129]. However, the analysis by Gaut et al.[129] was limited to a small proportion of grass diversity, and extending the sampling could change its outcome. The study of Salamin and Davies[123] failed to show any link between a number of other characters and speciation rate including the ability to resist drought, ability to tolerate saline environments, open versus forest habitat and bisexual versus monoecious breeding system. Other factors require investigation.

## 17.4 FUTURE PERSPECTIVES

### 17.4.1 THEORETICAL CONSIDERATIONS FOR THE PRODUCTION OF LARGE PHYLOGENETIC TREES OF THE GRASSES

The task of producing a complete c. 10,000 species phylogenetic tree for the grasses is immense and presents numerous problems. First, all species need to be collected, correctly identified and vouchered. Second, sufficient data need to be collected for phylogenetic reconstruction, although this task has become relatively straightforward with modern automated sequencing. Finally, these data have to be analysed and visualised in an appropriate way. The first task is the most challenging and will require the longest time and major international collaboration. Good phylogenetic practice therefore depends on good taxonomy and vice versa; the two are inextricably linked. There is a paucity of suitably trained grass taxonomists, and they have limited time and resources. However, assuming that the plants can be collected, and the data generated, we can ask whether there are other theoretical impediments to the process. One major concern is whether methods of phylogenetic reconstruction can accommodate large datasets.

Increasing the number of taxa sampled[130–134] and the number of characters[135–137] in a dataset generally results in more reliable phylogenetic trees. This is partly because such datasets tend to reduce random error and sampling bias (both properties of finite datasets) and sometimes inconsistency (an asymptotic property[138]). Furthermore, Källersjö et al.[133] and Savolainen and Chase[50] describe examples where homoplasy, present in large datasets, improves local phylogenetic signal. In a parsimony framework, multiple successive substitutions along a branch cannot be observed, and saturated characters can rapidly confound the tree search. Although large datasets contain potentially more homoplastic changes, the large number of branches will spread these multiple changes throughout the tree, making them locally informative, and therefore increasing their inferential qualities[50,84]. Model based methods are less prone to such problems even with smaller trees, because a sufficient model will take into account the hidden multiple changes of each character.

Most phylogenetic reconstructions of the grasses have included a relatively small number of species. For example, the combined analyses of the GPWG[4] contained 61 genera and 62 species (respectively, c. 8% and c. 0.6% of the total). If we accept that large phylogenetic trees are desirable, it is also important to ask whether existing methods of phylogenetic reconstruction can accommodate matrix sizes that approximate to the size of the grass family and include multiple gene regions.

Many multi-gene analyses[95,139], including the GPWG[4] analyses, have incorporated missing data because complete large datasets are often not practically producible. The influence of missing data on phylogenetic inference has been discussed[84,140]. On theoretical grounds it would be expected that increasing missing data could reduce accuracy of phylogenetic inference by increasing the number of optimal solutions found and creating uncertainty in the placement of some taxa relative to others[84]. It is therefore not clear how much missing data can be accommodated in phylogenetic reconstruction, and the most conservative option is to produce complete maximum combined matrices[84,141].

Perhaps the largest phylogenetic reconstructions have been made within the higher plants. The first 'large tree' was by Chase et al.[142] who included 499 seed plants in a parsimony analysis of *rbcL* sequences. The number of taxa included in analyses of plant groups has steadily grown, and one study in particular included 2,538 *rbcL* sequences in a parsimony jackknifing analysis[143]. Other large multi-gene analyses have been made[95,144]. The results of these empirical studies are encouraging and could satisfy us that matrix sizes up to approximately 2,500 taxa can be analysed using existing methods such as maximum parsimony (MP) and NJ. Advances in phylogenetic reconstruction methods should also improve the situation. Methods that can search the tree space more efficiently have been proposed to overcome the computationally intensive task of finding an optimal tree. For example, parsimony jackknifing[145] estimates well supported groups that will guide the search through the tree space. Bayesian methods[146,147], which are powerful and versatile approaches, use a random walk through the tree space guided by the posterior probability of each tree to infer credibility intervals on the topology, branch lengths and model parameters used. In comparison, the more traditional way of estimating support by calculating bootstrap percentages is problematic with matrices containing thousands of taxa. Even reducing the complexity of the search during the bootstrap, which was shown to give comparable results than more extensive tree searches[148], will require enormous computer power. Parallelisation of the process on larger computer clusters will reduce the computation time, but the process will remain tedious.

We can conclude, therefore, that it should be suitable to use existing methods of phylogenetic reconstruction to analyse reliably a large multi-gene matrix of Poaceae, and that these analyses should provide better estimates of phylogenetic pattern than analyses of smaller matrices. For studies of tens of thousands of taxa we move into uncharted territory. We have to rely on results of simulation studies to test the utility of methods such as parsimony to accurately reconstruct phylogeny from such data sets. Salamin et al.[137] used Monte Carlo simulations to assess the accuracy of MP and NJ methods to retrieve model trees using 13,000 taxa (the number of angiosperm genera and close to the number of grass species). The results were encouraging because even with relatively inefficient heuristic search options a high percentage of nodes on the model tree were correctly inferred (80% with parsimony). NJ was more problematic because computing the distance matrix for such large matrices is more computationally demanding than a MP search with more than 5,000 characters. The number of characters that will be needed to construct complete and robust generic- or species-level trees of the grasses is unknown. However, the simulation studies of Salamin et al.[137] with a 13,000-taxon tree found a sharp improvement in phylogenetic accuracy when character data sets were increased in size between 5,000 and 10,000 informative characters. Increasing the number of characters beyond this had a slower impact on phylogenetic reconstruction accuracy.

### 17.4.2 PRACTICAL CONSIDERATIONS: DATA AVAILABILITY FOR SUPERMATRIX AND SUPERTREE RECONSTRUCTIONS

Disregarding the theoretical considerations listed above, it is worth noting how far we still are from a complete generic-level tree of the grasses from a practical perspective. Sequence data on the grasses is accumulating at a rapid rate. A preliminary examination, in December 2004, of sequence availability in DDBJ/EMBL/GenBank looked promising, as there were 68,153 sequences deposited for the grasses. However, if we examine the subfamily distribution of these sequences (data not shown) we find that most (over 98%) have been produced for Ehrhartoideae (68.6%), Pooideae (15.6%) and Panicoideae (13.9%). These contain the most important cereal crops that have been subject to intense genomic study. The sequencing of the c. 400 mbp of the *Oryza* genome (Ehrhartoideae) has been completed[88,89,149]. Furthermore, these figures represent the number of entries in DDBJ/EMBL/GenBank (including replicated information and pseudogenes) and will therefore be an overestimation of what could be used in a phylogenetic analysis.

For maximum sized multi-gene analyses it is desirable to combine data from some combination of the most frequently sequenced gene regions (with good taxonomic sampling across the grass family). The prime candidate regions for combination are therefore from the top 10 most sequenced gene regions. The number of grass genera and species sequenced for each of these regions is given in Figure 17.4. The maximum number of species and genera sequenced for any particular DNA region is 577 and 162, respectively, for a nuclear ribosomal DNA region (ITS). The plastid gene *ndhF* is the next best represented gene region (354 species, 162 genera). Assuming even complete overlap of taxa, the maximum sized data set for two regions would be limited to 162 genera and 354 species. The reality is far worse than this ideal scenario, as there is often poor overlap of taxa between genes. Combined datasets will therefore have to accommodate missing data, and targeted sequencing needs to be conducted to maximise future dataset size.

### 17.4.3 DIVIDE-AND-CONQUER

An alternative approach to the direct analyses of large multi-gene region datasets for phylogenetic reconstruction is to use some sort of 'divide-and-conquer' strategy to build trees from individual data matrices and later assemble them on the basis of taxonomic overlap with other such trees (Wilkinson and Cotton, *Chapter 5*[9,83,84,150]). From a theoretical stance, these meta-analysis techniques may be less favourable than direct large multi-region analyses for phylogenetic reconstruction, but they may offer an adequate solution to the problem of constructing large trees when we

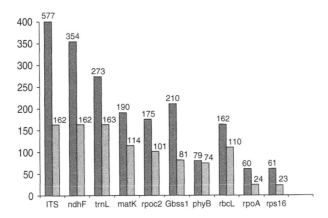

**FIGURE 17.4** Data availability for single and multi-gene analyses. Number of grass species and genera sequenced for each of the top 10 most frequent gene regions in DDBJ/EMBL/GenBank (December 2004). Light grey bars represent number of genera and dark grey bars represent number of species.

consider the data availability problems of the grasses[9,86]. However, the analysis of truly large multi-region analyses (supermatrices) may also ultimately depend on some sort of divide-and-conquer strategy (the decomposition of the tree into subproblems and the recombining of these into an overall solution; Wilkinson and Cotton, *Chapter 5*). The reliability of supertree methods has been questioned[151] and debated at length[83,152]. However, many of these reliability issues, such as poor quality data, data duplication and data accountability, have been addressed (Wilkinson and Cotton, *Chapter 5*[85,152]). Several empirical studies are also adding support to the validity of the supertree approach[9,153]. For example, Salamin et al.[9] used the results of single gene region analyses from the GPWG[4] to generate supertrees that were congruent with trees generated by combining these data into multi-region analyses. They also constructed a supertree of the grass family (containing 395 genera) that was broadly congruent with previous studies.

Disregarding practical considerations, some sort of divide-and-conquer strategy will be required to piece together large phylogenetic trees of the grasses, at least in the short term. Appropriate data are simply not available for large multi-gene phylogenetic trees. If data become available, supertree phylogenetic divide-and-conquer methods are still likely to be required to handle the scale of the computational problem presented by the supermatrices. The source trees in the supertrees of Salamin et al.[9] and Hodkinson et al.[86] do contain some duplicated data (the source data are not totally independent). We are currently generating new supertrees that remove this problem by reconstructing optimal single gene region trees from all available sequence data for the top ten genes shown in Figure 17.4 (primary data analyses of single regions) and then combining these trees using a range of meta-analysis methods into supertrees (Kinney et al., in prep.) that also incorporate support statistics such as bootstrapping. This approach removes the problem of duplication and non-independence of data that can occur when taking topologies directly from published literature. We are also comparing how these methods compare to supermatrix methods using the same data (Kinney et al., in prep.).

## 17.5 CONCLUSIONS

The grasses represent a small but highly significant piece of the tree of life[154–156] and lessons learnt by grass systematists and taxonomists should help guide those researchers studying less well known groups of organisms. The future of many aspects of grass systematics will depend critically on how well classical taxonomy can be meshed with phylogenetics and informatics. The collection of grass species, correct identification of specimens, maintenance of high-quality collections and associated literature are likely to be the major limiting factors in the production of large phylogenetic trees (or the use of these sequences for DNA barcoding or taxonomy). Technical advances in molecular biology and digital archiving of sequence information have already occurred. There is a need to reach an international consensus over taxonomic nomenclature and accepted names, but this will need major coordinated action at an international level. We can also conclude that, despite the great accumulation of sequence data for the grasses and the advances that have been made in phylogenetic theory and phylogenetic reconstruction in the grasses, we are still a long way from complete species-level or even generic-level trees of the family. Targeted sequencing should focus on filling in gaps in existing data and improving taxonomic sampling across the family. These trees and associated supertrees will have, amongst other things, great utility for macroevolutionary studies of grasses.

## ACKNOWLEDGEMENTS

This work was supported by the Irish Higher Education Authority, Enterprise Ireland (SC2003/437) and the Swiss National Science Foundation (81AN-068367).

## REFERENCES

1. Clayton, W.D. and Renvoize, S.A., *Genera Graminum: Grass Genera of the World,* Her Majesty's Stationery Office, London. 1986.
2. Mabberley, D.J., *The Plant Book,* 2nd ed., Cambridge University Press, England, 1997.
3. Archibold, O.I.V., *Ecology of World Vegetation,* Chapman and Hall, London, 1995, chap. 1.
4. GPWG. Phylogeny and subfamilial classification of the grasses (Poaceae), *Ann. Missouri Bot. Gard.,* 88, 373, 2001.
5. Hodkinson, T.R. et al., Phylogenetics of *Miscanthus, Saccharum* and related genera (Saccharinae, Andropogoneae, Poaceae) based on DNA sequences from ITS nuclear ribosomal DNA and plastid *trnL* intron and *trnL-F* intergenic spacers, *J. Plant Res.,* 115, 381, 2002.
6. Hodkinson, T.R et al., The use of DNA sequencing (ITS and *trnL-F*), AFLP and fluorescent *in situ* hybridization to study allopolyploid *Miscanthus* (Poaceae), *Amer. J. Bot.,* 89, 279, 2002.
7. Purvis, A., Using interspecies phylogenies to test macroevolutionary hypotheses, in *New Uses for New Phylogenies,* Harvey, P.H. et al., Eds., Oxford University Press, Oxford, 1996, 153.
8. Barraclough, T.G. and Nee, S., Phylogenetics and speciation, *Trends Ecol. Evol.,* 16, 391, 2001.
9. Salamin, N., Hodkinson T.R., and Savolainen, V., Building supertrees: an empirical assessment using the grass family (Poaceae), *Syst. Biol.,* 51, 136, 2002.
10. Graybeal, A., Is it better to add taxa or characters to a difficult phylogenetic problem? *Syst. Biol.,* 47, 9, 1998.
11. Rannala, B. et al., Taxon sampling and the accuracy of large phylogenies, *Syst. Biol.,* 47, 702, 1998.
12. Hillis, D.M. et al., Is sparse taxon sampling a problem for phylogenetic inference? *Syst. Biol.,* 52, 124, 2003.
13. Felsenstein, J., Phylogenies and the comparative method, *Am. Nat.,* 125, 1, 1985.
14. Harvey, P.H. and Pagel, M.D., *The Comparative Method in Evolutionary Biology,* Oxford University Press, London, 1991, chap. 1.
15. Pagel, M., Inferring the historical patterns of biological evolution, *Nature,* 401, 877, 1999.
16. Renvoize, S. and Clayton, W.D., Classification and evolution of the grasses, in *Grass Evolution and Domestication,* Chapman, G.P., Ed., Cambridge University Press, Cambridge, 1992, 3.
17. Watson, L. and Dallwitz, M.J., *The Grass Genera of the World,* CAB International, Wallingford, 1992.
18. Clark, L.G., Zhang W., and Wendel, J.F., A phylogeny of the grass family (Poaceae) based on *ndhF* sequence data, *Syst. Bot.,* 20, 436, 1995.
19. Godfray, H.C.J., Challenges for taxonomy, *Nature,* 417, 17, 2002.
20. Wheeler, Q.D., Transforming taxonomy, *The Systematist,* 22, 3, 2003.
21. Powell, D.J., *Publishing Output to 2020,* The British Library, 2004.

22. Wheeler, Q.D., Lipscomb, D., and Platnick, N., Terascale taxonomy: cyber-infrastructure and the Linnaean legacy, in *Proceedings of the Fourth Biennial Conference of the Systematics Association,* Trinity College Dublin, Ireland, 14, 2003.

23. Bisby, F.A., et al., Taxonomy, at the click of a mouse, *Nature,* 418, 367, 2002.

24. Knapp, S. et al., Taxonomy needs evolution, not revolution, *Nature,* 419, 559, 2002.

25. Blackmore, S., Biodiversity update: progress in taxonomy, *Science* 298, 365, 2002.

26. Sharp, D. and Simon, B.K., *AusGrass: Grasses of Australia,* ABRS Identification Series, interactive CD ROM, ABRS and EPA, Queensland, 2002.

27. Clayton, W.D., The logarithmic distribution of angiosperm families, *Kew Bull.,* 29, 271, 1974.

28. Clayton, W.D., Chorology of the genera of Gramineae, *Kew Bull.,* 30, 111, 1975.

29. Clayton, W.D., The genus concept in practice, *Kew Bull.,* 38, 149, 1983.

30. Frodin, D.G., History and concepts of big plant genera, *Taxon,* 53, 753, 2004.

31. Barker, N.P. et al., The paraphyly of *Cortaderia* (Danthonioideae; Poaceae): evidence from morphology, chloroplast and nuclear DNA sequence data, *Ann. Missouri Bot. Gard.,* 90, 1, 2003.

32. Ingram, A.L. and Doyle, J.J., Is *Eragrostis* (Poaceae) monophyletic? Insights from nuclear and plastid sequence data, *Syst. Bot.,* 29, 545, 2004.

33. Gómez-Martínez, R. and Culham, A., Phylogeny of the subfamily Panicoideae with emphasis on the tribe Paniceae: evidence from the chloroplast *trnL-F* cpDNA region, in *Grasses: Systematics and Evolution,* Jabobs, S.W.L. and Everett, J.E., Eds., CSIRO, Collingwood, Victoria, 2000, 136.

34. Duvall, M.R., Noll, J.D., and Minn, A.H., Phylogenetics of Paniceae (Poaceae), *Amer. J. Bot.,* 88, 1988, 2001.

35. Giussani, L.M. et al., A molecular phylogeny of the grass subfamily Panicoideae (Poaceae) shows multiple origins of $C_4$ photosynthesis, *Amer. J. Bot.,* 88, 1993, 2001.

36. Doust, A.N. and Kellogg, E.A., Inflorescence diversification in the panicoid 'bristle grass' clade (Paniceae, Poaceae): evidence from molecular phylogenies and developmental morphology, *Amer. J. Bot.,* 89, 1203, 2002.

37. Kellogg, E.A. et al., Taxonomy, phylogeny and inflorescence development of the genus *Ixophorus* (Panicoideae: Poaceae), *Intern. J. Pl. Sci.,* 165, 1089, 2004.

38. Spangler, R.E. et al., Andropogoneae evolution and generic limits in *Sorghum* (Poaceae) using *ndhF* sequences, *Syst. Bot.,* 24, 267, 1999.

39. Spangler, R.E., Taxonomy, *Sarga, Sorghum* and *Vacoparis* (Poaceae: Andropogoneae), *Aust. Syst. Bot.,* 16, 279, 2003.

40. Dillon, S.L., Lawrence, P.K., and Henry, R.J., The use of ribosomal ITS to determine phylogenetic relationships within *Sorghum, Pl. Syst. Evol.,* 230, 97, 2001.

41. Yu, P., Comparative Reproductive Biology of Two Vulnerable and Two Common Grasses in *Bothriochloa* Kuntze and *Dicanthium* Willem, Ph.D. thesis, University of New England, Armidale, Australia, 1999.

42. Yu, P., Prakash, N. and Whalley, R.D.B., Sexual and apomictic seed development in the vulnerable grass *Bothriochloa biloba* S.T. Blake, *Aust. J. Bot.,* 51, 75, 2003.

43. Steussy, T., The role of creative monography in the biodiversity crisis, *Taxon,* 42, 313, 1993.

44. Tautz, D. et al., DNA points the way ahead in taxonomy, *Nature,* 418, 479, 2002.

45. Lipscomb, D., Platnick, N., and Wheeler, Q., The intellectual content of taxonomy: a comment on DNA taxonomy, *Trends Ecol. Syst.,* 18, 65, 2003.

46. Tautz D. et al., A plea for DNA taxonomy, *Trends Ecol. Syst.,* 18, 70, 2003.

47. Wheeler, Q.D., Taxonomic triage and the poverty of phylogeny, *Phil. Trans. Royal Soc. Lond. B,* 359, 571, 2004.

48. Chase, M.W. et al., Land plants and DNA barcodes: short-term and long-term goals, *Phil. Trans. R. Soc. B.,* 360, 1889, 2005.

49. Kristiansen, K.A. et al., DNA taxonomy: the riddle of *Oxychloë* (Juncaceae), *Syst. Bot.,* 30, 284, 2005.

50. Savolainen, V. and Chase, M.W., A decade of progress in plant molecular phylogenetics, *Trends Genet.,* 19, 717, 2003.

51. Palmer, J.D., Soltis, D.E., and Chase, M.W., The plant tree of life: an overview and some points of view, *Amer. J. Bot.,* 91, 1437, 2004.

52. APG, An ordinal classification for the families of flowering plants, *Ann. Missouri Bot. Gard.,* 85, 531, 1998.

53. Chase, M.W., Fay, M.F., and Savolainen, V., Higher-level classification in the angiosperms: new insights from the perspective of DNA sequence data, *Taxon*, 49, 685, 2000.
54. APG, An update of the angiosperm phylogeny group classification for the orders and families of flowering plants: APG II, *Bot. J. Linn. Soc.*, 141, 399, 2003.
55. Davis, J.I., et al., A phylogeny of the monocots, as inferred from *rbcL* and *atpA* sequence variation, and a comparison of methods for calculating jacknife and bootstrap values, *Syst. Bot.*, 29, 467, 2004.
56. Simpson, D.A. et al., Phylogenetic relationships in Cyperaceae subfamily Mapanioideae inferred from pollen and plastid DNA sequence data, *Amer. J. Bot.*, 90, 1071, 2003.
57. Bremer, K., Gondwanan evolution of the grass alliance of families (Poales), *Evolution*, 56, 1374, 2002.
58. Cronquist, A., *An Integrated System of Classification of Flowering Plants*, Columbia University Press, New York. 1981.
59. Linder, H.P. and Ferguson, I.K., On the pollen morphology and phylogeny of the Restionales and Poales, *Grana*, 24, 65, 1985.
60. Campbell, C.S. and Kellogg, E.A., Sister group relationships of the Poaceae, in *Grass Systematics and Evolution*, Soderstrom, T.R. et al., Eds., Smithsonian Institution Press, Washington, D.C., 1987, 217.
61. Briggs, B.G. and Johnson, L.A.S., Hopkinsiaceae and Lyginiaceae, two new families of Poales in Western Australia, with revisions of *Hopkinsia* and *Lyginia, Telopea*, 8, 477, 2000.
62. Rudall, P.A. et al., Evolution of reproductive structures in grasses (Poaceae) inferred by sister-group comparison with their putative closest living relatives, Ecdeiocoleaceae, *Amer. J. Bot.*, 92, 1432, 2005.
63. Doyle, J.J. et al., Chloroplast DNA inversions and the origin of the grass family (Poaceae), *Proc. Natl. Acad. Sci. USA*, 89, 7722, 1992.
64. Soreng, R.J. and Davis, J.I., Phylogenetics and character evolution in the grass family (Poaceae): simultaneous analysis of morphological and chloroplast DNA restriction site character sets, *Bot. Rev.*, 64, 1, 1998.
65. Hilu, K.W., Alice, L.A., and Liang, H.P., Phylogeny of Poaceae inferred from *matK* sequences, *Ann. Missouri Bot. Gard.*, 86, 835, 1999.
66. Briggs, B.G., et al., A molecular phylogeny of Restionaceae and allies, in *Grasses: Systematics and Evolution*, Jacobs, S.W.L. and Everett, J.E., Eds., CSIRO Collingwood, Victoria, 2000, 661.
67. Michelangeli, F.A., Davis J.I., and Stevenson, D.W., Phylogenetic relationships among Poaceae and related families as inferred from morphology, inversions in the plastid genome, and sequence data from the mitochondrial and plastid genomes, *Amer. J. Bot.*, 90, 93, 2003.
68. Hamby, R.K. and Zimmer, E.A., Ribosomal RNA sequences for inferring phylogeny within the grass family (Poaceae), *Pl. Syst. Evol.*, 160, 29, 1988.
69. Doebley, J. et al., Evolutionary analysis of the large subunit of carboxylase (*rbcL*) nucleotide sequence among the grasses (Gramineae), *Evolution*, 44, 1097, 1990.
70. Davis, J.I. and Soreng, R.J., Phylogenetic structure in the grass family (Poaceae) as inferred from chloroplast DNA restriction site variation, *Amer. J. Bot.*, 80, 1444, 1993.
71. Mason-Gamer, R.J., Weil, C.F., and Kellogg, E.A., Granule-bound starch synthase: structure, function, and phylogenetic utility, *Mol. Biol. Evol.*, 15, 1658, 1998.
72. Hsiao, C. et al., A molecular phylogeny of the grass family (Poaceae) based on the sequences of nuclear ribosomal DNA (ITS), *Aus. Syst. Bot.*, 11, 667, 1999.
73. Mathews, S. and Sharrock, R.A., The phytochrome gene family in grasses (Poaceae): a phylogeny and evidence that grasses have a subset of the loci found in dicot Angiosperms, *Mol. Biol. Evol.*, 13, 1141, 1996.
74. Mathews, S., Tsai, R.C., and Kellogg, E.A., Phylogenetic structure in the grass family (Poaceae): evidence from the nuclear gene phytochrome B, *Amer. J. Bot.*, 87, 96, 2000.
75. Liang, H. and Hilu, K.W., Application of the *matK* gene sequences to grass systematics, *Canad. J. Bot.*, 74, 125, 1995.
76. Barker, N.P., Linder, H.P., and Harley, E.H., Polyphyly of Arundinoideae (Poaceae): evidence from *rbcL* sequence data, *Syst. Bot.*, 20, 423, 1995.
77. Duvall, M.R. and Morton, B.R., Molecular phylogenetics of Poaceae: an expanded analysis of *rbcL* sequence data, *Molec. Phylogenet. Evol.*, 5, 353, 1996.
78. Zhang, W.P. Phylogeny of the grass family (Poaceae) from *rpl16* intron sequence data, *Molec. Phylogenet. Evol.*, 15, 135, 2000.

79. Barker, N.P., Linder, H.P., and Harley, E.H., Sequences of the grass-specific insert in the chloroplast *rpoC2* gene elucidate generic relationships of the Arundinoideae (Poaceae), *Syst. Bot.,* 23, 327, 1999.
80. Nadot, S., Bajon, R., and Lejeune B., The chloroplast gene *rps4* as a tool for the study of Poaceae phylogeny, *Pl. Syst. Evol.,* 191, 27, 1994.
81. Clark, L.G. et al., The Puelioideae, a new subfamily of Poaceae, *Syst. Bot.,* 25, 181, 2000.
82. Soreng, R.J. and Davis, J.I., Phylogenetic structure in Poaceae subfamily Pooideae as inferred from molecular and morphological characters: misclassification vs. reticulation, in *Grasses: Systematics and Evolution,* Jacobs, S.W.L. and Everett. J.E., Eds., CSIRO Collingwood, Victoria, 2000, 61.
83. Bininda-Emonds, O.R.P., Gittleman, J.L., and Steel, M.A., The (super)tree of life: procedures, problems, and prospects, *Ann. Rev. Ecol. Syst.,* 33, 265, 2002.
84. Sanderson, M.J., and Driskell, A.C., The challenge of constructing large phylogenetic trees, *Trends Plant Sci.,* 8, 374, 2003.
85. Wilkinson, M. et al., The shape of supertrees to come: tree shape related properties of fourteen supertree methods, *Syst. Biol.,* 54, 419, 2005.
86. Hodkinson, T.R. et al., Large trees, supertrees and diversification of the grass family, in *Aliso 23 (Grasses: Systematics and Evolution Vol. 3),* Columbus, J.T. et al., Eds., Allen Press, KS, USA, in press.
87. Salamin, N., Large trees, supertrees and the grass phylogeny, Ph.D. thesis, University of Dublin, Trinity College Dublin, 2001.
88. Goff, S.A. et al., A draft sequence of the rice genome (*Oryza sativa* L. ssp. *japonica*), *Science,* 296, 79, 2002.
89. Yu, J., et al., A draft sequence of the rice genome (*Oryza sativa* ssp. *indica*), *Science,* 296, 79, 2002.
90. Matsuoka, Y. et al., Whole chloroplast genome comparison of rice, maize and wheat: implications for chloroplast gene diversification and phylogeny of cereals, *Mol. Biol. Evol.,* 19, 2084, 2002.
91. Ogihara, Y. et al., Structural features of a wheat plastome as revealed by complete sequencing of chloroplast DNA, *Molec. Genet. Genomics,* 266, 740, 2002.
92. Eisen, J.A. and Fraser, C.M., Phylogenomics: intersection of evolution and genomics, *Science,* 300, 1706, 2003.
93. Friis, E.M., Pedersen K.R., and Crane, P.R., Fossil evidence of water lilies (Nymphaeales) in the Early Cretaceous, *Nature,* 410, 357, 2001.
94. Sun, G. et al., Archaefructaceae, a new basal angiosperm family, *Science,* 296, 899, 2002.
95. Soltis, P.S., Soltis, D.E., and Chase, M.W., Angiosperm phylogeny inferred from multiple genes as a tool for comparative biology, *Nature,* 402, 402, 1999.
96. Linder, H.P., The evolutionary history of the Poales/Restionales: a hypothesis, *Kew Bull.,* 42, 297, 1987.
97. Crepet, W.L. and Feldmann, G.D., The earliest remains of grasses in the fossil record, *Amer. J. Bot.,* 78, 1010, 1991.
98. Prasad, V. et al., Dinosaur coprolites and early evolution of grasses and grazers, *Science,* 310, 1177, 2005.
99. Poinar, G.O. and Columbus, J.T., Adhesive grass spikelet with mammalian hair in Dominican amber: first fossil evidence of epizoochory, *Experientia,* 48, 906, 1992.
100. Jacobs, B.F., Kingston, J.D., and Jacobs, L.L., The origin of grass-dominated ecosystems, *Ann. Missouri Bot. Gard.,* 86, 590, 1999.
101. Kellogg, E.A., The grasses: a case study in macroevolution, *Ann. Rev. Ecol. Syst.,* 31, 217, 2000.
102. Gaut, B.S. and Doebley, J.F., DNA sequence evidence for the segmental allotetraploid origin of maize, *Proc. Natl. Acad. Sci. USA,* 94, 6809, 1997.
103. Sanderson, M.J., Estimating absolute rates of molecular evolution and divergence times: a penalized likelihood approach, *Mol. Biol. Evol.,* 19, 101, 2002.
104. Donoghue, P.C.J. and Smith, M.P., *Telling the Evolutionary Time: Molecular Clocks and the Fossil Record,* Systematics Association Series 66, CRC Press, Florida, 2003.
105. Magallón, S.A. and Sanderson, M.J., Angiosperm divergence times: the effect of genes, codon positions, and time constraints, *Evolution,* 59, 1653, 2005.
106. Sanderson, M.J., A nonparametric approach to estimating divergence times in the absence of rate constancy, *Mol. Biol. Evol.,* 14, 1218, 1997.
107. Thorne, J.L., Kishino, H., and Painter, I.S., Estimating the rate of evolution of the rate of molecular evolution, *Mol. Biol. Evol.,* 15, 1647, 1998.
108. Cutler, D.J., Understanding the overdispersed molecular clock, *Genetics,* 154, 1403, 2000.

109. Cutler, D.J., Estimating divergence times in the presence of an overdispersed molecular clock, *Mol. Biol. Evol.*, 17, 1647, 2000.

110. Bremer, K., Early cretaceous lineages of monocot flowering plants, *Proc. Natl. Acad. Sci. USA*, 97, 4704, 2000.

111. Gaut, B.S., Evolutionary dynamics of grass genomes, *New Phytol.*, 154, 15, 2002.

112. Piperno, D.R. and Sues, H-D., Dinosaurs dined on grass, *Science*, 310, 1126, 2005.

113. Audley-Charles, M.G., Dispersal of Gondwanaland: relevance to evolution of the angiosperms, in *Biogeographical Evolution of the Malay Archipelago*, Whitmore, T.C., Ed., Clarendon Press, Oxford, 1987, 5.

114. Ronquist, F., Dispersal-vicariance analysis: a new approach to the quantification of historical biogeography, *Syst. Biol.*, 46, 195, 1997.

115. Ronquist, F., DIVA ver. 1.1, http://www.ebc.uu.se/systzoo/research/diva/diva.html, 1996.

116. Chan, K.M.A. and Moore, B.R., SYMMETREE: whole-tree analysis of differential diversification rates, *Bioinformatics*, 21, 1709, 2004.

117. Moore, B.R., Chan, K.M.A., and Donoghue, M.J., Detecting diversification rate variation in supertrees, in *Phylogenetic Supertrees: Combining Information to Reveal the Tree of Life*, Bininda-Emonds, O.R.P., Ed., Kluwer Academic Publishers, Dordrecht, 2004, 487.

118. Chan, K.M.A. and Moore, B.R., Whole-tree methods for detecting differential diversification rates, *Syst. Biol.*, 51, 855, 2002.

119. Burger, W.C., Why are there so many kinds of flowering plants? *Bioscience*, 31, 577, 1981.

120. Maynard Smith, J. and Szathmáry, E., *The Major Transitions in Evolution*, Freeman, Oxford, UK, 1995, 346.

121. Weiblen, G.D., Oyama, R.K., and Donoghue, M.J., Phylogenetic analysis of dioecy in monocotyledons, *Am. Nat.*, 155, 46, 2000.

122. Slowinski, J.B. and Guyer, C.G., Testing whether certain traits have caused amplified diversification: an improved method based on a model of random speciation and extinction, *Am. Nat.*, 142, 1019, 1993.

123. Salamin, N., and Davies, T.J., Using supertrees to investigate species richness in grasses and flowering plants, in *Phylogenetic Supertrees: Combining Information to Reveal the Tree of Life*, Bininda-Emonds, O.R.P. Ed., Kluwer Academic Publishers, Dordrecht, 2004, 461.

124. Goudet, J., An improved procedure for testing the effects of key innovations on rate of speciation, *Am. Nat.*, 153, 549, 1999.

125. Bousquet, J., et al., Extensive variation in evolutionary rate of *rbcL* gene-sequences among seed plants, *Proc. Natl. Acad. Sci. USA*, 89, 7844, 1992.

126. Barraclough, T.G., Harvey, P.H., and Nee, S., Rate of *rbcL* gene sequence evolution and species diversification in flowering plants (angiosperms), *Proc. R. Soc. Lond. B.*, 263, 589, 1996.

127. Savolainen, V. and Goudet, J., Rate of gene sequence evolution and species diversification in flowering plants: a re-evaluation, *Proc. R. Soc. Lond. B.*, 265, 603, 1998.

128. Barraclough, T.G. and Savolainen, V., Evolutionary rates and species diversity in flowering plants, *Evolution*, 55, 677, 2001.

129. Gaut, B.S. et al., Comparisons of the molecular evolutionary process at *rbcL* and *ndhF* in the grass family (Poaceae), *Mol. Biol. Evol.*, 14, 769, 1997.

130. Hillis, D.M., Inferring complex phylogenies, *Nature*, 383, 130, 1996.

131. Hillis, D.M., Taxonomic sampling, phylogenetic accuracy, and investigator bias, *Syst. Biol.*, 47, 3, 1998.

132. Soltis, D.E. et al., Inferring complex phylogenies using parsimony: an empirical approach using three large DNA data sets for angiosperms, *Syst. Biol.*, 47, 32, 1998.

133. Källersjö, M., Albert, V.A., and Farris, J.S., Homoplasy increases phylogenetic structure, *Cladistics-Int. J. Willi Hennig Soc.*, 15, 91, 1999.

134. Savolainen, V. et al., Phylogenetics of flowering plants based upon a combined analysis of plastid *atpB* and *rbcL* gene sequences, *Syst. Biol.*, 49, 306, 2000.

135. Erdos, P.L. et al., A few logs suffice to build (almost) all trees: part II, *Theor. Comp. Sci.*, 221, 77, 1999.

136. Bininda-Emonds, O.R.P. et al., Scaling of accuracy in extremely large phylogenetic trees, in *Pacific Symposium on Biocomputing 6*, Altman, R.B. et al., Eds., World Scientific Publishing Company, River Edge, New Jersey, 2001, 547.

137. Salamin, N., Hodkinson, T.R., and Savolainen, V., Towards building the tree of life: a simulation study for all angiosperm genera, *Syst. Biol.,* 54, 183, 2005.
138. Sanderson, M.J., et al., Error, bias, and long-branch attraction in data for two chloroplast photosystem genes in seed plants, *Mol. Biol. Evol.,* 17, 782, 2000.
139. Qiu, Y.-L. et al., Phylogeny of basal angiosperms: analysis of five genes from three genomes, *Int. J. Plant Sci.,* 161, S3, 2000.
140. Wiens, J.J., Does adding characters with missing data increase or decrease phylogenetic accuracy? *Syst. Biol.,* 47, 625, 1998.
141. Kearney, M., Fragmentary taxa, missing data, and ambiguity: mistaken assumptions and conclusions, *Syst. Biol.,* 51, 369, 2002.
142. Chase, M.W. et al., Phylogenetics of seed plants: an analysis of nucleotide-sequence from the plastid gene *rbcL, Ann. Missouri Bot. Gard.,* 80, 528, 1993.
143. Källersjö, M. et al., Simultaneous parsimony jackknife analysis of 2538 *rbcL* DNA sequences reveals support for major clades of green plants, land plants, seed plants and flowering plants, *Pl. Syst. Evol.,* 213, 259, 1998.
144. Savolainen, V. et al., Phylogeny reconstruction and functional constraints in organellar genomes: plastid versus animal mitochondrion. *Syst. Biol.,* 51, 638, 2002.
145. Farris, J.S. et al., Parsimony jackknifing outperforms neighbor-joining, *Cladistics,* 12, 1996.
146. Rannala, B. and Yang, Z.H., Probability distribution of molecular evolutionary trees: a new method of phylogenetic inference. *J. Mol. Evol.,* 43, 304, 1996.
147. Larget, B. and Simon, D.L., Markov Chain Monte Carlo algorithms for the Bayesian analysis of phylogenetic trees, *Mol. Biol. Evol.,* 16, 750, 1999.
148. Salamin N. et al., Assessing internal support with large phylogenetic DNA matrices, *Molec. Phylogenet. Evol.,* 27, 528, 2003.
149. Adam, D., Now for the hard ones, *Nature,* 408, 792, 2000.
150. Bininda-Emonds, O.R.P., Novel versus unsupported clades: assessing the qualitative support for clades in MRP supertrees, *Syst. Biol.,* 52, 839, 2003.
151. Gatesy, J. et al., Resolution of a supertree/supermatrix paradox, *Syst. Biol.,* 51, 652, 2002.
152. Bininda-Emonds, O.R.P. et al., Supertrees are a necessary not-so-evil: a comment on Gatesy et al., *Syst. Biol.,* 52, 724, 2003.
153. Purvis, A., A composite estimate of primate phylogeny, *Philos. Trans. R. Soc. Lond. B,* 348, 405, 1995.
154. Baldauf, S.L., The deep roots of eukaryotes, *Science,* 300, 1703, 2003.
155. Mace, G.M., Gittleman, J.L., and Purvis, A., Preserving the tree of life, *Science,* 300, 1707, 2003.
156. Pennisi, E., Modernizing the tree of life, *Science,* 300, 1692, 2003.

# 18 Collecting Strategies for Large and Taxonomically Challenging Taxa: Where Do We Go from Here, and How Often?

*T. M. A. Utteridge and R. P. J. de Kok*
South East Asia Regional Team, Herbarium, Royal Botanic Gardens, Kew, Richmond, Surrey, UK

## CONTENTS

## ABSTRACT

The major task of systematists is to document the planet's biodiversity. This has traditionally been done using morphological characters, especially for inventory work involving the description of new taxa (species or genera) and the production of checklists and Floras. We present an analysis of collections from a well collected area on the highly biodiverse tropical island of New Guinea. The species accumulation curve for this area reveals different collection patterns between the type of collector and the type of habitat visited. Future surveys should be based on databases of large collections (or large samples of collections, such as geographic or systematic subsets) and should use a combination of generalist and specialist collectors in the field to produce a comprehensive and rigorous sampling strategy.

## 18.1   INTRODUCTION

An enduring challenge in systematics is how to accurately estimate the number of species on the planet[1]. It is a question that is pertinent at practically all taxonomic scales of investigation and particularly for large species rich taxa, the topic of this book. This chapter assesses the best collecting strategies for sampling a large and taxonomically difficult taxon in one region on the tropical island of New Guinea.

The lack of taxonomic capacity is well documented, with various solutions proposed including, amongst others, DNA taxonomy and barcoding (for example, Blaxter[2]; Seberg and Peterson, *Chapter 3*) and the use of parataxonomists in inventory studies (Krell[3]). These solutions often assume that all life on Earth has already been sufficiently sampled and is just waiting in museums to be catalogued. However, this is not the case, and the systematic community is far from efficiently sampling the planet[4], especially as it undergoes drastic changes in the extent and composition of natural areas. This is a particular problem in the tropics, which have an incredibly high diversity but not enough taxonomic capacity to sufficiently sample and describe their diversity. Sampling strategies can broadly be grouped into two categories: unstructured and structured surveys, explored below.

### 18.1.1   Unstructured Surveys

Systematists generally perform unstructured surveys as they examine differences between organisms. As such, systematic botanists undertake random searches through a habitat, collecting whatever is useful for systematic purposes, that is, plants with flowers or fruits that can be compared with other collections. Once a species has been collected, it is unlikely that another specimen of the same species will be collected during the same trip. An area may be visited again if thought to be 'rich' (a term often based on the fact that not everything in flower/fruit could be collected in a single visit), or if it was not sampled due to little flowering or fruiting at the time of visit. Taxonomic sampling, especially of plants (that is, 'plant collecting'), involves collecting specimens encountered while moving through a large area during a limited time span without using any fixed reference points such as plots or transects. Collecting specimens for systematic purposes is usually undertaken via unstructured surveys, often resulting in Floras or checklists that have little if any data on such things as the number of individuals seen or their rarity; all that is noted are the differences between the different organisms.

### 18.1.2   Structured Surveys

Structured surveys are often carried out to explore ecological relationships between an area and the number of species found within it. Such survey techniques use a predefined methodology, such as Gentry's transect methodology (Phillips and Miller[5]) or permanent plots (for example, Newbery[6]). In such studies, the plots will be visited many times, and every organism within the study parameters will be collected, often resulting in sterile specimens that would have been disregarded by systematists.

### 18.1.3   A Case Study: New Guinea

New Guinea has recently been designated a 'wilderness area' by Conservation International[7]. Wilderness areas are defined as being greater than 10,000 km$^2$ in size and retaining 70% of their historical habitat. New Guinea is politically divided down its centre with the Indonesian half of Papua to the West, and Papua New Guinea on the East. Recently published collection densities for Papua are not available. If one uses Campbell's theory that a collecting density of 1 specimen/km$^2$ is sufficient to sample a given large area[8], Papua New Guinea is still undercollected at 0.46 specimen/km$^2$.

We wanted to examine a real life unstructured taxonomic inventory to try and reveal the presence of any sampling patterns and, if so, investigate how these could affect future sampling strategies. We studied a plant collecting programme in Mt. Jaya, New Guinea, organised by systematic botanists at the Royal Botanic Gardens (RBG), Kew, U.K., from 1997 to 2005. The questions addressed in this chapter are:

- What collecting patterns, if any, are present in an unstructured survey?
- Are we using our limited time and resources efficiently?
- Can collectors (generalist/specialist) be broadly categorised, and do they contribute to an inventory in different ways?
- How good are particular sampling strategies for collecting large and taxonomically difficult groups?
- Is there a most efficient (cost effective) way to conduct a systematic inventory?

We do not seek to examine classical species area patterns; we assume that habitats have different diversities and acknowledge that systematists rarely follow any particular experimental design or ecological methodology to produce comparative data during taxonomic collecting trips. We also wish to investigate collecting effort in regard to patterns of collections made during taxonomic trips, that is, a species accumulation curve independent of time and area, rather than examining species diversity (z values) between different habitats in New Guinea. We have only estimated the number of species, not their relative abundances, and this is not an attempt to estimate the number of species in the study area (for example, Rosenzweig et al.[9]).

## 18.2 METHODS

We analysed historical and contemporary herbarium collections from the Mt. Jaya area of New Guinea. Mt. Jaya at 4,884 m is the highest peak in South East Asia and has been the focus for scientific expeditions since 1913.

Recently, the company operating a copper mine on the mountain facilitated access to the area, and in 1997 a plant checklist project was instigated at RBG Kew involving several expeditions to the area and a programme of databasing all historical collections from Mt. Jaya held in herbaria around the world. Our current analysis used information from the RBG Kew database that holds 9,600 records and includes nearly all, if not all, historical collections and all contemporary RBG Kew collections from the study area.

Collections were assigned to 18 trips from 1913 onwards. Prior to the RBG Kew project, there had been 13 collecting trips to the region with five RBG Kew expeditions carried out during the period 1998–2000. Collections were assigned to the following habitats: lowland (0–1200 m); montane (>1200 m and <3000 m); and alpine (3000 m and over), following Hope's vegetation classification for the mountain[10,11].

Both historical and contemporary expeditions made collections in all habitats from mangrove to alpine. Collections from montane to alpine areas are currently being used as a basis for a checklist of the alpine flora of the area[12]. Utteridge has worked extensively on the project making collections in the area, processing specimens and writing taxonomic accounts[13-15], and has become well acquainted with the Mt. Jaya flora. We are confident, therefore, that identifications in the database to family and generic level are correct, as are species-level identifications for the alpine and montane zones.

The data were analysed at species level, with the first occurrence of each species recorded, and subsequent recollection of a species deleted from the database. Thus, if a set of collections was identified to genus only they were treated as a single species; this will significantly underestimate diversity for the lowland collections which have yet to be worked on. Plots were made using Microsoft Excel, as XY (scatter) plots; logarithmic trend lines were added using the options provided in Excel. The vertical lines on Figures 18.1 and 18.2 indicate the first time a RBG Kew expedition visited the area. Splitting the data into habitat type, we plotted species collection accumulation curves for the alpine and montane habitats. To investigate how different collectors contributed to the project, we excluded all collections made by specialists of their specialist groups; specialists often made general collections as well. Plots of the accumulation of genera and species were made for each collection event against the cumulative number of all collections up to and including that event. Finally, we examined collecting efficiency by calculating the number of collections it took before a new species was added to the inventory.

## 18.3 RESULTS

Only 7,945 of the 9,600 records could be used due to factors such as inadequate data, duplicated records and misidentifications. For all habitats in the study area, a total of 698 genera and 1,935 species were recorded. Collections from the lowlands were not analysed at habitat level because of the mosaic of habitats (such as mangrove, swamp forest, lowland rainforest and heath forest), several of which had been visited many times and others hardly at all[16]. In addition we found that many of the lowland collections had incomplete locality and collection data and were not named to the same accuracy as the montane and alpine collections. Two different patterns can be seen in the analysis: the effect of 'collector type' and collecting 'efficiency' through time.

### 18.3.1 COLLECTOR TYPE EFFECT

We examined the effect of 'collector type' by plotting collections made by generalist and specialist collectors (Figure 18.1). The majority of collecting trips to the project area were by general collectors. A total of 22 specialists visited the study site, collecting 14 taxa; only two specialists visited the study area before the RBG Kew project, with the majority of the collecting trips undertaken by general collectors. At the end of the project specialist collectors had contributed 10.6% of the genera to the collections (a total of 698 genera versus 624 genera without specialist collected taxa by the last expedition) and at species level this rose to 15.9% (a total of 1,935 species versus 1,627 species without specialist collected taxa at the last expedition). The differences between the generalist and specialist collections during the RBG Kew phase of collecting when the majority

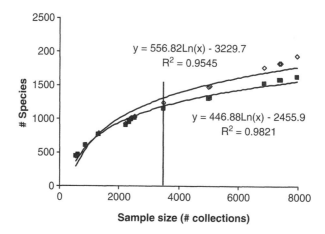

**FIGURE 18.1** Plot of all collections from the Mt. Jaya area. Cumulative number of collections (x-axis) and cumulative number of species (y-axis). Generalist collector curve shown in solid squares; specialist collector curve in open diamonds. The vertical line indicates the first RBG Kew expedition to the area.

of specialist collectors visited the area were significantly different for the number of species collected ($P = 0.05$ using a $t$-test), but not the number of genera ($P = 0.06$).

## 18.3.2 COLLECTING EFFICIENCY OVER TIME

Examination of historical collection patterns provides a measure of the 'efficiency' of collecting over time. The species to collection accumulation curves developed follow an asymptotic pattern (Figure 18.1). Historical collections made prior to RBG Kew's involvement constitute 31.8% of specimens but make up 53.4% and 59.8% of the species and genera, respectively. This equates to an efficiency of 1.3 collections made per new species added during the first expedition of the region, dropping slightly to 4.1 collections per new species by the end of the project.

The curve in the species accumulation plot for the entire dataset has yet to flatten out (Figure 18.1), reflecting the difference in diversity of, and collecting effort undertaken in the different habitats in the project area. Analysing the collections from the altitudinal zones separately reveals stark differences in their collecting histories and the relative efficiency of collecting in each zone. RBG Kew expeditions added an additional 104 species to the alpine region (18% of the total alpine species) and an additional 630 species to the montane region (60.5% of the montane species) (Figure 18.2). Plotting the $\log_{10}$ values reveals the differences in the slope between the collecting programmes in the different habitats (Figure 18.3). A null model of efficient collecting (every additional collection adds another new species to the inventory) has a slope of 1 (Figure 18.3); this reduces to 0.7 for the montane collections (Figure 18.3 solid squares) and 0.56 for the alpine collections (Figure 18.3 open diamonds).

Collecting efficiency in the alpine area (data not shown) was low because of the previous history of collecting, with the collecting programme taking 7.7 collections per new species added to the project for the first RBG Kew expedition, dropping sharply to 45 collections made per new species for the last expedition. This contrasts with the expeditions in the montane region (data not shown) where efficiency was much higher at 1.2 and 3.1 collections made per new species for the first and last Kew expeditions, respectively.

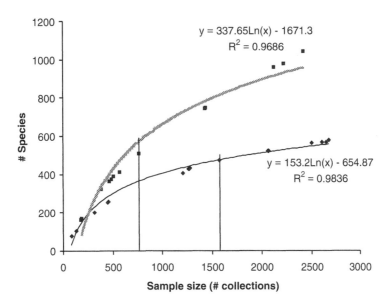

**FIGURE 18.2** Plot of collections from alpine and montane habitats of the Mt. Jaya area. Cumulative number of collections (x-axis) and cumulative number of species (y-axis). Alpine collections shown in solid diamonds; montane collections in solid squares. The vertical line indicates the first RBG Kew expedition to the area.

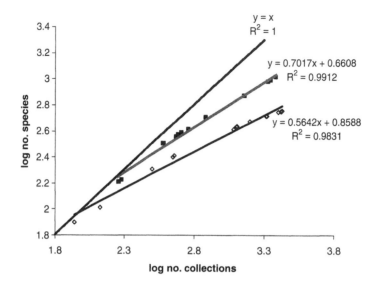

**FIGURE 18.3** Plot of collections from alpine and montane habitats of the Mt. Jaya area. Log$_{10}$ cumulative number of collections (x-axis) and log$_{10}$ cumulative number of species (y-axis). Alpine collections shown in open diamonds (slope = 0.56); montane collections in solid squares (slope = 0.70); null model of efficient collecting (see text) shown as the y = x line.

## 18.4 DISCUSSION

Collecting trips to the Mt. Jaya area have been unstructured surveys, with each individual trip collecting independently of each other. When the collections have been databased as a whole, however, the cumulative species versus cumulative collections data plot out as classic species accumulation curves (Figure 18.1). The number of new species added to the project initially rises sharply compared to the number of collections made and then starts to level out towards an asymptote. For the Mt. Jaya project this was shown most notably in the collecting plots from the alpine regions (Figure 18.2).

General collecting has taken place during all stages of this project, whilst specialist collecting was mainly limited to the RBG Kew trips. Specialist collectors collected many more species from their taxon of expertise. This pattern was mentioned by Prance and Campbell[17] in their analysis of incoming collections of Chrysobalanaceae, who concluded that "recent collecting is baling in a lot of herbarium material of common species, but it is not adequately covering rare species. This indicates the need for more informed and knowledgeable collectors, especially specialists in various plant families and less general collecting". This is shown in our analysis, which looked at 14 groups collected by 22 specialists (Figure 18.1). We interpret our data to mean that specialists recognise and collect species that generalist collectors overlook. In a well organised large-scale survey, general collecting has its place in the early stages. However, a phase where collecting efficiency sharply drops can soon be reached, such as in the alpine collecting pattern post RBG Kew involvement (Figure 18.2). This is either due to all species in an area being collected or the area only being visited by general collectors who are not getting rare species, especially in taxonomically difficult and species rich taxa. RBG Kew is a large institute which can call on both generalist and specialist botanists. At least two members of the RBG Kew team took part in every Mt. Jaya expedition and became knowledgeable of the flora; that is, they became good generalist collectors. However, they were still unable to collect the full range of species in several difficult and species rich taxa such as Araliaceae, Arecaceae, Cyperaceae, Ericaceae, Myrtaceae, Poaceae and pteridophytes. However, specialists on the field trips did recognise and collect these extra taxa.

It is important for inventory projects to initiate projects with generalist collectors and then use specialist collectors in the field. This process will include drawing on databases of taxonomists and inviting specialists from outside institutes to join their project where the expertise is not present in their own institute. We argue that it is not sufficient to use specialists only to name collections once the fieldwork has been done, because their expertise may include spotting taxa in the field, something a generalist may not do, with the result that any additional collections of specialist groups are an 'ad hoc bonus'.

Databasing specimens is a cost effective way of estimating the stage of an inventory. Expeditions to the tropics can be extremely expensive, and we have estimated that a two-month expedition involving a team of four people equates to a year's databasing. It has been difficult to estimate the cost of collecting each specimen in New Guinea, as so much manpower was contributed by the mining company without cost to the expedition. However, costs for a specimen have recently been calculated at the RBG Kew as approximately STG£5.50 to enter the herbarium (including accessioning and curation) and an additional STG£3 to identify a specimen[18]. During Kew's last expedition to the alpine region, 45 specimens were collected before a new species was added to the project's list, equating to herbarium costs of STG£247.50 for each new species. These costs are higher, by a considerable amount, than those calculated by Parnell[19] and Mann[20], suggesting that costs may vary considerably depending on factors such as the area of study undertaken and the number of duplicates collected. Effort would have been better directed in the montane region (with a slope of 0.7) to reduce that slope to that approaching that of the alpine region (with a slope of 0.56) (Figure 18.3).

We appreciate that much of what we present here is intuitive and already 'known', in people's minds at least. However, economic analyses of flora, checklist and similar projects carried out to fulfil systematic means are sparse (but see Parnell[19]; Mann[20]; Funk and Richardson[21]). As such, this simple analysis will allow predictions to be made as to the efficacy of a collecting programme in a particular locality, and the number of visits that should be made to an area.

We recommend that an inventory strategy should be put in place before large projects are undertaken. This should start with some understanding of the collecting history of the area, databasing as many (if possible, all) known herbarium collections from the area, regardless of age. These data should be analysed at the habitat level, to estimate the phase the inventory has reached. The collector history should be examined to see if specialist collectors had previously visited the inventory site. This will dictate future strategy, for example if general collecting is no longer necessary for an area, then specialist collectors should be used. Decisions on taxa not fully represented in the collections can be made from the database and will influence which specialists are required to complete the inventory. Institutes will also have to consider external collaboration.

For the Mt. Jaya project many expeditions were conducted in the alpine zone, partly because of the intrinsic desire of humans to get to the summits of large mountains, and partly to fulfil the needs of the project to produce a checklist of the area impacted by mining. However, it is important to know the collection history of areas with high peaks, especially as many of these areas are conserved without due regard for the lowland forests surrounding them. For example, in the Malaysian state of Sabah, many of the protected areas are on mountains with very few lowland areas protected.

## 18.5  CONCLUSIONS

- Generalists do not collect all species in an area, even if they know the flora well. A parataxonomist or generalist collector based in one area may not collect all species if undertaking unstructured surveys, even after a long time.
- Species rich and taxonomically difficult groups are systematically undercollected by generalists.

- Specialist collectors have a high value in fieldwork, as they are able to pick out rare species in the field.
- Databasing is a cost-effective way to guide the sampling strategy of an area.

## ACKNOWLEDGEMENTS

We would like to thank our colleagues at RBG Kew for discussion during the writing of this paper, especially Neil Brummitt, Stuart Cable, Aaron Davis, Helen Hopkins, Bob Johns, Justin Moat and Maria Vibe Norup.

## REFERENCES

1. Systematics Agenda, *Systematics Agenda 2000: Charting the Biosphere,* Systematics Agenda 2000, New York, 1994.
2. Blaxter, M., Counting angels with DNA, *Nature,* 421, 122, 2003.
3. Krell, F.-T., Parataxonomy vs. taxonomy in biodiversity studies: pitfalls and applicability of 'morphospecies' sorting, *Biodivers. Conserv.,* 13, 795, 2004.
4. Raven, P.H. and Wilson, E.O., A fifty-year plan for biodiversity surveys, *Science,* 258, 1099, 1992.
5. Phillips, O. and Miller, J.S., *Global Patterns of Plant Diversity: Alwyn H. Gentry's Forest Transect Data Set,* Missouri Botanical Garden Press, St. Louis, MO, 2002.
6. Newbery, D.M. et al., Primary forest dynamics in lowland dipterocarp forest at Danum Valley, Sabah, Malaysia, *Phil. Trans. R. Soc. Lond. B,* 354, 1763, 1999.
7. Mittermeier, R.A. et al., Wilderness and biodiversity conservation, *Proc. Nat. Acad. Sci. USA,* 100, 10309, 2003.
8. Campbell, D.G., The importance of floristic inventory in the tropics, in *Floristic Inventory of Tropical Countries: The Status of Plant Systematics, Collections, and Vegetation, plus Recommendations for the Future,* Campbell, D.G. and Hammond, H.D., Eds., New York Botanic Garden, New York, 1989, 5.
9. Rosenzweig, M.L. et al., Estimating diversity in unsampled habitats of a biogeographical province, *Conserv. Biol.,* 17, 864, 2002.
10. Hope, G.S., Vegetation, in *The Equatorial Glaciers of New Guinea,* Hope, G.S. et al., Eds., Balkema, Rotterdam, 1976, 112.
11. Hope, G.S., New Guinea Mountain Vegetation Communities, in *The Alpine Flora of New Guinea,* van Royen, P., Ed., J. Cramer, Vaduz, Liechtenstein, 1980, 153.
12. Johns, R.J., *A Guide to the Alpine and Subalpine Flora of Mt Jaya,* Royal Botanic Gardens, Kew, 2006.
13. Utteridge, T.M.A., Two new species of *Maesa (Myrsinaceae)* from Puncak Jaya, New Guinea: contributions to the Flora of Mt Jaya I, *Kew Bull.,* 55, 443, 2000.
14. Utteridge, T.M.A., The subalpine members of *Pittosporum (Pittosporaceae)* from Mt Jaya, New Guinea: contributions to the Flora of Mt Jaya II, *Kew Bull.,* 55, 699, 2000.
15. Utteridge, T.M.A., A new species of *Medusanthera* Seem. *(Icacinaceae)* from New Guinea: *Medusanthera inaequalis* Utteridge: contributions to the Flora of Mt Jaya IV, *Kew Bull.,* 56, 233, 2001.
16. Utteridge, T.M.A., personal observations, 1999–2000.
17. Prance, G.T. and Campbell, D.G., The present state of tropical floristics, *Taxon,* 37, 519, 1988.
18. Harvey, T., personal communication, 2005.
19. Parnell, J.A.N., The monetary value of herbarium collections, in *Biological Collections and Biodiversity,* Rushton, B.S., Hackney, P., and Tyrie, C.R., Eds., Linnean Society of London Special Publication 3, Samora Publishing, UK, 2001, 271.
20. Mann, D.G., The economics of botanical collections, in *The Value and Valuation of Natural Science Collections,* Nudds, J.R. and Pettitt, W., Eds., Geological Society, London, 1997, 68.
21. Funk, V.A. and Richardson, K.S., Systematic data in biodiversity studies: use it or lose it, *Syst. Biol.,* 51, 303, 2002.

# 19 Large and Species Rich Taxa: Diatoms, Geography and Taxonomy

*D. M. Williams and G. Reid*
Botany Department, The Natural History Museum, London, UK

## CONTENTS

## ABSTRACT

Diatoms are one of the largest, if not the largest, group of cryptogamic photosynthetic organisms, in terms of species diversity. This chapter address some of the basic issues concerning the history of alpha taxonomy and the study of diatom diversity, exploring future possibilities from the perspective of their geography.

## 19.1 INTRODUCTION

The thrust of this book is to address issues behind the systematics of large and species rich taxa. Whilst that topic is of much significance, it begs a number of questions, those that pertain to definitions and those that pertain to solutions. How large is large? What kinds of numbers allow a taxon to legitimately be described as 'species rich'? These are not trivial questions. Whatever yardsticks are applied, we would probably find no disagreement if we were to describe the group of organisms that are the focus of our concern as both extremely large and species rich, or at least potentially so in both cases, regardless of the precise definition of each term.

Our organisms of study are diatoms (Bacillariophyta), a large and diverse group of photosynthetic, single-celled eukaryotes, with their cells encased in a silica shell[1]. They are members of the heterokont (stramenochrome) algae[2]; their sister taxon has been identified as a group of recently described tiny flagellates, Bolidophyceae[3], which has no more than three to five currently recognised species[4,5]. The heterokonts belong within stramenopiles, themselves a remarkably large and diverse group of eukaryotic organisms[2]. Stramenopiles and stramenochromes are both names that have been applied to heterokont algae and their relatives[6,7]. The term 'strameno', meaning straw, refers

305

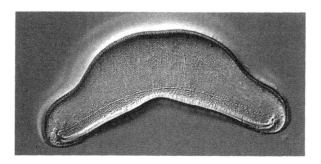

**FIGURE 19.1** Light micrograph of the Miocene fossil *Amphorotia americana* (Kain et Schultz) D.M. Williams & G. Reid. (From lectotype slide, BM-Adams D. 846, Atlantic City, NJ. For further details see Williams and Reid[65].)

to the characteristic tripartite flagella hairs, a synapomorphy uniting these taxa. The stramenochromes[8–10] are equivalent to the heterokont algae, whereas the stramenopiles include, as well as the heterokont algae, öomycetes, labyrithulids, thraustochytrids and certain other flagellate protozoa[2]. Surprisingly, the stramenopiles merit only a short two-page account in the recent *Assembling the Tree of Life* compendium[11], buried in the chapter on the relationships of green plants (Delwiche et al.[12]), though an illustration of the diatom *Cymbella cistula* (Hemprich et Ehrenb.) Kirchner does appear on the dust jacket.

Silica is inert, so when the organism dies the siliceous parts sink into the sediment and are often preserved, creating a splendid diatom fossil record (Figure 19.1), stretching from the present to the Cretaceous[13]. Reasonable estimates place the total number of diatom species (fossil and Recent) somewhere in the region of 200,000[14], roughly the same number of species known for all higher plants[15]. It is estimated that some 15,000 living species have so far been described, with another 5,000–8,000 species that are now extinct[13]. Should the estimates of diatom diversity at the species level be even approximately accurate, it is evident that there is a considerable amount of work that remains to be undertaken. Accounting for species diversity is often referred to as alpha taxonomy, the discovery and description of the basic units in classification[16], yet species diversity is not the only problem facing diatom taxonomists and systematists.

At present, diatom species are assigned to around 350 genera[17,18], with another approximately 150 genera for fossil groups[13]; only a handful (no more than 30) of higher taxa, classes, subclasses, orders and families, are recognised[1], or at least in use. Rarely do diatomists refer to a specimen's family, for example, as characters for higher taxa have rarely been documented. In short, diatoms are poorly known and poorly accounted for in their phylogenetic relationships (higher-level taxonomy), their current classification (numbers of higher taxa) and biodiversity estimates, regardless of the agreed vast number of species that remain to be described.

So much for definitions, solutions are of more immediate concern, particularly solutions accounting for diatom diversity, both in terms of documenting and understanding it. Whilst it is clear that many of the problems are simply practical, devising a strategy, should one wish to approach the problem in that way, requires further consideration.

Below we deal with the understanding of diatom diversity by examining three aspects. First, we briefly examine the history and understanding of alpha taxonomy as it has developed and how algal taxonomy progressed within these changing paradigms. Second, we present some more numbers for direct comparison within both the stramenopile clade as well as other plant and animal groups. Finally, we propose some options for dealing with large genera within diatoms, focusing studies on the geographical dimension, a poorly examined aspect of diatom studies, which, in some circles, is considered a pointless endeavour[19,20].

## 19.2 THERE ARE TAXA, AND THEN THERE ARE TAXA . . .

Thomas Henry Huxley, Darwin's bulldog and champion of English evolutionary understanding[21,22], aptly summed up the efforts required by taxonomists when late in life he wrote of the descriptive task in relation to biogeography: "I think there is no greater mistake than to suppose that distribution, or indeed any other large biological question, can be studied to good purpose by those who lack either the opportunity or the inclination to go through what they are pleased to term the drudgery of exhaustive anatomical, embryological, and physiological preparation"[23].

Drudgery it may be, but it is certainly necessary. Nevertheless, descriptive taxonomy does have a rather strange reputation. To many it appears unscientific and restricted to the domain of those peculiar persons who spend their time trying to capture in words the details of a particular specimen laid out before them, a curious hangover from Victorian times where endless, lengthy monographs were lovingly put together, recording the natural world in all its detail. Taxonomists have developed a strange reputation as well. The entomologist Wayne Moss offered the following: "Taxonomists have always had the reputation of being difficult. Intransigence may be rooted in the necessity of defending prolonged self-immersion in a taxon that others find a total bore; it is frustrating to have one's life work greeted with a yawn"[24].

Yet the common view concerning alpha taxonomy as second-class drudgery is simply mistaken, as it is the key to defining relationships in any group whatever its size. That common viewpoint is now fading as descriptive taxonomy is undergoing something of a renaissance[16]. First though, an historical overview is relevant, namely the history and origin of the term alpha taxonomy, usually invoked to categorise the activities of those whose preoccupation is species-level diversity. It is necessary, as the framework for much phycological taxonomy in the twentieth century rests within this ancient and defunct paradigm[25]. As diatomists, our historical focus is framed around British phycology[26].

The British Phycological Society was born at a meeting in January 1952 at the British Museum (Natural History), as it was then called. The programme for the meeting announced two presentations: the first by William Bertram Turrill (1890–1961)[27] from the Royal Botanic Gardens, Kew, England; the second by Robert Ross (1912–2005), a diatomist based at the Natural History Museum. The aim of that meeting was to initiate discussion on how to discover and name biological taxa. Turrill's task was to outline how one discovered the biological units worth naming; Ross's task was to outline the process of naming those units once discovered. The title of Ross's presentation was 'The international rules of botanical nomenclature and their application to the algae', a subject that occupied him throughout his life. Turrill's title was 'The integration of the older methods of classification with the more recent techniques of ecological and genetical research'. A version of Turrill's lecture was published in *Nature* with a slightly different title and advertised for sale in the first *Phycological Bulletin* at nine old pence a copy, post and packing free[28].

Turrill was a truly remarkable man. At the age of 18 he was appointed as Temporary Technical Assistant in the Herbarium of the Royal Botanic Gardens, Kew, eventually becoming Keeper of the Herbarium and Library in 1946, a position he held until his retirement in 1957. A year after retiring he was elected to the Royal Society and was awarded the Linnean Gold Medal, that society's most prestigious honour. He died suddenly in 1961. Turrill was many things but he was not a phycologist[27].

So why Turrill, and what did he have to say to British phycologists over 50 years ago? The headings from the *Nature* paper give some kind of clue: 'Characters, Special and General Classifications, Individuals, Species, Ecads, Genetics and Taxonomy, Karyology, Variation, Reticulation and Phylogeny'[28]. These topics summarised what Turrill liked to call the 'modern' viewpoint, as opposed to, one can imagine, the efforts of the antiquated herbarium taxonomists who poured over the dried and dusty specimens in their care (or, in this case, as he was Keeper of Kew Gardens, in Turrill's care).

Early in his career, Turrill became convinced that intensive and wide-ranging studies were required before any lasting improvements could be made to botanical classification, particularly at the species level. He tirelessly urged botanists to move away from just the morphological herbarium approach and advocated the investigation of ecological, cytological, genetical and chemical factors to enhance and supplement the taxonomy of organisms in general. In 1935 he published a short paper on 'The investigation of plant species'. He wrote: "Those who, having been trained to an appreciation of modern discoveries in ecology, cytology, genetics, and chemical factors, are trying to widen the basis of taxonomy, have undertaken a long, slow and perhaps thankless task."[29]. Well, maybe. But that was 1935, not 1952. Even earlier, Turrill published an essay on 'Species', which dealt with these very same issues[30]. So in 1952, 27 years on, the ideas were hardly modern, yet Turrill persisted with the viewpoint that morphology was limited. His 1935 paper gained a certain amount of significance for taxonomy as a whole.

For the first time Turrill mentioned the idea of alpha taxonomy: "It is suggested … that the time has come when the student of floras whose taxonomy on the old lines is relatively well known should attempt to investigate species by much more complete analyses of a wider range of characters than is the rule"[29]. 'On the old lines' was to be read as morphology, herbarium taxonomy. Turrill then delivered his coup de grace: "There is thus distinguished an alpha taxonomy and an omega taxonomy, the latter being an ideal which will probably never be completely realized"[29]. Turrill also noted: "The alpha taxonomist, however, really studies the constitution of the genus, not of the species; he 'gets inside' the genus but does not 'get inside' the species"[29].

Turrill was also a member of the group of biologists that in 1937 eventually became the Systematics Association[31]. Through the efforts of Julian Huxley, the Association was largely responsible for giving birth to what was called the 'New Systematics', another grand title given to this all-encompassing approach to taxonomy[32]. Of course, at the time it seemed the right thing to do, as if all information was of some equal merit. Turrill acknowledged that his omega classification was probably impossible to obtain, but what was it? No one really quite knew. Yet what did happen was that all the different facets of the new experimental approaches to taxonomy created a whole series of special classifications or, more accurately, artificial classifications, designed to represent each property being considered. Under this experimental umbrella were: Turesson and his genecology[33]; Gilmore and his demes, along with its complicated terminology[34,35]; Danser and his classificatory system[36]; and so on.

Gilmore and Turrill[37] outlined one possibility, where all the special classifications, those based on ecology, those based on genetics, those based on demes, could be combined into one general classification (Figure 19.2). At this point it appeared that alpha taxonomy (and to a certain extent, omega taxonomy) had vanished from consideration. Nevertheless, it is not difficult to see, with all these special classifications and the desire to produce from them a general classification, how

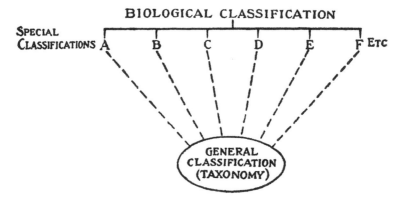

FIGURE 19.2 The relationship between special and general classifications. (Reproduction of illustration taken from Gilmour and Turrill[37].)

phenetics (classification using overall similarity[38,39]) became popular not only among those who embraced the 'New Systematics' but also those that needed to embrace a 'modern' approach to taxonomy but were unwilling to engage with the earlier literature[40].

Turrill's views stem from an even earlier preoccupation. Around the turn of the nineteenth century there was a general revolt from reconstructing phylogenies, at that time largely understood as the apparent unifying principle of morphology. The history of this episode has been captured by Boney's account of the 'Tansley Manifesto' affair[41], a general revolt against the more and more extravagant phylogenetic speculations of a previous generation, ending up in a revolt against morphology.

Phylogenetic speculations, in the form of ancestor-descendant sequences, were largely the inspiration of one man, Ernst Haeckel, who coined the word phylogeny among many others[42]. The revolt, set in motion for phycologists by Tansley and his associates, and pursued relentlessly by people like Turrill and Gilmore, added to the decline and drift away from morphology and its relation to classification, the effects of which are still with us today[16].

For Turrill, alpha taxonomy was concerned with morphology but as a starting point, so with the inclusion of all other forms of data it would finally lead to the omega taxonomy, or something close. Whilst this history betrays the effects of a certain kind of thinking in British botany, zoologists probably relied more on Mayr for their understanding[43]. Mayr, like Turrill, saw the issue as one of progress, from alpha taxonomy onwards: "In the first stage, often called *alpha taxonomy,* emphasis is on the description of new species and their preliminary arrangement in comprehensive genera. In *beta taxonomy* relationships are worked out more carefully on the species level and on that of the higher categories; emphasis is placed on the development of a sound classification. At the level of *gamma taxonomy* much attention is paid to intraspecific variation, to various sorts of evolutionary studies, and to a causal interpretation of organic diversity"[43].

Thus both zoologists and botanists understood progress to be from the simple exercise of description to the 'various sorts of evolutionary studies' that might seem of some greater interest. This prescription, although rendered into modern vocabulary, is still repeated today, as if complexity and more data offer serious solutions to the problem of classification[44].

Of course, in the 1970s and 1980s the theory of classification known as 'cladistics' became the dominant force for understanding the interrelationships of organisms, regardless of taxonomic rank. Cladistics was viewed by some as a critique[45] allowing a greater understanding of past endeavours, while others saw in it nothing of substance, and much of phycology has remained bathed in the ideas of Turrill and Mayr. So what is the significance of this for the study of 'large and species rich taxa', even in the face of a taxonomic renaissance[16]?

## 19.3 THERE ARE NUMBERS, AND THEN THERE ARE NUMBERS . . .

We noted above that there may be in the region of 15,000–20,000 diatom species so far described, with estimates of the total number anywhere in the region of 100,000 to 200,000 and beyond[46]. To get some idea of the extraordinary diversity in diatoms, the usual approach is to compare numbers of diatom species known and expected with those of other 'algal' groups. Yet it has been known for some time that 'algae' are paraphyletic[12], a nongroup having no reality in nature[47] and serving no real basis for comparison other than giving some overall impression of species level diversity across all organisms. Therefore, to give some further perspective, we have prepared approximate figures of numbers of species for the stramenochromes (Table 19.1), a selection of other 'algal' groups (Table 19.2) and some other organisms, plant and animal (Table 19.3).

Even from a superficial glance at the taxa in Table 19.1, it is evident that diatoms are by far the most diverse group, in terms of numbers of species described as well as numbers of species expected. Rows 7, 12 and 13, the phaeophytes, xanthophytes and chrysophytes, respectively, have around 2,000–2,500 species each. The chrysophytes (enclosed in square brackets) is the taxon most difficult to circumscribe at the moment, with many new taxa being removed from it, usually described

**TABLE 19.1**
**Estimated Numbers of Species in Subgroups of the Stramenochromes**

| Class | Species | Genera |
|---|---|---|
| Bacillariophyta | 15,000 (200,000) | c. 350–400 |
| Bolidophyceae | 2–5 | 1 |
| Chrysomerophyceae* | 6–7 | 5 |
| Dictyochophyceae | 200–400 | 1 |
| Eustigmatophyceae | 12 | 4 |
| Pelagophyceae | 3–5 | 3 |
| Phaeophyceae | 2,000 | 120 |
| Phaeothamniophyceae* | 20–30 | 16 |
| Pinguiophyceae | 1–12 | 5 |
| Raphidophyceae | 8–10 | 3–5 |
| Schizocladophyceae* | 1 | 1 |
| Xanthophyceae | 2,000 | 200 |
| [Chrysophyta | 2,400] | > 200 |

*Note:* For Bacillariophyta (diatoms), the number outside the brackets is the number of species so far described. Those marked with a * have been considered members of Chrysophyta but have recently been removed[48]. Thus figures for taxa marked * relate to the last entry in the table for Chrysophyta, enclosed in square brackets. In Dictyochophyceae approximately 2–7 species are extant, belonging to the same genus; the majority of species in this taxon are known only from fossils, arranged in 10–15 genera.

*Source:* See Andersen[2] for summary; see also Bailey et al.[91]; Andersen et al.[92]; Honda and Inouye[93]; Kawachi et al.[94–96]; O'Kelly[97].

**TABLE 19.2**
**Estimated Numbers of Species in a Selection of Other 'Algal' and Plant Groups**

| Group | Species |
|---|---|
| Bacillariophyta | 200,000 |
| Rhodophyta | 20,000 |
| 'Chlorophyta' | 120,000 |
| Charophyta | 20,000 |
| Cryptophyta | 1,200 |
| Dinophyta | 11,000 |
| Haptophyta | 2,000 |
| Xanthophyta | 2,000 |
| 'Bryophytes' | 15,000 |
| 'Ferns' | 13,025 |
| 'Gymnosperms' | 980 |
| 'Dicotyledons' | 199,350 |
| Monocotyledons | 59,300 |

*Note:* Single quotes indicate non-monophyletic or possibly non-monophyletic group.

*Source:* Data from Cracraft and Donoghue[11].

**TABLE 19.3**
**Estimated Numbers of Species in a Selection of Animal Groups**

| Group | Species |
| --- | --- |
| Bacillariophyta | 200,000 |
| Insects | 950,000 |
| Molluscs | 70,000 |
| Mammals | 4,842 |
| Birds | 9,932 |
| 'Reptiles' | 8,134 |
| 'Amphibians' | 5,578 |
| 'Fishes' | 26,018 |

*Note:* Single quotes indicate non-monophyletic or possibly non-monophyletic group.

*Source:* Data from Cracraft and Donoghue[11].

at the class level (those marked with an asterisk in Table 19.1, see Kristiansen[48]). Chrysophyta sensu lato[48] illustrate another problem when attempting to compare taxa, as many new genera, classes and even families are separated away when they include only one (or very few species), indicating a rather idiosyncratic approach to ranking, if not classification. For example, Kawai et al.[49] erected the new class Schizocladophyceae for the new species *Schizocladia ischiensis*. Comparison between the number of species and number of genera indicates that, again, idiosyncratic criteria are used in assigning rank.

Amongst a wider more varied choice of photosynthetic organisms (Table 19.2), potential numbers of diatom species compare best with dicotyledons, again a large number in the context of the plant kingdom. For example, there are between 11,000 and 13,000 known species of Charophytes but an estimated total of only 20,000; for dinoflagellates there are some 2,000–4,000 known species with an estimated 11,000 total; and for Rhodophytes (red algae) there are some 4,000–6,000 known species with an estimated total of 5,500–20,000. With the exception of the insects as a whole, expected diatom species are far in excess of most other groups (Table 19.3).

We offer two further comparisons of the numbers of diatom taxa to allow some idea of the resolution in terms of higher categories. Our first comparison is with Lepidoptera and angiosperms, as both have similar numbers of known species when compared to expected number of diatom species (Table 19.4). Under the assumption that diatom species will reach the 200,000 mark, then

**TABLE 19.4**
**Comparison between Some Higher Taxa of Lepidoptera and Angiosperms Relative to the Number of Species in Each Group (Number of Diatoms Described)**

| | Superfamilies | Families | Genera | Species |
| --- | --- | --- | --- | --- |
| Lepidoptera | 46 | — | 16,298 | 174,250 |
| Angiosperms | — | 454 | 13,479 | 250,000 |
| Bacillariophyta | — | 90 | 450 | 15,000 |

*Source:* For Lepidoptera, Kristensen[98]; for angiosperms, Brummitt[99]; for Bacillariophyta (diatoms), Round et al.[1].

**TABLE 19.5**
**Comparison between Described Species of Diatoms and Described Species of Mammals, Both Fossil and Recent**

|         |         | Genera | Species |                 |         | Genera | Species |
|---------|---------|--------|---------|-----------------|---------|--------|---------|
| Mammals | Extant  | 1,083  | 5,000   | Bacillariophyta | Extant  | 450    | 12,000  |
|         | Extinct | 4,076  | 20,000  |                 | Extinct | 150?   | c. 5,000 |

*Source:* For mammals, O'Leary et al.[50]; for Bacillariophyta (diatoms), Round et al.[1].

expected genera will be around 9,000 (at 450 times 20 for an even relative increase in numbers). Should this figure be anywhere near accurate, there are at least 8,000 genera needed to achieve some greater resolution.

The numbers of diatom species described so far can be compared with the number of species described in mammals. This comparison, seemingly strange, allows direct comparison between diatoms and a group of organisms with similar numbers of species but considerably better known (Table 19.5; O'Leary et al.[50]). Mammals have fewer species placed in many more genera, implying that the hierarchical structure in mammalian classification is better resolved (understood) and that resolution and structure in diatom classification is rather poor (not well understood).

## 19.4 THERE ARE NAMES, AND THEN THERE ARE NAMES . . .

Published diatom names have been collected together in VanLandingham's eight-volume *Catalogue of Diatoms*[51]. Whilst the catalogue is not a complete record, it is estimated that there are as many as 40,000–50,000 available names. The number of names added to the literature in the last four decades is probably around 10,000–15,000, with estimates suggesting that new taxa are being described at the rate of 50–150 per year, depending on the number of Floras of new or little studied regions that are published. A new catalogue of names is being prepared at the California Academy of Sciences in San Francisco (http://www.calacademy.org/research/diatoms).

One estimate is that there are 450 diatom genera described and accepted. The rate at which generic names appear is fairly uniform, with a mean of 5.5 names per year (Figure 19.3). Exceptions are found in two years. In 1844, 57 new generic names were introduced, primarily as a result of the publication of two large, comprehensive accounts from Ehrenberg[52] and Kützing[53], the former dealing mostly with fossils, the latter with living organisms. In 1990, a major revision of diatom taxonomy was presented in Round et al.[1], a book dealing with the biology and morphology of diatom genera, which included 44 new generic names, not all described for the first time in that book. Of note are two other peaks corresponding to the publications of revisions of diatom classification by Cleve[54] and Karsten[55] and towards the late 1990s many more genera were being routinely described as part of minor taxonomic revisions (Figure 19.3).

However, diatom genera, like most other taxa, are variable in size, with a relatively large number of recently described monotypic genera (as is the case in the Chrysophyta sensu stricto, and its new subgroups; see Table 19.1 and Kristiansen and Preisig[48]) as well as some with well over a thousand species.

Table 19.6 lists genera from VanLandingham's catalogue that have the largest number of species names, from the 9,000+ in *Navicula* Bory to the 850 in *Triceratium* Ehrenb. Table 19.6 includes 15 genera, of which nine have been revised since VanLandingham's catalogue was completed (marked with an asterisk), with most now separated into a series of smaller genera but invariably leaving a large residue of unplaced species. Thus, for example, while the number of species for the genus *Navicula* is very likely artificially high, it has functioned as a 'dustbin' for any bilaterally symmetrical, raphid diatom that lacks the features of the more precisely defined genera separated from or related

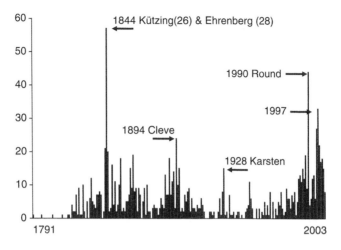

**FIGURE 19.3** Numbers of diatom genera described against time. (Data from Fourtanier and Kociolek[17,18].)

to it. Most 'dustbin' taxa turn out to be paraphyletic nongroups defined by the lack of characters. Nevertheless, since the publication of VanLandingham, *Navicula* has been more precisely defined[56], and some 30 new (or resurrected) genera are now commonly used for subgroups within the older genus. Yet it remains unknown whether any of these new subgroups are monophyletic in any demonstrable

## TABLE 19.6
## List of Diatom Genera with the Largest Number of Species

| Genera | Species | Freshwater | Marine | Fossil | Cosmopolitan |
|---|---|---|---|---|---|
| *Navicula** | 9,000+ | + | + | + | + |
| *Eunotia* | 2,170 | + | - | + | + |
| *Pinnularia* | 1,820 | + | + | + | + |
| *Nitzschia* | 1,745 | + | + | + | + |
| *Coscinodiscus** | 1,395 | − | + | + | + |
| *Synedra** | 1,200 | + | + | + | + |
| *Cymbella** | 1,180 | + | − | + | + |
| *Amphora* | 1,050 | + | + | + | + |
| *Cocconeis* | 1,000 | + | + | + | + |
| *Gomphonema* | 1,000 | + | - | + | + |
| *Achnanthes* * | 970 | + | + | + | + |
| *Fragilaria** | 925 | + | + | + | + |
| *Surirella** | 910 | + | + | + | + |
| *Melosira** | 900 | + | + | + | + |
| *Triceratium** | 850 | + | + | + | + |

*Note:* Genera marked with an * have been revised in the last 10 years and are now better circumscribed as a series of smaller genera. However, in most cases a large residue remains after taxa have been separated.

*Source:* Data based on the numbers of names (rather than taxa) in VanLandingham's *Catalogue of Diatoms*[51]; figures for *Synedra* and *Fragilaria* are supplemented with databases records kept at the Natural History Museum, London as part of ongoing research projects, and the figure for *Navicula* species is from California Academy of Science databases (Kociolek personal communication).

way, that is, have unique sets of synapomorphies (but see Cox and Williams[57,58], for a beginning). Additionally, many of the 'unique' groups of species have been placed not only in a new genus but also in new families, thereby obscuring rather than clarifying relationships (for example, *Cavinula* D.G. Mann and A. Stickle, six species previously in *Navicula,* in Cavinulaceae D.G. Mann).

Nevertheless, the number of new species added to *Navicula* continues to increase regardless of attempts to discover monophyletic subgroups within this large genus[57]. Somewhat ironically, Lange-Bertalot and Moser created the genus *Naviculadicta* Lange-Bertalot[59] to accommodate species that would have previously been described as part of *Navicula* but can now no longer be so because of its more precise definition. In other words, they deliberately created a paraphyletic group from another paraphyletic group, a taxonomic decision that is perhaps unique in systematics (see Kociolek[60] for further commentary).

Yet progress can be made. Between the two genera *Fragilaria* and *Synedra* there are over 2,000 names, covering freshwater, marine, fossil and Recent taxa (Table 19.6). Revision of both genera, beginning in 1986[61], has allowed the discovery and recognition of many better-defined genera from both a morphological perspective as well as their general ecological requirements (freshwater or marine) (Table 19.7; further details can be found in Williams[62]). Many of the remaining names can be examined (on a piecemeal basis if necessary) to see if they 'fit' the described genera. By fit, we

## TABLE 19.7
## Araphid Diatom Genera Described since 1986

| Genus | Date | Approximate Species Numbers | Freshwater(F)/Marine(M) |
|---|---|---|---|
| *Fragilaria* | 1819 | 925 | F |
| *Synedra* | 1830 | 1,200 | F |
| *Staurosira* | 1843 | 10–15 | F |
| *Catacombas* | 1986 | 4–6 | M |
| *Hyalosynedra* | 1986 | 2 | M |
| *Tabularia* | 1986 | 10–25 | M |
| *Ctenophora* | 1986 | 2 | M |
| *Neosynedra* | 1986 | 2 | M |
| *Fragilariforma* | 1988 | 10–25 | F |
| *Staurosirella* | 1987 | 10 | F |
| *Pseudostaurosira* | 1987 | 5–7 | F |
| *Punctastriata* | 1987 | 5–7 | F |
| *Tabulariopsis* | 1988 | 2 | M |
| *Cavitatus* | 1989 | 8–10 | M |
| *Martyana* | 1990 | 3–5 | M |
| *Psammosynedra* | 1993 | 2 | M |
| *Synedropsis* | 1994 | 1 | M |
| *Stauroforma* | 1996 | 2 | F |
| *Nanofrustulum* | 1999 | 4–8 | M |
| *Syndrella* | 2001 | 1 | F |
| *Belanostrum* | 2001 | 1 | F |
| *Pseudostaurosiropsis* | 2001 | 5 | F |
| *Sarcophagodes* | 2002 | 5 | F |

*Note:* Includes taxa that would have been previously placed in either *Fragilaria* or *Synedra* or else with species that would have been placed in either of those two genera when the more general circumscription applied.

*Source:* Data from Williams[61]; Williams and Round[100,101].

mean search for and identify synapomorphic characters, rather than some 'estimate' of similarity, overall or otherwise[63]. Such a search should apply to genera listed in Table 19.7 and, as such, would allow both detailed revisionary work to continue along with floristic work and species recognition. In other words, proper attention to the processes of classification is required, regardless of taxonomic rank, that is the identification of synapomorphic characters (see also Ruck and Kociolek[64] and Williams and Reid[65]). As a rule, once 'large' genera are broken up into smaller more precisely circumscribed units (monophyletic units), they become more sharply defined with respect to other parameters, such as their ecology. One aspect that remains contentious is whether the subdivisions correspond to geographical regions or areas. Put more bluntly: does biogeography matter with respect to diatoms[19]?

## 19.5 THERE IS BIOGEOGRAPHY, AND THEN THERE IS BIOGEOGRAPHY . . .

But there are no such things as water-babies. How do you know that? Have you been there to see? And if you had been there to see, and had seen none, that would not prove that there was none … . And no one has a right to say no water-babies exist till they have seen no water-babies existing; which is quite a different thing, mind, from not seeing water-babies; and a thing which nobody ever did, or perhaps ever will do you know that?

**Kingsley, C.**
*The Water-Babies, a Fairy Tale for a Land Baby, 1863*[66]

The term biogeography can be defined in different ways[67]. To some, biogeography is simply an extension of ecology[68]; to others it deals with the subject of migration and dispersal[69]; and to others it concerns itself with taxa, the areas they occupy and their relationships[70]; and still symposia are organised to address the question 'what is biogeography?'[71]. That such diversity of focus exists makes it somewhat problematic to discuss the issue relative to diatoms and their distribution. Often, one might read of diatoms, or of other 'protists', as having 'no biogeography', or lacking sufficient patterns of distribution to indicate any regional separation of significance[19].

Given the relevance and potential of biogeography for tackling the subject of 'large genera', as well as the more all encompassing world of evolutionary studies[72], we try and tackle the subject here from the viewpoint that biogeographic studies deal with the relationships of taxa and the areas they occupy, a viewpoint which requires thorough taxonomic work on a scale fit to address a particular problem. That is, if the focus is the distribution of freshwater diatoms around the Pacific rim (areas bordering the Pacific Ocean, Eastern Russia, Eastern China, Japan, Western United States, Central America, Western South America, Eastern South East Asia), then ideally one requires knowledge of endemic diatom species from at least three different areas within that region[73]. This raises the issue of endemism and what it might mean.

A paper of major significance in the development of biogeography was that of De Candolle[74], who, amongst other things, was the first to introduce the term endemism to biogeography. He related the idea to genera, defining endemic genera as those with many species confined to one region (De Candolle[74]; see Nelson[75] for translation). Since that time, the term endemism does not refer to any particular taxon or any particular area: any taxon (species, genus) might be endemic to Lake Baikal, East Russia or the Southern hemisphere. Finlay et al.[76] puzzle over the term 'endemic protist'. They offer the argument that because of their small size, many rare species may not be sampled, thus allowing them to be interpreted as endemics when, in fact, they are simply rare. Whilst the possibility of undersampling is a constant issue, it is hard to see how their argument could in any way be either tested or falsified. It is phrased in such a way as to qualify for what one might call a 'water-baby' theory, that not finding something will never be sufficient reason for believing it to not exist. Thus, one might always invoke undersampling for never having found specimens of rare species.

Finlay et al.[76] speak of typical definitions of cosmopolitanism, such as 'widely distributed'. They reject this in favour of 'one that thrives wherever its required habitat is realised'. But this refers to ecology rather than geography, and cosmopolitan, in biogeography, is a geographical term. De Candolle wrote of cosmopolitanism: "The first travellers always thought that they found plants of their home in far-away countries, and they delighted in giving the plants the same names. But as soon as specimens were bought back to Europe, the illusion dissipated for the vast majority of the plant species. Even when examination of dry specimens left some doubt, horticultural investigation generally removed the doubts, leaving only a very small number of species known to be cosmopolitan." (De Candolle[74]; see Nelson[75] for translation).

De Candolle understood cosmopolitanism to be the exception rather than the rule[75]. He referred to those who first put European names to 'foreign' plants only to discover that on inspection they differed enough to have their own identities. Much post World War II diatom taxonomy was guided, if not structured around, European and North America Floras[77–79]. Thus, levels of endemism were masked from recognition[79]. More recently, this focus has changed[61]. A brief examination of the Floras published in the series *Iconographia Diatomologica* yields some interesting figures (Table 19.8).

**TABLE 19.8**
**Numbers of Species in Total against Numbers Described for the First Time**

| Area | Total Species | New (%) | References |
|---|---|---|---|
| Africa | | | |
| Madagascar | 628 | 249 (40%) | Metzeltin and Lange-Bertalot[102], Spaulding and Kociolek[103] |
| South America | | | |
| Tropical South America (including Brazil, Guyana, Venezuela) | c. 700 | 202[a] (29%) | Metzeltin and Lange-Bertalot[104] |
| The Andes (including Ecuador, Venezuela, Chile, Tierra del Fuego) | 888 | 184[b] (21%) | Rumrich, Lange-Bertalot and Rumrich[80] |
| Uruguay | c. 850 | 102[c] (12%) | Metzeltin, Lange-Bertalot and Garcia-Rodriguez[105] |
| North America | | | |
| Cape Cod | c. 250 | 42[d] (17%) | Siver et al.[106] |
| Europe (Boreal) | | | |
| Siberia (Vaigach, Mestnyi and Matveev Islands–North West Siberia) | 490 | 48[e] (10%) | Lange-Bertalot and Genkal[107] |
| Australasia | | | |
| New Caledonia (South West Pacific) | 643 | 257 (40%) | Moser et al.[108,109], Moser[110] |
| Subantarctica | | | |
| Ile de la Possession | 220 | 57[f] (26%) | van de Vijer et al.[111] |

*Note:* Data from a selection of Floras primarily published in the *Iconographia Diatomologica* series. Few of the published Floras deal with noted or recognised biogeographical regions, other than those that coincide with an island (for example Madagascar). Rather, most describe collections made within political regions. Thus, while these data are of limited use in providing a general interpretation, they do provide an indication of the potential number of species yet to be described.

[a] According to Metzeltin and Lange-Bertalot[104], they described 202 new taxa, identified 131 as 'supposedly' new species. If these 131 are confirmed as new species the percentage endemic taxa rises to 47.6%, nearly half.

[b] According to Rumrich et al.[80], they described 84 new taxa, identified 59 as 'probably' new and a further 271 'morphodemes' requiring 'further evaluation'.

[c] Includes 5 species published in an earlier version of this Flora[105].

[d] This number includes taxa that could not be identified; they may not all be new.

[e] According to Lange-Bertalot and Genkal[107], they described 42 new taxa from 159 'taxonomically undefined'.

[f] The total number includes the 20 new species of *Stauroneis* described in a later monograph[112].

There may be much to criticise in the sampling regimes and taxonomic procedures adopted for each of these floras. For example, the Andes is not a region in the biological sense, as either side of this mountain range has quite distinct floras, and to include samples from Ecuador, Venezuela, Chile and Tierra del Fuego, may obscure rather than illuminate regional differences[80]. However, nearly all report levels of endemism from 10% of the total up to the remarkable figure of 40% (Table 19.8). Perhaps even for diatoms, cosmopolitanism may be the exception. Nevertheless, endemism needs to be understood as a hierarchical concept, with cosmopolitanism describing global distributions, if indeed anything can be truly global[81].

Rather than attempt to discover if there are any patterns of distribution to explain, Finlay et al.[19,76,82] proposed the term 'ubiquitous dispersal', a process, one imagines, intended to describe all 'protist' distribution. 'Ubiquitous dispersal' is a process meaning "random dispersal across all spatial scales, all the way up to the global scale"[76], and it "is essentially a 'neutral' process, driven by the absolute abundance of organisms"[76]. To bolster their argument — that it is the sheer quantity of organisms as the driving force — they refer to an analogy: "Millions of people indulge every week [in the lottery], but the grim truth is that each individual has a vanishingly low probability of winning the big prize. Almost every week, however, one or a few randomly self-selecting individuals do win because of the vast number of individuals taking part raises the probability that someone has to win."[76]

Analogies do not usually bear close examination. Yet the form of argument — probabilities that something will happen somewhere — is not new, extending back to Darwin's notion of dispersal as an explanation for disjunct distributions, a process that Finlay and his colleagues take for granted as the driving force. George Gaylord Simpson, for example, made the same kind of argument in 1952 with respect to geological time:

> If the probability that some member of a population will cross a barrier is .000001 in any one year, in a large population this means that the probability for any one designated individual is almost infinitesimally small, so much that it would seem absolutely impossible to even the best qualified observer in the field. Yet during the course of a million years the event would be probable, $p = 0.63$, again. In the course of 10 million years the event would become so extremely probable as to be, for most practical purposes, certain, $p = 0.99995$[83].

Both Finlay and Simpson argue not for 'ubiquitous dispersal' but a version of 'improbable dispersal'[75], effectively suggesting that given enough specimens (Finlay and Simpson) or enough time (Simpson), a water-baby will indeed be found (see also Lund[84] and Wilkinson[85]). Such arguments appeal to probabilities rather than facts, the latter being the number of endemic diatoms recognised and their nonrandom distribution[65,73,79,86]. Such arguments also ignore the fact that, while life evolves on Earth, so the Earth does too, both, in fact, together[87]. Alexander du Toit (1878–1948) was one of the first persons to defend (and promote) continental drift; Simpson, initially, saw no use for continental movements to explain organism distributions. Du Toit wrote in response to one of Simpson's papers: "The notion of random, and sometimes two-way 'rafting' across the wide oceans … evinces, however, a weakening of the scientific outlook, if not a confession of doubt from the viewpoint of organic evolution"[88] (also see McCarthy[89]).

From an earlier diatom perspective Ehrenberg had considered the distribution of organisms to be a problem worthy of consideration, especially to explain the disjunct distribution of fossil 'Infusoria' around Pacific coastal regions: "… the Rocky Mountains are a more powerful barrier between the two sides of America, than the Pacific Ocean between America and China; the infusorial forms of Oregon and California being wholly different from those of the east side of the mountains, while they are partly identical with Siberian species"[90].

For example, our work on Lake Baikal has focused on the genus *Eunotia* Ehrenb. Among the many species present in the Lake was *E. clevei,* a rather large and unusual species for the genus. Intensive examination of the Lake Baikal flora and surrounding areas yielded a number of species endemic to the lake, all most closely related to each other (by virtue of synapomorphies) and best separated into a new genus[65]. Further examination of more specimens yielded not wider distributions but more species, some occurring in other large, deep water lakes (for example, Lake Hovsgol), others extending into South East Asia and (often) into marine waters[65]. Furthermore, a number of extinct fossil specimens added yet more species, with a West Coast North American–Chinese South East Asian distribution suggesting a trans-Pacific relationship, one already found for *Tetracyclus* and offering a test for Ehrenberg's earlier proposals[73].

## 19.6 SUMMARY

Although estimates for numbers of species differ, diatoms are a large group, most likely exceeding 200,000 species (fossil and Recent). Their taxonomic hierarchy is poorly resolved, many genera having more than 1,000 available names, with few families circumscribed in such a way as to assist identification. The adoption of cladistic approaches to classification has been slow; even now, few taxa are demonstrably monophyletic: that is, form groups supported by synapomorphies (homologies). Because the relationship between taxon and synapomorphy (homology) is so well established, cladistic approaches seem more than reasonable, in spite of the perceived difficulty in finding appropriate synapomorphic resemblances.

Since the 1950s, geographical data have rarely been considered useful information, apart from the largely nomenclatural tradition of noting a specimen's location when found. Thus, we suggest the most reasonable way forward for diatom taxonomy when considering 'large' genera (taxa) is to adopt a regional approach and attack the problem from both a taxonomic (cladistic) and geographical perspective, allowing testable theories of relationships among the organisms investigated and the areas they occupy.

## REFERENCES

1. Round, F.E., Crawford, R.M., and Mann, D.G., *The Diatoms: Biology and Morphology of the Genera*, Cambridge University Press, Cambridge, 1990.
2. Andersen, R.A., Biology and systematics of Heterokont and Haptophyte algae, *Amer. J. Bot.*, 91, 1508, 2004.
3. Guillou, L. et al., *Bolidomonas:* a new genus with two species belonging to a new algal class, the Bolidophyceae (Heterokonta), *J. Phycol.*, 35, 368, 1999.
4. Guillou, L. et al., Diversity and abundance of Bolidophyceae (Heterokonta) in two Oceanic regions, *App. Environ. Microb.*, 65, 4528, 1999.
5. Kühn, S., Medin, M., and Eller, G., Phylogenetic position of the parasitoid nanoflagellate *Pirsonia* inferred from nuclear-encoded small subunit ribosomal DNA and a description of *Pseudopirsonia* n. gen. and *Pseudopirsonia mucosa* (Drebes) comb. nov., *Protist*, 155, 143, 2004.
6. Patterson, D.J., Stramenopiles: chromophyte from a protistan perspective, in *The Chromophyte Algae: Problems and Perspectives*, Green, J.C., Leadbeater, B.S.C., and Diver, W.L., Eds., Clarendon Press, Oxford, 1989, 357.
7. Leipe, D.D. et al., 16S-like rRNA sequences from *Developayella elegans, Labyrinthuloides haliotidis*, and *Proteromonas lacerate* confirm that the stramenopiles are a primarily heterotrophic group, *Eur. J. Protistol.*, 32, 449, 1996.
8. Leipe, D.D. et al., The stramenophiles from a molecular perspective: 16S-like rRNA sequences from *Labyrinthuloides minuta* and *Cafeteria roenbergensis, Phycologia*, 33, 369, 1994.
9. Patterson, D.J., The diversity of eukaryotes, *Am. Nat.*, 154, 96, 1999.
10. Ben Ali A., et al., Phylogenetic relationships among algae based on complete large-subunit rRNA sequences, *Int. J. Syst. Evol. Micr.*, 51, 737, 2001.
11. Cracraft, J., and Donoghue, M.J., Eds., *Assembling the Tree of Life*, Oxford University Press, Oxford, 2004.
12. Delwiche, C.F. et al., Algal evolution and the early relation of green plants, in *Assembling the Tree of Life*, Cracraft, J., and Donoghue, M.J., Eds., Oxford University Press, Oxford, 2004, 121.

13. Nikolaev, V.A. and Harwood, D.M., Morphology and taxonomic position of the Late Cretaceous diatom genus *Pomphodiscus* Barker and Meakin, *Micropaleontology,* 46, 167, 2000.
14. Mann, D.G. and Droop, S.J.M., Biodiversity, biogeography and conservation of diatoms, *Hydrobiologia,* 336, 19, 1996.
15. Frodin, G.G., History and concepts of big plant genera, *Taxon,* 53, 753, 2004.
16. Wheeler, Q.D., Taxonomic triage and the poverty of phylogeny, *Phil. Trans. R. Soc. Lond. B,* 359, 571, 2004.
17. Fourtanier, E. and Kociolek, J.P., Catalogue of the diatom genera, *Diatom Res.,* 14, 1, 1999.
18. Fourtanier, E. and Kociolek, J.P., Addendum to 'Catalogue of the diatom genera', *Diatom Res.,* 18, 245, 2003.
19. Finlay, B.J., Monaghan, E.B., and Maberly, S.C., Hypothesis: the rate and scale of dispersal of freshwater diatom species is a function of their global abundance, *Protist,* 153, 261, 2002.
20. Finlay, B.J. and Fenchel, T., Cosmopolitan metapopulations of free-living microbial eukaryotes, *Protist,* 155, 237, 2004.
21. Desmond, A., *Huxley: The Devil's Disciple,* Michael Joseph, London, 1994.
22. Desmond, A., *Huxley: Evolution's High Priest,* Michael Joseph, London, 1997.
23. Huxley, T.H., The Gentians: notes and queries, *J. Linn. Soc.,* 24, 101, SM 4, 612, 1888.
24. Moss, W., Taxa, taxonomists, and taxonomy, in *Numerical Taxonomy,* J. Felsenstein, Ed., NATO ASI series G, Ecological Sciences 1, 72, 1983.
25. Williams, D.M and Ebach, M.C., *The Foundations of Comparative Biology,* Kluwer Academic-Plenum Publishers, submitted.
26. Williams, D.M., Classification, Collections, Diatoms and Biogeography, in *The British Phycological Society 50th Jubilee Meeting, Programme and Abstracts,* 6, 2002.
27. Hubbard, C.E., William Turrill, *Biographical Memoirs of Fellows of the Royal Society,* 17, 689, 1971.
28. Turrill, W.B., Some taxonomic aims, methods and principles: their possible application to the algae, *Nature,* 169, 388, 1952.
29. Turrill, W.B., The investigation of plant species, *Proc. Linn. Soc. Lond.,* 147, 104, 1935.
30. Turrill, W.B., Species, *J. Bot.,* 63, 359, 1925.
31. Winsor, M.P., The English debate on taxonomy and phylogeny, 1937–1940, *Hist. Phil. Life Sci.,* 17, 227, 1995.
32. Huxley, T.H., *Evolution: The Modern Synthesis,* G. Allen and Unwin, London, 1942.
33. Turesson, G., The species and the variety as ecological units, *Hereditas,* 3, 100, 1922.
34. Gilmour, J.S.L. and Heslop-Harrison, J., The deme terminology and the units of micro-evolutionary change, *Genetica,* 27, 147, 1954.
35. Walters, S.M., Experimental and orthodox taxonomic categories and the deme terminology, *Plant Syst. Evol.,* 167, 35, 1989.
36. Danser, H.B., Über die begriffe komparium, kommiskuum und konvivien und über die Entstehungsweise der konvivien, *Genetica,* 11, 399, 1929.
37. Gilmour, J.S.L. and Turrill, W.B., The aim and scope of taxonomy, *Chronica Botanica,* 6, 217, 1941.
38. Sokal, R.R. and Sneath, P.H.A., *Principles of Numerical Taxonomy,* W.H. Freeman and Co., San Francisco, 1963.
39. Sneath, P.H.A., and Sokal, R.R., *Numerical Taxonomy. The Principles and Practices of Numerical Classification,* W.H. Freeman and Co., San Francisco. 1973.
40. Winsor, M.P., Species, demes, and the omega taxonomy: Gilmour and The New Systematics, *Biol. Philos,* 15, 349, 2000.
41. Boney, A.D., The 'Tansley Manifesto' affair, *New Phytol.,* 118, 3, 1991.
42. Williams, D.M., Haeckel, E., and Agassiz, L., Trees that bite and their geographical dimension, in *What is Biogeography?* Ebach, M.C. and Tangey, R., Eds., CRC Press, 2006.
43. Mayr, E., *Principles of Systematic Zoology,* McGraw Hill, New York, 1969.
44. Dayrat, B., Towards integrative taxonomy, *Biol. J. Linn. Soc.,* 85, 407, 2005.
45. Nelson, G., Species and taxa: systematics and evolution, in *Speciation and Its Consequences,* Otte, D. and Endler, J.A., Eds., Sinauer Associates, Sunderland, MA, 1989, 60.
46. Poulin, M. and Williams, D.M., Conservation of diatom biodiversity: a perspective, *Proceedings of the 15th International Diatom Symposium,* 161, 2002.
47. Ebach, M.C. and Williams, D.M., Classification, *Taxon,* 53, 791, 2004.

48. Kristensen, J. and Preisig, H.R., Encyclopedia of Chrysophyte genera, *Bibliotheca Phycologia,* 110, 1, 2001.

49. Kawai, H. et al., *Schizocladia ischiensis:* a new filamentous marine Chromophyte belonging to a new class, Schizocladiophyceae, *Protist,* 154, 211, 2003.

50. O'Leary, M.A. et al., Building the mammalian sector of the tree of life: combining different data and a discussion of divergence times for placental mammals, in *Assembling the Tree of Life,* Cracraft, J. and Donoghue, M.J., Eds., Oxford University Press, Oxford, 2004, 490.

51. VanLandingham, S.L., *Catalogue of the Fossil and Recent Genera and Species of Diatoms and Their Synonyms,* Cramer, Lehre, 1967–1979.

52. Ehrenberg, C.G., *Verbreitung und Einfluss des mikroskopischen Lebens in Süd- und Nord-Amerika,* Abhandlungen der Königliche Akademie der Wissenschaften zu Berlin (1841), 1844, 291.

53. Kützing, F.T., *Die kieselschaligen Bacillarien oder Diatomeen,* Nordhausen, 152 S., 30 Taf. 1844, Auflage 2, 1865.

54. Cleve, P.T., Synopsis of naviculoid diatoms, *Kongliga Svenska Vetenskaps-Akademiens Handlingar,* 26, 1, 1894.

55. Karsten, G., Abteilung Bacillariophyta (Diatomeae), in *Die naturlichen Pflanzenfamilien, Peridineae (Dinoflagellatae), Diatomeae (Bacillariophyta), Myxomycetes,* Engler, A. and Prantl, K., Eds., Wilhelm Engelmann, Leipzig, 2, 105, 1928.

56. Cox, E.J., Studies on the diatom genus *Navicula* Bory: the typification of the genus, *Bacillaria,* 2, 137, 1979.

57. Cox, E.J., and Williams, D.M., Systematics of naviculoid diatoms: the interrelationships of some taxa with a stauros, *Eur. J. Phycol.,* 35, 273, 2003.

58. Cox, E.J., and Williams, D.M., Systematics of Naviculoid diatoms (Bacillariophyta): a preliminary analysis of protoplast and frustule characters for family and order level classification, *Syst. Biodivers.,* 4, 2006.

59. Lange-Bertalot, H. and Moser, G., *Brachysira:* Monoraphie der Gattung, *Bibliotheca Diatomologica,* 32, 1, 1994.

60. Kociolek, J.P., Comment: taxonomic instability and the creation of *Naviculadicta* Lange-Bertalot in Lange-Bertalot and Moser, a new catch-all genus of diatoms, *Diatom Res.,* 11, 219, 1996.

61. Williams, D.M., Comparative morphology of some species of *Synedra* with a new definition of the genus, *Diatom Res.,* 1, 131, 1986.

62. Williams, D.M., Some notes on the classification of *Fragilaria, Synedra* and their sub-groups, in *Microalgal Biology, Evolution and Ecology,* Crawford, R.M., Moss, B., Mann, D.G., and Preisig, H.R., Eds., Nova Hedwigia Beihefte 130, Gebrüder Borntraeger Verlagsbuchhandlung, Science Publishers, Stuttgart, 2006.

63. Kociolek, J.P., Historical constraints, species concepts and the search for a natural classification, *Diatom,* 13, 3, 1997.

64. Ruck, E.C. and Kociolek, J.P., Preliminary phylogeny of the family Surirellaceae (Bacillariophyta), *Bibliotheca Diatomologica,* 50, 1, 2004.

65. Williams, D.M. and Reid, G., *Amphorotia* nov. gen., a new genus in the family Eunotiaceae (Bacillariophyceae), based on *Eunotia clevei* Grunow in Cleve et Grunow, *Diatom Monographs,* 6, 1, 2005.

66. Kingsley, C., *The Water-Babies, A Fairy Tale for a Land Baby,* Macmillan, London, 1863.

67. Ebach, M.C., Forum on biogeography: introduction, *Taxon,* 53, 889, 2004.

68. Walter, H.S., Understanding places and organisms in a changing world, *Taxon,* 53, 905, 2004.

69. Avise, J.C., What is the field of biogeography, and where is it going? *Taxon,* 53, 893, 2004.

70. Parenti, L.R. and Humphries, C.J., Historical biogeography, the natural science, *Taxon,* 53, 899, 2004.

71. Ebach, M.C. and Tangey, R., *What is Biogeography?* CRC Press, 2006.

72. Williams, D.M. and Ebach, M.C., The reform of palaeontology and the rise of biogeography: 25 years after 'Ontogeny, Phylogeny, Paleontology and the Biogenetic law' (Nelson 1978), *J. Biogeogr.,* 31, 685, 2004.

73. Williams, D.M., Fossil species of the diatom genus *Tetracyclus* (Bacillariophyta, 'ellipticus' species group): morphology, interrelationships and the relevance of ontogeny, *Phil. Trans. R. Soc., Lond. B,* 351, 1759, 1996.

74. Candolle, A.P., Géographie botanique, in *Dictionnaire des Sciences Naturelles,* XVIII, Strasbourg and Paris, 1820.

75. Nelson, G., From Candolle to Croizat: comments on the history of biogeography, *J. Hist. Biol.,* 11, 269, 1978.
76. Finlay, B.J., Esteban, G.F., and Fenchel, T., Protist diversity is different? *Protist,* 155, 15, 2004.
77. Hustedt, F., Bacillariophyta (Diatomeae), in *Die Süsswasser-Flora Mitteleuropas,* 2nd ed., Pascher, A., Ed., G. Fischer, Jena, 1930, 10.
78. Patrick, R. and Reimer, C.W., The diatoms of the United States exclusive of Alaska and Hawaii, *Monographs of the Natural Sciences of Philadelphia,* 13, 1, 1966.
79. Kociolek, J.P. and Spaulding, S.A., Freshwater diatom biogeography, *Nova Hedwigia,* 71, 223. 2000.
80. Rumrich, U., Lange-Bertalot, H., and Rumrich, M., Diatoms of the Andes (from Venezuela to Patagonia/Tierra del Fuego), *Annotated Diatom Micrographs, Iconographia Diatomologica,* 9, 1, 2000.
81. Williams, D.M., On diatom endemism and biogeography: *Tetracyclus* and Lake Baikal endemic species, *Proceedings of the 17th International Diatom Symposium,* 433, 2004.
82. Finlay, B.J. and Clarke, K.J., Ubiquitous dispersal of microbial species, *Nature,* 400, 828, 1999.
83. Simpson, G.G., Probabilities of dispersal in geological time, *Bull. Am. Mus. Nat. Hist.,* 99, 163, 1952.
84. Lund, J.W.G., *Annual Report of the Freshwater Biological Association,* 70, 43, 2002.
85. Wilkinson, D.M., Dispersal, cladistics and the nature of biogeography, *J. Biogeogr.,* 30, 1779, 2003.
86. Williams, D.M., Diatom biogeography: some preliminary considerations, *Proceedings of the 13th International Diatom Symposium,* BioPress Ltd, 311, 1994.
87. Nelson, G. and Platnick, N., *Systematics and Biogeography: Cladistics and Vicariance,* Columbia University Press, New York, 1981, 567.
88. Du Toit, A., Tertiary mammals and continental drift: a rejoinder to George G. Simpson, *Am. J. Sci.,* 242, 145, 1944.
89. McCarthy, D., Biogeography and scientific revolutions, *The Systematist,* 25, 3, 2005.
90. Ehrenberg, G.C., On infusorial deposits on the River Chutes in Oregon, *Am. J. Sci.,* 2nd ser., 9, 140, 1850.
91. Bailey, J.C. et al., Phaeothamniophyceae classis nova.: a new lineage of chromophytes based upon photosynthetic pigments, *rbcL* sequence analysis and ultrastructure, *Protist,* 149, 245, 1998.
92. Andersen, R.A., Potter, D., and Bailey, C.J., *Pinguiococcus pyrenoidosus* gen. et sp. nov. (Pinguiophyceae), a new marine coccoid alga, *Phycol. Res.,* 50, 57, 2002.
93. Honda, D. and Inouye, I., Ultrastructure and taxonomy of a marine photosynthetic stramenopile *Phaeomonas parva* gen. et sp. nov. (Pinguiophyceae) with emphasis on the flagellar apparatus architecture, *Phycol. Res.,* 50, 75, 2002.
94. Kawachi, M., Noël, M.H., and Andersen, R.A., Re-examination of the marine 'chrysophyte' *Polypodochrysis teissieri* (Pinguiophyceae), *Phycol. Res.,* 50, 91, 2002.
95. Kawachi, M. et al., *Pinguiochrysis pyriformis* gen. et sp. nov. (Pinguiophyceae), a new picoplanktonic alga isolated from the Pacific Ocean, *Phycol. Res.,* 50, 49, 2002.
96. Kawachi, M. et al., The Pinguiophyceae *classis nova,* a new class of photosynthetic stramenopiles whose members produce large amounts of omega-3 fatty acids, *Phycol. Res.,* 50, 31, 2002.
97. O'Kelly, C.J., *Glossomastix chrysoplasta* n. gen., n. sp. (Pinguiophyceae), a new coccoidal, colony-forming golden alga from southern Australia, *Phycol. Res.,* 50, 67, 2002.
98. Kristensen, N.P., Ed., *Lepidoptera, Moths and Butterflies: Vol. 1 Evolution, Systematics, and Biogeography,* W. de Gruyter, Berlin, New York, 1999, 1.
99. Brummitt, R.K., *Vascular Plant Families and Genera,* Royal Botanic Gardens, Kew, 1992, 804.
100. Williams, D.M and Round, F.E., Revision of the genus *Fragilaria, Diatom Res.,* 2, 267, 1987.
101. Williams, D.M. and Round, F.E., Revision of the genus *Synedra* Ehrenb., *Diatom Res.,* 1, 313, 1986.
102. Metzeltin, D. and Lange-Bertalot, H., Diatoms from the 'Island Continent' Madagascar, *Annotated Diatom Micrographs, Iconographia Diatomologica,* 11, 1, 2002.
103. Spaulding, S.A., and Kociolek, J.P., Freshwater diatoms (Bacillariophyceae), in *Natural History of Madagascar,* Goodman, S. and Benstead, J., Eds., University of Chicago Press, 276, 2003.
104. Metzeltin, D. and Lange-Bertalot, H., Diversity–taxonomy–geobotany: tropical diatoms of South America I: about 700 predominantly rarely known or new taxa representative of the neotropical flora, *Annotated Diatom Micrographs, Iconographia Diatomologica,* 5, 1, 1998.
105. Metzeltin, D., Lange-Bertalot, H., and Garcia-Rodriguez, F., Diatoms of Uruguay, *Annotated Diatom Micrographs, Iconographia Diatomologica,* 15, 1, 2004.

106. Siver, P.A. et al., Diatoms of North America. The freshwater flora of Cape Cod, *Annotated Diatom Micrographs, Iconographia Diatomologica*, 14, 1, 2005.

107. Lange-Bertalot, H., and Genkal, S.I., Phytogeography–diversity–taxonomy: diatoms from Siberia I: islands in the Arctic Ocean (Yugorsky–Shar Strait), 2nd corrected printing, *Annotated Diatom Micrographs, Iconographia Diatomologica*, 6, 1, 1999.

108. Moser, G., Steindorf, A., and Lange-Bertalot, H., Neukaledonien. Diatomeenflora einer Tropeninsel, revision der collection Maillard und untersuchung neuen materials, *Bibliotheca Diatomologica*, 32, 1, 1995.

109. Moser, G., Lange-Bertalot, H., and Metzeltin, D., Island of endemics, New Caledonia: a geobotanical phenomenon, *Bibliotheca Diatomologica*, 38, 1, 1998.

110. Moser, G., Die Diatomeenflora von Neukaledonien–Systematik–Geobotanik — Ökologie: Ein Fazit, *Bibliotheca Diatomologica*, 43, 1, 1999.

111. van de Vijver, B., Frenot, Y., and Beyens, L., Freshwater Diatoms from Ile de la Possession (Crozet Archipelago, Subantarctica), *Bibliotheca Diatomologica*, 46, 1, 2002.

112. Van de Vijver, B., Beyens, L., and Lange-Bertalot, H., The genus *Stauroneis* in the Arctic and (sub-) Antarctic regions, *Bibliotheca Diatomologica*, 51, 1, 2004.

# 20 Systematics of the Species Rich Algae: Red Algal Classification, Phylogeny and Speciation

*J. Brodie*
Botany Department, The Natural History Museum, London, UK

*G. C. Zuccarello*
School of Biological Sciences, Victoria University of Wellington, New Zealand

## CONTENTS

## ABSTRACT

The algae are a non-monophyletic group of highly numerous organisms which exhibit extremely diverse morphologies. It is estimated that there are >350,000 species of algae globally, although only a fraction of this number have been described. Here we use the red algae (Rhodophyta) to demonstrate the approaches that have been taken in their identification and classification. The red algae are an ancient and morphologically highly diverse group with about 5,800 described species. Different approaches have been employed in their classification, including anatomical, biochemical and physiological studies, but molecular studies have also had a profound impact on our understanding of their evolution. Ordinal classification has increased from four orders recognised in the nineteenth century to 30 currently recognised. Based primarily on morphological observations, some orders have considerably more species than others; for example, the largest is Ceramiales with c. 2,300 species, with some of its genera being very large (>200 species). We explore in the red algae what factors, many of them unique to this group, may have led to high levels of speciation (reproductive isolation). In red algae, levels of genetic uniqueness are shown to be correlated with reproductive isolation and not always with morphological distinctness. There is also evidence that red algal populations are highly differentiated over small distances. In addition, red algae have unique reproductive systems that may lead to the easy acquisition of reproductive isolation. In order to obtain a greater understanding of the causes and mechanisms of reproductive isolation in red algae, we propose that a new line of research targeted at reproductive incompatibility be explored. We conclude that the continuation of a multifaceted approach, including molecular techniques, population studies and cell biology remains necessary to illuminate evolution in the red algae.

## 20.1  INTRODUCTION

The algae are an artificial, non-monophyletic, grouping of highly numerous organisms exhibiting extremely diverse morphologies which have been grouped together on the basis of their ability to photosynthesise and that they are not higher plants (see Williams and Reid, *Chapter 19*). They have been variously classified from Linnaeus'[1] concept in 1753 of a few genera within a subdivision of the class Cryptogamia, which also included the ferns, mosses and fungi, through some 15 genera[2] to the latest classifications in which we find the eukaryotic algae placed among several of the major lineages of the organisms on Earth[3,5]. Keeling[4] has tentatively identified five supergroups of eukaryotes where the distribution of the plants reflects a history of endosymbiosis. Algae are found in four of these five groups. Green and red algae and glaucophytes are in Primoplantae, which arose from primary cyanobacterial symbiosis. Dinoflagellates, phaeophytes, chrysophytes, bacillariophytes and haptophytes are placed in Chromalveolates and possibly have plastids as a result of red algal secondary symbiosis. Chlorarachniophytes are in Rhizaria and euglenophytes in Excavates, and both these algal groups are thought to have evolved 'plant like' attributes through secondary symbiosis via green algal endosymbionts (Palmer et al.[5], and see Hodkinson and Parnell, *Chapter 1*, for a phylogenetic tree of these groups). If we are to continue to refine the tree of life and resolve its structure at different levels of classification, then it is necessary to continue to identify and describe algae, to understand the interrelationships of algal groups and to understand some of the mechanisms that lead to this great diversity.

It has been estimated that there could be over 350,000 species of algae[6] (Table 20.1). To put this into context, despite this figure being subject to large margins of error, there are c. 300,000 higher plant species (bryophytes, pteridophytes, gymnosperms and angiosperms). However, the number of species estimated for each algal group varies enormously. While some groups of algae, such as the brown algae, appear to be relatively well documented taxonomically, the number of described species for others, such as the diatoms (silicated unicells) falls well below the estimated total (Table 20.1). Whatever the ultimate reality, evidence suggests that there is still an enormous number of species to be identified, particularly in some groups. For example, diatoms require much further study to determine their diversity (see Williams and Reid, *Chapter 19*), and the same is true for algae that occur in terrestrial habitats such as on bark, leaves and rocks[6], and for minute planktonic organisms that have only been recognised in the last quarter century[7]. Whatever the true figure of algal diversity, we are faced with a bewildering number of species and therefore need to consider the ways in which species identification of large taxa can be studied. We also need to consider the implications of the additional knowledge of all new species at all levels of classification in relation to the topology of the tree of life.

The red algae (phylum Rhodophyta) have approximately 5,800 described species[8], making them a species rich group within the algae. In this chapter, we examine how this group of algae has been studied in order to identify and classify species. The first part of the chapter defines the red algae and how phycologists have tried to identify and classify species using both traditional and modern techniques. The second part examines the higher classification and describes how this has developed over the last 250 years. The third part explores work in defining red algal species and factors that may have lead to these high levels of diversity. In this chapter we concentrate on the subphylum

**TABLE 20.1**
**The Eukaryote Algal Groups and Estimated Numbers of Species in Each Division**

| Division | Subdivision | Algal Groups | Estimated No. of Described Species | Estimated No. of Species |
|---|---|---|---|---|
| Excavates | | Euglenophytes (euglenoids) | 959 | 800 |
| Rhizaria | | Chlorarachniophytes | 12 | 12 |
| Primoplantae | | Charophytes (stoneworts) | 1,602 | 20,000 |
| | | Chlorophytes (green algae) | 3,215 | 120,000 |
| | | Rhodophytes (red algae) | 5,781 | 20,000 |
| | | Glaucophytes | 4 | 13 |
| Chromalveolates | Chromists | Phaeophytes (brown algae) | 1,718 | 2,000 |
| | | Chrysophytes (golden algae) | 2,400 | 2,400 |
| | | Bacillariophytes (diatoms) | 6,423 | 200,000 |
| | | Haptophytes (coccolithophorids) | 510 | 2,000 |
| | | Cryptophytes (cryptomonads) | 85 | 1,200 |
| | Alveolates | Dinophytes (dinoflagellates) | 1,240 | 11,000 |

*Source:* Classification based on Keeling[4].

Eurhodophytina, which reflects the new red algal classification system of Saunders and Hommersand[9] and is the most species-rich red algal group.

## 20.2  THE RED ALGAE

The red algae are characterised by having plastids that contain phycobiliproteins (phycoerythrin, phycocyanin and allophycocyanin) as accessory pigments and by the absence of centrioles and flagella[10]. They are an ancient and morphologically highly diverse group of primarily marine organisms. The oldest known fossil of a red alga is *Bangiomorpha pubescens* Butterfield, and this is dated at 1.2 billion years old[11]. This is a highly significant fossil, because it is the oldest taxonomically resolved eukaryote in the fossil record. Although there are almost 5,800 described red algal species, estimates are as high as 20,000 species[6]. The exact number of red algae remains uncertain, because species are extremely difficult to identify on morphological grounds alone. So although the number of known species might not seem great in relation to, for example, some flowering plant groups such as orchids (Orchidaceae) or composites (Asteraceae), it is a species rich group of algae. As any survey of the marine seaweed floras demonstrates[12–15] Rhodophyta is the most species rich macroalgal division, with more species than the green and brown seaweeds combined. Furthermore, it contains several genera with large numbers of species.

This diversity is reflected in the complexity and diversity of red algal morphology and anatomy. Conspicuous morphological features with which species can be readily identified are often lacking in the red algae; morphology can vary between individuals of the same species, but it can also be highly convergent. This is further compounded by the complexities of red algal life histories. This has inevitably forced phycologists to look for characters in addition to morphology to distinguish between species to draw conclusions about taxonomic relationships. Different approaches undertaken include anatomical, biochemical and physiological studies, but it is molecular studies over the last decade that have profoundly changed our notions of red algal classification (and indeed seaweed classification) at all levels.

### 20.2.1  HIGHER CLASSIFICATION OF RED ALGAE

The history of red algal systematics is well described in Dixon and Irvine[16], who provided an insight into the difficulties presented and the approaches that were taken to classify red algae from the origins of the current classification started by Linnaeus in the eighteenth century to the time that they were writing. The difficulty of identifying and classifying red algae as a consequence of morphological variation is well illustrated from the beginning of such studies. Linnaeus'[1] class Cryptogamia in which all algae were placed, was created for the groups of species which did not display phanerogamic reproductive structures and was included in just one of 24 classes (the other 23 being devoted to the phanerogams). Linnaeus' classification accepted 14 genera of algae, but essentially only *Chara, Conferva, Fucus* and *Ulva* contained organisms now accepted as algae[17]. Species of red algae were referred to on the basis of their morphological forms (*Conferva,* slender and filamentous; *Fucus,* fleshy or cartilaginous thalli; *Ulva,* flat, membranous thalli).

In the nineteenth century, colour was used by both Lamouroux[18] and Harvey[19] to split up the algae. In 1813 Lamouroux[18] established the Floridée for what we now know as species belonging to the red algae, and from which Florideophyceae is derived. In 1836 Harvey[19] divided the algae into Rhodospermae (red algae), Melanospermae (brown algae), Chlorospermae (green algae) and Diatomaceae (diatoms). The difficulty of classifying the red algae based on morphology and colour is well illustrated when considering what were originally thought of as red algae. Harvey's[19] Rhodospermae was largely accurate, with the exception of *Porphyra* and *Bangia,* which were placed in the green algae until Berthold[20] in 1882 aligned them with the red algae. The ability to distinguish the red seaweeds from the browns and greens can still be a problem for the beginner today.

By the start of the twentieth century the two subclasses of the red algae, Bangiophycidae (as Bangioideae) and Florideophyceae (as Florideae) were established and have remained a subject for debate since that time. At the supraordinal level, relationships had remained virtually unchanged since 1900 until Magne[21] proposed a new scheme in 1989 based on morphological characters and reproductive systems. He proposed three subclasses for Rhodophyceae: Archaeorhodophycidae (without sporangia), Metarhodophycidae (having only a portion of the parent cell converted to a sporangium) and Eurhodophycidae (having sporangia), which includes Bangiales and Florideophyceae. Magne's hypothesis was an attempt to infer monophyletic lineages but failed to take into account the diversity within and between lineages.

The next major revision was by Saunders and Hommersand[9], who proposed a classification based on recent and traditional evidence. In this classification, summarised in Figure 20.1, all the species encompassed traditionally in Rhodophyta have been placed in the subkingdom of Rhodoplantae, with two phyla: Cyanidiophyta and Rhodophyta. Rhodophyta are split into three subphyla: Rhodellophytina, Metarhodophytina and Eurodophytina. Bangiophyceae and Floridae are raised to class status within Eurodophytina. Floridae are further subdivided into four subclasses; Hildenbrandiophycidae, Nemaliophycidae, Ahnfeltiophycidae and Rhodomeniophycidae. There is a lot of work to be done to refine this classification, as acknowledged by Saunders and Hommersand[9], but for the first time in the history of red algal classification, it should be a better reflection of evolutionary history and diversity.

## 20.2.2 ORDERS OF RED ALGAE

Since the last part of the nineteenth century, the number of orders of red algae has gone from four[22] to thirty[9]. The first and, as noted by Dixon and Irvine[16], 'crucial' step in creating the basis for red algal systematics over the next 100 years or so was the recognition in Florideophyceae (now classified as Floridae) of the importance of the development of the carposporophyte, in particular the recognition of the 'auxiliary cell' and its position and time of formation. This is the cell that receives the diploid nucleus from the carpogonium (egg), either directly or through specialised filaments, and gives rise to tissue that develops on the female plant after fertilisation and which ultimately produces diploid sporangia. There were other alterations to the ordinal classification, notably the creation of Ceramiales by Oltmanns[23], where the auxiliary cell is formed after fertilisation. Guiry[24,25] used comparative anatomy to create Palmariales. Detailed reproductive developmental studies have proved invaluable in red algal classification, notably Fredericq and

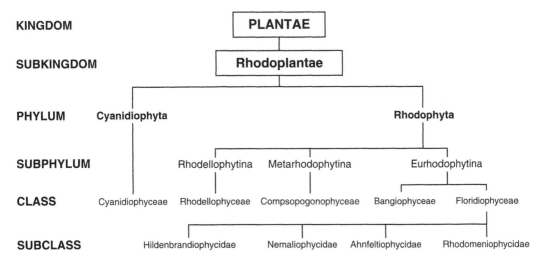

**FIGURE 20.1** Summary of higher level classification of red algae (Rhodoplantae). (From Saunders and Hommersand[9]. With permission.)

Hommersand[26], whose meticulous study of *Gracilaria verrucosa* (Hudson) Papenfuss, using aceto-iron-hematoxylin-chloral hydrate[27] to stain nuclei, enabled them to erect the order Gracilariales.

However, the next major breakthrough was the work of Pueschel and Cole[28], who examined the fine structure of pit plugs (occlusions in the small pore between cells following cell division) in red algae. The structure of these pit plugs was found to be stable and a useful systematic character in resolving monophyletic groups (although not further phylogenetic relationships; see Garbary and Gabrielson[29]). As a consequence of their observations, Pueschel and Cole[28] were able to provide further support for orders already in existence and to propose some other orders (Table 20.2).

Garbary and Gabrielson[29] used cladistic analysis to attempt to resolve phylogenetic relationships amongst the red algae. However, a major advance has been the use of molecular DNA sequence data, and this has enabled a much clearer understanding of phylogenetic relationships. For the red algae in particular, after nearly 250 years, our understanding of all levels of classification has been revolutionised, particularly where molecular data has been used in conjunction with traditional techniques. It is this that has enabled Saunders and Hommersand[9] to almost double the number of orders from those presented by Garbary and Gabrielson[29]. No doubt the new classification will change with time, but the changes proposed because of molecular studies are arguably the most fundamental that we have seen since Linnaeus created Cryptogamia.

### 20.2.3 SPECIATION IN RED ALGAE

A comparison of the numbers of species reported for each red algal order (Table 20.3) shows that some orders have considerably more species than others. Ceramiales has the largest number of species at c. 2,300, an order of magnitude more than those with the next highest numbers. Conversely, some orders have very few species, such as Rhodogorgonales with two species, or Pihiellales with one. It is possible to interpret these findings in a number of ways: first that these are artificial groupings based on the acquisition of what are considered 'ordinal' characters by relatively recent taxa; second that these species numbers will be greatly revised as new taxonomic studies are continued; and third that this really does reflect successful speciation in some groups and not others. For Rhodophyta as a whole, there are a handful of genera which are much more species rich than all the others. *Polysiphonia* is the largest genus with >200 species, followed by *Ceramium* with c. 170 species, then *Gracilaria* with c. 160 species, *Batrachospermum* with c. 130 species, *Laurencia* with c. 120 species and *Porphyra* with c. 110 species[8]. With the exception of the freshwater genus

**TABLE 20.2**
**Classifications of Red Algae to Show the Increase in the Number of Recognised Orders over the Last Half Century**

| Fritsch (1945)[66] | Kylin (1956)[67] | Dixon (1973)[17] | Dixon and Irvine (1977)[16] | Pueschel and Cole (1982)[28] | Garbary and Gabrielson (1990)[29] | Saunders and Hommersand (2004)[9] |
|---|---|---|---|---|---|---|
| BANGIOIDEAE[a] | BANGIOIDEAE | BANGIOPHYCEAE | | RHODOPHYCEAE | RHODOPHYCEAE | BANGIOPHYCEAE |
| | Porphyridiales | Porphyridiales | | | Porphyridiales | Porphyridiales 1, 2 and 3 |
| | Goniotrichales | | | | | |
| | Compsopogonales | Compsopogonales | | | Compsopogonales | Compsopogonales |
| | Rhodochaetales | Rhodochaetales | | | Rhodochaetales | Rhodochaetales |
| | | | | | | Erythropeltidales |
| Bangiales | Bangiales | Bangiales | | | Bangiales | |
| FLORIDEAE | FLORIDEAE | FLORIDIOPHYCEAE | | FLORIDIOPHYCEAE | | FLORIDIOPHYCEAE |
| | | | | | Acrochaetiales | Acrochaetiales |
| Nemalionales | Nemalionales | Nemalionales | Nemaliales[b] | Nemaliales | | Nemaliales |
| | | | Palmariales | Palmariales | Palmariales | Palmariales |
| | | | | Corallinales | Corallinales | Corallinales |
| Gelidiales | Gelidiales | | | Gelidiales | Gelidiales | Gelidiales |
| | | | | Hildenbrandiales | Hildenbrandiales | Hildenbrandiales |
| | | | | Batrachospermales | Batrachospermales | Batrachospermales |
| | | | | Bonnemaisoniales | Bonnemaisoniales | Bonnemaisoniales |
| Gigartinales | Gigartinales | Gigartinales | Gigartinales | Gigartinales | Gigartinales | Gigartinales |

| | | | | |
|---|---|---|---|---|
| Cryptonemiales | Cryptonemiales | Cryptonemiales | Cryptonemiales | Rhodymeniales |
| Rhodymeniales | Rhodymeniales | Rhodymeniales | Rhodymeniales | Ceramiales |
| Ceramiales | Ceramiales | Ceramiales | Ceramiales | Balbianiales |
| | | | | Balliales |
| | | | | Colaconematales |
| | | | | Rhodogorgonales |
| | | | | Thoreales |
| | | | Gracilariales | Gracilariales |
| | | | Ahnfeltiales | Ahnfeltiales |
| | | | | Pihiellales |
| | | | | Bangiales |
| | | | | Halymeniales |
| | | | | Nemastomatales |
| | | | | Plocamiales |
| | | | | Cyanidiales |

[a] Names in capital letters are classes.

[b] See Christensen[68] regarding change from Nemalionales to Nemaliales.

*Source:* Data from Garbary and Gabrielson[29] and Saunders and Hommersand[9].

**TABLE 20.3**
**Number of Species of Red Algae for Each Order**

| Orders of Red Algae | Numbers of Species |
|---|---|
| BANGIOPHYCEAE | |
| Porphyridiales | 19-23(?) |
| Compsopogonales | 12 |
| Rhodochaetales | 1 |
| Erythropeltidales | 45 |
| FLORIDEOPHYCEAE | |
| Acrochaetiales | 275 |
| Nemaliales | 201 |
| Palmariales | 43 |
| Corallinales | 564 |
| Gelidiales | 166 |
| Hildenbrandiales | 17 |
| Batrachospermales | 147 |
| Bonnemaisoniales | 34 |
| Gigartinales | 920 |
| Rhodymeniales | 315 |
| Ceramiales | 2,300 |
| Balbianiales | 3 |
| Balliales | 6 |
| Colaconematales | 19 |
| Rhodogorgonales | 2 |
| Thoreales | 15 |
| Gracilariales | 222 |
| Ahnfeltiales | 9 |
| Pihiellales | 1 |
| Bangiales | 124 |
| Halymeniales | 241 |
| Nemastomatales* | |
| Plocamiales | 40 |
| Cyanidiales | 5 |

ᵃ In Gigartinales in Guiry et al.[8]
*Source:* Data from Guiry et al.[8]

*Batrachospermum,* they are all marine genera. These genera are all characteristically common and widely distributed. They also present taxonomic problems because of their variable morphology and because generic circumscriptions have been in various states of flux.

Many of the large genera have been the subject of DNA sequence studies in recent years and have been split into smaller genera as new phylogenetic hypotheses have revealed monophyletic groups that share previously unappreciated synapomorphies (*Laurencia*[30,31] and *Polysiphonia*[32]). Furthermore, many genera that were thought to consist of few species have been shown to reveal many highly divergent lineages, such as *Bangia*[33,34].

What has driven this great diversity and high numbers of species? The question has to be qualified with the observation that most of these recognised 'species' are still based on morphological, and often typological, concepts of species. Very few studies have taken the concept of species beyond this alpha taxonomic description, but what has been done reveals that diversity at the species level is probably even greater than previously recognised.

Let us propose that we consider an entity as a species in the genetic and evolutionary sense when certain criteria either have been demonstrated or are at least strongly suspected. The main criterion is some form of reproductive isolation[35]. It is important that this is tested experimentally or that distinct entities are found in sympatry that remain distinct. The strength of the reproductive isolation cannot easily be predicted a priori and needs experimental, ecological and demographic studies to be performed. Another criterion is that they should show some level of genetic uniqueness. However, the level of distinctness needed is not easy to determine globally. The genetic markers used in these analyses need to be critically explored. Nearly all phylogenetic studies in red algae involve sets of genetic markers which, though useful at higher levels, may not be appropriate for providing information on reproductive processes in speciating and newly speciated algae[36]. Yet genetic uniqueness, percentage of genetic divergence and phylogenetic topology have been used to indicate species status in some algae (see example in Saunders and Lehmkuhl[37]). More importantly, in the few algal groups studied, levels of genetic divergence have been shown to be correlated with reproductive isolation[38,39] and not always with morphological distinctness.

Reproductive isolation is an important criterion in many species definitions, particularly in producing populations that have unique evolutionary trajectories. Reproductive isolation has been criticised mainly because it is often not feasible to test, and, if used in the strict sense (100% reproductive isolation for species status), would exclude many well recognised species that can form viable hybrids with other species[35]. Reproductive isolation is often a consequence of population isolation, with divergent populations accumulating mutations that cause gamete incompatibility or hybrid sterility/inviability. An understanding of red algal population genetics is important if we are to gain insights into how populations differentiate and hence speciate. Also important in our understanding of speciation in red algae is knowledge into factors that cause prezygotic isolation (incompatibilities of gametes to fuse, inability of gametic nuclei to fuse) or postzygotic isolation such as improper zygote (carposporophyte) division and inviability of meiosis.

### 20.2.4   POPULATION STRUCTURE OF RED ALGAE

Red algae are the only group of macroalgae completely lacking flagellated stages. Despite this, spores are released into the water column and theoretically can disperse relatively large distances, with spore survival being in the order of days[40]. Until recently population studies in red algae have been few due to the lack of appropriate genetic markers to address these questions and difficulties with methodologies applied to these organisms. Allozyme studies, which have been available for over 40 years and which have proven very useful in many organisms, have been used sparingly in algae[41,42] both for technical reasons and because levels of allozyme variation have been too low to adequately address population genetic questions[42]. Recent studies using DNA markers (microsatellites, RFLP, organellar haplotypes) indicate that red algal populations are highly differentiated at the level of a few to tens of kilometres[43–45] and that the distribution of genotypes can be very patchy even at local scales[46]. These studies indicate that the ability of red algal populations to become genetically isolated at larger scales is great, even without obvious geographic barriers to dispersal. These factors will lead to the accumulation of genetic change that could contribute to speciation.

### 20.2.5   REPRODUCTIVE ISOLATING MECHANISMS

Red algae are also unique among macroalgae in their reproductive structures. Besides the lack of flagellated stages several other factors make their gametes, syngamy/karyogamy and early zygote stages unique and areas in which the potential of reproductive incompatibility is increased relative to other organisms. Red algae have male and female gametes that possess walls or are at least partially walled. Due to their lack of flagellated stages, gamete union is guided by chance, although once it occurs, walled gametes must stick. This interaction involves a cell-to-cell recognition mechanism involving unique carbohydrate-protein (lectin) interactions[47–49] found in the cell walls.

Once sperm-to-egg contact has been made, enzymes are needed for the digestion of the two cell walls before the respective membranes and cytoplasms can fuse. In both these interactions, genetic changes in isolated populations can lead to incompatibility when populations come into contact again. The extended egg receptive area means that once a sperm nucleus has entered the egg cytoplasm it must travel down the egg to the egg nucleus. This interaction involves actin filaments within the extended egg structure and motor proteins (myosin) on the sperm nuclei[50,51], another area where incompatibilities can arise.

Also unique in red algae is the fate of the zygote nucleus. In contrast to most organisms in which the zygote is released to produce the diploid individual, the red algal zygote is amplified while still attached to the female gametophyte before it releases diploid spores (zygote copies). The process by which this occurs is unique to different groups of red algae and is used in higher-level classification, but is complex enough that the dividing zygote is usually considered an alternate stage of the life cycle (the carposporophyte, in a three life history stage sexual cycle). This unique development is also an area in which incompatibility is manifested, even before a free living diploid is formed. This has been shown in some experiments in which red algae from different locations, and/or genetic types, were artificially hybridised. Carposporophytes began forming and then aborted forming 'pseudocystocarps' (partially formed carposporophytes plus surrounding female tissue) (Brodie et al.[52]; Zuccarello and West[53]; Kamiya et al.[54] and references therein), suggesting that the diploid hybrid nucleus was not able to properly coordinate carposporophyte development. Many of these processes are only now starting to be investigated using molecular methods, but all these unique red algal attributes indicate that there is greater scope for red algae to develop incompatible reproductive interactions and hence to lead to reproductively isolated species.

## 20.2.6 EXAMPLES OF SPECIATION STUDIES IN RED ALGAE

There are few studies in which reproductive isolation and genetic data have been used to address questions of speciation. An excellent summary of premolecular-era studies on hybridisation in red algae is by Guiry[55]. Several problem groups were indicated in this summary, including one where morphologically similar algae were shown not to hybridise. One of these species was *Mastocarpus stellatus* (Stackhouse) Guiry, a common intertidal red algae in the North Atlantic. Work by Guiry and West[56] showed not only that certain cultured isolates were not able to hybridise with each other but that there was a North–South distribution of these breeding groups. A Southern breeding group was found in Spain and Portugal plus in areas of the Brittany coast in Northern France, while a Northern breeding group was found in Britain and Ireland, but also Brittany. In fact, populations existed in France in which both breeding groups were found.

This work was then extended by Zuccarello et al.[57] using molecular markers. Two organellar markers, one from plastid DNA, the other from mitochondrial DNA, were used to assess genetic differences between breeding group samples. The plastid marker was an intergenic spacer between the large and small subunit of the ribulose bisphosphate carboxylase/oxygenase gene (the rubisco spacer). This marker was first developed and used in *Gracilaria*[58,59] but has proven to be useful for untangling the intraspecies relationships of many red algae (Wattier and Maggs[60] and references therein). The mitochondrial marker is also an intergenic spacer between the cytochrome oxidase subunit 2 and cytochrome oxidase subunit 3 genes (*cox2-3* spacer), is 'universal' (amplifies in many florideophycean red algae) and shows greater intra-species variation than the rubisco spacer[61].

These molecular markers confirmed that the breeding groups identified by Guiry and West[56] were genetically distinct. For the rubisco spacer the breeding groups varied by as little as two base pairs (bp), while for the mitochondrial *cox2-3* spacer they differed by at least 15 bp. Increased sampling again has shown that these two breeding groups are parapatric with mixed populations found in Brittany France and South East England. These two breeding groups could therefore have been isolated during the last glacial maximum (c. 20,000 years before present), leading to their reproductive isolation and genetic divergence, when there was a southern refugium in Spain and a northern refugium around the English Channel[62]. This scenario would mean that reproductive isolation occurred between the interglacial cycles before the population could have reconnected (several tens of thousands of years). We have no knowledge of whether this is considered a fast or slow rate of reproductive isolation acquisition in algae.

These data indicate that even low levels of genetic divergence in a commonly used nonfunctional intraspecies marker, the rubisco spacer, are correlated with reproductively isolated individuals. How far this can be projected is difficult to say, but in the few studies that have used rubisco spacer data and hybridisation experiments a similar pattern is seen. In *Caloglossa postiae* M. Kamiya and R. J. King, samples collected from Japan and Australia do not differ in rubisco spacer sequence and only produce pseudocystocarps when artificially hybridised[63]. In a study of *Spyridia filamentosa* (Wulfen) Harvey in which only a limited amount of hybridisations were possible a difference of only 7 bp in the rubisco spacer correlated with reproductive isolation between two samples[38]. Also in *Caloglossa vieillardii* (Kützing) Setchell samples differing by 13 bp are also reproductively isolated (data not shown).

The only other well documented example within red algae in which molecular studies and hybridisation data have been used to understand speciation and the prevalence of cryptic species is within the *Bostrychia radicans/B. moritziana* complex[39,53,64,65]. This cosmopolitan tropical species group is composed of seven distinct genetic lineages that are highly divergent and morphologically indistinguishable[39]. These lineages are also reproductively isolated, representing cryptic species with either narrow or wide-ranging distributions, with lineages being sympatric in certain populations[39,65]. Along the East coast of the United States, two lineages are found. Within one of these lineages different chloroplast and mitochondrial haplotypes are found along the East coast, with one haplotype found in Northern populations (haplotype B; New York, Virginia, North Carolina) and another in more Southern populations (haplotype C; Georgia, Florida). Both haplotypes are found in intermediate areas (South Carolina)[39]. These two haplotypes differ by 4 bp differences in the rubisco spacer and are reproductively isolated. Haplotype C is 1 bp different from samples of the same species complex found in Pacific Mexico (haplotype A), and yet these samples show intermediate levels of reproductive isolation (diploids are formed but they do not go through meiosis)[39]. These data indicate that even within phylogenetically well supported lineages, reproductive isolation can be present, increasing the number of evolutionary lineages within this red algal group.

The increasing number of molecular studies at the species/population level in red algae will increase our knowledge of genetic variation, evolution and genetic structure within species. This work must go beyond the standard molecular markers to incorporate more unlinked nuclear markers that will give clues to hybridisation within and between populations. These studies must still be combined with time-consuming algal culturing and hybridisation studies. Only with knowledge of reproductive isolation within alga groups can we couple our ever-increasing molecular data with an important reproductive parameter. A new line of research should be targeted at reproductive incompatibility in red algae. Combining our knowledge of sister species or semi-isolated populations with an understanding of the fertilisation process and cellular/physiological areas of incompatibility will lead to a greater understanding of the causes and mechanisms of reproductive isolation in red algae.

## 20.3  CONCLUSIONS

In conclusion, molecular data will continue to illuminate the evolutionary relationships within the red algae at multiple levels. The use of multiple molecular markers and better taxon sampling will lead to more natural higher-level taxonomies. A population focus in species studies (greater sampling within and between populations) will lead to a better understanding of the tempo and history of speciation in these organisms. Concentration on cell biological techniques may unravel the complex physiological process that occur in the unique prezygotic and postzygotic processes that are crucial in establishing and maintaining distinct species in this diverse and important algal group.

## REFERENCES

1. Linnaeus, C., *Species plantarum, exhibentes plantas rite cognitas, ad genera relatas, cum differentiis specificis, nominibus trivialibus, synonymis selectis, locis natalibus digestas, Tomas I, Cryptogamia,* Salvi, Stockholm, Sweden, 1753, 1061.
2. Hoek, C. van den, Mann, D.G. and Jahns, H.M., *Algae: An Introduction to Phycology,* Cambridge University Press, Cambridge, 1995, 623.
3. Baldauf, S.L., The deep roots of eukaryotes, *Science,* 300, 1703, 2003.
4. Keeling, P.J., Diversity and evolutionary history of plastids and their hosts, *Am. J. Bot.,* 91, 1481, 2004.
5. Palmer, J.D., Soltis, D.E., and Chase, M.W., The plant tree of life: an overview and some points of view, *Am. J. Bot.,* 91, 1437-1445, 2004.
6. World Conservation Monitoring Centre, *Global Biodiversity: Status of the Earth's Living Resources,* Chapman and Hall, 1992.
7. Moon-van der Staay, S.Y., de Wachter, R., and Vaulot, D., Oceanic 18S rDNA sequences from picoplankton reveal unsuspected eukaryotic diversity, *Nature,* 409, 607, 2001.
8. Guiry, M.D., Rindi, F., and Guiry, G.M., *AlgaeBase version 4.0,* National University of Ireland, Galway, http://www.algaebase.org, 2005.
9. Saunders, G.W. and Hommersand, M.H., Assessing red algal supraordinal diversity and taxonomy in the context of contemporary systematic data, *Am. J. Bot.,* 91, 1494, 2004.
10. Woelkerling, W.J., An introduction, in *Biology of the Red Algae,* Cole, K.M. and Sheath, R.G., Eds., Cambridge University Press, Cambridge, 1990, chap. 1.
11. Butterfield, N.J., *Bangiomorpha pubescens* n. gen., n. sp.: implications for the evolution of sex, multicellularity, and the Mesoproterozoic/Neoproterozoic radiation of eukaryotes, *Paleobiology,* 26, 386, 2000.
12. Brodie, J. and Irvine, L.M., *Seaweeds of the British Isles Volume 1 Part 3B Bangiophycidae,* Intercept, Hampshire, 2003, 167.
13. Littler, D.S., *Marine Plants of the Caribbean: A Field Guide from Florida to Brazil,* Smithsonian Institution Press, Washington, DC, 1989, 263.
14. Womersley, H.B.S., *The Marine Benthic Flora of Southern Australia Part IIIC,* State Herbarium of South Australia, South Australia, 1998, 535.
15. Silva, P.C., Basson, P.W., and Moe, R.L., Catalogue of the benthic marine algae of the Indian Ocean, *Univ. Calif. Publs. Bot.,* 79, 1.
16. Dixon, P.S. and Irvine, L.M., *Seaweeds of the British Isles Volume 1 Rhodophyta Part 1: Introduction, Nemaliales, Gigartinales,* British Museum (Natural History) London, 1977, 252.
17. Dixon, P.S., *Biology of Rhodophyta,* Oliver and Boyd, Edinburgh, 1973, 285.
18. Lamouroux, J.V.F., Essai sur les genres de la famille de Thalassiophytes non articulées, *Annales du Muséum (National) d'Histoire Naturelle (Paris),* 20, 21, 1813.
19. Harvey, W.H., Algae, in *Flora Hibernica Vol. 2,* Mackay, J.T., Ed., Curry, Dublin, Ireland, 1836, 157.
20. Berthold, G., *Fauna and Flora des Golfes von Neapel VIII Die Bangiaceen des Golfes von Neapel,* Leipzig, 1882, 1.
21. Magne, F., Classification et Phylogénie des Rhodophycées, *Cryptog., Algol.,* 10, 101, 1989.
22. Schmitz, F. and Hauptfleisch, P., Rhodophyceae, in *Die Naturlichen Pflanzenfamilien, vol. 1(2),* Engler, A. and Prantl, K., Eds., Englemann, Leipzig, 1896/7, 298
23. Oltmanns, F., *Morphologie und Biologie der Algen 1,* Fischer, Jena, 1904, 733.
24. Guiry, M.D., A preliminary consideration of the taxonomic position of *Palmaria palmata* (Linnaeus) Stackhouse = *Rhodymenia palmata* (Linneaus) Greville, *J. Mar. Biol. Assoc. UK,* 54, 509, 1974.
25. Guiry, M.D., The importance of sporangia in the classification of the Florideophyceae, in *Modern Approaches to the Taxonomy of Red and Brown Algae, Systematics Association Special Volume 10,* Irvine, D.E.G. and Price, J.H., Eds., Academic Press, London, 1978, 111.
26. Fredericq, S. and Hommersand, M.H., Proposal of the Gracilariales ord. nov. (Rhodophyta) based on an analysis of the reproductive development of *Gracilaria verrucosa, J. Phycol.,* 25, 213, 1989.
27. Wittmann, W., Aceto-iron-haematoxylin-chloral hydrate for chromosome staining, *Stain Technol.,* 40, 161, 1965.
28. Pueschel, C.M. and Cole, K.M., Rhodophycean pit plugs: an ultrastructural survey with taxonomic implications, *Am. J. Bot.,* 69, 703, 1982.

29. Garbary, D.J. and Gabrielson, P.W., Taxonomy and evolution, in *Biology of the Red Algae,* Cole, K.M. and Sheath, R.G., Eds., Cambridge University Press, Cambridge, 1990, chap. 18.
30. Nam, K.W., Maggs, C.A., and Garbary, D.J., Resurrection of the genus *Osmundea* with an emendation of the generic delineation of *Laurencia* (Ceramiales, Rhodophyta), *Phycologia,* 33, 384, 1994.
31. Nam, K.W. et al., Taxonomy and phylogeny of *Osmundea* (Rhodomelaceae, Rhodophyta) in Atlantic Europe, *J. Phycol.,* 36, 759, 2000.
32. Choi, H.G. et al., Phylogenetic relationships of *Polysiphonia* (Rhodomelaceae, Rhodophyta) and its relatives based on anatomical and nuclear small-subunit rDNA sequence data, *Can. J. Bot.,* 79, 1465, 2001.
33. Broom, J.E.S., Farr, T.J., and Nelson, W.A., Phylogeny of the *Bangia* flora of New Zealand suggests a southern origin for *Porphyra* and *Bangia* (Bangiales, Rhodophyta), *Mol. Phylogenet. Evol.,* 31, 1197, 2004.
34. Müller, K.M., Cannone, J.J., and Sheath, R.G., A molecular phylogenetic analysis of the Bangiales (Rhodophyta) and description of a new genus and species, *Pseudobangia kaycoleia, Phycologia,* 44, 146, 2005.
35. Coyne, J.A. and Orr, H.A., *Speciation,* Sinauer Associates, Sunderland, MA, 2004, 545.
36. Small, R.L., Cronn, R.C., and Wendel, J.F., Use of nuclear genes for phylogeny reconstruction in plants, *Aus. Syst. Bot.,* 17, 145, 2004.
37. Saunders, G.W. and Lehmkuhl, K.V., Molecular divergence and morphological diversity among four cryptic species of *Plocamium* (Plocamiales, Florideophyceae) in northern Europe, *Eur. J. Phycol.,* 40, 293, 2005.
38. Zuccarello, G.C., Sandercock, B., and West, J.A., Diversity within red algal species: variation in world-wide samples of *Spyridia filamentosa* (Ceramiaceae) and *Murrayella periclados* (Rhodomelaceae) using DNA markers and breeding studies, *Eur. J. Phycol.,* 37, 403, 2002.
39. Zuccarello, G.C. and West, J.A., Multiple cryptic species: molecular diversity and reproductive isolation in the *Bostrychia radicans/B. moritziana* complex (Rhodomelaceae, Rhodophyta) with focus on North American isolates, *J. Phycol.,* 39, 948, 2003.
40. Hoffmann, A.J., The arrival of seaweed propagules at the shore: a review, *Bot. Mar.,* 30, 151, 1987.
41. Sosa, P.A. and Lindstrom S.C., Isozymes in macroalgae (seaweeds): genetic differentiation, genetic variability and applications in systematics (Review), *Eur. J. Phycol.,* 34, 427, 1999.
42. Valero, M. et al., Concepts and issues of population genetics in seaweeds, *Cah. Biol. Mar.,* 42, 53, 2001.
43. Engel, C.R., Destombe, C., and Valero, M., Mating system and gene flow in the red seaweed *Gracilaria gracilis*: effect of haploid-diploid life history and intertidal rocky shore landscape on fine scale genetic structure, *Heredity,* 92, 289, 2004.
44. Faugeron, S. et al., Hierarchical spatial structure and discriminant analysis of genetic diversity in the red alga *Mazzaella laminarioides* (Gigartinales, Rhodophyta), *J. Phycol.,* 37, 705, 2001.
45. Wright, J.T., Zuccarello, G.C., and Steinberg, P.D., Genetic structure in the subtidal red alga *Delisea pulchra, Mar. Biol.,* 136, 439, 2000.
46. Zuccarello, G.C. et al., Population structure and physiological differentiation of haplotypes of *Caloglossa leprieurii* (Rhodophyta) in a mangrove intertidal zone, *J. Phycol.,* 37, 235, 2001.
47. Kim, G.H. and Fritz, L., Gamete recognition during fertilization in a red alga, *Antithamnion nipponicum, Protoplasma,* 174, 69, 1993.
48. Kim, G.H., Lee, I.K., and Fritz, L. Cell-cell recognition during fertilization in a red alga, *Antithamnion sparsum* (Ceramiaceae, Rhodophyta), *Plant Cell Physiol.,* 37, 621, 1996.
49. Kim, G.H. and Kim, S.H., The role of F-actin during fertilization in the red alga *Aglaothamnion oosumiense* (Rhodophyta), *J. Phycol.,* 35, 806, 1999.
50. Wilson, S.M., Pickett-Heaps, J.D., and West, J.A., Fertilization and the cytoskeleton in the red alga *Bostrychia moritziana* (Rhodomelaceae, Rhodophyta), *Eur. J. Phycol.,* 37, 509, 2002.
51. Wilson, S.M., West, J.A., and Pickett-Heaps, J.D., Time-lapse videomicroscopy of fertilization and the actin cytoskeleton in *Murrayella periclados* (Rhodomelaceae, Rhodophyta), *Phycologia,* 42, 638, 2003.
52. Brodie, J., Guiry, M.D., and Masuda, M., Life history, morphology and crossability of *Chondrus ocellatus* forma *ocellatus* and *C. ocellatus* forma *crispoides* (Gigartinales, Rhodophyta) from the north-western Pacific, *Eur. J. Phycol.,* 28, 183, 1993.
53. Zuccarello, G.C. and West, J.A., Hybridization studies in *Bostrychia*: 1. *B. radicans* (Rhodomelaceae, Rhodophyta) from Pacific and Atlantic North America, *Phycol. Res.,* 43, 233, 1995.

54. Kamiya, M., Zuccarello, G.C., and West, J.A., Evolutionary relationships of the genus *Caloglossa* (Delesseriaceae, Rhodophyta) inferred from large-subunit ribosomal RNA gene sequences, morphological evidence and reproductive compatibility, with description of a new species from Guatemala, *Phycologia*, 42, 478, 2003.

55. Guiry, M.D., Species concepts in marine red algae, in *Progress in Phycological Research 8,* Round, F.E. and Chapman, D.J., Eds., Biopress Ltd, Bristol, 1992, chap. 5.

56. Guiry, M.D. and West, J.A., Life history and hybridization studies on *Gigartina stellata* and *Petrocelis cruenta* (Rhodophyta) in the north Atlantic, *J. Phycol.,* 19, 474, 1983.

57. Zuccarello G.C. et al., A molecular re-examination of speciation in the intertidal red alga *Mastocarpus stellatus* (Gigartinales, Rhodophyta) in Europe, *Eur. J. Phycol.,* 40, 337, 2005.

58. Destombe, C. and Douglas, S.E., Rubisco spacer sequence divergence in the rhodophyte alga *Gracilaria verrucosa* and closely related species, *Curr. Genet.,* 19, 395, 1991.

59. Destombe, C., Correction, *Curr. Genet.,* 22, 173, 1992.

60. Wattier, R. and Maggs, C.A., Intraspecific variation in seaweeds: the application of new tools and approaches, *Adv. Bot. Res.,* 35, 171, 2001.

61. Zuccarello, G.C. et al., A mitochondrial marker for red algal intraspecific relationships, *Mol. Ecol.,* 8, 1443, 1999.

62. Provan, J., Wattier, R.A., and Maggs, C.A., Phylogeographic analysis of the red seaweed *Palmaria palmata* reveals a Pleistocene marine glacial refugium in the English Channel, *Mol. Ecol.,* 14, 793, 2005.

63. Kamiya, M. et al., Reproductive and genetic distinction between broad and narrow entities of *Caloglossa continua* (Delesseriaceae, Rhodophyta), *Phycologia*, 38, 356, 1999.

64. Zuccarello, G.C. and West, J.A., Hybridization studies in *Bostrychia* 2: correlation of crossing data and plastid DNA sequence data within *B. radicans* and *B. moritziana* (Ceramiales, Rhodophyta), *Phycologia*, 36, 293, 1997.

65. Zuccarello, G.C., West, J.A., and King, R.J., Evolutionary divergence in the *Bostrychia moritziana/B. radicans* complex (Rhodomelaceae, Rhodophyta): molecular and hybridization data, *Phycologia*, 38, 34, 1999.

66. Fritsch, F.E., *Structure and Reproduction of the Algae 2,* Cambridge University Press, Cambridge, 1945, 939.

67. Kylin, H., *Die Gattungen der Rhodophyceen,* Gleerup, Lund, 1956, 673.

68. Christensen, T., Two new families and some names and combinations in the algae, *Blumea*, 15, 91, 1967.

# Index

f = figure; t = table

# Systematics Association
# Publications

1. Bibliography of Key Works for the Identification of the British Fauna and Flora, 3rd edition (1967)[†]
   *Edited by G.J. Kerrich, R.D. Meikie and N. Tebble*

2. Function and Taxonomic Importance (1959)[†]
   *Edited by A.J. Cain*

3. The Species Concept in Palaeontology (1956)[†]
   *Edited by P.C. Sylvester-Bradley*

4. Taxonomy and Geography (1962)[†]
   *Edited by D. Nichols*

5. Speciation in the Sea (1963)[†]
   *Edited by J.P. Harding and N. Tebble*

6. Phenetic and Phylogenetic Classification (1964)[†]
   *Edited by V.H. Heywood and J. McNeill*

7. Aspects of Tethyan biogeography (1967)[†]
   *Edited by C.G. Adams and D.V. Ager*

8. The Soil Ecosystem (1969)[†]
   *Edited by H. Sheals*

9. Organisms and Continents through Time (1973)[†]
   *Edited by N.F. Hughes*

10. Cladistics: A Practical Course in Systematics (1992)[*]
    *P.L. Forey, C.J. Humphries, I.J. Kitching, R.W. Scotland, D.J. Siebert and D.M. Williams*

11. Cladistics: The Theory and Practice of Parsimony Analysis (2nd edition)(1998)[*]
    *I.J. Kitching, P.L. Forey, C.J. Humphries and D.M. Williams*

[*] Published by Oxford University Press for the Systematics Association
[†] Published by the Association (out of print)

## SYSTEMATICS ASSOCIATION SPECIAL VOLUMES

1. The New Systematics (1940)
   *Edited by J.S. Huxley (reprinted 1971)*

2. Chemotaxonomy and Serotaxonomy (1968)[*]
   *Edited by J.C. Hawkes*

3. Data Processing in Biology and Geology (1971)[*]
   *Edited by J.L. Cutbill*

4. Scanning Electron Microscopy (1971)*
   *Edited by V.H. Heywood*

5. Taxonomy and Ecology (1973)*
   *Edited by V.H. Heywood*

6. The Changing Flora and Fauna of Britain (1974)*
   *Edited by D.L. Hawksworth*

7. Biological Identification with Computers (1975)*
   *Edited by R.J. Pankhurst*

8. Lichenology: Progress and Problems (1976)*
   *Edited by D.H. Brown, D.L. Hawksworth and R.H. Bailey*

9. Key Works to the Fauna and Flora of the British Isles and Northwestern Europe, 4th
   edition (1978)*
   *Edited by G.J. Kerrich, D.L. Hawksworth and R.W. Sims*

10. Modern Approaches to the Taxonomy of Red and Brown Algae (1978)
    *Edited by D.E.G. Irvine and J.H. Price*

11. Biology and Systematics of Colonial Organisms (1979)*
    *Edited by C. Larwood and B.R. Rosen*

12. The Origin of Major Invertebrate Groups (1979)*
    *Edited by M.R. House*

13. Advances in Bryozoology (1979)*
    *Edited by G.P. Larwood and M.B. Abbott*

14. Bryophyte Systematics (1979)*
    *Edited by G.C.S. Clarke and J.G. Duckett*

15. The Terrestrial Environment and the Origin of Land Vertebrates (1980)
    *Edited by A.L. Pachen*

16  Chemosystematics: Principles and Practice (1980)*
    *Edited by F.A. Bisby, J.G. Vaughan and C.A. Wright*

17. The Shore Environment: Methods and Ecosystems (2 volumes)(1980)*
    *Edited by J.H. Price, D.E.C. Irvine and W.F. Farnham*

18. The Ammonoidea (1981)*
    *Edited by M.R. House and J.R. Senior*

19. Biosystematics of Social Insects (1981)*
    *Edited by P.E. House and J.-L. Clement*

20. Genome Evolution (1982)*
    *Edited by G.A. Dover and R.B. Flavell*

21. Problems of Phylogenetic Reconstruction (1982)
    *Edited by K.A. Joysey and A.E. Friday*

22. Concepts in Nematode Systematics (1983)*
    *Edited by A.R. Stone, H.M. Platt and L.F. Khalil*

23. Evolution, Time and Space: The Emergence of the Biosphere (1983)*
    *Edited by R.W. Sims, J.H. Price and P.E.S. Whalley*

24. Protein Polymorphism: Adaptive and Taxonomic Significance (1983)*
    *Edited by G.S. Oxford and D. Rollinson*

25. Current Concepts in Plant Taxonomy (1983)*
    *Edited by V.H. Heywood and D.M. Moore*

26. Databases in Systematics (1984)*
    *Edited by R. Allkin and F.A. Bisby*

27. Systematics of the Green Algae (1984)*
    *Edited by D.E.G. Irvine and D.M. John*

28. The Origins and Relationships of Lower Invertebrates (1985)‡
    *Edited by S. Conway Morris, J.D. George, R. Gibson and H.M. Platt*

29. Infraspecific Classification of Wild and Cultivated Plants (1986)‡
    *Edited by B.T. Styles*

30. Biomineralization in Lower Plants and Animals (1986)‡
    *Edited by B.S.C. Leadbeater and R. Riding*

31. Systematic and Taxonomic Approaches in Palaeobotany (1986)‡
    *Edited by R.A. Spicer and B.A. Thomas*

32. Coevolution and Systematics (1986)‡
    *Edited by A.R. Stone and D.L. Hawksworth*

33. Key Works to the Fauna and Flora of the British Isles and Northwestern Europe, 5th edition (1988)‡
    *Edited by R.W. Sims, P. Freeman and D.L. Hawksworth*

34. Extinction and Survival in the Fossil Record (1988)‡
    *Edited by G.P. Larwood*

35. The Phylogeny and Classification of the Tetrapods (2 volumes)(1988)‡
    *Edited by M.J. Benton*

36. Prospects in Systematics (1988)‡
    *Edited by J.L. Hawksworth*

37. Biosystematics of Haematophagous Insects (1988)‡
    *Edited by M.W. Service*

38. The Chromophyte Algae: Problems and Perspective (1989)‡
    *Edited by J.C. Green, B.S.C. Leadbeater and W.L. Diver*

39. Electrophoretic Studies on Agricultural Pests (1989)‡
    *Edited by H.D. Loxdale and J. den Hollander*

40. Evolution, Systematics, and Fossil History of the Hamamelidae (2 volumes)(1989)‡
    *Edited by P.R. Crane and S. Blackmore*

41. Scanning Electron Microscopy in Taxonomy and Functional Morphology (1990)‡
    *Edited by D. Claugher*

42. Major Evolutionary Radiations (1990)‡
    *Edited by P.D. Taylor and G.P. Larwood*

43. Tropical Lichens: Their Systematics, Conservation and Ecology (1991)‡
    *Edited by G.J. Galloway*

44. Pollen and Spores: Patterns and Diversification (1991)‡
    *Edited by S. Blackmore and S.H. Barnes*

45. The Biology of Free-Living Heterotrophic Flagellates (1991)‡
    *Edited by D.J. Patterson and J. Larsen*

46. Plant–Animal Interactions in the Marine Benthos (1992)‡
    *Edited by D.M. John, S.J. Hawkins and J.H. Price*

47. The Ammonoidea: Environment, Ecology and Evolutionary Change (1993)‡
    *Edited by M.R. House*

Systematics Association Publications

69. Neotropical Savannas and Seasonally Dry Forests: Plant Diversity, Biogeography and Conservation (2006)
    *Edited by R.T. Pennington, G.P. Lewis and J.A. Rattan*
70. Biogeography in a Changing World (2006)
    *Edited by M.C. Ebach and R.S. Tangney*
71. Pleurocarpous Mosses: Systematics & Evolution (2006)
    *Edited by A.E. Newton and R.S. Tangney*

*Published by Academic Press for the Systematics Association
†Published by the Palaeontological Association in conjunction with Systematics Association
‡Published by the Oxford University Press for the Systematics Association
**Published by Chapman & Hall for the Systematics Association

51

9 780367 389581